GAME THEORY: STRATEGIES, EQUILIBRIA, AND THEOREMS

GAME THEORY: STRATEGIES, EQUILIBRIA, AND THEOREMS

INGRID N. HAUGEN
AND
ANNA S. NILSEN
EDITORS

Nova Science Publishers, Inc.
New York

For permission to use material from this book please contact us:
Telephone 631-231-7269; Fax 631-231-8175
Web Site: http://www.novapublishers.com

LIBRARY OF CONGRESS CATALOGING-IN-PUBLICATION DATA

Game theory : strategies, equilibria, and theorems / Ingrid N. Haugen and Anna S. Nilsen, editors.
p. cm.
ISBN 978-1-60456-844-8 (hardcover)
1. Game theory. I. Haugen, Ingrid N., 1973- II. Nilsen, Anna S., 1955-
QA269.G34 2008
519.3--dc22 2008022979

Published by Nova Science Publishers, Inc. ✢ New York

CONTENTS

PREFACE

Game theory is the research field to analyze the whole system as a group of many components interacting with each other. Considering the terms in game theory, the idea could be mapped into human's rationality in social science or the forces of nature affecting all of natural phenomenon in fundamental level, and the relationship could be described with many sub-theories of game theory like non-cooperative game or cooperative game.

Most of events and situations could be projected by the eye of game theory on the aspect that there always exist conflicts and cooperation on the interactions in them. Game theory was originated form economics, one of social science, but it applies to not just social system but also the realm of nature. For example we could see the molecular world of materials with game theoretic viewpoint. Molecules in material are connected or interacted with each other by several different kinds of physical forces like metallic bond, ionic bond, van der Waals forces, etc. This book provides leading-edge research from around the world on this field.

Ultimately information needs to be written, stored and read by physical devices. The study of the quantum mechanical aspects of information processing is known as quantum information theory. There is much current interest in utilizing some of the intrinsic features of quantum mechanics, such as superposition, interference and entanglement, to create algorithms that work on a quantum computer in completely novel ways in comparison to classical algorithms. At the intersection of the theory of quantum information and game theory is a new area of investigation known as quantum game theory. The authors outline the idea of quantum bits (qubits), their interference and entanglement, and the allowable operations on them. Then the authors discuss the various protocols for quantizing well known classical games. At this early stage in the theory there is no consensus on precisely how this is done.

Chapter 1 reviews the work carried out in the model of Eisert et al. [Eisert, Wilkins, and Lewenstein, *Phys. Rev. Lett.* **83** (1999) 3077]. Despite early claims that this model can lead to new equilibria in 2×2 games that "solve" dilemmas like the famous Prisoners' Dilemma, the authors emphasize that this is an artifact of an artificial restriction on the players' allowed quantum actions. Access to the full set of quantum operations enriches the players' choices, but in 2×2 games does not lead to any new insights. The situation is more fruitful in multi-player games. In certain circumstances the power of quantum entanglement can be exploited to produce equilibria that are more efficient than those that can be obtained classically, and that have no classical analogue. Additionally, in some systems there is a correspondence between the payoffs for these equilibria and fundamental quantum properties of the initial state, as measured by violations of the Bell inequalities.

Generalizations of the Eisert protocol are discussed as well as other approaches to quantum games. Recent experimental implementations of some simple quantum games are presented.

Chapter 2 introduces the new concept of reversibility in game theory including (1) game-theoretic playing reversibility and (2) reversibility in error correction of game decision communication. Game-theoretic multiple-valued quantum computation (MVQC) is also introduced. Game theory (GT) involves the study of competition and cooperation, without regard to the particular entities (agents) involved, and issues of rationality associated with such phenomena. Reversibility property in GT can be important in situations where: (1) an outside observer needs to know reversely the (correct) paths that lead to specific payoffs on a game's extensive form, and (2) modeling the (maximin) dynamics using low-power consuming circuits as reversibility is a main requirement for low-power circuit design of future technologies such as in quantum computing (QC). Error correction of communicated decisions in two-player games and reversible error correction of communicated batch of decisions in multi-player games are important tasks in situations where noise from a third player(s) or from the communication channel exists. Quantum decision trees (QDTs) are also introduced as quantum representations for applying MVQC to games' dynamics.

Quantum games have proposed a new point of view for the solution of the classical problems and dilemmas in game theory. It has been shown that are more efficient than classical games and provide a saturated upper bound for this efficiency. Certain quantization relationships can be proposed with the objective that a game can be generalized into a quantum domain where the linear superposition of actions is allowed. This quantization let us describe and solution problems originated by conflicting or cooperative behaviors among the members of a system from the point of view of quantum mechanical interactions. This leads us to interesting aspects which only can be observed through the quantization of a game like the possibility of the entanglement between players, the definition of a socioeconomical temperature in a system and the analysis of a game through elements of quantum information theory.

Although both systems analyzed are described through two theories apparently different (quantum mechanics and game theory) both are analogous and thus exactly equivalents. The quantum analogue of the replicator dynamics is the von Neumann equation. The classical equilibrium concepts in game theory can be also generalized through a maximum entropy approach in the so called Collective Welfare Principle. Nature is a game in where its players compete for the equilibrium of the system that they are members. They act as a whole besides individuals like they obey a rule in where they prefer to work for the collective besides the individual welfare. If it is maximized the welfare of the individual above the collective welfare the system gets unstable and eventually it collapses.

Quantum mechanics (and physics) could be used to explain more correctly biological and economical processes (econophysics). A special consequence of the relationships between quantum mechanics and game theory is analyzed in Chapter 3. It is shown that the so called "globalization" process (i.e., the inexorable integration of markets, currencies, nation-states, technologies and the intensification of consciousness of the world as a whole) has a behavior exactly equivalent to a system that is tending to a maximum entropy state i.e., to its state of equilibrium. This let us predict the apparition of big common markets and strong common currencies that will reach the "equilibrium" by decreasing its number until they get a state characterized by only one common currency and only one big common community around the world.

In Chapter 4, a general class of mixed equilibrium problems involving non - differentiable functions is considered. The authors describe some recent advances in the theory and solution methods for such mixed equilibrium and variational inequality problems under monotonicity type assumptions. The methods are based on descent and combined relaxation approaches and involve different linesearch procedures. As a result, they do not require a priori information on the problem. Their numerical implementation is further simplified by making use of specific features of the problems, in particular, their decomposable structure. Then methods' parts can be executed in parallel, making them suitable for rather large problems. Some additional examples of applications are also presented.

In Chapter 5, a discrete two-dimensional political competition model has been proposed and addressed with geometric strategies that find the equilibrium positions, if they exist, and ensure their uniqueness. To adapt the problem to various political landscapes one simply assumes that the distribution of voters is not uniform. This complexity can be represented by assigning an appropriate weight to each position of a voter in the policy plane (the authors assume a finite number of voters). The case in which all the voters have the same weight, and the general case, when they have different weights is studied. It has been proved that, in both cases, except for the particular case where all voters are aligned along a single line of the plane, an equilibrium, if it exists, is attained only when both parties choose to offer the same policy to their voters. That is to say, the two parties will converge to essentially the same political program in order to maximise the number of voters. To avoid the uniqueness of the equilibrium position a definition of equilibrium weaker than the classical one is provided. As a result, a "region of equilibrium" appears. In this area, the parties can operate in a situation of "almost" equilibrium, in which they are not necessarily required to adopt the same policy. Finally, the maximum number of positions of weak equilibrium when the two parties choose positions of the voters is stated, and some examples where this maximum number is attained are presented.

The Goore Game (GG) introduced in [1] has the fascinating property that it can be resolved in a completely distributed manner with no inter-communication between the players. The game has recently found applications in many domains, including the field of sensor networks and Quality-of-Service (QoS) routing. In actual implementations of the solution, the players are typically replaced by Learning Automata (LA). In this Chapter, the authors shall first briefly survey the field of LA and report the LAbased solutions to the GG. The problem with the existing reported approaches is that the accuracy of the solution achieved is intricately related to the number of players participating in the game - which, in turn, determines the resolution. In other words, an arbitrary accuracy can be obtained only if the game has an infinite number of players, which renders a practical solution infeasible. Thus, the authors shall describe some of the recent advances, and show how they can attain an unbounded accuracy for the GG by utilizing no more than three stochastic learning machines, and by recursively pruning the solution space to guarantee that the retained domain contains the solution to the game with a probability as close to unity as desired. Chapter 6 contains the formal algorithms, the proofs of the respective convergence results, and simulation results demonstrating its power. It also conjectures on how the solution can be applied to some of the application domains.

In Chapter 7, the authors study strategic product line designs under duopoly. While the availability of a rich variety of products may attract consumers, it may also lead to confusion in the purchase decision-making process resulting in consumer dissatisfaction. Firms are thus motivated to offer their products with ICT tools in order to supply consumers with sufficient communication about them. The authors present a simple game theoretic model to analyze the impact on product line decisions of the cost of developing these tools, which may even be affected even by the rival's product line. The authors show that the ICT progress in solving consumers' confusion due to a firm's own products enlarges the profit gap between asymmetric firms. In contrast, it is shown that the ICT progress in solving consumers' confusion arising from a comparison with the rival's products reduces the profit gap between them. The cost asymmetry between two firms makes both of these opposite effects more robust.

The objective of Chapter 8 is to propose and analyze a solution concept for the class of transferable utility cooperative games which generalizes the notion of core and is stable in relation to a set of individual goals, in the sense that the players can guarantee the achievement of certain goals in any coalition they join. In other words, individual players impose their own goals in any cooperation group.

The authors also investigate which are the best individual goals that can be attained in a game and provide a characterization of the allocations associated to those goals as the solutions of a multiobjective linear problem. Finally, the authors explore the potential of goal programming approaches to obtain compromise allocations for the transferable utility game. These allocations are such that the excesses the players obtain can not be improved simultaneously. In particular, the authors apply a minimax reference point approach, which is specially appropriate for finding equitable solutions when a group of decision makers is involved in deciding the most preferred solutions as is the case in n-person cooperative games.

Budget-constrained and financially motivated members of independent groups participated in a series of two-stage contests to win a single, commonly valued, and exogenously determined prize. In Chapter 9 the authors present and test an equilibrium model that, in addition to the utility of receiving the prize, incorporates 1) a non-pecuniary utility of winning each stage of the contest, and 2) allows for misperception of the probability of winning, which is determined by Tullock's contest success function. The equilibrium solution accounts for the major finding of excessive aggregate expenditures in stage 1 of the contest. The authors then test a Cognitive Hierarchy model that attributes individual differences in stage 1 expenditures to different levels of depth of reasoning. Although the explanatory power of this model is limited, it emphasizes the importance of the non-pecuniary utility of winning in accounting for the excessive stage 1 expenditures.

Chapter 10 examines the effectiveness of sophisticated capital budgeting practices. The theoretical applications of sophisticated capital budgeting practices (defined as the use of real option reasoning and/or game theory decision rules) have been well documented; however, empirical evidence on the effectiveness of these sophisticated capital budgeting practices is scarce. The empirical results from a survey of Dutch organizations suggest that the use of sophisticated capital budgeting practices is not necessarily associated with higher performance. However, the results also indicate that particular industries may benefit from sophisticated capital budgeting practices. Additional research may identify the specific characteristics that affect the effectiveness of sophisticated capital budgeting practices for individual companies.

The aim of Chapter 11 is to give a new class of functions called θ-β-I-irresolute functions in ideal topological space. Some characterizations and several basic properties of this class of functions are obtained.

Chapter 12 studies diversity by temporal oscillations in plant communities with a differential timing of reproduction.

Background and Aims: Species can coexist at non-equilibrium circumstances, for instance by oscillations in population densities or chaos, caused by non-linear responses of species to their environment. The authors analyzed whether plant genotypes that vary in their timing of reproduction can coexist under equilibrium or non-equilibrium circumstances when competing for light.

Methods: The authors used a game theoretical approach, based on a biologically mechanistic model of plant growth.

Key Results: In our model, the genotype switching to reproduction slightly later than its competitor attained a higher fitness. This caused a succession from early switching genotypes to those switching later to reproductive investment. However, there were cyclic opportunities for extinct genotypes that switch early to reproduction to re-establish and grow into the community. The cause was that genotypes that switched very late produced relatively very little seed because of an overinvestment in vegetative growth; especially when competing against individuals of the same genotype. Because the very early switch genotypes could establish, circumstances were such that other extinct switch genotypes could re-enter the vegetation as well. In this way the diversity of genotypes was maintained over time by temporal oscillations of genotype abundances.

Conclusions: The authors show that within a model, an externally undisturbed plant community can produce its own temporal cyclic or chaotic disturbances to promote diversity, rather than converge to a stable equilibrium when competing for light. Cyclic fluctuations in species composition can occur in a model community of plants sharing the same growing season and that are limited just by light as a single resource.

A well-known phenomenon (popularized by Malcolm Gladwell's book) in real world social networks is the existence of 'tipping points'. That is, thresholds that once surpassed lead to a whole new configuration of the social structure. Political upheavals, sudden fads and the fast adaptation of innovations are just examples of this. The authors intend to provide some clues on how this might happen, but instead of assuming that it is an unintended consequence of random actions the authors consider here a game theoretic framework in which rational agents make decisions aimed to maximize their payoffs. Starting with a framework very much like Bala & Goyal's (2000) the authors consider a finite society in which agents are endowed with some amount of a private but reproducible good (*information*) that upon contact can be copied or transmitted from one agent to another. While there is a cost of establishing a connection, there are also gains in accessing new information. The difference between these two yields the payoff of a connection. Rational agents will behave strategically and the Nash equilibria will provide the *network architecture*.

As it is well known from Erdös and Renyi (1959) seminal treatment of random graphs, new connections may lead to phase transitions in the density of the graph. That is, jumps in the number of clusters from many to a single major one. While for social networks the framework of random graph is not quite cogent, similar results may arise varying the nature of the probability distribution on potential connections (Newman et al., 2002).

In Chapter 13 the authors will show how the same is true in our non-probabilistic, gametheoretic framework. By slight changes in the information carried by individual agents (representing the influence of non-social sources) the equilibrium networks may vary suddenly. The authors will see that in the end, if each agent has an information endowment larger than the cost of establishing connections to her, a minimally connected network becomes the unique outcome. That means that a highly organized structure arises when everyone is "valuable". On the other hand, if the value is too low for every agent (i.e. there is no gain in connecting to others) the only efficient outcome is the empty network. In the middle, the authors will show, there exist some critical agents to which most of the others will want to establish contact and yield components in a disconnected network.

Game theory has been applied to many different kinds of fields including economics, politics, even engineering since it was developed and proposed by John Von Neumann and John Forbes Nash. It was originally applied to economic problems to analyze the interactions between players in competitive situations, but it has evolved as a general tool which is useful for modeling various kinds of systems including not just social but also physical ones if they have distributed schemes composed of interactive components. On the aspect of application, Chapter 14 introduces two different examples analyzed and modeled by game theoretic approach. Two examples have totally different qualities although both are the problems in electric power industry. One is a typical application of game theory to economic problems in electricity market representing social system, and the other is the application to SCADA communication and its security problem in the area of physical system which does not have independent decision making entities. By comparing two different cases based on different systems the authors intended to show the flexible application of game theory to solving the problems.

In: Game Theory: Strategies, Equilibria and Theorems
Editors: Ingrid N. Haugen et al, pp. 1-40
ISBN 978-1-60456-844-8

Chapter 1

Review of Quantum Game Theory

Adrian P. Flitney
School of Physics, University of Melbourne, Parkville VIC 3010, Australia

Abstract

Ultimately information needs to be written, stored and read by physical devices. The study of the quantum mechanical aspects of information processing is known as quantum information theory. There is much current interest in utilizing some of the intrinsic features of quantum mechanics, such as superposition, interference and entanglement, to create algorithms that work on a quantum computer in completely novel ways in comparison to classical algorithms. At the intersection of the theory of quantum information and game theory is a new area of investigation known as quantum game theory. We outline the idea of quantum bits (qubits), their interference and entanglement, and the allowable operations on them. Then we discuss the various protocols for quantizing well known classical games. At this early stage in the theory there is no consensus on precisely how this is done.

We review the work carried out in the model of Eisert et al. [Eisert, Wilkins, and Lewenstein, *Phys. Rev. Lett.* **83** (1999) 3077]. Despite early claims that this model can lead to new equilibria in 2×2 games that "solve" dilemmas like the famous Prisoners' Dilemma, we emphasize that this is an artifact of an artificial restriction on the players' allowed quantum actions. Access to the full set of quantum operations enriches the players' choices, but in 2×2 games does not lead to any new insights. The situation is more fruitful in multi-player games. In certain circumstances the power of quantum entanglement can be exploited to produce equilibria that are more efficient than those that can be obtained classically, and that have no classical analogue. Additionally, in some systems there is a correspondence between the payoffs for these equilibria and fundamental quantum properties of the initial state, as measured by violations of the Bell inequalities.

Generalizations of the Eisert protocol are discussed as well as other approaches to quantum games. Recent experimental implementations of some simple quantum games are presented.

1. Introduction

"Landauer based his research on a simple rule: information is physical. That is, information is registered by physical systems such as strands of DNA, neu-

rons and transistors; in turn the ways in which systems such as cells, brains and computers can process information is governed by the laws of physics. Landauer's work showed that the apparently simple and unproblematic statement of the physical nature of information had profound consequences."
— Seth Lloyd on Rolf Laundauer [1]

There is currently an intense interest in the nature of information in the quantum domain. The "quantum weirdness" in the behavior of subatomic particles that can, for example, be in multiple states at the same time, has long been a subject of fascination to physicists and of perplexity to non-physicists. Feynman was the first to consider the idea of a computer that could advantageously exploit quantum behavior [2]. In 1994 Shor deduced how quantum entanglement and superposition could be utilized for the difficult problem of prime factorization of integers [3,4]. Two years later Grover came up with an algorithm for a fast database search that used a quantum superposition to examine multiple states simultaneously, and as a result could find an item in an unsorted date quadratically faster than the best classical search algorithms [5]. Once the problem of error detection and correction had been addressed [6], the idea of a practical quantum computer began to be taken seriously.

As early as 1980 Blaquiere made the connection between quantum (wave) mechanics and game theory [7]. The idea of quantum cryptography, which can be considered in game-theoretic terms, was initiated a little later [8]. Mermin's ideas on multi-party entanglement [9, 10] can be viewed as an n-person game that can be won with certainty using quantum resources but with a probability no better than $\frac{1}{2} + \frac{1}{2^{(n/2)}}$ classically. The field of quantum game theory finally came to the general attention of theorists in 1999 with publications by Eisert *et al.* [11] and Meyer [12], and excited interest in the popular science press [13–15]. The model of Eisert *et al.* remains the standard for the exploration of 2×2 games, and with extensions, to larger game spaces.

In the literature on quantum theory there are two main uses of the term "quantum games." In one, known as cooperative games or non-local games [16] two or more agents in separated locations try to achieve some joint task with little or no communication, by first sharing an entangled quantum state. The maximum possible success probability of completing the task is the value of such a game. One compares the success probability using quantum resources with that obtainable classically to determine if there is an advantage in using quantum entanglement. The term *pseudo-telepathy* [17] is also used to describe a similar type of non-local cooperative task where quantum entanglement serves to eliminate the need for communication between the parties, thus giving the appearance of "telepathy." These are cooperative tasks for remote players and not games in the von Neumann sense. In a larger body of literature, the term "quantum games" refers to competitive von Neuman games played with quantum resources. Games fitting this latter definition will be the sole concern of this review. This is not to say that the non-local tasks are unimportant. Indeed, in many cases, they are relevant to the prospective applications of quantum computers or quantum communication devices. However, they are more of concern to quantum theorists rather than game theorists, to which this current review is aimed.

In this chapter we review the work to date on quantum game theory with particular emphasis on the properties of games in the Eisert protocol. Other work in the field of quantum games will be outlined with a bias towards more recent ideas. Interested readers are encouraged to view the source references for details of the alternative schemes. An earlier review was published in Ref. [18]. I have tried to be complete but inevitably there will be omissions for which I apologize in advance.

This chapter is divided as follows. A short introduction to the necessary notation and concepts of elementary quantum mechanics is given in Sec. 2., which can be skipped by those familiar with the field. Section 3. discusses the original protocol of Eisert, with subsections detailing the various investigations into 2×2 quantum games in this protocol. Generalizations to multiple players are considered in Sec. 4. and to iterated quantum games in Sec. 5. Various other models of quantum games are discussed in Sec. 6. The controversy over the nature of quantum games is discussed in Sec. 7. Finally, a brief outline of the physical implementation of quantum games is given in Sec. 8.

2. Introductory Concepts of Quantum Mechanics

A good introduction to all aspects of quantum computation is provided in several recent books. The bible remains the excellent work by Nielsen and Chuang [19]. Below I summarize some concepts from elementary quantum mechanics that I hope are sufficient for the non-specialist to follow the remainder of the chapter.

The Dirac notation for a quantum state labeled by ψ is the *ket* $|\psi\rangle$. The state $|\psi\rangle$ belongs to a complex vector space known in quantum mechanics as a *Hilbert space*. If $\{|\phi_i\rangle, i = 1, \ldots, N\}$ is an orthonormal basis for an N-dimensional Hilbert space, then $|\psi\rangle$ may be decomposed as

$$|\psi\rangle = \sum_{i=1}^{N} c_i |\phi_i\rangle, \tag{1}$$

where the c_i are complex numbers.

The *bit* is the fundamental unit of information in classical computation, taking the values 0 or 1. It's quantum analog is the quantum bit or *qubit*. Amongst its possible values are $|0\rangle$ or $|1\rangle$, known as the *computational basis* states. However, a qubit may also be a convex linear combination, or *superposition*

$$|\psi\rangle = \alpha|0\rangle + \beta|1\rangle, \tag{2}$$

with the normalization condition $|\alpha|^2 + |\beta|^2 = 1$. The physical interpretation of (2) is that a measurement of $|\psi\rangle$ in the computational basis (that is, if we examine a qubit to determine if it is $|0\rangle$ or $|1\rangle$) will return $|0\rangle$ with probability $|\alpha|^2$ and $|1\rangle$ with probability $|\beta|^2$. The quantum state (2) can be represented by the vector

$$\begin{bmatrix} \alpha \\ \beta \end{bmatrix}. \tag{3}$$

Such a superposition is not simply a classical ensemble of its component states. It is **not** the case that $|\psi\rangle$ is actually $|0\rangle$ or $|1\rangle$ but we do not yet know which. Instead each component of the superposition is simultaneously present. The superposition (2) is often referred to as a *coherent superposition* to emphasize the existence of coherence between the components. *Coherence* can be thought of as a measure of the "quantumness" of a state.

Multiple qubits, each in their own two-dimensional Hilbert space are written, for example, as $|0\rangle \otimes |1\rangle \equiv |01\rangle$. The notation $\langle\phi|$, known as the *bra* ϕ, refers to the Hermitian conjugate of $|\phi\rangle$. For example, the Hermitian conjugate of (2) is

$$\langle\psi| = \alpha^* \langle 0| + \beta^* \langle 1|, \tag{4}$$

or

$$\begin{bmatrix} \alpha^* & \beta^* \end{bmatrix}, \tag{5}$$

where * refers to complex conjugation. The overlap between two states is written as the *bra-ket* $\langle\phi|\psi\rangle$. That is, the probability of a measurement[1] revealing $|\phi\rangle$ and $|\psi\rangle$ in the same state is given by $|\langle\phi|\psi\rangle|^2$. The two states are orthogonal if this value is zero. For an orthonormal basis of an N-dimensional Hilbert space, $\{|\phi_i\rangle, i = 1, \ldots, N\}$, we have $|\langle\phi_i|\phi_j\rangle|^2 = \delta_{ij}$, where δ_{ij} is the usual Kronecker delta.

In this work we are generally only concerned with the simplest of Hilbert spaces, those spanned by $\{|0\rangle, |1\rangle\}^{\otimes N}$. When we have two or more qubits they may exist in an *entangled* state such as

$$|\psi\rangle = \frac{|0\rangle_{\mathrm{A}} \otimes |0\rangle_{\mathrm{B}} + |1\rangle_{\mathrm{A}} \otimes |1\rangle_{\mathrm{B}}}{\sqrt{2}} \equiv \frac{|0_{\mathrm{A}}0_{\mathrm{B}}\rangle + |1_{\mathrm{A}}1_{\mathrm{B}}\rangle}{\sqrt{2}}. \tag{6}$$

The peculiarity of quantum mechanics is encapsulated in such states. Even though both qubits A and B are indeterminate, with each having a 50/50 chance of being found in the $|0\rangle$ or $|1\rangle$ state when measured in the computational basis, a measurement on both qubits will always yield the same value. This is true even if the physical systems representing the qubits are separated. It is as if a measurement on one qubit *instantaneously* determines[2] the state of a distant qubit. This "spooky action at a distance" lead Einstein, Podolsky and Rosen (EPR) in a famous paper [20] to conclude that either quantum mechanics is inconsistent with *local, realism* or does not provide a complete description of nature. Subsequent experiments have confirmed quantum mechanical predictions [21]. The usual philosophical position is to give up *locality*: in quantum mechanics measurements on spatially separated systems that have been prepared together can affect one another. The measurement correlations in the quantum state (6) are stronger than can exist in *any* classical system [22].

Classical computation proceeds by means of logic gates acting on bits that are transmitted down wires. The only non-trivial gate on a single bit is the NOT gate that transforms $0 \leftrightarrow 1$ (Fig. 1). The quantum NOT gate acts in the same way on the computational basis states $|0\rangle$ and $|1\rangle$ and acts linearly on a superposition:

$$\alpha|0\rangle + \beta|1\rangle \to \alpha|1\rangle + \beta|0\rangle. \tag{7}$$

The action of the quantum NOT operator $\hat{X}$ can be represented by the matrix

$$\hat{X} \equiv \begin{bmatrix} 0 & 1 \\ 1 & 0 \end{bmatrix}, \tag{8}$$

[1] We are not going to be concerned with what constitutes a measurement in quantum mechanics, since this is an unsolved problem. A common sense definition of a measurement will suffice.

[2] It is too much of digression to discuss this in detail, but note that this instantaneous action at a distance cannot be used to transmit useful information, and so there is no conflict with Einstein's theory of Special Relativity.

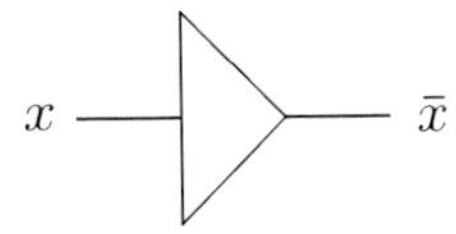

Figure 1. A classical NOT gate.

where the hat over a symbol indicates that it is an operator. Equation (8) is the matrix representation of the operator in the computational basis. This vector/matrix notation will be used through out the chapter. In quantum mechanics there are other important single qubit gates, The phase flip

$$\hat{Z} \equiv \begin{bmatrix} 1 & 0 \\ 0 & -1 \end{bmatrix}, \tag{9}$$

has the effect of flipping the relative phase between the $|0\rangle$ and $|1\rangle$ components of a superposition. A combination of bit- and phase-flips is carried out by the operator

$$\hat{Y} \equiv \begin{bmatrix} 0 & -\mathrm{i} \\ \mathrm{i} & 0 \end{bmatrix}. \tag{10}$$

The matrices $\{\hat{X}, \hat{Y}, \hat{Z}\}$ are collectively known as Pauli matrices and together with the 2×2 identity matrix $\hat{I}$ form a basis for the space of 2×2 unitary matrices.

In quantum mechanics multiplying a state by an arbitrary phase $\mathrm{e}^{\mathrm{i}\alpha}$ has no physical effect: the probabilities of measurement are not affected since these are proportional to $|\cdot|^2$. However, the relative phase between the components of a superposition is important as we shall see later. An arbitrary phase difference between the $|0\rangle$ and $|1\rangle$ states can be applied by the phase operator

$$\hat{\mathcal{P}}(\alpha) \equiv \begin{bmatrix} 1 & 0 \\ 0 & \mathrm{e}^{\mathrm{i}\alpha} \end{bmatrix}. \tag{11}$$

The *Hadamard* gate

$$\hat{H} \equiv \frac{1}{\sqrt{2}} \begin{bmatrix} 1 & 1 \\ 1 & -1 \end{bmatrix}, \tag{12}$$

changes the computational basis states into a superposition half way between $|0\rangle$ and $|1\rangle$. That is, in operator notation:

$$\begin{aligned} \hat{H}|0\rangle &= (|0\rangle + |1\rangle)/\sqrt{2}, \\ \hat{H}|1\rangle &= (|0\rangle - |1\rangle)/\sqrt{2}. \end{aligned} \tag{13}$$

It is important to note that all operators act linearly on a superposition. In a quantum circuit the operators are represented as in Fig. 2, which shows the action of the Hadamard gate. Here the "wires" represent any medium through which a qubit can be propagated.

In general, any single qubit operation can be represented by a 2×2 unitary matrix. Since the overall phase is not physically relevant three real numbers are required to parametrize an arbitrary single qubit operator. In this work we shall choose the form

$$\hat{M}(\theta, \alpha, \beta) = \begin{bmatrix} \cos(\theta/2)\, \mathrm{e}^{\mathrm{i}\alpha} & \mathrm{i}\sin(\theta/2)\, \mathrm{e}^{\mathrm{i}\beta} \\ \mathrm{i}\sin(\theta/2)\, \mathrm{e}^{-\mathrm{i}\beta} & \cos(\theta/2)\, \mathrm{e}^{-\mathrm{i}\alpha} \end{bmatrix}, \tag{14}$$

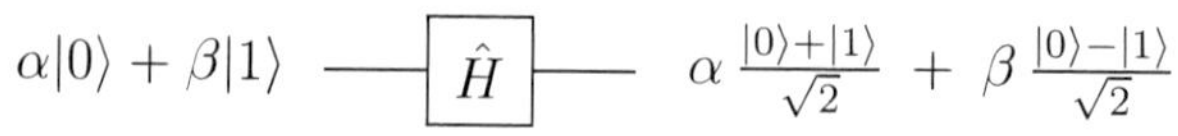

Figure 2. A Hadamard gate.

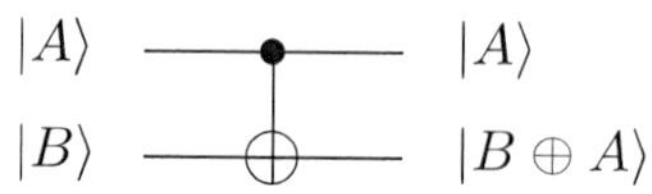

Figure 3. A controlled-NOT gate, where the binary operator $\oplus$ represents addition modulo two.

where the rotation angle $\theta \in [0, \pi]$ and the phases $\alpha,\ \beta \in [-\pi, \pi]$.

One operator that is of particular use is the projection operator onto the state $|\phi\rangle$:

$$\hat{P} = |\phi\rangle\langle\phi|. \tag{15}$$

When applied to a state $|\psi\rangle$ it has the effect of projected out only the component parallel with $|\phi\rangle$. For example $\hat{P}_0 \equiv |0\rangle\langle 0|$ applied to the superposition (2) results in the state $\alpha|0\rangle$. Multi-qubit projection operators, such as $|00\rangle\langle 00|$, are also possible.

In classical computation it is known that the two-bit gate NAND is universal. That is, any computation can be performed by combinations of single-bit gates (NOT gates) and NAND gates. In quantum computation, since all gates can be represented by unitary operators, they are necessarily reversible. Hence, there is no quantum analogue to the NAND gate, or of the other two-bit gates, AND, OR, and XOR. However, there is a two-qubit gate that together with the set of single-qubit gates forms a universal set. Any multiple qubit gate can be formed from the combination of single qubit gates and controlled-NOT or CNOT gates which flip the second qubit if the first, or control qubit, is in the $|1\rangle$ state, as indicated in Fig. 3.

Quantum parallelism coupled with the phenomenon of *interference* gives us the ability to extract more information from a process than would be possible classically. The simplest example of this is Deutsch's algorithm [23] which we sketch below. Suppose we are given a black box that determines the function $f : \{0, 1\} \rightarrow \{0, 1\}$. To answer the question, "Does $f(0) = f(1)$?" using only classical methods requires two evaluations of the function by the black box. In an information theoretic sense this seems excessive since we only wanted a single bit of information about the function. Now, consider the quantum circuit in Fig. 4. At the heart of the circuit is a two-qubit unitary gate that evaluates f on the first qubit and adds it modulo two to the second, as indicated. By applying f to a superposition of the $|0\rangle$ and $|1\rangle$ states we achieve parallelism: we effectively compute f on both possibilities

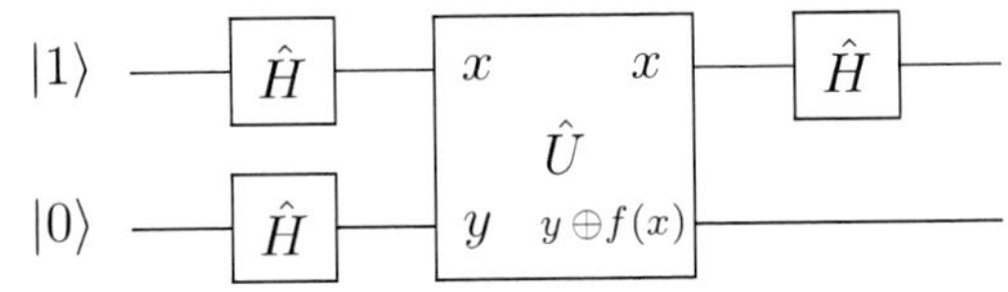

Figure 4. Deutsch's algorithm to answer whether the binary function $f : \{0, 1\} \rightarrow \{0, 1\}$ is constant, using only a single evaluation of f.

simultaneously. A little calculation reveals that the final state can be written as

$$|\psi_{\text{final}}\rangle = \pm|0\rangle \otimes \frac{|0\rangle - |1\rangle}{\sqrt{2}} \qquad \text{if } f(0) = f(1)$$
$$|\psi_{\text{final}}\rangle = \pm|1\rangle \otimes \frac{|0\rangle - |1\rangle}{\sqrt{2}} \qquad \text{if } f(0) \neq f(1). \tag{16}$$

By examining the first qubit we can now tell if $f(0) = f(1)$. For the details of the calculations the interested reader is referred to section 1.4.3 of Ref. [19].

States such as the superposition (2) are known as *pure states*. Sometimes it is necessary to consider a probabilistic mixture of states, or a *mixed states* $\sum p_i|\psi_i\rangle$, where $\sum p_i = 1$. Vector notation like (3) is no longer sufficient. Instead such states can be represented by a *density matrix*

$$\rho = \sum p_i|\psi_i\rangle\langle\psi_i|. \tag{17}$$

The action of applying an operator $\hat{A}$ to this state is a new density matrix $\hat{A}\rho\hat{A}^\dagger$. The ith diagonal element of the density matrix represents the probability of finding the state $|\psi_i\rangle$ upon taking a suitable measurement.

3. The Eisert Protocol for 2×2 Quantum Games

3.1. Introduction

For the purposes of this chapter, quantum game theory is the playing of a von Neumann competitive game between two or more parties using quantum resources. For the most part we shall consider the strategic form of the game. A simple demonstration of the power of quantum superpositions in a game-theoretic context was published by Meyer in 1999 [12]. The game of Penny-Flip proceeds as follows. A coin is placed in a box with heads face up. First Alice, then Bob, then Alice again secretly choose to flip the coin or leave it as is. Then the coin is revealed and if it shows heads Alice wins, tails Bob wins. The game is simply solved: the Nash equilibrium (NE) strategy for both sides is to choose between flipping and not flipping with equal probability. Both players then have an equal chance of winning.

Meyer supposed that instead we play the game with a quantum "coin" and that Alice is permitted any unitary operation on the coin, while Bob continues to choose between the classical moves. Surprisingly, with this arrangement Alice can ensure she wins 100% of the time against any strategy by Bob. We shall write $|0\rangle$ for the heads state and $|1\rangle$ for tails to avoid confusion with the Hadamard operator $\hat{H}$. First she operates on the coin with the Hadamard operator to produce the state $(|0\rangle + |1\rangle)/\sqrt{2}$. This state is unaltered by any combination of $\hat{I}$ and $\hat{X}$ chosen by Bob. Alice then applies $\hat{H}$ again, returning the state to $|0\rangle$ with certainty. In this example Alice benefited from her increased strategic space, and from the fact that she had two moves to Bob's one. A second (classical) move by Bob would return the game to one balanced between Alice and Bob despite Alice's quantum powers.

3.2. The Eisert Model

Consider a classical game with two pure strategies. A player's move may be encoded as a single bit. A strategy is then a prescription of how to choose the value of the player's bit. If the players start with a bit in the 0 state the two pure strategies can be considered as operators on this bit: the identity operator, leaving the choice at 0, or the bit-flip operator changing the choice to 1. Mixed strategies correspond to a convex linear combination of the identity and bit-flip operations. Once the final state of the players' bits have been determined the payoffs can be assigned in accordance with the payoff matrix. It is this description of a player's actions in terms of operators acting on information states that forms the basis of the idea of Eisert *et al.* [11].

When a strategic game is translated into the quantum realm the bit representing the player's move is replaced by a qubit, with the computational basis states $|0\rangle$ and $|1\rangle$ corresponding to the classical pure strategies 0 and 1. We suppose the players have instruments that can manipulate the state of their qubit. In the protocol of Eisert *et al.* [11, 24] a pure quantum strategy is the choice of unitary operator that the player applies to their qubit, while a mixed quantum strategy is a convex linear combination of unitaries. Strictly, a strategy is prescription for choosing the operator(s) but in what follows I will generally refer to the player's operator as their strategy. Hopefully no confusion will arise. The players are first given a qubit prepared in some known state. Each player then has the opportunity of executing their strategy by applying their chosen unitary operator. The state of the qubit pair is then measured in some basis and the payoffs are assigned according to the classical payoff matrix. In Eisert's protocol the qubits are prepared in an entangled state by applying to an initial state $|\psi_{\text{in}}\rangle = |00\rangle$ the entangling operator

$$\begin{aligned}\hat{J}(\gamma) &= \cos\left(\frac{\gamma}{2}\right) \hat{I} \otimes \hat{I} + \mathrm{i}\sin\left(\frac{\gamma}{2}\right) \hat{X} \otimes \hat{X} \\ &= \begin{bmatrix} \cos\frac{\gamma}{2} & 0 & 0 & \mathrm{i}\sin\frac{\gamma}{2} \\ 0 & \cos\frac{\gamma}{2} & \mathrm{i}\sin\frac{\gamma}{2} & 0 \\ 0 & \mathrm{i}\sin\frac{\gamma}{2} & \cos\frac{\gamma}{2} & 0 \\ \mathrm{i}\sin\frac{\gamma}{2} & 0 & 0 & \cos\frac{\gamma}{2} \end{bmatrix}. \end{aligned} \tag{18}$$

where $\hat{I}$ is the single qubit identity operator, $\hat{X}$ is the bit-flip operator (8), and $\gamma \in [0, \pi/2]$ is a measure of the degree of entanglement. Maximal entanglement is achieved by setting $\gamma = \pi/2$, since $\hat{J}(\pi/2)|00\rangle = (|00\rangle + \mathrm{i}|11\rangle)/\sqrt{2}$. An unentangled game results if $\gamma = 0$ and is equivalent to the classical game. In the original paper of Eisert *et al.* [11] a slightly different form of the entangling operator was used. This only has the result of rotating the final quantum state (prior to measurement) in the complex plane by some phase factor. As noted Sec. 2., this has no physically measurable effects. That is, there is no difference in

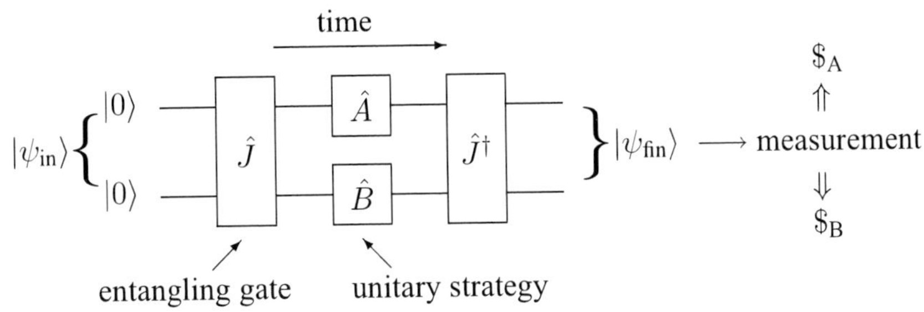

Figure 5. Eisert's protocol for a 2×2 quantum game, where $\hat{A}$ and $\hat{B}$ are the (unitary) strategies applied by Alice and Bob and $\hat{J}$ is an entangling gate.

the expected payoffs between the two schemes. The form of (18) is chosen since it is more easily generalizable to multiple players (see Sec. 4.). After the players have completed their moves, the operator $\hat{J}^\dagger$ is applied to the state, where the $\dagger$ corresponds to taking the Hermitian conjugate, and then a measurement is made in the computational basis. These two steps are equivalent to taking a measurement on the pair of qubits in the orthonormal basis $\{(|00\rangle + \mathrm{i}|11\rangle)/\sqrt{2}, (|01\rangle + \mathrm{i}|10\rangle)\sqrt{2}, (\mathrm{i}|01\rangle + |10\rangle)/\sqrt{2}, (\mathrm{i}|00\rangle + |11\rangle)/\sqrt{2}\}$. With this formalism, the classical moves correspond to $\hat{I}$ and $\mathrm{i}\hat{X}$. If both players select operators of the form of (14) with $\alpha = \beta = 0$ a classical game results where each player's choice of θ determines their mixture between the moves $|0\rangle$ ($\theta = 0$) and $|1\rangle$ ($\theta = \pi$). However, when we permit non-zero values of the phases α and β new and interesting results can be obtained. To summarize: we generate a final state of a pair of qubits, one for each player, by

$$|\psi_{\text{fin}}\rangle = \hat{J}^\dagger(\gamma)(\hat{A} \otimes \hat{B})\hat{J}(\gamma)|00\rangle, \tag{19}$$

where $\hat{A}$ and $\hat{B}$ represent Alice and Bob's strategies, respectively. The final state $|\psi_{\text{fin}}\rangle$ is then measured in the computational basis, and payoffs are assigned according to the values of the players qubits (which are now either in the $|0\rangle$ or $|1\rangle$ state) and the classical payoff matrix. The protocol is shown schematically in Fig. 5. Since quantum mechanics is inherently probabilistic, even when both players choose a pure quantum strategy the final state can be a mixture of different results. The quantities of interest to us are the *expectation values*, or average values taken over a large number of identical trails, of the payoffs to Alice and Bob. These may be computed by

$$\langle \$ \rangle = \sum_{i,j=1}^{2} \$_{ij} |\langle ij|\psi_f\rangle|^2, \tag{20}$$

where $\$_{ij}$ is the payoff to Alice (or Bob) associated with the classical game outcome $ij,\ i,j \in \{0,1\}$.

3.3. Similar Models for 2×2 Quantum Games

A modification of the above protocol was published by Marinatoo and Weber [25]. This can be said to be a classical game played with quantum resources. An initial entangled state

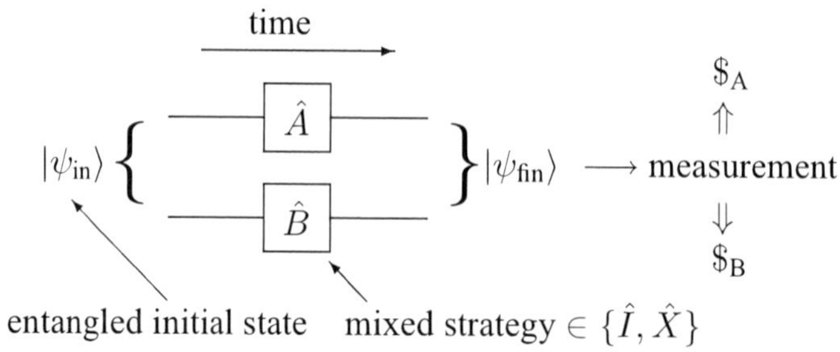

Figure 6. Marinatto and Weber's protocol for a 2×2 quantum game. Alice and Bob's strategies are a mixture of the classical moves $\hat{I}$ and $\hat{X}$.

is prepared with one qubit passed to each of the players. The players are restricted to mixed classical strategies with probability p of selecting the identity operator and probability $1-p$ of choosing the bit-flip operator $\hat{X}$. After the players have executed their strategy the state of the qubits is measured in the computational basis and payoffs are awarded as before. There is no "unentangling" operator $\hat{J}^\dagger$ in this protocol. The classical game is again a subset of the quantum protocol but in this case it is achieved by preparing the initial state $|00\rangle$. The fully entangled initial state $(|00\rangle + |11\rangle)/\sqrt{2}$ gives the maximum quantum effects. The game is still played with quantum resources (qubits) and coherence needs to be maintained until the final measurement. A schematic of this protocol is given in Fig. 6. This scheme did not gain as much popularity as that of Eisert *et al.* However, in both, quantum solutions may be artifacts of the restrictions on the allowable quantum actions of the players, and classicality of the appropriate parts of the physical system reproduces the classical game. The schemes need to be judged in terms of the insight they provide.

A generalization of Eisert's protocol with two different γ's:

$$|\psi_{\text{fin}}\rangle = \hat{J}^\dagger(\gamma_2)(\hat{A} \otimes \hat{B})\hat{J}(\gamma_1)|00\rangle, \tag{21}$$

was proposed by Nawaz and Toor [26]. This includes as a subset the protocols of Eisert's ($\gamma_1 = \gamma_2$) and Marinatto ($\gamma_2 = 0$), in the latter case with the additional restriction that $\hat{A}$ and $\hat{B}$ can only be probabilistic mixtures of $\hat{I}$ and $\hat{X}$. An extension of the Eisert protocol to the consideration of games with more than two classical pure strategies is easily made by replacing the players' qubits by qunits, n-state quantum systems, and replacing the 2×2 unitary operators by $n \times n$ ones.

Lee and Johnson [27] propose a general formalism for quantum games as a triple (N, Ω, P), where N is the number of players, $P : \Omega \rightarrow \mathbb{R}^N$ such that $P_k(\cdot)$ is the payoff function of the kth player, and $\Omega = \otimes_{k=1}^N \Omega_k$, $\Omega_k =$ strategy space of the kth player. With this formalism quantum versions of the minimax theorem and the Nash equilibrium theorem can then be demonstrated. The authors also show that the class of quantum games is strictly larger than the class of classical ones. Their formalism was then applied to a game-theoretic approach to the problems of quantum state estimation and quantum cloning [28].

3.4. Quantum versus Classical Players

One of the first things to be considered in quantum games was the advantage of a player with access to a full set of quantum strategies over one that is restricted to the classical strategies, represented by $\hat{I}$ and $\mathrm{i}\hat{X}$. This property can be illustrated by considering the well known 2×2 game of Prisoners' Dilemma, typically characterized by the payoff matrix

	Bob : C	Bob : D
Alice : C	$(3,3)$	$(0,5)$
Alice : D	$(5,0)$	$(1,1)$

(22)

where the first number in each parenthesis corresponds to Alice's payoff and the second to Bob's. The strategies C and D correspond to cooperation and defection, respectively. The literature on this game is extensive. For an accessible introduction see, for example, Ref. [29].

In this game the dilemma arises since there is a conflict between the NE strategy of "always defect" and the Pareto optimal (PO) strategy of "always cooperate." The former is the rational choice in a one-shot game since it is the dominant strategy, yielding a higher payoff than cooperation what ever the opponent plays. However, if both players had chosen to cooperate instead of defect they would have both received payoffs of three rather than one. The PO result represents the best result for the players as a whole. It is just such conflicts between individual and group "rationality" in our every day life and in the life on nations that lead to much of the misery through out the world. In a repeated Prisoners' Dilemma cooperation can emerge. Early work in this area was performed by Axelrod [30, 31].

In the quantum extension of Prisoners' Dilemma $|0\rangle$ will be associated with cooperation and $|1\rangle$ with defection. The strategy "always cooperate" is represented by the operator $\hat{M}(0,0,0)$ while "always defect" is represented by $\hat{M}(\pi,0,0)$. Operators $\hat{M}_{\mathrm{cl}}(\theta) \equiv \hat{M}(\theta,0,0)$ correspond to mixed classical strategies. We can now consider a quantum game of Prisoners' Dilemma between a quantum player Bob, with access to the full set of unitary operators, and a classical Alice, who is restricted to the set of $\hat{M}_{\mathrm{cl}}$. When we have full entanglement ($\gamma = \pi/2$) it is most advantageous for Bob to play Eisert's "miracle" move [11]

$$\hat{M}(\frac{\pi}{2},\frac{\pi}{2},0) = \frac{\mathrm{i}}{\sqrt{2}}\begin{bmatrix} 1 & 1 \\ 1 & -1 \end{bmatrix} \tag{23}$$

that yields a payoff of $\langle \$_{\mathrm{B}} \rangle = 3 + 2\sin\theta_{\mathrm{A}}$ for Bob while leaving Alice with only $\langle \$_{\mathrm{A}} \rangle = (1 - \sin\theta_{\mathrm{A}})/2$ when she plays $\hat{M}_{\mathrm{cl}}(\theta_{\mathrm{A}})$. In this case the dilemma is resolved in favor of the quantum player. The payoffs are dependent on the entanglement parameter γ. There is a critical value $\sin\gamma = 1/\sqrt{5}$ below which (23) is no longer optimal and the quantum player should revert to the classical dominant strategy of $\hat{M}(\pi,0,0)$ as shown in Fig. 7. The critical value of γ that signifies the phase-like transition from the quantum to the classical regions depends on the precise values of the payoffs in (22) [32, 33].

Quantum versions of other well studied 2×2 games have been considered in the literature. In particular these include various formulations of the Battle of the Sexes [25, 34] and Chicken [24,25,35]. There are different miracle moves for the quantum player depending upon which final result is optimal for them. The advantage obtainable by the quantum

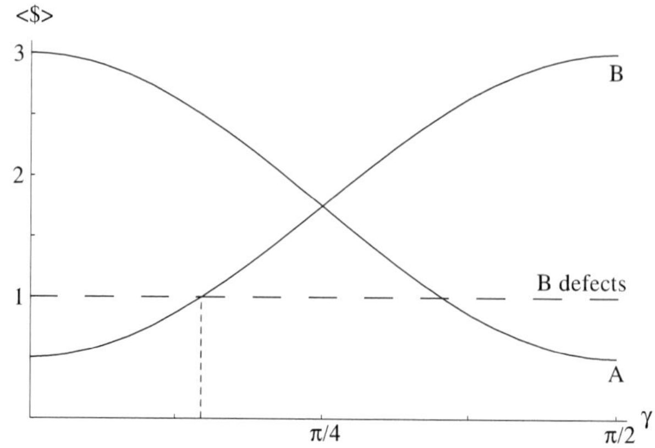

Figure 7. The expected payoffs for a classical Alice (A) playing "always defect" and a quantum Bob (B) versus the entanglement parameter γ. The solid lines give the payoffs when Bob plays his "miracle" move (23) while the dashed line gives the payoffs to both players when Bob defects. Below an entanglement of $\arcsin(1/\sqrt{5})$ (short dashes) Bob does best against a defecting Alice to also defect. Figure adapted from Ref. [11].

player against a classical one in quantum versions of the game of Chicken, Deadlock, Stag Hunt, and Battle of the Sexes is discussed in Refs. [36, 37]. Thresholds in the entanglement parameter below which the quantum player cannot obtain an advantage occur in all these games except the Battle of the Sexes.

3.5. New "Equilibria" in Two-Parameter Strategic Spaces

It is more interesting to consider the case where both players have the same strategic space. There was initially some excitement in the popular physics literature that quantum game theory may provide "solutions" to the Prisoners' Dilemma and other game-theoretic problems [13, 14]. However, this excitement was premature. In the original paper by Eisert *et al.* [11] the strategic space of the players was restricted to operators of the form

$$\hat{M}(\theta, \alpha) = \begin{bmatrix} e^{i\alpha} \cos\frac{\theta}{2} & -\sin\frac{\theta}{2} \\ \sin\frac{\theta}{2} & e^{-i\alpha}\cos\frac{\theta}{2} \end{bmatrix}, \tag{24}$$

with $\theta \in [0, \pi]$ and $\alpha \in [0, \pi/2]$. That is, (14) with $\beta \equiv 0$. When entanglement is maximal ($\gamma = \pi/2$) this restriction produces a new preferred strategy for both players in quantum Prisoners' Dilemma[3]

$$\hat{C}' \equiv \hat{M}(0, \frac{\pi}{2}, 0) = \begin{bmatrix} i & 0 \\ 0 & i \end{bmatrix}, \tag{25}$$

with a new NE of $\hat{C}' \otimes \hat{C}'$. This new NE has the benefit of also being PO with a payoff of three to both players, the same as mutual cooperation. The situation when entanglement is less than maximal is analyzed by Du *et al.* [32]. With the payoffs of (22), $\hat{C}' \otimes \hat{C}'$ is NE provided $\sin^2\gamma \geq \frac{2}{5}$, while for $\sin^2\gamma < \frac{1}{5}$ the game behaves classically with $\hat{D} \otimes \hat{D}$ the

[3] In Ref. [11] and other papers this strategy is referred to as $\hat{Q}$ but the strategy is cooperation with an additional phase, hence the symbol used in this chapter.

only NE. In the intermediate region there are two asymmetric NE, $\hat{C}' \otimes \hat{D}$ and $\hat{D} \otimes \hat{C}'$, with a payoff of $5\sin^2\gamma$ to the player choosing $\hat{C}'$ and $5\cos^2\gamma$ to the defector.

The game of Chicken can be characterized by the payoff matrix

	Bob : C	Bob : D
Alice : C	$(3,3)$	$(1,4)$
Alice : D	$(4,1)$	$(0,0)$

(26)

A similar NE versus entanglement structure emerges, with the boundary between the "quantum" region with the $\hat{C}' \otimes \hat{C}'$ NE being at $\sin^2\gamma = \frac{1}{3}$. The intermediate region extends down to $\gamma = 0$ and only at this value does the game behave fully classically (as it must in the absence of entanglement).

There appears to be no physical justification for selecting the particular subset of unitary operators with $\beta \equiv 0$, though this has not stopped a number of authors from exploring the properties of quantum games restricted to this, so called, two-parameter strategic space [38–43]. If an alternative subset of unitary operators, those of the form of (14) but with $\alpha \equiv 0$ are chosen, different consequences result. Such a strategic space was used in Ref. [44] to explore the NE versus entanglement structure of two- and three-player quantum Prisoners' Dilemma, though the authors failed to comment on the divergence of their results from earlier ones for two-player Prisoners' Dilemma [32, 33] achieved with $\beta \equiv 0$. A full comparison of the results for these two-parameter strategic spaces in various two-player quantum games and multi-player quantum Prisoners' Dilemma is given in Ref. [45]. Some of the main results are summarized in Fig. 8.

3.6. Three-Parameter Strategic Spaces and the Loss of the New "Equilibria"

In response to Eisert's original paper on 2×2 quantum games [11] it was observed by Benjamin and Hayden that the two-parameter space (24) in the original paper is not the most general, and is not closed under composition [46]. For an arbitrary three-parameter strategy $\hat{A} = \hat{M}(\theta, \alpha, \beta)$ there always exists $\hat{B} = \hat{M}(\theta, \alpha, -\frac{\pi}{2} - \beta)$ such that

$$(\hat{A} \otimes \hat{I})\frac{1}{\sqrt{2}}(|00\rangle + \mathrm{i}|11\rangle) = (\hat{I} \otimes \hat{B})\frac{1}{\sqrt{2}}(|00\rangle + \mathrm{i}|11\rangle). \tag{27}$$

That is, on the maximally entangled state any unitary operation carried out by Alice on her qubit is equivalent to a unitary operation carried out by Bob on his. Thus for any strategy $\hat{M}(\theta, \alpha, \beta)$ chosen by Alice, Bob can undo its effect and apply his preferred strategy $\hat{D}$, say, by playing $\hat{D}\hat{M}(\theta, -\alpha, \frac{\pi}{2} - \beta)$. Hence every pure quantum strategy has an effective counter strategy and there can be no equilibrium amongst pure quantum strategies. This is a general feature of two-party entanglement and carries over to $n \times n$ games.

If mixed quantum strategies are allowed then there is an infinite set of NE [24]. Consider quantum Prisoners' Dilemma with four allowed strategies $\{\hat{C} \equiv \hat{I}, \hat{D} \equiv \mathrm{i}\hat{X}, \mathrm{i}\hat{Z}, -\mathrm{i}\hat{Y}\}$. The strategy $\hat{D}$ is the optimal counter to $\hat{C}$, $\mathrm{i}\hat{Z}$ is the optimal counter to $\hat{D}$, $-\mathrm{i}\hat{Y}$ is the best reply to $\mathrm{i}\hat{Z}$, and finally $\hat{C}$ is optimum against $-\mathrm{i}\hat{Y}$. One mixed NE is for Alice to select $\hat{C}$ or $\mathrm{i}\hat{Z}$ each with probability $\frac{1}{2}$ while Bob plays $\hat{D}$ or $-\mathrm{i}\hat{Y}$ each with probability $\frac{1}{2}$. There is a continuous set of such NEs, namely where both Alice and Bob each play one of a pair of

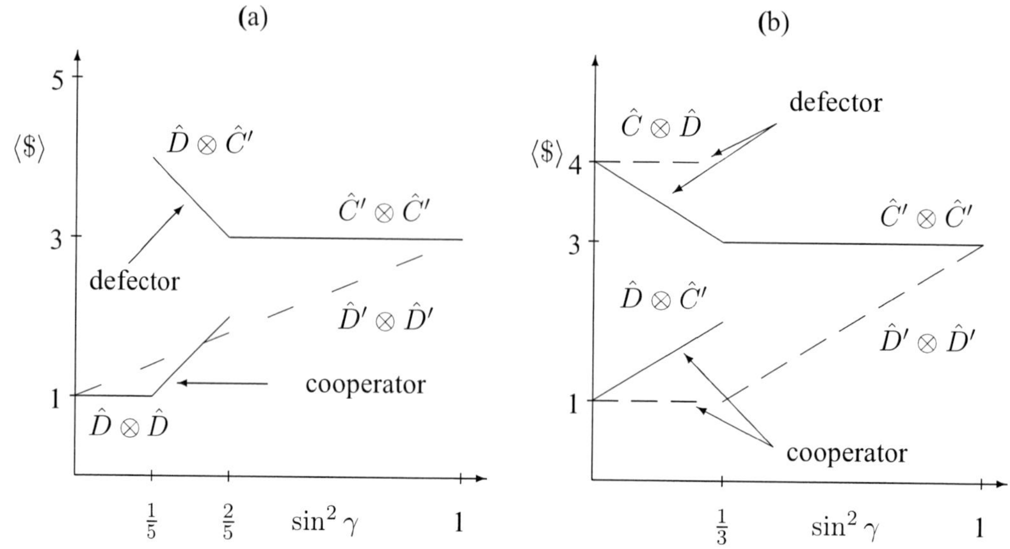

Figure 8. The Nash equilibrium payoffs for a quantum version of (a) Prisoners' Dilemma (b) Chicken versus entanglement as measured by $\sin^2\gamma$. The solid lines are for the two-parameter strategic space with $\beta \equiv 0$ and the dashed lines are those for $\alpha \equiv 0$. The strategy $\hat{C}'$ is a cooperative-like strategy that asymptotes to cooperation as $\gamma \to 0$ while $\hat{D}'$ is a defection-like strategy that asymptotes to defection as $\gamma \to 0$. Figures adapted from Ref. [45].

moves with equal probability [18]:

$$\hat{A}_1 = \hat{M}(\theta, \alpha, \beta), \qquad \hat{A}_2 = \hat{M}(\theta, \frac{\pi}{2} + \alpha, \frac{\pi}{2} + \beta), \tag{28a}$$

$$\hat{B}_1 = \hat{M}(\pi - \theta, \frac{\pi}{2} + \beta, \alpha), \qquad \hat{B}_2 = \hat{M}(\pi - \theta, \pi + \beta, \frac{\pi}{2} + \alpha). \tag{28b}$$

Since we have an infinite set of mixed strategy NEs we must appeal to psychological factors to decide which, if any, would occur in practice.

For all outcomes one player scores the maximum payoff of 5 while the other scores zero. The average payoff to both players is 2.5, superior to that of the classical NE of mutual defection. If we adjust the payoffs in (22) it is possible to get the average payoff for the mixed strategy quantum NE to be above (as here) or below the classical NE payoff.

The situation for multi-player games is more interesting, as we shall see in Sec. 4.. Here new equilibria can arise amongst pure quantum strategies that have no classical analogue and have expected payoffs that are more efficient than their classical counter-parts.

3.7. Decoherence in Quantum Games

A major hurdle in the physical implementation of quantum computers is the maintenance of coherence during the calculations without which the special features of quantum computation are lost. Interaction of a quantum system with the environment leads to environment-induced decoherence, usually referred to simply as *decoherence*. It is this process of decoherence that induces the transformation from the quantum state, that may involve "strange" properties like superposition or entanglement, to the classical, where all properties are localized and well defined. Decoherence is a major subject in quantum mechanics and will be touched on only briefly here. Recent overviews on the subject can be found in Refs. [47–49].

Although we can simulate quantum games on a classical digital computer, provided the number of players are small, the real test of the theory is in the physical implementation of the games. Any such implementation needs to consider the effects of decoherence. Mathematically decoherence leads to non-unitary dynamics, the precise form of which will depend on the physical implementation. In the density matrix notation for quantum states, the off-diagonal elements of the density matrix are reduced by decoherence, moving the state from a pure to a mixed quantum state.

Quantum error correction [6] or decoherence free subspaces [50] permit quantum computation in the presence of noise, though both techniques have the disadvantage of expanding the required number of qubits.

Johnson [51] has studied a three-player quantum game in which "noise" can cause the initial state to be flipped from $|000\rangle$ to $|111\rangle$ while Özdemir *et al.* [42] have considered various 2×2 quantum games where the initial state is corrupted by the presence of bit-flip errors. In the latter case the authors restricted the strategic space to the two-parameter space of (24). In both cases it was found that above a certain threshold of noise the players were impeded by quantum effects: the classical game yielded a more efficient solution.

The first studies of decoherence in the execution of a quantum game rather than in the initial state considered Meyer's quantum Penny Flip game [52] and quantum Prisoners' Dilemma [53]. A systematic treatment of decoherence in 2×2 quantum games in the Eisert protocol is given in Refs. [37,54], which was later extended to include multi-player games, in particular the quantum Minority game, in Ref. [55]. In these references decoherence during computation is considered. For optical implementations of quantum games coherence times are long—photons do not interact strongly with their environment unless absorbed—and it is more useful to instead consider errors in the production of the initial state [56,57]. We shall briefly review both methods below using the density matrix notation for a quantum state.

Consider decoherence during the calculation of a quantum game. Suppose we have a (mixed) quantum state represented by the density matrix ρ. A process of decoherence can be represented by the function $D(\rho, p)$ that decoheres ρ with probability p. For example, the effect on a two-qubit density matrix of a bit flip error with probability p on either qubit can be calculated by

$$D(\rho, p) = (\sqrt{p}\hat{X} + \sqrt{1-p}\,\hat{I})^{\otimes 2} \rho\, (\sqrt{p}\hat{X} + \sqrt{1-p}\,\hat{I})^{\otimes 2}. \tag{29}$$

The Eisert protocol for two-player games can be modified to account for decoherence by

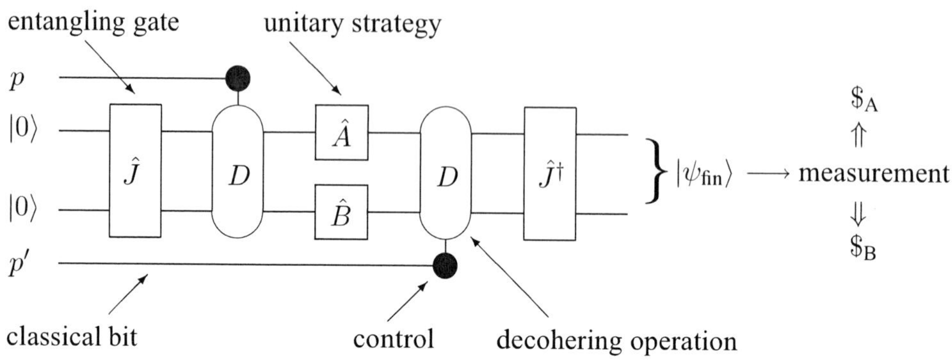

Figure 9. A two-person quantum game in the Eisert protocol with decoherence. The central horizontal lines represent the flow of qubits from left to right, $\hat{A}$ and $\hat{B}$ are the strategies of Alice and Bob, respectively, and $\hat{J}$ is the entangling gate (18). The top and bottom lines are classical random bits with a probability p or p', respectively, of being one. When they are one they switch on the related decoherence function D. Adapted from Ref. [54].

calculating the final density matrix through the following sequence of steps:

$$
\begin{aligned}
\rho_{\rm in} &= |\psi_{\rm in}\rangle\langle\psi_{\rm in}| && \text{initial state} \\
\rho_1 &= \hat{J}\rho_0\hat{J}^\dagger && \text{entanglement} \\
\rho_2 &= D(\rho_1, p) && \text{partial decoherence} \\
\rho_3 &= (\hat{A}\otimes\hat{B})\rho_2(\hat{A}\otimes B)^\dagger && \text{players' moves} \\
\rho_4 &= D(\rho_3, p') && \text{partial decoherence} \\
\rho_{\rm fin} &= \hat{J}^\dagger\rho_4\hat{J} && \text{preparation for measurement,}
\end{aligned}
\tag{30}
$$

where p and p' are some probabilities $\in [0,1]$. We assume that the probability of decoherence is the same for all qubits and that there are no correlated errors, though some work has been done on quantum games with correlated noise [58]. A schematic of the arrangement is shown in Fig. 9.

The effect of decoherence on the properties of quantum games can be computed with the above scheme, substituting various methods of decoherence for the function $D(\rho, p)$. Typically, in a quantum computer one considers bit-flip and phase-flip error or a combination of both. It is also useful to consider de-phasing by an arbitrary phase [i.e. multiplication by the phase shift operator (11)].

In the restricted (two-parameter) strategy space the NE for quantum Prisoners' Dilemma discussed in Sec. 3.5. is found to still hold but with reduced payoffs as decoherence increases [53]. The advantage of a quantum player over a classical one (Sec. 3.4.) is reduced by decoherence as expected but does not entirely disappear until decoherence is maximum ($p = p' = 1$), in which case the payoffs to the two players are equal [37, 54]. The classical game is not reproduced. By contrast, in a quantum version of a truel (a three person duel) increasing the level of decoherence smoothly alters the regions of parameter space corresponding to different preferred strategies towards the classical result [59].

The loss of advantage is similar to that which occurs as the entanglement parameter γ is reduced. Shuai and Mao-Fa study the loss of advantage of a quantum player over a classical player when only the quantum player is subject to decoherence [60]. This approach appears to be flawed since the classical player is still playing a quantum game using a qubit and hence will be subject to decoherence. It is of course possible that the two players are affected differently, a possibility that has not been studied.

In a multi-player quantum Minority game (Sec. 4.3.) dephasing reduces the NE payoff smoothly towards the classical result, as does bit- and phase-flip errors provided $p < \frac{1}{2}$ [55]. Once p exceeds $\frac{1}{2}$ the NE strategy changes, but this knowledge has no practical value since error rates of such magnitude would render a quantum computer completely dis-functional.

When considering entangled photon states there is almost no decoherence once the photons are created but allowances must be made for some loss of fidelity in the preparation of the state. In Ref. [56] a four photon entangled state is prepared. The loss of fidelity in the initial state is modeled as completely random noise by representing the initial state by the density matrix

$$\rho_{\text{in}} = p\,|\psi_{\text{in}}\rangle\langle\psi_{\text{in}}| + \frac{1-p}{16}\Sigma_{ijk\ell=0,1}\,|ijk\ell\rangle\langle ijk\ell|, \tag{31}$$

where $p \in [0,1]$ is a measure of the fidelity, and $|\psi_{\text{in}}\rangle$ is the desired initial state. For the four-player quantum Minority game the NE payoff to the players is equal to the classical value ($\frac{1}{8}$) plus a quantum excess proportional to p [61].

3.8. Evolutionarily Stable Strategies

Game theory was first applied to the problem of evolutionary dynamics when Maynard-Smith and Price introduced the concept of evolutionarily stable strategies (ESS) [62, 63]. An ESS is a refinement of the concept of a NE. A strategy s is evolutionarily stable against s' if it performs better against the mixed strategy $(1-\epsilon)s + \epsilon s'$ for sufficiently small $\epsilon > 0$. The strategy s is ESS if it is evolutionarily stable against all $s' \neq s$. In practical terms this means that the strategy s is a NE that is stable against small mutations. In a population setting where there is a group of agents engaging in repeated pair-wise encounters an ESS is an equilibrium that resists invasion by a small group playing a mutant strategy.

Various studies have been made of ESS in the quantum setting [38, 64–66] with a recent review being given in Ref. [67]. It is not clear if nature exploits quantum games at the microscopic level but the idea at least is not far fetched and has been mooted by some distinguished physicists [11, 68]. In addition, examining ESS in quantum games may provide an alternative view point in considering evolutionary quantum algorithms.

In the Eisert protocol ESS in quantum Prisoners' Dilemma have been studied [38]. When players are restricted to the two-parameter space (24) a quantum strategy $\hat{M}(\theta,\alpha)$ can invade a population playing $\hat{D}$ provided $\alpha > \arcsin(1/\sqrt{5})$. This is to be expected since the space of quantum strategies is much enlarged compared to the classical subspace. The strategy $\hat{C}'$ is ESS in the two-parameter space.

When considering quantum games in the protocol of Marinatto and Weber the strategies that are ESS depends on the choice of initial state. If a strategy is an ESS of the quantum

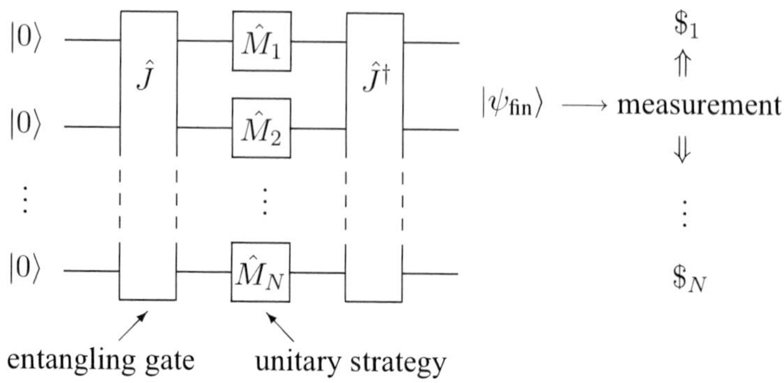

Figure 10. Eisert protocol for an N-person quantum game, where $\hat{M}_k$ is the strategy of the kth player and $\hat{J}$ is an entangling gate.

game for all initial states then it is an ESS of the classical game [38]. It is possible for a NE to gain or lose the property of evolutionary stability by suitable choice of (entangled) initial state. For example, in a quantum version of the well know children's game Rock-Scissors-Paper the classical mixed strategy NE (choosing each of the three options with equal probability) can become ESS when a particular class of entangled initial states is chosen [69].

4. Multi-player Quantum Games

4.1. General Considerations for Multi-player Quantum Games

An extension of the Eisert protocol to more than two players was first made by Benjamin and Hayden [70]. The multi-player protocol is a natural extension of the two-player one. The initial state is now N qubits $|00\ldots 0\rangle$ entangled by a generalization of (18)

$$\hat{J}(\gamma) = \cos\left(\frac{\gamma}{2}\right)\hat{I}^{\otimes N} + \mathrm{i}\sin\left(\frac{\gamma}{2}\right)\hat{X}^{\otimes N} \tag{32}$$

to produce the N-qubit Greenberger-Horne-Zeilinger (GHZ) [10,71] state

$$\frac{1}{\sqrt{2}}(|00\ldots 0\rangle + \mathrm{i}|11\ldots 1\rangle). \tag{33}$$

The final game state is computed by

$$|\psi_{\text{fin}}\rangle = \hat{J}^\dagger(\gamma)(\hat{M}_1 \otimes \ldots \otimes \hat{M}_N)\hat{J}(\gamma)|0\ldots 0\rangle, \tag{34}$$

where the $\hat{M}_k$, $k = 1, \ldots, N$ are the players' strategies. The protocol is shown schematically in Fig. 10.

This scheme has been used to study three player quantum Prisoners' Dilemma [40,44], but with a restricted strategy space. Of greater interest, quantum versions of a multi-player Public Goods game [72] and Minority game [55,70,73,74] have been studied. Unlike the two player games, in the multi-player context with the full set of unitary operators available to all players new equilibria can arise that have no classical analogue. These two examples are discussed in the next two subsections.

In addition, we can ask what type of initial entanglement is required to reproduce the classical game as a subset of the quantum game by restricting the players to a pair of unitary operators each? Only if such reproducibility is possible can we compare the results of the quantum and classical versions of the game. Such a study has been carried out by Shimamura *et al.* [75, 76]. For $N \geq 3$ players they find that an initial GHZ state is always satisfactory but for other forms of entanglement[4] this is generally not the case. In particular, with an Eisert-like protocol for multi-player games the use of W states, such as the three particle state $W_3 = (|001\rangle + |010\rangle + |100\rangle)/\sqrt{3}$, are generally ruled out according to their criteria.

4.2. Quantum Public Goods Game

In the Public Goods game each of N players indexed by $k = 1, \ldots, N$ voluntarily contribute an amount c_k to a public good from their private holdings of y_k and then all share in the benefits derived form the public good. Formally, the game can be described as follows:

x	amount of public good
y_k	initial private holding of player k
c_k	contribution of player k to the public good
$Q_k(x, y)$	utility of player k
$g(C)$	production function of the public good as a function of total contribution $C = \sum_k c_k$

The simple example used by Chen *et al.* [72] sets $Q_k(x, y) = x + y$ and $g(C) = aC/N$ for some constant a. We further suppose that $y_k = 1 \; \forall k$. If $a < 1$ there is no gain by contributing to the public good and the NE (and PO) outcome is for all players to contribute nothing, while for $a > N$ the NE outcome is for all players to contribute their entire holdings. It is the intermediate case $1 < a < N$ that is socially interesting. The NE solution is still zero contribution but this is not Pareto efficient. One PO outcome is $c_k = 1$ but in this solution, players do better by defecting and contributing nothing, while still sharing in the public good produced by the others. This is the well known free rider problem. In the quantum model of this game the actions corresponding to the qubit states $|0\rangle$ and $|1\rangle$ are taken to be cooperation (contribute $c_k = 1$) and defection ($c_k = 0$), respectively. Chen *et al.* consider three cases:

- full entanglement between the N parties using (32)
- players having multiple qubits, with two-particle entanglement between each pair of players using (18)

[4] The subject of multi-particle entanglement is a complex one and well beyond this review. Suffice to say that for $N > 2$ particles there are many forms of entanglement with different properties. The GHZ state, or equivalent, is by all measures maximally entangled.

- each player having two qubits with entanglement by (18) between nearest neighbors only

In the case of full entanglement there is no pure strategy quantum NE, however, there are mixed strategy NE. One such equilibrium is where all players select between the pair of strategies

$$\hat{M}(0,0,0) = \begin{bmatrix} 1 & 0 \\ 0 & 1 \end{bmatrix},$$
$$\hat{M}(0,\frac{\pi}{2},0) = \begin{bmatrix} \mathrm{i} & 0 \\ 0 & -\mathrm{i} \end{bmatrix}, \tag{35}$$

each with probability $\frac{1}{2}$. The expected payoff to each player is then $(1+a)/2$. Though less than the PO payoff, it still exceeds the classical payoff of 1.

In the case of two-party entanglement between each pair, every player has $N-1$ qubits. For example, the four-player game requires the twelve qubit initial state

$$\begin{aligned} |\psi_{\text{in}}\rangle &= |q_{\text{R}}^{(4)} q_{\text{L}}^{(1)} q_{\text{R}}^{(1)} q_{\text{L}}^{(2)} q_{\text{R}}^{(2)} q_{\text{L}}^{(3)} q_{\text{R}}^{(3)} q_{\text{L}}^{(4)} q_{\text{M}}^{(1)} q_{\text{M}}^{(3)} q_{\text{M}}^{(2)} q_{\text{M}}^{(4)}\rangle \\ &= \frac{1}{8}(|00\rangle + i|11\rangle)^{\otimes 6}, \end{aligned} \tag{36}$$

where $q^{(k)}$ is a qubit belonging to player k, entangled with the player to the left (subscript L), right (R) or middle (M). There is the question of how to interpret a player's multi-qubit state. If we suppose that the number of zeros in the state determines the extent of the player's contribution, the equilibrium payoffs for the pairwise entangled game are the same as that for the fully entangled game. If instead we suppose that a player only contributes all or nothing, choosing the former if any of their $N-1$ qubits is 0, we arrive at an NE with expected payoff

$$\langle \$ \rangle = a - 2^{1-n}(a-1). \tag{37}$$

Given that $1 < a < N$ this is only slightly below the maximum possible average payoff of a.

In the case of nearest neighbor entanglement, the players are listed in some arbitrary order in a circular arrangement. An entangled pair of qubits is provided between each successive pair of players. Such an arrangement requires $2N$ qubits. For example, the initial state for a four-player game is the eight qubit state

$$\begin{aligned} |\psi_{\text{in}}\rangle &= |q_{\text{R}}^{(4)} q_{\text{L}}^{(1)} q_{\text{R}}^{(1)} q_{\text{L}}^{(2)} q_{\text{R}}^{(2)} q_{\text{L}}^{(3)} q_{\text{R}}^{(3)} q_{\text{L}}^{(4)}\rangle \\ &= \frac{1}{4}(|00\rangle + i|11\rangle)^{\otimes 4}. \end{aligned} \tag{38}$$

Using the all-or-nothing interpretation of a player's two qubits the NE payoff to all players is $(1+3a)/4$. Again this is an improvement over the classical NE but unlike the case with pairwise entanglement between each pair, the payoff does not improve with N.

It is important to note that an improved NE can be found in the case of two-party entanglement, since multi-qubit entangled states are difficult to produce and susceptible to the effects of decoherence.

4.3. Quantum Minority Game

"It is not worth an intelligent man's time to be in the majority. By definition, there are enough people to do that."
— G.H. Hardy

The Minority game is perhaps the simplest multi-player game. Each of N players must independently select between one of two options '0' and '1' with those that choose the minority option receiving a payoff of one, while the others receive nothing. If the numbers are balanced no player scores. Consideration of the game arose as a means of studying adaptive reasoning in a multi-agent model. It has been used to study the dynamics of multi-agent markets [77,78] where the two options correspond to buy and sell. A recent review is provided by Moro [79].

In a one-shot game the players can do no better than selecting the mixed strategy of choosing each of the two options with probability $\frac{1}{2}$. The payoff to each player is then

$$\langle \$ \rangle = \sum_{k=1}^{\lfloor (N-1)/2 \rfloor} {}^{N}C_k \, \frac{2k}{N} \left(\frac{1}{2} \right)^N . \tag{39}$$

In the classical game the number of players N is assumed to be odd to avoid the problem of equal numbers choosing the two options. Since N is generally large this is not of practical significance. The interest lies in how the agents adapt their strategies in a sequence of repeated games based upon past history.

By contrast, a one-shot quantum game proves interesting when N is even. For small even N the probability of getting a balance between the two choices is a difficulty. By exploiting quantum entanglement the players in a quantum Minority game can reduce the probability of getting a balanced result. Benjamin and Hayden were the first to study the three and four player quantum Minority game [70] showing that in the three player case entanglement provides no benefit but in the four player case a new quantum equilibrium arose that was more efficient than that achievable in a one-shot classical game. In our notation, the NE strategy for the four player quantum Minority game found by Benjamin and Hayden is

$$\begin{aligned} \hat{s}_{\rm NE} &= \frac{1}{\sqrt{2}} \cos(\frac{\pi}{16})(\hat{I} + {\rm i}\hat{X}) \; - \; \frac{1}{\sqrt{2}} \sin(\frac{\pi}{16})({\rm i}\hat{Y} + {\rm i}\hat{Z}) \\ &= \hat{M}(\frac{\pi}{2}, \frac{-\pi}{16}, \frac{\pi}{16}). \end{aligned} \tag{40}$$

When all four players choose this strategy the result is a NE with an expected payoff of $\frac{1}{4}$, the maximum average payoff possible since in a four player game only one player can be in the minority. We note that the $\hat{J}^{\dagger}$ operator is unnecessary in this game since it only permutates between states with the same set of payoffs to all the players [70].

Subsequently, the result of Benjamin and Hayden were extended to arbitrary even N [73] and to the consideration of decoherence [55] and quantum coalitions [74]. For arbitrary even N, the probability that $|\psi_{\rm fin}\rangle$ contains states for which the numbers selecting the 0 and 1 options are balanced can be made to vanish if all players select the strategy

$$\hat{s}_\delta = \hat{M}(\frac{\pi}{2}, -\delta, \delta), \qquad \delta = \frac{(4n+1)\pi}{4N}, \quad n = 0, \pm 1, \pm 2, \ldots \tag{41}$$

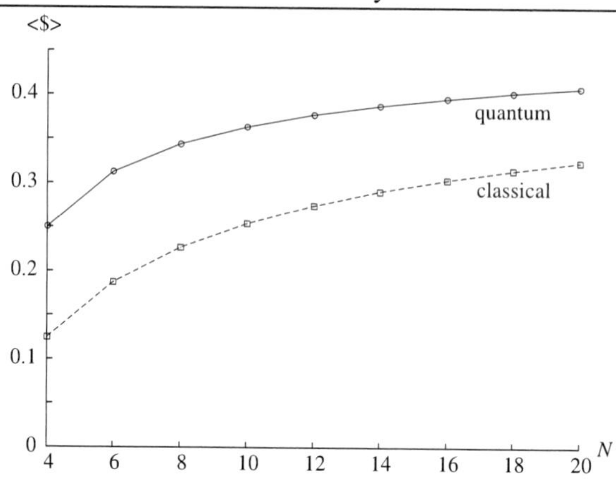

Figure 11. Nash equilibrium payoffs for even integer numbers of players N for the fully entangled quantum case (circles) and the classical case (boxes). The lines are a guide to the eye only and do not indicate payoffs for the game for N other than even integers. Note that the quantum payoff curve is the same as that for the classical game equilibrium with N odd but shifted one unit (in N) to the right. Adapted from Ref. [55].

The vanishing of the balanced states is optimal for all players since these are the ones for which no player scores. Each value of δ gives a NE for the N even player quantum Minority game. In addition, for each δ there is a continuum of symmetric NE strategies of the form $\hat{M}(\pi/2, \eta - \delta, \eta + \delta)$. However, the strategy with $\eta = 0$ and $n = 0$ is the simplest and therefore is a focal or Schelling point [80] that attracts the player's attention. Figure 11 shows the quantum and classical NE payoffs for even N. For odd N, the quantum game provides no advantage over the classical. It is interesting to note that the quantum NE payoff for arbitrary even N is the same as the classical NE payoff for $N - 1$.

Recently the quantum Minority game with other forms of initial entanglement has been considered [57]. In the four player game with an initial state that is a mix of the GHZ state $(|0000\rangle + |1111\rangle)/\sqrt{2}$ and EPR pairs $(|01\rangle + |10\rangle)/\sqrt{2}$, an exciting correspondence has been found between the symmetric PO payoffs and the violation of Bell inequalities which measure fundamental quantum properties of entangled states, thus providing a link between quantum game theory and fundamentals of quantum mechanics [57, 81]. It has also been shown that two party entanglement is insufficient to produce any benefit to the players [57], in contrast to the situation for the multi-player quantum Public Goods game in the previous section.

5. Iterated Quantum Games

5.1. Formalism for Repeated Quantum Games

The subject of repeated, or iterated, quantum games has only rarely been considered. An early attempt in this direction was made by Iqbal and Toor [82] who considered a two round quantum Prisoners' Dilemma in Marinatto and Weber's protocol. In this protocol quantum effects arise by making the initial state a particular entangled state, in this case requiring

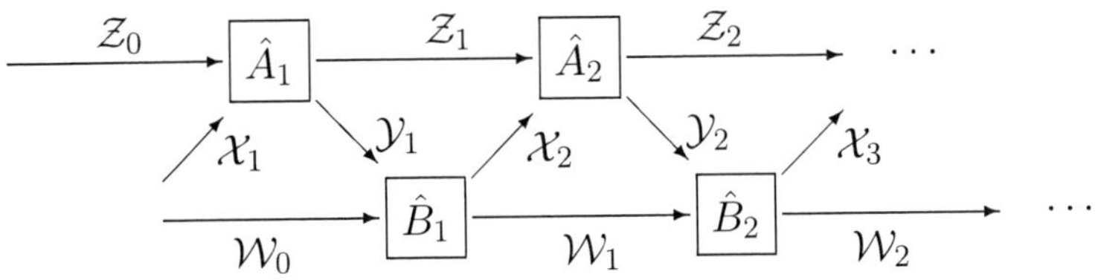

Figure 12. An iterated quantum game, from Ref. [83]. Alice's strategy consists of a string of operators $\hat{A}_j$ acting on $\mathcal{Z}_{j-1} \otimes X_j$, while Bob's strategy consists of a string of operators $\hat{B}_j$ acting on $W_j \otimes Y_j$, where the $\mathcal{X}_j, Y_j, Z_j, W_j$ are vector spaces (see text).

four qubits. Computations are simplified by only allowing strategies to be mixtures of the classical strategies $\hat{I}$ or $\hat{X}$.

A powerful general scheme for repeated quantum games was recently introduced by Gutoski and Watrous [83]. This scheme can model both the one-shot quantum games of the Eisert protocol and iterated quantum games. The two players have strategy sets $\{\hat{A}_1, \hat{A}_2, \ldots,\}$ and $\{\hat{B}_1, \hat{B}_2, \ldots\}$, with each operator being a linear mapping between two vector spaces:

$$\begin{aligned} \hat{A}_j &: \mathcal{X}_j \otimes \mathcal{Z}_{j-1} \rightarrow \mathcal{Y}_j \otimes \mathcal{Z}_j, \\ \hat{B}_j &: \mathcal{Y}_j \otimes \mathcal{W}_{j-1} \rightarrow \mathcal{X}_{j+1} \otimes \mathcal{W}_j. \end{aligned} \tag{42}$$

The $\mathcal{Z}_j$ and $\mathcal{W}_j$ are vector spaces representing the private "memory" of the players after j iterations of the game, while $\mathcal{X}_j$ and $\mathcal{Y}_j$ are vector spaces representing the information transfered between the players at time step j, as indicated in Fig. 12.

The scheme has been used to provide an alternate proof for Kitaev's bound for strong coin flipping [84], a task where by two parties at separate locations want to flip a coin but do not trust each other, as well as to study zero-sum quantum games. The generality and firmer mathematical basis of Ref. [83] should lead to more applications and cross overs to other areas of quantum information.

5.2. Iterated Quantum Games as Quantum Walks

In classical repeated games the players' strategies are contingent upon some finite history of previous game results. A two-dimensional random walk can be used as an aid to model an iterated game where the position of the walker in the x- and y-directions is used to represent the cumulative payoffs to the two players. Such a scheme for modeling repeated quantum games as a history-dependent discrete quantum walk with variable step size was introduced by Abal *et al* [85]. Every play of the game results in one step of the walker, with the step size in the x- and y-directions representing the payoffs to Alice and Bob, respectively. The players' choice is contingent upon the result of the most recent game, in a two-dimensional generalization of a history-dependent quantum walk [86]. Unfortunately the scheme in Ref. [85] is not a true quantum analogue of repeated classical games since the move of the second player (Bob) depends not upon Alice's move in the previous game but her move in the current one. At some computational cost, this flaw can be removed by using ancillary qubits.

Below we summarize the model of Abal *et al.* for iterated quantum games and show how it is a special case of the generalized repeated quantum games of Ref. [83]. We also present an outline of some modifications to the model that provides some improvements.

In iterative games, the strategy chosen by the players can be contingent upon the results of previous games, that are in turn determined by the strategies previously chosen by the players. Thus, in an iterative quantum game we desire that the operators chosen by the players depend upon the existing values of all the player qubits. Players can still only manipulate their own qubit. The existing protocol for quantum walks in one-dimension[5] can be used to update the position of a walker. The position in the x-direction will represent the cumulative payoff to Alice, and in the y-direction to Bob. Since the walker is quantum mechanical and can be in a superposition of states there is no need to take a measurement in order to assign payoffs. That is, with position eigenstates $|x\rangle \otimes |y\rangle$, $x, y = 0, \pm 1, \pm 2, \ldots$ the conditional shift operator

$$\begin{aligned} \hat{\Omega} = \Sigma_{x,y} \; \big[& |x + \$^{\rm A}_{00}, y + \$^{\rm B}_{00}\rangle\langle x, y| \otimes |00\rangle\langle 00| \\ & + |x + \$^{\rm A}_{01},\; y + \$^{\rm B}_{01}\rangle\langle x, y| \otimes |01\rangle\langle 01| \\ & + |x + \$^{\rm A}_{10},\; y + \$^{\rm B}_{10}\rangle\langle x, y| \otimes |10\rangle\langle 10| \\ & + |x + \$^{\rm A}_{11},\; y + \$^{\rm B}_{11}\rangle\langle x, y| \otimes |11\rangle\langle 11| \big] \,, \end{aligned} \tag{43}$$

where $\$^{\rm X}_{ij}$ is the payoff to player X for the game outcome ij, is used to update the position of the walker. If necessary, payoffs are scaled so that they are all integers. The players' choice of unitary operator is dependent upon the existing value of the two qubits. For example, for Alice we can write

$$\begin{aligned} \hat{A} = & (a_0|00\rangle + a_1|10\rangle)\langle 00| + (a_2|01\rangle + a_3|11\rangle)\langle 01| \\ & + (a_4|00\rangle + a_5|10\rangle)\langle 10| + (a_6|01\rangle + a_7|11\rangle)\langle 11| \end{aligned}$$

for complex coefficients a_j, and similarly for Bob. The requirements of unitarity reduce the set of indpendent real parameters to seven for each player. A repeated Prisoners' Dilemma with restrictions on the strategy space of the players was investigated in the above scheme.

A connection between this scheme and the generalized scheme for iterated quantum games of Sec. 5.1. can be made as follows. A step by the walker proceeds in two phases. Firstly, the players make their choices and then the conditional shift operator is applied to the walker position. The $\mathcal{Z}_j$ and $\mathcal{W}_j$ cannot be the position spaces for the walker in the x- and y-directions since these can become entangled. Rather all the information is contained in the vector spaces $\mathcal{X}_j$ and $\mathcal{Y}_j$ that are $\mathcal{H}_x \otimes \mathcal{H}_y \otimes \mathcal{H}_{\rm A} \otimes \mathcal{H}_{\rm B}$, where $\mathcal{H}_x$ and $\mathcal{H}_y$ are the Hilbert spaces of particle position in the x- and y-directions, and $\mathcal{H}_{\rm A}$ and $\mathcal{H}_{\rm B}$ are the Hilbert space of Alice and Bob's qubits. In the format of Fig. 12, one step of the iterated quantum game of Ref. [85] can be represented as in Fig. 13.

[5]For an introduction to quantum walks see Ref. [87].

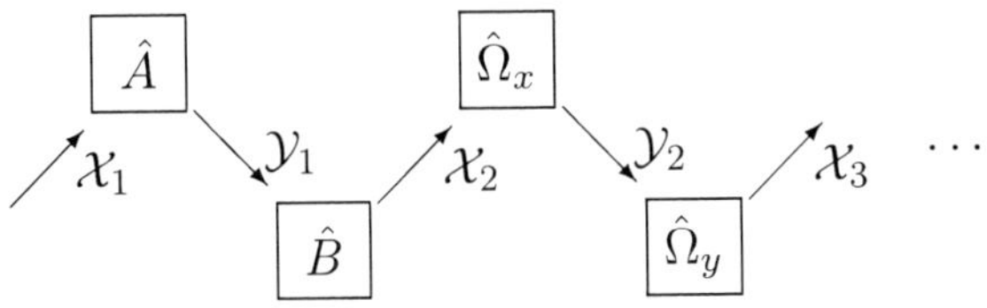

Figure 13. A single step in the iterated quantum game as a quantum walk model of Ref. [85] arranged in the format of the generalized repeated quantum game of Ref. [83]. The $\mathcal{X}_j$ and $\mathcal{Y}_j$ both represent the Hilbert spaces of the walker position and of the player qubits. The qubits are updated first by the player operators $\hat{A}$ and $\hat{B}$, followed by the conditional shift operators $\hat{\Omega}_x$ and $\hat{\Omega}_y$ that updates the walker position.

5.3. Suggested Improvement to the Model of Iterated Quantum Games as Quantum Walks

One difficulty with the above scheme is that the game is no longer a simultaneous move game, in keeping with the classical description of strategic games in normal form. Rather, the order in which Alice and Bob apply their operators matter: Alice's decision is contingent upon the moves played in the previous game while Bob's depends upon Alice's move in the current game and Bob's in the previous.

Below we suggest a means of circumventing this problem by introducing an ancilla or ancillas to the scheme, as shown in Fig. 14. Using a CNOT and an ancillary qubit Bob's decision can now be made contingent upon Alice's move in the previous game. For subsequent games further ancilla are required, or the single ancilla must be reset to $|0\rangle$ before the next step. By adopting the first idea, some of the interference that we would normally see in a two-dimensional quantum walk is removed since the set of ancilla hold a record of Alice's past moves. This is analogous to the loss of interference seen in multi-coin one-dimensional quantum walks as the number of coins in increased [88]. The use of a single ancilla that is reset to $|0\rangle$ after each move results in some loss of coherence since we must trace over the ancilla states after each move.

6. Other Models of Quantum Games

6.1. Introduction

There have been a variety of other quantum game-theoretic investigations. These include games that do not involve entanglement [89, 90], games of incomplete information [91], and continuous variable quantum games [92]. An early model of zero-sum quantum games was proposed by Boukas [93], who seemed unaware of the previous publications in the area. In this little known paper, the concept of a classical player, corresponding to a simple random variable on a finite cardinality probability space, is extended to a quantum player, corresponding to a self-adjoint operator on a Hilbert space, and the formalism is used to demonstrate a quantum version of von Neumann's minimax theorem. In quantum computational models, entanglement is used as a measure of whether a game is "really quantum."

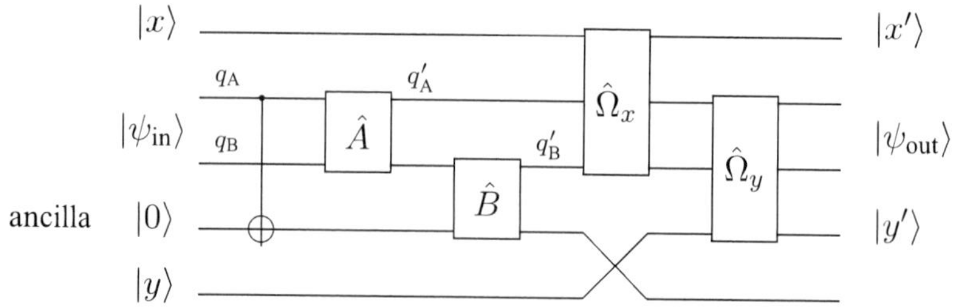

Figure 14. Schematic of one step in a two-player iterated quantum game with input state $|\psi_{\text{in}}\rangle$ and final state $|\psi_{\text{out}}\rangle$. The latter becomes the input state for the next game in the sequence. Each iteration uses an ancilla with initial state $|0\rangle$. Alice and Bob's strategies are performed by the operators $\hat{A}$ and $\hat{B}$, respectively. The operators $\hat{\Omega}_x$ and $\hat{\Omega}_y$ determine the payoffs to Alice and Bob and hence update the position of the walker in the x- and y-directions, respectively.

Ref. [94] uses a different but equally non-intuitive property of quantum information, the non-distributive nature of logical propositions, to establish a connection between quantum logic and quantum games. More speculatively, an attempt is being made to develop a new representation of game theory that encompasses both classical and quantum games [95,96]. These models have generally only been taken up by a single author, or group of authors, so their impact on the field is limited at this stage.

6.2. Quantum Correlation Games

In an attempt to circumvent some of the objections to quantum games [97,98] (see Sec. 7.) an attempt has been made to develop models of quantum games taking Einstein-Podolsky-Rosen type experiments [20] with particle spins as their underlying basis [99–103]. In two-player games, each player performs a measurement of the spin of one of a pair of particles with their strategy determined by the choice of measurement axis, as indicated in Fig. 15. Classical games are reproduced if the particles are classically correlated but when there is quantum correlations—entanglement—results are obtained that cannot be realized in a classical game setting.

6.3. Quantum Market Games

Various market phenomena such as the buying and selling of commodities, bargaining, and auctions have been modeled with a quantum game-like formalism [104–113]. This should be distinguished from attempts to use the mathematical machinery of quantum mechanics and quantum field theory to solve classical financial market problems [114,115]. In the new quantum market games, transactions are described in terms of projective operations acting on Hilbert spaces of strategies of traders. A superposition of trading actions is represented by a quantum strategy. The authors claim that outcomes can be achieved not realizable by classical means [107]. Furthermore, some features of quantum mechanics can be used to model aspects of market behavior: traders observe the actions of other players and adjust their actions accordingly so there is non-commutativity of bidding [104], maximal capital

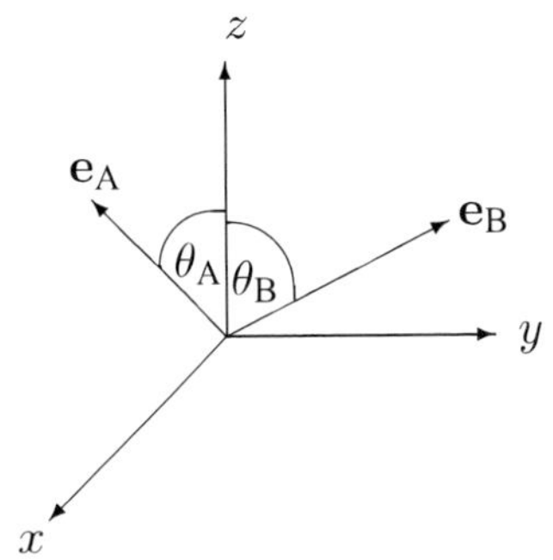

Figure 15. In a quantum correlation game, Alice and Bob choose axes $\mathbf{e}_{\mathrm{A}}$ in the xz plane and $\mathbf{e}_{\mathrm{B}}$ in the yz plane at angles to the θ_{A} and θ_{B} to the z-axis, respectively, about which to make a measurement of their particle's spin. The results of the measurement are then mapped onto the payoffs for the players. When the pair of particles are initialized in an entangled state quantum effects can result.

flow at a given price corresponds to entanglement between buyers and sellers [106], and so on. It remains to be seen whether these ideas will be taken up by other theorists. In addition, the technological possibility of clearing actual markets with quantum computers remains science fiction at this stage of experimental development. There has, however, been some recent investigations of quantum auctions [116, 117]. In these non-winning bids need never be revealed, and entanglement can be used to arrange correlations between buyers for different parts of a contract.

6.4. Quantum Monty Hall

The quantum Monty Hall problem clearly demonstrates the non-uniqueness of the quantization of a game. The Monty Hall problem is based around an old TV game show situation where the optimal strategy is surprising and counter-intuitive. The problem came to prominence in the early 1990s when it gave rise to a lively debate in the popular literature [118, 119], perplexing many non-mathematicians and occasionally some mathematicians as well! The problem can be described as follows: the host ('Monty Hall') secretly selects one of three doors behind which to hide a prize. The contestant makes an initial choice of door. Monty Hall now opens one of the other two doors demonstrating that the prize is not behind that door. The contestant is then given the option of remaining with their initial choice or of switching their selection to the untouched door. At first thought it seems that both of the unopened doors would have an equal probability of hiding the prize but this is not so. The optimal strategy is always to switch doors [120, 121], doubling the chance of winning from $\frac{1}{3}$ to $\frac{2}{3}$.

This problem was bought to the attention of several quantum game theorists in 2001–2 and as a result three different quantization schemes for the problem were published. In Ref. [122] the authors have one particle, representing the prize, in one of three boxes. They make the game fair between the contestant and the host by allowing the host to use an ancillary particle entangled with the prize particle. By making a suitable measurement on the ancilla after the contestant's initial guess and the opening of one box, the probability of placement of the prize particle can be redistributed (without touching the particle!) evenly between the remaining two boxes. A second group quantized the original game also using an ancillary particle [123] and have a simulation of the quantum game on the web [124].

In Refs. [37, 125] a quantum version of the game is presented without the use of ancillas and one that has the most in common with the Eisert protocol. In this scheme the system is described by three qutrits [6] $|\psi\rangle = |oba\rangle$, with o, b, and a representing the opened door, and the contestant's and the host's choices, respectively. Quantum effects depend on the selection of the initial state $|\psi_{\text{in}}\rangle$. When the initial state is unentangled the classical game is reproduced. Interest comes when the initial state is prepared with some form of entanglement. When the initial state is maximally entangled

$$|\psi_{\text{in}}\rangle = \frac{1}{\sqrt{3}}(|000\rangle + |111\rangle + |222\rangle), \tag{44}$$

if both players have access to the full set of unitary operators the classical expected payoffs are reproduced (the contestant wins $\frac{1}{3}$ of the time by not switching and $\frac{2}{3}$ of the time by switching). If the host is restricted to just the identity or bit-flip operators (classical moves) then the contestant can win with certainty, while if the contestant is so restricted while the host has access to the full set of quantum operations then the host is able to choose a strategy that gives the contestant a winning probability of $\frac{1}{2}$, whether they switch or not, like in the model of Ref. [122].

6.5. Quantum Games with Nonfactorizable Joint Probabilities

If we have two independent classical coins, each of which has a 50/50 chance of coming up heads (H) or tails (T) then the probabilities of finding each of the four combinations HH, HT, TH, and TT must be $\frac{1}{4}$. If the "coins" are quantum mechanical two-state systems we can create an entangled state for which this is no longer true: although the "coins" can be unbiased and non-interacting, their results can be correlated due to the special nature of the state preparation. The probabilities of the combinations HH, HT, TH, or TT may not then be factorizable into products of probabilities for the individual coins. Motivated by the idea of making quantum games more accessible to non-physicists, Iqbal and Cheon have proposed using nonfactorizable joint probabilities directly to create a quantum game [126], rather than referring to quantum states with their associated mathematics. In their proposal each player would be given two coins, with different biases. The selection of one coin or the other are the two "pure" strategies. This choice is equivalent to selecting one of two possible directions of spin measurement in an EPR experiment. All coins are returned to a referee who "flips" them. The referee rewards the players according to the their strategies, the underlying statistics of the coins determined from a sequence of tosses, and the payoff matrix of the game. If the probabilities of the various combinations of H and T are factorizable then this is equivalent to a classical game where each coin is the same as a particular mixed strategy (depending on the coin's bias). Certain contraints are imposed in order to ensure that the classical game corresponds to factorizable joint probabilities. These remain valid even when the joint probabilities become nonfactorizable. In the latter case, new game

[6] A qutrit is three-state generalization of a qubit.

behavior may arise. In this scheme no new NE occur in a 2×2 Prisoners' Dilemma but with three players new results can arise by selection of a suitable set of nonfactorizable probabilities [127].

6.6. Quantum Parrondo's Games

Parrondo's games or Parrondo's paradox are not strategic games in the von Neumann sense. They arise in gambling-like situations with one agent playing against a bank. Parrondo and others demonstrated that it is possible to have two games that are losing when played individually but when played in some alternating combination they form an overall winning game [128]. This necessarily involves some form of coupling between the games, for example, through the player's capital [129] or through the history dependent rules [130]. The history dependent Parrondo's game has been translated directly into the quantum realm by replacing the tossing of biased coins by unitarty operations on qubits [131]. There is coupling through the history dependent rules and through interference effects when a suitable initial state is chosen. Meyer and Blumer [132, 133] have constructed a quantum version of the capital dependent Parrondo's game where the role of the player's capital is taken by the position of a quantum particle moving between discrete sites in a one-dimensional lattice under the influence of some potential. Quantum interference can result in ratcheting in one direction. A history-dependent quantum walk can also be used to create quantum Parrondo-like effects [86].

7. Are Quantum Games "Quantum"?

There was early criticism of quantum games by van Enk and others [97,98] suggesting that the games are not truly quantum mechanical. It is true that all quantum games can be simulated on a classical computer, however, any classical implementation scales exponentially with the size of the Hilbert space, unlike a quantum implementation [134]. The efficiency of quantum games was also emphasized by Lee and Johnson [27]. This is of particular importance when studying multi-player games where the size of the Hilbert space increases rapidly with the number of players.

Van Enk considers quantum Prisoners' Dilemma with the two-parameter strategic space to be equivalent to a new classical game with three strategies $\hat{C}$, $\hat{D}$ and $\hat{C}'$. As a result the $\{\hat{C}', \hat{C}'\}$ NE does not address the dilemma in the original game [98]. Other criticisms of the two-parameter strategic spaces have also been raised [45, 46] (Sec. 3.5.–3.6.) casting uncertainty on the results obtained for this case. In addition, the sharing of an entangled state blurs the distinction between cooperative and non-cooperative games. The criticisms have some merit, but apart from the issue of efficient implementation of quantum games, it is difficult to see how the new equilibria in multi-player games, such as those in the Public Goods and Minority games (Sec. 4.2.–4.3.), can arise simply by adding extra strategies to the classical game. Iqbal and Weigert have tried to circumvent van Enk's criticism by playing a game in an EPR-like setting that utilizes quantum correlations [100–102] (Sec. 6.). Finally, the main reason for studying quantum games is not as another model for classical games but as a model for competitive scenarios involving quantum information or quantum control.

8. Physical Implementation of Quantum Games

There is currently intense interest in the implementation of quantum computers though many challenges remain [135]. There are many proposed physical implementations including solid state (silicon), optical, nuclear magnetic resonance, ion traps, superconducting devices etc. For a recent overview see Ref. [136]. Since most quantum games involve only a few qubits they are ideal candidates on which to test new technologies. This section will only give a brief overview of the physical implementations since it is intended for game theorists and not physicists.

An early experimental realization of a quantum game was made by Du *et al.* [39] in a two qubit nuclear magnetic resonance device. They implemented quantum Prisoners' Dilemma in Eisert's protocol with two-parameter strategies (Sec. 3.5.), studying the effect of varying the degree of entanglement (i.e. varying γ). Their results were consistent with the theoretical predictions.

Over the past few years there have been a number of proposals for implementing quantum games in optical systems [137–139]. Here, individual photons of light produced by a pulsed laser act as the qubits. It is generally the polarization of the photon that is used to encode the value of the qubit with, say, horizontal polarization representing $|0\rangle$ and vertical representing $|1\rangle$. In Ref. [137] a pair of entangled photons was produced by spontaneous parametric down-conversion, a process where by a photon is split by a nonlinear crystal into two photons, each with half the energy and hence twice the wavelength, of the original photon. Some way towards implementing a quantum Prisoners' Dilemma in the Eisert protocol was made. In this protocol it is necessary to distinguish between each of the four Bell states

$$|\phi_\pm\rangle = \frac{1}{\sqrt{2}}(|0\rangle_{\rm A}|0\rangle_{\rm B} \pm |1\rangle_{\rm B}|1\rangle_{\rm B}),$$
$$|\psi_\pm\rangle = \frac{1}{\sqrt{2}}(|1\rangle_{\rm A}|0\rangle_{\rm B} \pm |1\rangle_{\rm B}|0\rangle_{\rm B}), \tag{45}$$

in order to implement the $\hat{J}^\dagger$ operator. Although Ref. [137] did not implement the quantum strategies a method for complete Bell state analysis of a pair of photons was constructed. Subsequently, at the same institution, a four player quantum Minority game was implemented optically with results consistent with theoretical predictions for a certain fidelity of preparation of the initial state [61].

Following a proposal by Paternostro *et al.* [140] quantum Prisoners' Dilemma was implemented in a four photon cluster state [7] by a group at the University of Vienna [141]. A schematic of the arrangement of qubits is given in Fig. 16. The experimental arrangement made it necessary to restrict the strategy space to the set $\{C, D, q(\alpha)\}$, where $\alpha \in [0, \pi]$ gives a range of quantum strategies including $\hat{C}' \equiv q(\pi/2)$.

Apart from the current interest in optical implementations of quantum games there is a proposal for a general scalable framework for executing two-choice multi-player quantum games with ion traps [142]. The authors have considered implementations for 2×2 quantum games in the Eisert protocol and for a four player quantum Minority game, analyzing both

[7] A cluster state is a type of highly entangled multiple qubit state generated in a lattice of qubits, in this case a 2×2 array of photons.

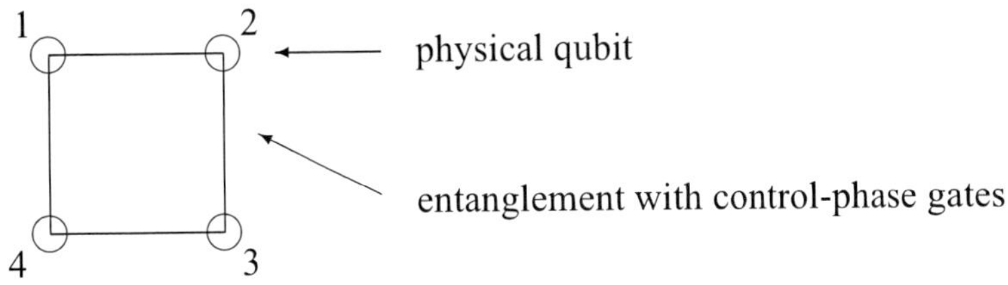

Figure 16. A schematic of a four qubit cluster state used to implement quantum Prisoners' Dilemma in an optical system. The circles represent physical qubits. The adjoining lines create entanglement between neighboring pairs using a control-phase gate.

the effect of the degree of initial entanglement and of decoherence. It remains to be seem if any experimentalists will take up the challenge of this implementation.

9. Conclusion

Existing classical game-theoretic problems have been the starting point for quantum game theory. Classical strategic games have been translated into the quantum domain by changing classical information (bits) into quantum information (qubits). A classical mixed strategy can then be modeled by a superposition of states and new quantum effects can be obtained by introducing entanglement between the strategies of different players. The original game is a subset of the quantum game and is obtained either by appropriate restrictions to the players' strategic space or by a suitable selection of initial state, depending on the model. The quantization process changes the nature of the game and in that sense quantum games do not address problems of the original classical game. Nevertheless, the quantum game-theoretic models demonstrate what can be achieved by the expansion into the quantum realm and may be interesting for their own sake. More importantly, quantum game theory is the appropriate language when dealing with competition or conflict situations in quantum computing and quantum communication where the information *is* quantum.

There is no unique method for quantizing a game and many models have been proposed. The dominant model is that due to Eisert *et al.* and a number of issues have been explored in this model. While in two-player games there is debate about the validity of the new equlibria that have been obtained, in multi-player quantum games new Nash equilibria can arise that have no classical analogue and that are more efficient than those of the underlying classical game.

With recent developments in generalized models of quantum games there is hope that the theory made be placed on a firmer foundation and may see applications in other areas of quantum information theory. Requiring only a few qubits, quantum games provide an ideal avenue for experimentalists to test their abilities and a number of physical implementations of quantum games have been carried out with expectations of more in the near future.

References

[1] S. Lloyd, "Obituary: Rolf Laundauer," *Nature* **400**, 720 (1999).

[2] R. P. Feynman, "Simulating physics with computers," *Int. J. Theor. Phys.* **21**, 467–88 (1982).

[3] P. Shor, "Algorithms for quantum computation: discrete logarithms and factoring," *Proc. 35th Annual Symp. on Foundations of Computer Science*, pp. 124–34 (1994).

[4] P. Shor, "Polynomial-time algorithms for prime factorization and discrete logarithms on a quantum computer," *SIAM J. Sci. Statist. Comput.* **26**, 1484 (1997).

[5] L. K. Grover, "A fast quantum mechanical algorithm for database search," *Proc. 28th Annual ACM Symp. on the Theory of Computing (STOC)*, pp. 212–9 (1996).

[6] J. Preskill, "Reliable quantum computing," *Proc. R. Soc. (Lond.) A* **454**, 385–410 (1998).

[7] A. Blaquiere, "Wave mechanics as a two-player game," in *Dynamical Systems and Microphysics* pp. 33–69 (1980).

[8] S. Wiesner, "Conjugate coding," *SIGACT News* **15**, 78–88 (1983).

[9] N. D. Mermin, "Quantum mysteries revisited," *Am. J. Phys.* **58**, 731–4 (1990).

[10] N. D. Mermin, "Extreme quantum entanglement in a superposition of macroscopically distinct states," *Phys. Rev. Lett.* **65**, 2838–40, (1990).

[11] J. Eisert, M. Wilkins, and M. Lewenstein, "Quantum games and quantum strategies," *Phys. Rev. Lett.* **83**, 3077–88 (1999).

[12] D. A. Meyer, "Quantum strategies," *Phys. Rev. Lett.* **82**, 1052–5 (1999).

[13] E. Klarreich, "Playing by quantum rules," *Nature* **414**, 244–5 (2001).

[14] A. Cho, "Multiple choice," *New Scientist* **173**, Jan. 5th, pp. 13–6 (2002).

[15] C. F. Lee and N. F. Johnson, "Let the quantum games begin," *Physics World* **15**, 25–9 (2002).

[16] R. Cleve, P. Hoyer, B. Toner, and J. Watrous, "Consequence and limits of nonlocal strategies," *Proc. of the 19th Annual Conference on Computational Complexity*, pp. 236–49 (2004).

[17] G. Brassard and A. Broadbent and Alain Tapp, "Quantum pseudo-telepathy," *Found. Phys.* **35**, 1877–1907 (2005).

[18] A. P. Flitney and D. Abbott, "An introduction to quantum game theory," *Fluct. Noise Lett.* **2**, R175–87 (2002).

[19] M. A. Nielsen and I. Chuang, "Quantum Computation and Quantum Information," Cambridge University Press, Cambridge, UK (2000).

[20] A. Einstein, B. Podolsky, and N. Rosen, "Can quantum-mechanical description of physical reality be considered complete?," *Phys. Rev.* **47**, 777–80 (1935).

[21] A. Aspect, Dalibard, and G. Roger, "Experimental test of Bell's inequalities using time-varying analyzers," *Phys. Rev. Lett.* **49**, 1804–7 (1982).

[22] J. S. Bell, "On the Einstein Podolsky Rosen paradox," *Physics* **1**, 195–200 (1964).

[23] D. Deutsch and R. Jozsa, "Rapid solutions of problems by quantum computation," *Proc. R. Soc. (Lond.) A* **439**, 553–8 (1992).

[24] J. Eisert and M. Wilkins, "Quantum games," *J. Mod. Opt.* **47**, 2543–56 (2000).

[25] L. Marinatto and T. Weber, "A quantum approach to games of static information," *Phys. Lett. A* **272**, 291–303 (2000).

[26] A. Nawaz and A. H. Toor, "Generalized quantization scheme for non-zero sum games" *J. Phys. A* **37** 11457–63 (2004).

[27] C. F. Lee and N. F. Johnson, "Efficiency and formalism of quantum games," *Phys. Rev. A* **67**, 022311 (2003).

[28] C. F. Lee and N. F. Johnson, "Game-theoretic discussion of quantum state estimation and cloning," *Phys. Lett. A* **319**, 429–33 (2003).

[29] W. Poundstone, *Prisoner's Dilemma*, Doubleday, New York (1992).

[30] R. Axelrod, "The evolution of cooperation," *Science* **211**, 1390–6 (1981).

[31] R. Axelrod and W. Hamilton, *The Evolution of Cooperation*, Basic Books, New York (1984).

[32] J. Du, X. Xu, H. Li, X. Zhou, and R. Han, "Entanglement playing a dominating role in quantum games," *Phys. Lett. A* **289**, 9–15 (2001).

[33] J. Du, H. Li, X. Xu, X. Zhou, and R. Han, "Phase transition-like behavior of quantum games," *J. Phys. A* **36**, 6551–62 (2003).

[34] A. Nawaz and A. H. Toor, "Dilemma and quantum battle of the sexes," *J. Phys. A* **37**, 4437–43 (2004).

[35] S. C. Benjamin, "Comment on 'A quantum approach to static games of complete information'," *Phys. Lett. A* **277**, 180 (2000).

[36] A. P. Flitney and D. Abbott, "Advantage of a quantum player against a classical one in 2 × 2 quantum games," *Proc. R. Soc. (Lond.) A* **459**, 2463–74 (2003).

[37] A. P. Flitney, "Aspects of Quantum Games," PhD thesis, The University of Adelaide (2005).

[38] A. Iqbal and A. H. Toor, "Evolutionarily stable strategies in quantum games," *Phys. Lett. A* **280**, 249–56 (2001).

[39] J. Du, H. Li, X. Xu, M. Shi, J. Wu, and R. Han, "Experimental realization of quantum games on a quantum computer," *Phys. Rev. Lett.* **88**, 137902 (2002).

[40] J. Du, X. Xu, H. Li, X. Zhou, and R. Han, "Playing Prisoners' Dilemma with quantum rules," *Fluct. Noise Lett.* **2**, R189–203 (2002).

[41] J. Shimamura, S. K. Özdemir, F. Morikoshi, and N. Imoto, "Quantum and classical correlations between players in game theory," *Int. J. Quant. Inf.* **2**, 79–89 (2004).

[42] S. K. Özdemir, J. Shimamura, N. Imoto, "Quantum advantage does not survive in the presence of a corrupt source: optimal strategies in simultaneous move game," *Phys. Lett. A* **325**, 104–11 (2004).

[43] S. K. Özdemir, J. Shimamura, F. Morikoshi, N. Imoto, "Dynamics of a discoordination game with classical and quantum correlations," *Phys. Lett. A* **333**, 218–31 (2004).

[44] J. Du, H. Li, X. Xu, M. Shi, and X. Zhou, "Entanglement enhanced multiplayer quantum games," *Phys. Lett. A* **302**, 229–33 (2002).

[45] A. P. Flitney and L. C. L. Hollenberg, "Nash equilibria in quantum games with generalized two-parameter strategies," *Phys. Lett. A* **363**, 381–8 (2007).

[46] S. C. Benjamin and P. M. Hayden, "Comment on 'Quantum games and quantum strategies' " *Phys. Rev. Lett.* **87**, 069801 (2001).

[47] W. H. Zurek, "Decoherence, einselection and the quantum origins of the classical," *Rev. Mod. Phys.* **75**, 713–75 (2003).

[48] M. Schlosshauer, "Decoherence, the measurement problem, and interpretations of quantum mechanics," *Rev. Mod. Phys.* **76**, 1267–305 (2004).

[49] M. Schlosshauer, *Decoherence and the quantum-to-classical transition*, Springer, Berlin-Heidelberg-New York (2007).

[50] D. A. Lidar and K. B. Whaley, "Decoherence free subspaces and subsystems," in Irreversible Quantum Dynamics, *Springer Lecture Notes in Physics* **622**, eds. F. Benatti and R. Floreanini, Springer, Berlin, pp. 83–120 (2003).

[51] N. F. Johnson, "Playing a quantum game with a corrupted source," *Phys. Rev. A* **63**, 020302(R) (2001).

[52] J-L. Chen, L. C. Kwek, and C. H. Oh, "Noisy quantum game," *Phys. Rev. A* **65**, 052320 (2002).

[53] L. K. Chen, H. Ang, D. Kiang, L. C. Kwek, and C. F. Lo, "Quantum prisoner dilemma under decoherence," *Phys. Lett. A* **316**, 317–23 (2003).

[54] A. P. Flitney and D. Abbott, "Quantum games with decoherence," *J. Phys. A* **38** 449–59 (2005).

[55] A. P. Flitney and L. C. L. Hollenberg, "Multiplayer quantum Minority game with decoherence," *Quant. Inf. Comput.* **7**, 111–26 (2007).

[56] H. Weinfurter and M. Żukowski "Four-photon entanglement from down-conversion," *Phys. Rev. A* **64**, 010102(R) (2001).

[57] A. P. Flitney and L. C. L. Hollenberg, "Quantum Minority game with various forms of entanglement," in *Proc. SPIE Symp. on Microelectronics, MEMS, and Nanotechnology* (2007).

[58] A. Nawaz and A. H. Toor, "Quantum games with correlated noise," *J. Phys. A* **39** 9321–8 (2006).

[59] A. P. Flitney and D. Abbott, "Quantum two and three person duels," *J. Optics B* **6** S860–6 (2004).

[60] C. Shuai and F. Mao-Fa, "The effect of quantum noise on the restricted quantum game," *Chinese Phys.* **15**, 60–5 (2006).

[61] C. Schmid, A. P. Flitney, W. Wieczorek, N. Kessel, H. Weinfurter, and L. C. L. Hollenberg, in preparation.

[62] J. Maynard Smith and G. R. Price, "The logic of animal conflict," *Nature* **246**, 15–8 (1973).

[63] J. Maynard Smith, *Evolution and the Theory of Games*, Cambridge University Press, Cambridge, UK (1982).

[64] A. Iqbal and A. H. Toor, "Entanglement and dynamic stability of Nash equilibrium in a symmetric quantum game," *Phys. Lett. A* **286**, 245–50 (2001).

[65] A. Iqbal and A. H. Toor, "Darwinism in quantum systems," *Phys. Lett. A* **294**, 261–70 (2002).

[66] A. Iqbal and A. H. Toor, "Stability of mixed Nash equilibrium in symmetric quantum games," *Commun. Theor. Phys.* **42**, 335–8 (2004).

[67] A. Iqbal and T. Cheon, "Evolutionary stability in quantum games," to be published in *Quantum Aspects of Life*, eds. D. Abbott, P. Davies, and A. Pati, World Scientific 2008 [*E-print*: quant-ph/0707.1413 (2007)].

[68] P. C. W. Davies, "Does quantum mechanics play a non-trivial role in life?," *BioSystems* **78**, 69–79 (2004).

[69] A. Iqbal and A. H. Toor, "Quantum mechanics gives stability to Nash equilibrium," *Phys. Rev. A* **65**, 022036 (2002).

[70] S. C. Benjamin and P. M. Hayden, "Multiplayer quantum games," *Phys. Rev. A* **64**, 030301(R) (2001).

[71] D. M. Greenberger, M. A. Horne, and A. Zeilinger, "Going beyond Bell's theorem," in *Bell's Theorem, Quantum Theory and Conceptions of the Universe*, ed. M. Kafatos, Kluwer Academic, Dordrecht, pp. 69–72 (1989).

[72] K. Y. Chen, T. Hogg and R. Beausoleil, "Quantum treatment of public goods economics," *Quant. Inf. Proc.* **1**, 449–69 (2002).

[73] Q. Chen, Y. Wang, J.-T. Liu, and K-L. Wang", "N-player quantum minority game," *Phys. Lett. A*, **327**, 98–102 (2004).

[74] A. P. Flitney and A. D. Greentree, "Coalitions in the quantum Minority game: Classical cheats and quantum bullies," *Phys. Lett. A* **362** 132–7 (2007).

[75] J. Shimamura, S. K. Özdemir, F. Morikoshi, and N. Imoto, "Entangled states that cannot reproduce original classical games in their quantum version," *Phys. Lett. A* **328**, 20–5 (2004).

[76] S. K. Özdemir, J. Shimamura, and N. Imoto", "A necessary and sufficient condition to play games in quantum mechanical settings," *New J. Phys.* **9**, 43 (2007).

[77] D. Challet and Y. C. Zhang, "Emergence of cooperation and organization in an evolutionary game," *Physica A* **246**, 407–18 (1997).

[78] D. Challet, M. Marsili, and Y-C Zhang, "Modeling market mechanism with minority game," *Physica A* **276**, 284–315 (2000).

[79] E. Moro, "The Minority game: an introductory guide," in *Advances in Condensed Matter and Statistical Physics*, eds. E. Korutcheva and R. Cuerono, Nova Science Publishers Inc., Hauppauge, NY (2004).

[80] T. C. Schelling *The Strategy of Conflict*, Harvard University Press, Cambridge, MA (1960).

[81] A. P. Flitney, M. Schlosshauer, C. Schmid, W. Laskowski, and L. Hollenberg, *E-print:* quant-ph/0803.0579 (2008).

[82] A. Iqbal and A. H. Toor, "Quantum repeated games," *Phys. Lett. A* **300**, 537–42 (2002).

[83] G. Gutoski and J. Watrous, "Towards a general theory of quantum games," *Proc. 39th Symp. on Theory of Computing (STOC)* pp. 565–74 (2007) [*E-print:* quant-ph/0611234].

[84] A. Kitaev, "Quantum coin-flipping," in *Proc. 6th Workshop on Quantum Information Processing (QIP)* (2003).

[85] G. Abal, R. Donangelo, and H. Fort, "Conditional quantum walk and iterated quantum games," to appear in *Physica A* [*E-print:* quant-ph/0607143].

[86] A. P. Flitney, D. Abbott and N. F. Johnson, "Quantum walks with history dependence," *J. Phys. A*, **30**, 7581–91 (2004).

[87] J. Kempe, "Quantum random walks: an introductory overview," *Contemp. Phys.* **44**, 307–27 (2003).

[88] T. A. Brun, H. A. Carteret and A. Ambainis, "Quantum walks driven by many coins," *Phys. Rev. A* **67**, 052317 (2003).

[89] J. Du, X. Xu, H. Li, M. Shi, X. Zhou and R. Han, "Multi-player and multi-choice quantum game," *Chinese Phys. Lett.* **19**, 1221–1224 (2002).

[90] H. Liu, "Generalized quantum games with Nash equilibrium," *Commun. Theor. Phys.* **41**, 553–6 (2002).

[91] Y.-J. Han, Y.-S. Zhang, and G-C. Guo, "Quantum game with incomplete information," *Fluct. Noise Lett.* **2**, L263–71 (2002).

[92] H. Li and J. Du and S. Massar, "Continuous-variable quantum games," *Phys. Lett. A* **306**, 73–8 (2002).

[93] A. Boukas, "Quantum formulation of classical two-person zero-sum games," *Open Sys. & Information Dyn.* **7**, 19–32 (2000).

[94] A. A. Grib and G. Parfionov, "Can the game be quantum," in *Notes of Scient. Sem. of St. Petersburg's Branch of Steklov Mathematical Institute of Russian Academy of Sciences* **291**, pp. 1–24 (2002).

[95] J. Wu, "Hamiltotian formalism of game theory," *E-print*: quant-ph/0501088 (2005).

[96] J. Wu, "Theory of games on quantum objects," *E-print*: quant-ph/0503094 (2005).

[97] S. J. van Enk, "Quantum and classical game strategies," *Phys. Rev. Lett.* **84**, 789 (2000).

[98] S. J. van Enk and R. Pike, "Classical rules in quantum games," *Phys. Rev. A* **66**, 024306 (2002).

[99] A. Iqbal, "Quantum games with a multi-slit electron diffraction set-up," *Nuov. Cimen. S. I. Fis. B* **118**, 463–8 (2003).

[100] A. Iqbal and S. Weigert, "Quantum correlation games," *J. Phys. A* **37**, 5873–85 (2004).

[101] A. Iqbal, "Quantum correlations and Nash equilibria of a bi-matrix game," *J. Phys. A* **37**, L353–9 (2004).

[102] A. Iqbal, "Playing games with EPR type experiments," *J. Phys. A* **38**, 249–56 (2005).

[103] A. Iqbal, "Investigations in quantum games using EPR-type set-ups," PhD thesis, University of Hull, UK (2006) [*E-print*: quant-ph/0604188].

[104] E. W. Piotrowski and J. Sładowski, "Quantum-like approach to financial risk: quantum anthropic principle," *Acta. Phys. Pol. B* **32**, 3873 (2001).

[105] E. W. Piotrowski and J. Sładowski, "Quantum bargaining games," *Physica A* **308**, 391–401 (2002).

[106] E. W. Piotrowski and J. Sładowski, "Quantum market games," *Physica A* **312**, 208–16 (2002).

[107] E. W. Piotrowski and J. Sładowski, "Quantum English auctions," *Physica A* **318**, 1101–6 (2002).

[108] E. W. Piotrowski and J. Sładowski, "Interference of quantum market strategies," *Physica A* **318**, 516–28 (2002).

[109] E. W. Piotrowski, "Fixed point theorem for simple quantum strategies in quantum market games," *Physica A* **324**, 196–200 (2003).

[110] E. W. Piotrowski and J. Sładowski, "Trading by quantum rules—quantum anthropic principle," *Int. J. Theor. Phys.* **42**, 1101–6 (2003).

[111] "E. W. Piotrowski and J. Sładowski, "Quantum game theory in finance," *Quant. Finance* **4**, 1–7 (2004).

[112] E. W. Piotrowski and J. Sładowski, "Quantum computer: an appliance for playing market games," *Int. J. Quantum Inf.* **2**, 495–509 (2004).

[113] E. W. Piotrowski and J. Sładowski, "Quantum diffusion of prices and profits," *Physica A* **345**, 185–95 (2005).

[114] K. Ilinski, *The Physics of Finance: Gauge Modeling in Non-equilibrium Pricing*, John Wiley and Sons Ltd., Chichester (2001).

[115] B. E. Baaquie, "Quantum field theory of treasury bonds," *Phys. Rev. E* **64**, 016121 (2001).

[116] T. Hogg and K.-Y. Chen, "Quantum auctions," *Int. J. Quantum Inf.* **5**, 751–80 (2007).

[117] K.-Y. Chen and T. Hogg, "Experiments with probabilistic quantum auctions," *E-print:* quant-ph/0707.4195 (2007).

[118] M. vos Savant, "Ask Marilyn," *Parade*, Dec. 2nd 1990, Feb. 17th 1991.

[119] M. vos Savant *The Power of Logical Thinking*, St Martin's Press, New York (1996).

[120] M. vos Savant, "Ask Marilyn," *Parade*, Sep. 9th 1990.

[121] L. Gillman, "The car and the goats," *The Am. Math. Monthly* **99**, 3–7 (1992).

[122] C-F. Li, Y-S. Zhang, and Y-F. Huang, "Quantum strategies of quantum measurement," *Phys. Lett. A* **280**, 257–60 (2001).

[123] G. M. D'Ariano *et al.*, "The quantum Monty Hall problem," *Quant. Inf. Comput.* **2**, 355-66 (2002).

[124] http://www.imaph.tu-bs.de/qi/monty/

[125] A. P. Flitney and D. Abbott, "Quantum version of the Monty Hall problem," *Phys. Rev. A* **65**, 062318 (2002).

[126] A. Iqbal and T. Cheon, "Constructing quantum games from non-factorizable joint probabilities," *Phys. Rev. E* **76**, 061122 (2007).

[127] A. Iqbal and T. Cheon, "Constructing multi-player quantum games from non-factorizable joint probabilities," in *Proc. SPIE Symp. on Microelectronics, MEMS, and Nanotechnology* [*E-print*: quant-ph/0712.3701] (2007).

[128] G. P. Harmer and D. Abbott, "Losing strategies can win by Parrondo's paradox," *Nature* **402**, 864 (1999)

[129] G. P. Harmer and D. Abbott "Parrondo's paradox," *Stat. Sci.* **14**, 206–13 (1999).

[130] J. M. R. Parrondo, G. Harmer, and D. Abbott, "New paradoxical games based on Brownian ratchets," *Phys. Rev. Lett.* **85**, 5226–9 (2000).

[131] A. P. Flitney and J. Ng and D. Abbott, "Quantum Parrondo's games," *Physica A* **314**, 35–42 (2002).

[132] D. A. Meyer and H. Blumer, "Parrondo games as lattice gas automata," *J. Stat. Phys.* **107**, 225–39 (2002).

[133] D. A. Meyer and H. Blumer, "Quantum Parrondo games: biased and unbiased," *Fluct. Noise Lett.* **2**, L257–62 (2002).

[134] D. A. Meyer, "Meyer replies," *Phys. Rev. Lett.* **84**, 790 (2000).

[135] D. Abbott *et al.*, "Dreams versus reality: plenary debate session on quantum computing," *Quant. Inf. Proc.* **2**, 449–72 (2003).

[136] J. Eisert and M. M. Wolf, "Quantum computing," in *Handbook of Innovative Computing*, Springer, Berlin-Heidelberg-Frankfurt (2004).

[137] C. Schuck, "Experimental implementation of a quantum game," Diploma thesis, Universität München, Germany (2003).

[138] J. Lu, L. Zhou, and L.-M. Kuang, "Linear optics implementation for quantum game with two players," *Phys. Lett. A* **330**, 48–53 (2004).

[139] C. Shuai and F. Mao-Fa, "Linear optics implementation for quantum game under quantum noise," *Chinese Phys.* **15**, 276–80 (2006).

[140] M. Paternostro, M. S. Tame, and M. S. Kim, "Hybrid cluster state proposal for a quantum game," *New J. Phys.* **7**, 226 (2005).

[141] R. Prevedel, A. Stefanov, P. Walther, and A. Zeilinger, "Experimental realization of a quantum game on a one-way quantum computer," *New J. Phys.* **9**, 205 (2007).

[142] I. M. Buluta, S. Fujiwara, and S. Hasegawa, "Quantum games in ion traps," *Phys. Lett. A* **358**, 100–4 (2006).

Reviewed by Dr Azhar Iqbal, Dept. of Electrical and Electronic Engineering, The University of Adelaide, Adelaide, SA 5005, Australia.

In: Game Theory: Strategies, Equilibria and Theorems
Editors: Ingrid N. Haugen et al, pp. 41-82
ISBN: 978-1-60456-844-8

Chapter 2

REVERSIBLE ERROR CORRECTION IN DECISION COMMUNICATION WITHIN QUANTUM GAME-THEORETIC BIJECTIVITY

Anas N. Al-Rabadi*

Computer Engineering Department, The University of Jordan, Jordan & The Office of Graduate Studies and Research (OGSR), Portland State University, U.S.A.

Abstract

This research introduces the new concept of reversibility in game theory including (1) game-theoretic playing reversibility and (2) reversibility in error correction of game decision communication. Game-theoretic multiple-valued quantum computation (MVQC) is also introduced. Game theory (GT) involves the study of competition and cooperation, without regard to the particular entities (agents) involved, and issues of rationality associated with such phenomena. Reversibility property in GT can be important in situations where: (1) an outside observer needs to know reversely the (correct) paths that lead to specific payoffs on a game's extensive form, and (2) modeling the (maximin) dynamics using low-power consuming circuits as reversibility is a main requirement for low-power circuit design of future technologies such as in quantum computing (QC). Error correction of communicated decisions in two-player games and reversible error correction of communicated batch of decisions in multi-player games are important tasks in situations where noise from a third player(s) or from the communication channel exists. Quantum decision trees (QDTs) are also introduced as quantum representations for applying MVQC to games' dynamics.

Keywords: Game Theory, Noise, Error-Correcting Codes, Error-Control Coding, Coding, Modeling, Low-Power Computing, Circuits and Systems, Quantum Circuits, Quantum Computing, Reversible Logic, Reversible Circuits, Low-Power Circuits and Systems, Low-Power VLSI Circuit Design.

---

* E-mail address: alrabadi@yahoo.com; http://web.pdx.edu/~psu21829/

1. Introduction

The notion of implementing game theory (GT) using quantum computing has been suggested by various authors, e.g., [8,16,28,29,39,46,49]. Researchers have succeeded in implementing many-valued (*m*-valued) logic gates using many-valued quantum systems (e.g., [42]). This research investigates the issues of reversibility in game theory including: (1) playing reversibility and (2) reversibility in error correction in game decision communication. Also, this research investigates the the implementation of games' maximin dynamics using many-valued quantum computing (MVQC).

Quantum computing (QC) is a method of computation that uses a dynamic process governed by the Schrödinger Equation (SE) [4,42,44]. Research in QC gained momentum, for its application within the context of GT (e.g., [8,16,28,29,39,46,49]), when it has been shown that the Prisoner's Dilemma (PD) game, which is not solvable in a single iteration and can be classically solved in several ways such as (1) iterated games, (2) non-myopic rationality, and (3) meta-games, can also be solved using quantum computing (QC) method.

Due to the anticipated failure of Moore's law around the year 2020, quantum computing will hopefully play an increasingly crucial role in building more compact and less power consuming computers [4,44]. Due to this fact, and because all quantum computer gates (i.e., building blocks) should be reversible [4,44], reversible computing will have an increasingly more existence in the future design of regular, compact, and universal circuits and systems. (k, k) reversible circuits are circuits that have the same number of inputs k and outputs k and are one-to-one mappings between vectors of inputs and outputs, thus the vector of input states can be always uniquely reconstructed from the vector of output states. A (k, k) conservative circuit has the same number of inputs k and outputs k and has the same number of values (states) in inputs and outputs (e.g., the same number of ones and twos in inputs and outputs for ternary) [4,44]. The importance of the conservativeness property stems from the fact that this property reflects the physical law of energy preservation: no energy can be created or destroyed, but can be transformed from one form to another. Thus, conservative logic will incorporate the fundamental law of energy preservation into the logic design of systems.

Other motivations for pursuing the possibility of implementing game dynamics using reversible logic (RL) and QC would include items such as: (1) *power*: the fact that, theoretically, the internal computations in RL systems consume no power. It is shown in [31] that a necessary (but not sufficient) condition for not dissipating power in any physical circuit is that all system circuits must be built using fully reversible logical components. For this reason, different technologies have been studied to implement reversible logic in hardware, such as adiabatic CMOS [54], optical [48], and quantum [4,44]. Fully reversible digital systems will greatly reduce the power consumption (theoretically eliminate) through three conditions: (i) *logical reversibility*: the vector of input states can always be uniquely reconstructed from the vector of output states, (ii) *physical reversibility*: the physical switch operates backwards as well as forwards, and (iii) the use of *"ideal-like" switches* that have no parasitic resistances; (2) *size*: since the newly emerging quantum computing technology must be reversible [4,9,15,31,44], the current trends related to more dense hardware implementations are heading towards 1 Angstrom (atomic size), at which quantum mechanical effects have to be accounted for; and (3) *speed*: if the properties of superposition and entanglement of quantum mechanics can be usefully employed in the GT context, significant computational and modeling speed enhancements can be expected [4,44].

In general, in decision communications between two playing entities (agents; parties), noise exists and corrupts sent messages (decisions), and thus noisy corrupted messages will be received. Therefore, error correction of communicated decisions in two-player games and reversible error correction of communicated batch of decisions in multi-player games are highly important tasks in situations where noise occurs. The corrupting noise may be sourced from either: (1) a third entity(s) or (2) from the communication channel. To solve this highly important problem, that is to extract the correct message from the noisy erroneous counterpart, various coding schemes, used principally in communications theory, have been proposed that work optimally for specific types of statistical distributions of noise [1,2,3,5,10,11,12,13,17,18,19,21,23,25,26,27,30,32,33,34,36,37,38,40,41,47,50,53,55,56,58,59,60,62,63,64,65,66,69,71]. Thus, one main aim of this research is the implementation of reversibility in error correction in game decision communication.

The main contributions of this research are as follows: (1) reversibility application in playing games to generate reversible game theory (RGT), (2) reversible error correction in decision communication within game theory context, (3) multiple-valued quantum computation (MVQC) of games' (maximin) dynamics, and (4) introducing quantum decision trees (QDTs) as quantum representations for applying MVQC to games' dynamics.

Basic background in game theory, error-control coding, reversible logic and quantum computing is presented in Section 2. Logic reversibility in game theory is introduced in Section 3. Reversible error correction in games is introduced in Section 4. The implementation of games using multiple-valued (*m*-valued) quantum computing and the associated QC tree-based representations are introduced in Section 5. Conclusion and future work are presented in Section 6.

2. Fundamentals

This Section presents basic background in the topics of game theory, error-correction coding, reversible logic, and quantum computing. The fundamentals presented in this section will be utilized in the development of the new results introduced in Sections 3 – 5.

2.1. Game Theory

Game theory (GT) is an important modeling method which is used in several areas like economics, business, finance, marketing, biology, engineering, anthropology, social psychology, politics, and philosophy [6,7,14,20,22,24,35,43,51,52,61,67,70].

While the context of decision theory (DT) is one player games against nature, the context of GT is the decision making and moves based on payoffs (utilities) for two players or more than two players [6,7,14,22,35,43,67,70]. Game theory deals with situations that are: (1) pure conflict (strictly competitive; zero sum games (ZSG)) where what one player wins is equal to what the other player loses and the total pay (utility) is constant, (2) partial conflict (competition and cooperation; non-zero sum games (NZSG)) where the total pay (utility) is variable, and (3) pure cooperation. While a ZSG has a theory that leads to a solution, a NZSG has no general solution and often exhibits paradoxical features.

Problem representation in GT can be ordinal utility versus interval (cardinal) utility, and extensive (tree-based) form where time is explicit versus normal (matrix; table) form, using a payoff matrix (table), where time is only implicit. Typically, some information is lost by going from the extensive form to the normal form. Information in GT can be: (1) perfect information that means that one knows always where he is on tree in the extensive form, (2) complete information where the payoff table is known, or (3) incomplete information where the payoff table is not known. Principle of rationality in GT states that (1) one maximizes (or secures some minimum) utility by some appropriate decision rule, and (2) one assumes the other player is doing likewise and acts accordingly for one's own utility.

A strategy in GT is a decision or a sequence of decisions, and a dominant strategy is a strategy that is optimal no matter what the other player does. A strategy that solves a game without probabilistic choices is called pure strategy, and a strategy that uses probability to solve a game is called mixed strategy, i.e., if a player in a game chooses among two or more strategies at random according to specific probabilities then this choice is called a "mixed strategy." If a pure strategy fails to solve the game then one uses mixed strategy.

A players' solution to a ZSG can be a maximin: get the best of the worst possible outcomes, where a dominant strategy is automatically maximin. A joint (mutual) maximin means that both the row and column players use maximin. In ZSG, if the column (row) player does maximin on his own payoffs, he will choose the same strategy as if he does minimax on the row (column) players' payoffs (while the value (outcome payoff) would be different).

A saddle point (SP) is a cell which is the maximum utility in its column and minimum utility in its row and is a cell which neither the row player nor the column player wants to move from. An equilibrium point (EP) is the cell(s) that game dynamics settle in. In ZSG, saddle points are the same as equilibrium points, but this is not necessarily true in a NZSG. Pareto optimal (PO) cell(s) is the cell that makes everyone happy and moving from it will hurt at least one player, i.e., is a cell that you remain at if any player vetoes (all) movements to (all) other cells (or in other words it is the cell that moving away from will hurt at least one player; the cell that if both players would be willing to move to is identical to the original cell). A non-Pareto optimal (NPO) cell is a cell that a player moves from since no player vetoes that movement to another cell(s), i.e., if you can move to at least one another cell then the current cell is NPO.

Hierarchy of 2-player ZSGs can be: (1) if both players have dominant strategy then each player uses dominant strategy; (2) if one player has dominant strategy then one uses the dominant strategy and the other acts appropriately; (3) if no player has dominant strategy but the game has a saddle point then both players do pure maximin (and in this case maximin equals minimax); and (4) if the game has no saddle point then both players do mixed maximin.

In ZSG, a game is called unstable if the joint (mutual) maximin cell is not the same as EP cell, otherwise the game is stable. Figure 1 illustrates common games in GT where a lighter color means maximin, a darker color means EPs, po is Pareto optimal, and npo is non-Pareto optimal.

		B		
		S	~S	
A	S	^{po}S,S	^{po}T,B	T
	~S	^{po}B,T	^{npo}W,W	W
		T	W	

(a)

		B		
		C	D	
A	C	^{po}S,S	^{po}W,B	W
	D	^{po}B,W	^{npo}T,T	T
		W	T	

(b)

		B		
		G	O	
A	G	^{po}2,1	^{npo}-1,-1	.4
	O	^{npo}-1,-1	^{po}1,2	.6
		.2	.8	

(c)

		B		
		B_1	B_2	
A	A_1	^{po}B,B	^{npo}W,S	W
	A_2	^{npo}S,W	^{npo}T,T	T
		W	T	

(d)

		B		
		J	~J	
A	J	^{npo}T,T	^{po}B,S	T
	~J	^{po}S,B	^{npo}W,W	W
		T	W	

(e)

		B		
		B_1	B_2	
A	A_1	^{po}S,S	^{po}B,W	S
	A_2	^{po}W,B	^{npo}T,T	W
		S	W	

(f)

Figure 1. Games (going left-to-right and top-to-down): (a) ordinal-utility pure-strategy chicken game, (b) ordinal-utility pure-strategy prisoner's dilemma (PD) game, (c) ordinal-utility pure-strategy assurance game, (d) interval-utility mixed-strategy leader (battle of the sexes) game, (e) ordinal-utility pure-strategy hero game, and (f) ordinal-utility pure-strategy convergence game.

2.2. Error Correction

In game-theoretic context, noise usually exists and may be generated from: (1) third agent(s) or (2) from the channel in which playing decisions are communicated. Such noise corrupts sent decisions from one end and thus noisy corrupted messages are received on the other end. To solve the problem of extracting a correct message from its corrupted counterpart, noise must be modeled [13,45,68] and accordingly an appropriate encoding / decoding communication schemes must be implemented [1,2,3,5,10,11,12,13,17,18,19,21,23,25,26,27, 30,32,33,34,36,37,38,40,41,47,50,53,55,56,58,59,60,62,63,64,65,66,69,71]. Various coding schemes have been proposed and one important family is the convolutional codes [17,50,62,63,64]. Figure 2 illustrates the modeling of 2-player games with noise and the solution to the noise problem using an encoder / decoder scheme.

Each of the two players' sides in the system seen in Figure 2 consists of three major parts: (1) encoding (e.g., generating a convolutional code using a convolutional encoder) to generate an encoded sent decision (message), (2) channel noise, and (3) decoding (e.g., generating the correct convolution code using the corresponding decoding algorithm (cf. Viterbi algorithm)) to generate the decoded correct received decision (message).

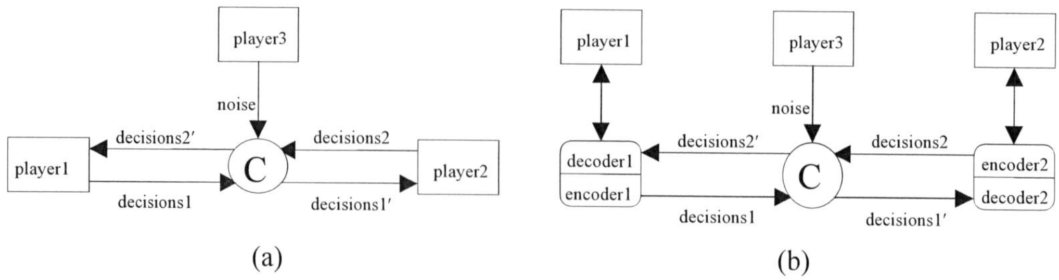

Figure 2. Modeling 2-player games with noise: (a) model of a noisy game where C is the channel, and (b) model of the solution to the noise problem using encoder / decoder schemes. In this model, decision1′ = decision1 + noise and decision2′ = decision2 + noise.

In general, in block coding, the encoder receives a k-bit message block and generates an n-bit code word, and therefore code words are generated on a block-by-block basis, and the whole message block must be buffered before the generation of the associated code word. On the other hand, message bits are received serially rather than in blocks where it is undesirable to use a buffer. In such case, one uses convloutional coding, in which a convolutional coder generates redundant bits by using modulo-2 convolutions.

The binary convolutional encoder can be seen as a finite state machine (FSM) consisting of an M-stage shift register with interconnections to n mouulo-2 adders and a multiplexer to serialize the outputs of the adders, in which an L-bit message sequence generates a coded output sequence of length $n(L + M)$ bits.

Definition 1. For an L-bit message sequence, M-stage shift register, n mouulo-2 adders, and a generated coded output sequence of length $n(L + M)$ bits, the code rate r is calculated as:

$$r = \frac{L}{n(L+M)} \text{ bits / symbol}$$

and for the typical case of $L >> M$, the code rate reduces to $r \approx \frac{1}{n}$ bits/symbol.

Definition 2. The constraint length of a convolutional code is the number of shifts over which a single message bit can influence the encoder output. Thus, for an encoder with an M-stage shift register, the number of shifts required for a message bit to enter the shift register and then come out of it is equal to $K = M + 1$. Thus, the encoder constraint length is equal to K.

A binary convolutional code can be generated with code rate $r = \frac{k}{n}$ by using k shift registers, an n modulo-2 adders, an input multiplexer, and an output multiplexer. An example of a convolutional encoder with constraint length = 3 and rate = ½ is the one shown in Figure 3.

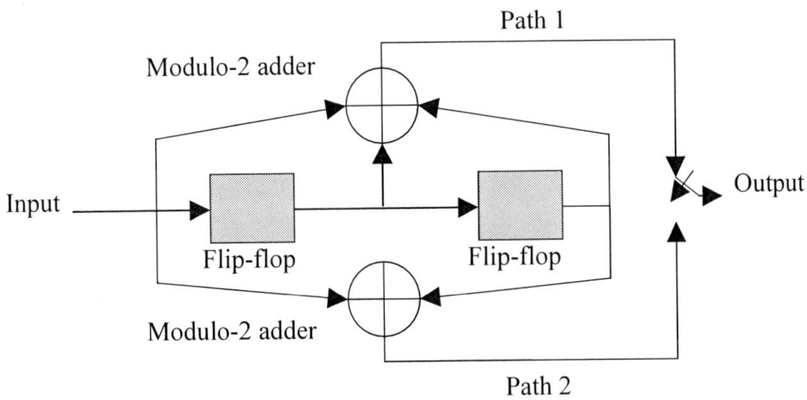

Figure 3. Convolutional encoder with constraint length = 3 and rate = ½ . The flip-flop is a unit-delay element, and the modulo-2 adder is the logic Boolean difference (XOR) operation.

The convolutional codes generated by the encoder in Figure 3 are part of what is generally called nonsystematic codes. Each path connecting the output to the input of a convolutional encoder can be characterized in terms of the impulse response which is defined as the response of that path to "1" applied to its input, with each flip-flop of the encoder set initially to "0". Equivalently, we can characterize each path in terms of a generator polynomial defined as the unit-delay transform of the impulse response. More specifically, the generator polynomial is defined as:

$$g(D) = \sum_{i=0}^{M} g_i D^i \tag{1}$$

where g_i is the generator coefficients $\in \{0, 1\}$, and the generator sequence $\{g_0, g_1, \ldots, g_M\}$ composed of generator coefficients is the impulse response of the corresponding path in the convolutional encoder, and D is the unit-delay variable.

Example 1. For the convolutional encoder in Figure 3, path1 impulse response is (1, 1, 1), and path2 impulse response is (1, 0, 1). Thus, according to Equation (1), the following are the corresponding generating polynomials, respectively, where addition is performed in modulo-2 addition arithmetic:

$$g_1(D) = 1\cdot D^0 + 1\cdot D^1 + 1\cdot D^2 = 1 + D + D^2$$

$$g_2(D) = 1\cdot D^0 + 0\cdot D^1 + 1\cdot D^2 = 1 + D^2$$

For a message sequence (10011), the following is the D-domain polynomial representation:

$$m(D) = 1\cdot D^0 + 0\cdot D^1 + 0\cdot D^2 + 1\cdot D^3 + 1\cdot D^4 = 1 + D^3 + D^4$$

As convolution in time domain is transformed into multiplication in the D-domain, path1 output polynomial and path2 output polynomial are as follows, respectively:

$$c_1(D) = g_1(D)m(D) = (1 + D + D^2)(1 + D^3 + D^4) = 1 + D + D^2 + D^3 + D^6$$

$$c_2(D) = g_2(D)m(D) = (1 + D^2)(1 + D^3 + D^4) = 1 + D^2 + D^3 + D^4 + D^5 + D^6$$

Therefore, the output sequences of paths 1 and 2 are as follows, respectively:

Output sequence of path1: (1111001)
Output sequence of path2: (1011111)

The resulting encoded sequence from the convolutional encoder in Figure 3 is obtained by multiplexing the two output sequences of paths 1 and 2 as follows:

$$c = (11, 10, 11, 11, 01, 01, 11)$$

Example 2. For the convolutional encoder in Figure 3, the following are examples of encoded messages:

$$m_1 = (11011) \rightarrow c_1 = (11010100010111)$$

$$m_2 = (00011) \rightarrow c_2 = (00000011010111)$$

$$m_3 = (01001) \rightarrow c_3 = (00111011111011)$$

In general, a message sequence of length L bits results in an encoded sequence of length equals to $n(L + K - 1)$ bits. Usually a terminating sequence of $(K - 1)$ zeros called the tail of the message is appended to the last input bit of the message sequence in order for the shift register to be restored to its zero initial state.

The structural properties of the convolutional encoder (e.g., Figure 3) can be represented graphically in several equivalent representations (cf. Figure 4) using: (1) code tree, (2) trellis, and (3) state diagram. The trellis contains $(L + K)$ levels where L is the length of the incoming message sequence and K is the constraint length of the code. Therefore, the trellis form is preferred over the code tree form because the number of nodes at any level of the trellis does not continue to grow as the number of incoming message bits increases, but rather it remains constant at 2^{K-1}, where K is the constraint length of the code. Figure 4 shows the various graphical representations for the convolutional encoder in Figure 3.

Therefore, any encoded output sequence can be generated from the corresponding input message sequence using the following methods: (1) circuit of the convolutional encoder, (2) polynomial generator, (3) code tree, (4) trellis, and (5) state diagram.

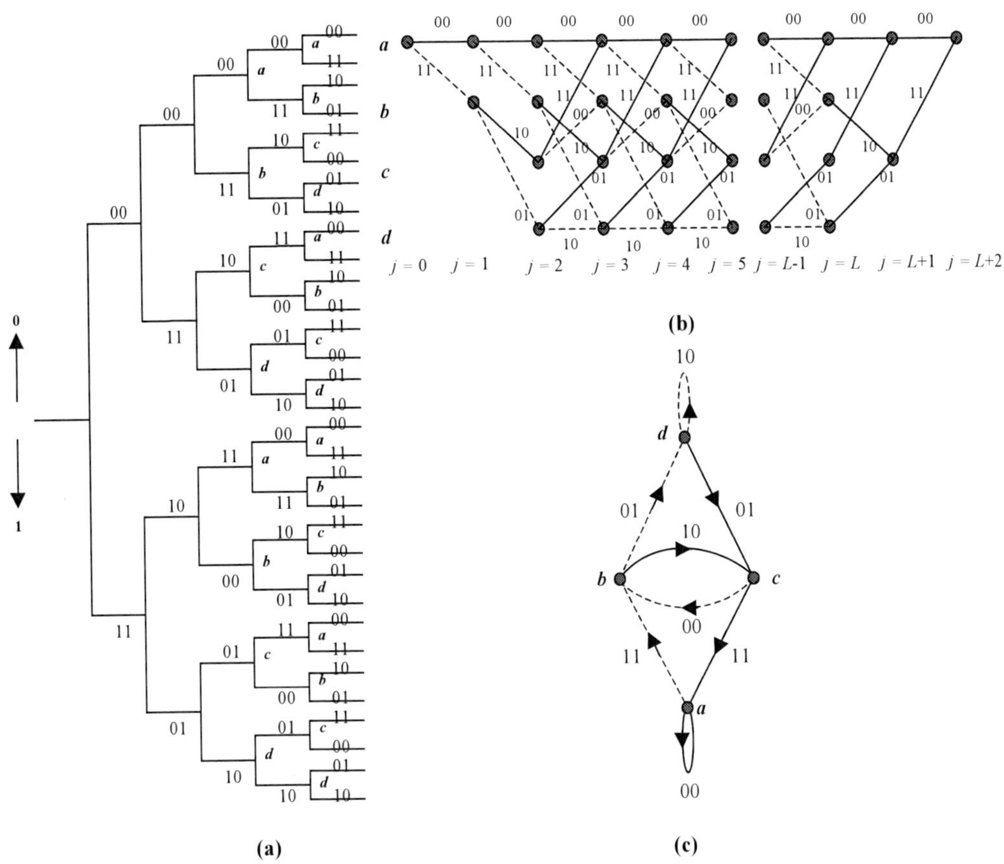

Figure 4. Various representations for the circuit of the convolutional encoder in Figure 3: (a) code tree, (b) trellis, and (c) state diagram. Solid line is the input of value "0" and the dashed line is the input of value "1". The binary label on each branch is the encoder's output as it moves from one state to another. The state encoding of the states can be as $\{a = 00, b = 10, c = 01, d = 11\}$.

An important decoder that uses the trellis representation to correct received erroneous messages is the Viterbi decoding algorithm [17,50,62,63,64]. The Viterbi algorithm is a dynamic programming algorithm which is used to find the maximum-likelihood sequence of hidden states, which results in a sequence of observed events particularly in the context of hidden Markov models (HMMs) [17,50]. The Viterbi algorithm forms a subset of information theory [1,13], and has been extensively used in a wide range of applications including speech recognition, keyword spotting, computational linguistics, bioinformatics, and in communications including digital cellular, dial-up modems, satellite, deep-space and wireless LAN communications.

The Viterbi algorithm is a maximum-likelihood decoder which is optimum for a noise type which is statistically characterized as an Additive White Gaussian Noise (AWGN). This algorithm operates by computing a metric for every possible path in the trellis representation. The metric for a specific path is computed as the Hamming distance between the coded sequence represented by that path and the received sequence. For a pair of code vectors c_1 and

c_2 that have the same number of elements, the Hamming distance $d(c_1, c_2)$ between such a pair of code vectors is defined as the number of locations in which their respective elements differ. In the Viterbi algorithm context, the Hamming distance is computed by counting how many bits are different between the received channel symbol pair and the possible channel symbol pairs, in which the results can only be "0", "1" or "2". Therefore, for each node (i.e., state) in the trellis the Viterbi algorithm compares the two paths entering the node. The path with the lower metric is retained and the other path is discarded. This computation is repeated for every level j of the trellis in the range $M \le j \le L$, where $M = K - 1$ is the encoder's memory and L is the length of the incoming message sequence. The paths that are retained are called survivor or active paths. In some cases, applying the Viterbi algorithm leads to the following difficulty: when the paths entering a node (state) are compared and their metrics are found to be identical then a choice is made by making a guess (i.e., flipping a fair coin). The Viterbi algorithm is a maximum likelihood sequence estimator and the following procedure illustrates the steps of this algorithm.

Algorithm Viterbi

---

1. Initialization: Label the left-most state of the trellis (i.e., all zero state at level 0) as 0
2. Computation step $j + 1$: Let $j = 0, 1, 2, \ldots,$ and suppose at the previous j the following is done:
   a. All survivor paths are identified
   b. The survivor paths and its metric for each state of the trellis are stored

   Then, at level (clock time) $j + 1$, compute the metric for all the paths entering each state of the trellis by adding the metric of the incoming branches to the metric of the connecting survivor path from level j. Thus, for each state, identify the path with the lowest metric as the survivor of step $j + 1$, therefore updating the computation
3. Final step: Continue the computation until the algorithm completes the forward search through the trellis and thus reaches the terminating node (i.e., all zero state), at which time it makes a decision on the maximum-likelihood path. Then, the sequence of symbols associated with that path is released to the destination as the decoded version of the received sequence

---

Example 3. Suppose that the resulting encoded sequence from the convolutional encoder in Figure 3 is as follows:

$$c = (0000000000)$$

Now suppose a noise corrupts this sequence, and the noisy received sequence is as follows:

$$c' = (0100010000)$$

Using the Viterbi algorithm, the following is the resulting step-by-step illustration to produce the survivor path which generates the correct sent message $c = (0000000000)$.

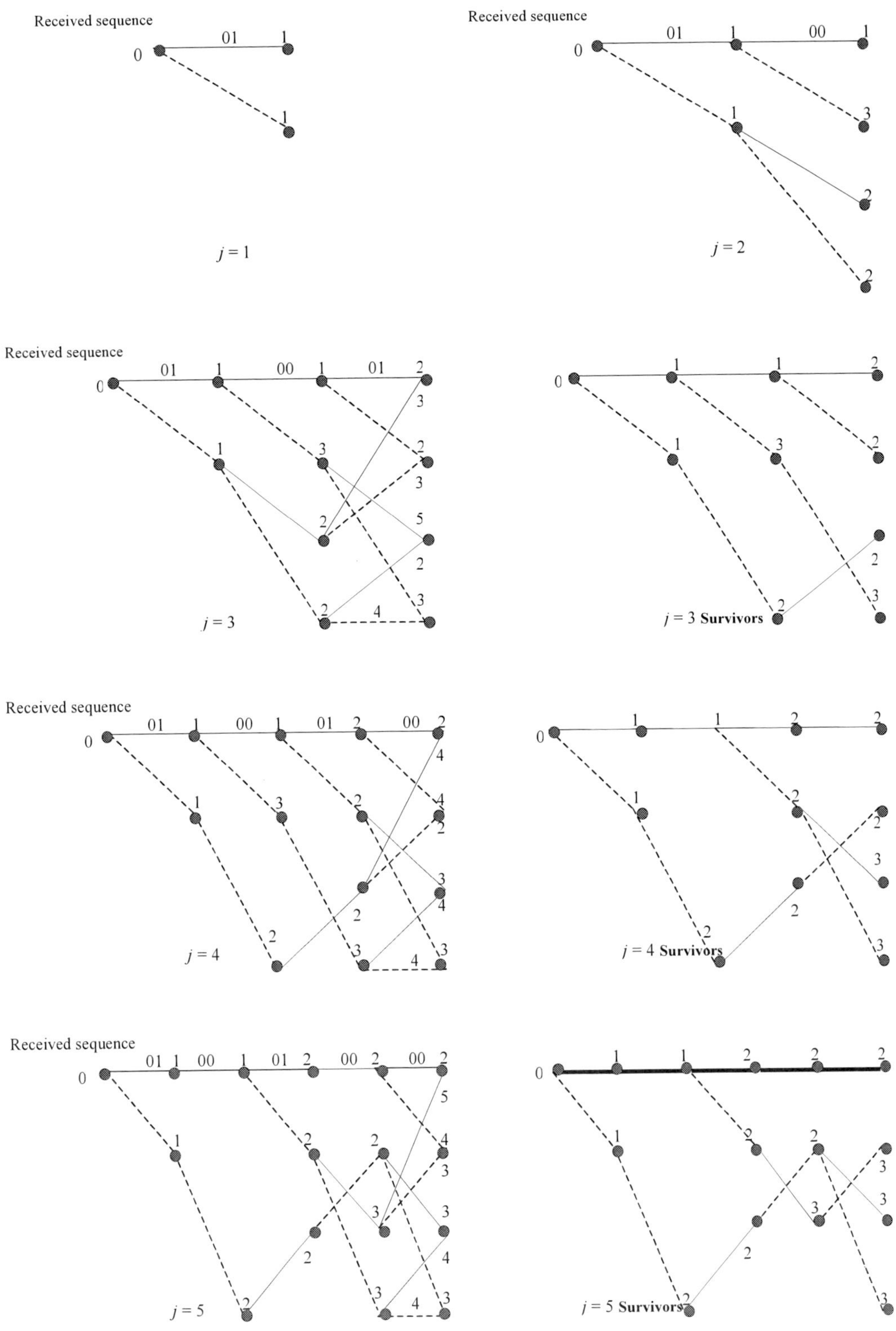

Figure 5. The illustration of the steps of the Viterbi algorithm when applied for Example 3, where the bold path is the survivor path.

Example 4. For the convolutional encoder in Figure 3, path1 impulse response is (1, 1, 1), and path2 impulse response is (1, 0, 1). Thus, the following are the corresponding generating polynomials, respectively:

$$g_1(D) = 1 \cdot D^0 + 1 \cdot D^1 + 1 \cdot D^2 = 1 + D + D^2$$

$$g_2(D) = 1 \cdot D^0 + 0 \cdot D^1 + 1 \cdot D^2 = 1 + D^2$$

For a message sequence (101), the following is the D-domain polynomial representation:

$$m(D) = 1 \cdot D^0 + 0 \cdot D^1 + 1 \cdot D^2 = 1 + D^2$$

As convolution in time domain is transformed into multiplication in the D-domain, the path1 output polynomial and path2 output polynomial are as follows, respectively, where addition is performed in modulo-2 arithmetic:

$$c_1(D) = g_1(D)m(D) = (1 + D + D^2)(1 + D^2) = 1 + D + D^3 + D^4$$

$$c_2(D) = g_2(D)m(D) = (1 + D^2)(1 + D^2) = 1 + D^4$$

Therefore, the output sequences of paths 1 and 2 are as follows, respectively:

Output sequence of path1: (11011)
Output sequence of path2: (10001)

The resulting encoded sequence from the convolutional encoder in Figure 3 is obtained by multiplexing the two output sequences of paths 1 and 2 as follows:

$$c = (11, 10, 00, 10, 11)$$

Now suppose a noise corrupts this sequence, and the noisy received sequence is as follows:

$$c' = (01, 10, 10, 10, 11)$$

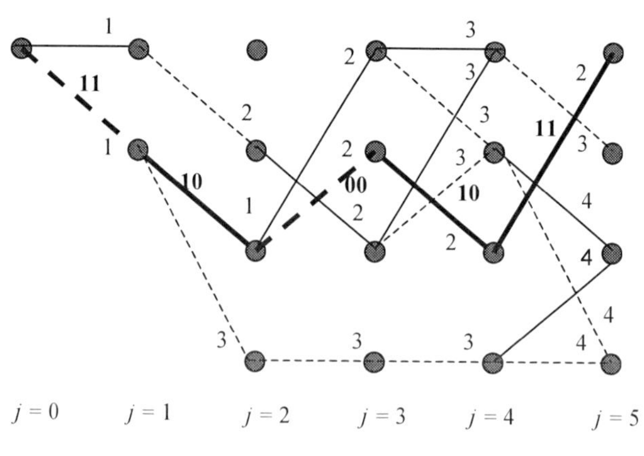

Figure 6. The resulting survivors of the Viterbi algorithm when applied for Example 4, where the bold path is the survivor path.

Using the Viterbi algorithm, the following is the resulting survivor path which generates the correct sent message $c = (11, 10, 00, 10, 11)$.

A difficulty with the application of the Viterbi algorithm occurs when the received sequence is very long. In this case the Viterbi algorithm is applied to a truncated path memory using a decoding window of length greater or equal five times the convolutional code constraint length K, in which the algorithm operates on a frame-by-frame of the received sequence each of length $l \geq 5K$. The decoding decisions made in this way are not a truly maximum likelihood, but they can be made almost as good provided that the decoding window is long enough. Another difficulty is the number of errors; for example, in case of three errors, the Viterbi algorithm when applied to a convolutional code of $r = ½$ and $K = 3$ cannot produce a correctable decoded message from the incoming erroneous message. Exceptions are triple-error patterns that spread over a time span $> K$.

2.3. Reversible Logic

An (n, k) reversible circuit is a circuit that has n number of inputs and k number of outputs and is one-to-one mapping between vectors of inputs and outputs, thus the vector of input states can be always uniquely reconstructed from the vector of output states [4,9,15,31,44,57]. Thus, a (k, k) *reversible* map is a *bijective* function which is both (1) *injective* ("one-to-one" or "(1:1)") and (2) *surjective* ("onto"). (Such bijective systems are also known as: *equipollent*, *equipotent*, and *one-to-one correspondence*.) The *auxiliary* outputs that are needed *only* for the purpose of reversibility are called "garbage" outputs. These are auxiliary outputs from which a reversible map is constructed (cf. Example 5).

Geometrically, achieving reversibility leads to value space-partitioning that leads to spatial partitions of unique values. Algebraically and in terms of systems representation, reversibility leads to multi-input multi-output (MIMO) bijective maps (i.e., bijective functions). An algorithm called reversible Boolean function (RevBF) that produces a reversible form from an irreversible Boolean function is as follows [4]. (We assume the Boolean function is specified via a table, as in Example 5 below.)

Algorithm RevBF

1. To achieve (k, k) reversibility, add sufficient number of auxiliary output variables such that the number of outputs equals the number of inputs. Allocate a new column in the mapping's table for each auxiliary variable
2. For construction of the first auxiliary output, assign a constant C_1 to half of the cells in the corresponding table column (e.g., zeros), and the second half as another constant C_2 (e.g., ones). For convenience, one may assign C_1 to the first half of the column, and C_2 to the second half of the column (cf. Table 1, column Y_1)
3. For the next auxiliary output, **If** non-reversibility still exists, **Then** assign for *identical* output tuples (irreversible map entries) values which are half zeros and half ones (cf. Table 1, last two entries of column Y_2), and then assign a constant for the remainder that are already reversible (cf. top two entries of Y_2)
4. **Do** step 3 until all map entries are reversible

Example 5. The standard two-variable Boolean OR: $Y = a + b$ is irreversible. The following table lists the mapping components:

a	b	Y
0	0	0
0	1	1
1	0	1
1	1	1

Applying the above RevBF algorithm, the following is *one* possible reversible two-variable Boolean OR:

Table 1. (2, 3) Reversible map for Boolean OR.

a	b	Y	Y_1	Y_2
0	0	0	**0**	**0**
0	1	1	**0**	**0**
1	0	1	**1**	**0**
1	1	1	**1**	**1**

Using the RevBF algorithm, the construction of the reversible map in Table 1 is obtained as follows: since Y is irreversible, assign auxiliary ("garbage") output Y_1 and assign the first half of its values the constant "0" and the second half another constant "1". Since the new map is still irreversible, assign a new garbage output Y_2 and assign the 3rd cell value to "0" and the last cell value to "1".

2.4. Quantum Computing

Quantum computing (QC) is a method of computation that uses a dynamic process governed by the Schrödinger Equation (SE) [4,44]. The one-dimensional time-dependent SE (TDSE) takes the following general form:

$$-\frac{(h/2\pi)^2}{2m}\frac{\partial^2|\psi\rangle}{\partial x^2}+V|\psi\rangle=i(h/2\pi)\frac{\partial|\psi\rangle}{\partial t} \quad (2)$$

or

$$H|\psi\rangle=i(h/2\pi)\frac{\partial|\psi\rangle}{\partial t} \quad (3)$$

where h is Planck's constant ($6.626\cdot10^{-34}$ J·s), $V(x,t)$ is the potential, m is particle's mass, i is the imaginary number, $|\psi(x,t)\rangle$ is the quantum state, H is the Hamiltonian operator ($H = -[(h/2\pi)^2/2m]\nabla^2 + V$), and ∇^2 is the Laplacian operator. While the above holds for all physical systems, in the quantum computing (QC) context, the time-independent SE (TISE) is normally used [4,44]:

$$\nabla^2 \psi = \frac{2m}{(h/2\pi)^2}(V - E)\psi \tag{4}$$

where the solution $|\psi\rangle$ is an expansion over orthogonal basis states $|\phi_i\rangle$ defined in Hilbert space **H** as follows:

$$|\psi\rangle = \sum_i c_i |\phi_i\rangle \tag{5}$$

where the coefficients c_i are called *probability amplitudes*, and $|c_i|^2$ is the probability that the quantum state $|\psi\rangle$ will collapse into the (eigen) state $|\phi_i\rangle$. The probability is equal to the inner product $|\langle\phi_i \mid \psi\rangle|^2$, with the unitary condition $\sum|c_i|^2 = 1$.

In QC, a linear and unitary operator $\Im$ is used to transform an input vector of quantum bits (qubits) into an output vector of qubits [4,44]. In two-valued QC, a qubit is a vector of bits defined as follows:

$$qubit\_0 \equiv |0\rangle = \begin{bmatrix} 1 \\ 0 \end{bmatrix}, qubit\_1 \equiv |1\rangle = \begin{bmatrix} 0 \\ 1 \end{bmatrix} \tag{6}$$

A two-valued quantum state $|\psi\rangle$ is a superposition of quantum basis states $|\phi_i\rangle$ such as those defined in Equation (6). Thus, for the orthonormal computational basis states $\{|0\rangle, |1\rangle\}$, one has the following quantum state:

$$|\psi\rangle = \alpha|0\rangle + \beta|1\rangle \tag{7}$$

where $\alpha\alpha^* = |\alpha|^2 = p_0 \equiv$ the probability of having state $|\psi\rangle$ in state $|0\rangle$, $\beta\beta^* = |\beta|^2 = p_1 \equiv$ the probability of having state $|\psi\rangle$ in state $|1\rangle$, and $|\alpha|^2 + |\beta|^2 = 1$. The calculation in QC for multiple systems (e.g., the equivalent of a register) follow the tensor product ($\otimes$) [4]. For example, given two states $|\psi_1\rangle$ and $|\psi_2\rangle$ one has the following QC:

$$\begin{aligned} |\psi_{12}\rangle &= |\psi_1\psi_2\rangle = |\psi_1\rangle \otimes |\psi_2\rangle \\ &= (\alpha_1|0\rangle + \beta_1|1\rangle) \otimes (\alpha_2|0\rangle + \beta_2|1\rangle) \\ &= \alpha_1\alpha_2|00\rangle + \alpha_1\beta_2|01\rangle + \beta_1\alpha_2|10\rangle + \beta_1\beta_2|11\rangle \end{aligned} \tag{8}$$

A physical system, describable by the following equation [4,44]:

$$|\psi\rangle = c_1 |Spinup\rangle + c_2 |Spindown\rangle \tag{9}$$

(e.g., the hydrogen atom), can be used to physically implement a two-valued QC. Another common alternative form of Equation (9) is:

$$|\psi\rangle = c_1 \left|+\frac{1}{2}\right\rangle + c_2 \left|-\frac{1}{2}\right\rangle \tag{10}$$

Many-valued QC (MVQC) can also be accomplished [4,42]. For the three-valued QC, the *qubit* becomes a 3-dimensional vector *qudit* (quantum discrete digit), and in general, for MVQC the qudit is of dimension "many". For example, one has for 3-state QC (in Hilbert space **H**) the following qudits:

$$qudit\_0 \equiv |0\rangle = \begin{bmatrix} 1 \\ 0 \\ 0 \end{bmatrix}, qudit\_1 \equiv |1\rangle = \begin{bmatrix} 0 \\ 1 \\ 0 \end{bmatrix}, qudit\_2 \equiv |2\rangle = \begin{bmatrix} 0 \\ 0 \\ 1 \end{bmatrix} \tag{11}$$

A three-valued quantum state is a superposition of three quantum orthonormal basis states (vectors). Thus, for the orthonormal computational basis states $\{|0\rangle, |1\rangle, |2\rangle\}$, one has the following quantum state:

$$|\psi\rangle = \alpha|0\rangle + \beta|1\rangle + \gamma|2\rangle \tag{12}$$

where $\alpha\alpha^* = |\alpha|^2 = p_0 \equiv$ the probability of having state $|\psi\rangle$ in state $|0\rangle$, $\beta\beta^* = |\beta|^2 = p_1 \equiv$ the probability of having state $|\psi\rangle$ in state $|1\rangle$, $\gamma\gamma^* = |\gamma|^2 = p_2 \equiv$ the probability of having state $|\psi\rangle$ in state $|2\rangle$, and

$$|\alpha|^2 + |\beta|^2 + |\gamma|^2 = 1.$$

In general, for an *n*-valued logic, a quantum state is a superposition of *n* quantum orthonormal basis states (vectors). Thus, for the orthonormal computational basis states $\{|0\rangle, |1\rangle, \ldots, |n-1\rangle\}$, one has the following quantum state:

$$|\psi\rangle = \sum_{k=0}^{n-1} c_k |q\rangle_k \tag{13}$$

where $$\sum_{k=0}^{n-1} c_k c_k^* = \sum_{k=0}^{n-1} |c_k|^2 = 1.$$

The calculation in QC for many-valued multiple systems follow the tensor product in a manner similar to the one demonstrated for two-valued QC in Equation (8).

A physical system comprising trapped ions under multiple laser excitations can be used to reliably implement MVQC [42]. A physical system in which an atom (particle) is exposed to a specific potential field (function) $V(x)$ can also be used to implement MVQC (two-valued being a special case) [4,44]. In such an implementation, the (resulting) distinct energy states are used as the orthonormal basis states. The latter is illustrated in Example 6 below which is an example of implementing MVQC by exposing a particle to a potential field V where the *distinct energy states* are used as the *orthonormal basis states*.

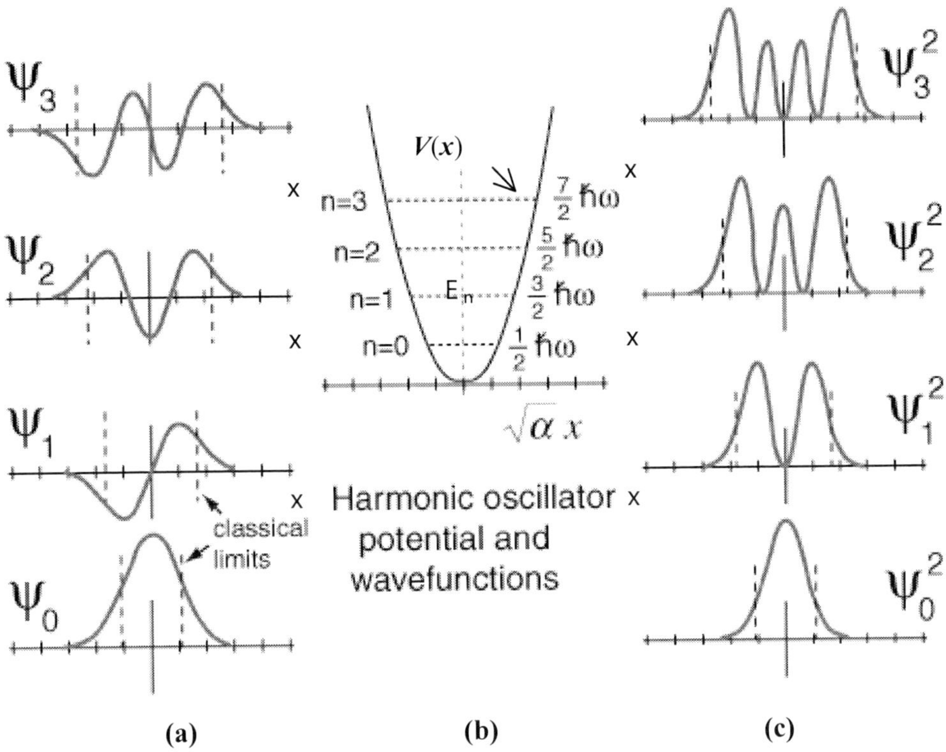

Figure 7. Harmonic oscillator (HO) potential and wavefunctions: (a) wavefunctions for various energy levels (subscripts), (b) spring potential $V(x)$ and the associated energy levels E_n, and (c) probabilities for measuring particle m in each energy state (E_n).

Example 6. We assume the following *constraints*: (1) spring potential $V(x) = (1/2)\, kx^2$, where m is a particle, $k = m\omega^2$ is spring constant, and ω is angular frequency ($= 2\pi \cdot$ frequency), and (2) boundary conditions. Also, assuming the solution of the TISE in Equation (4) for these constraints is of the following form (i.e., the Gaussian function):

$$\psi(x) = Ce^{-\alpha\frac{x^2}{2}}$$

Where $\alpha = \dfrac{m\omega}{h/2\pi}$. The general solution for the wave function $|\psi\rangle$ (for a spring potential) is:

$$C = \left[\frac{\alpha}{\pi}\right]^{1/4} \frac{1}{\sqrt{2^n n!}} H_n(\sqrt{\alpha}x)$$

where $H_n(x)$ are the Hermite polynomials. This solution leads to the sequence of evenly spaced energy levels (eigenvalues) E_n characterized by a quantum number n as follows:

$$E_n = (n + \frac{1}{2})(h/2\pi)\omega$$

The distribution of the energy states (eigenvalues) and their associated probabilities are shown in Figure 7.

3. Reversible Game Theory (RGT)

In game theory, decision path reversibility can be of fundamental importance in cases where a third neutral party needs to know the *temporal* path that leads to specific outcomes between two competitive parties. We define the following: (1) the column and row players' choices as inputs; (2) the corresponding utilities (payoffs) as the outputs; and (3) the path of player's decisions (choices) to specific payoffs (utilities) as a decision path. A game is called logically *irreversible* if an outside observer cannot reconstruct the input states from the output states using the game's extensive form, otherwise the game is called logically *reversible*.

To solve the problem of irreversibility in GT, i.e., to produce a reversible extensive form of the game from its irreversible counterpart, one can use the following new algorithm called reversible decision game theory (RDGT).

Algorithm RDGT

---

1. Encode numerically the column and row players' choices. These will be the inputs
2. Encode numerically the payoffs (utilities) in the payoff table (matrix). These will be the outputs
3. Represent the encoded inputs and outputs from steps (1) and (2) as a map (look-up-table (LUT))
4. **If** the encoded map is reversible, **Then** goto 6
5. **Else**, apply the algorithm RevBF on the payoffs
6. **End**

Example 7. The extensive form, for the row and column players in the game in Figure 8a, is shown in Figure 8b.

		B	
		C_1	C_2
A	R_1	1,3	1,3
	R_2	1,1	2,3

(a)

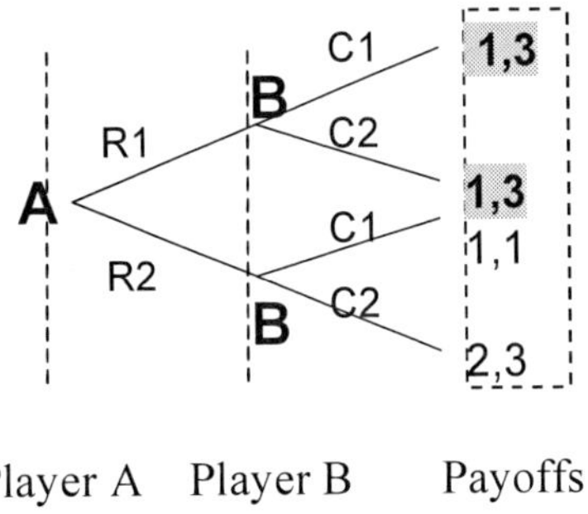

(b)

Figure 8. Irreversible game: (a) normal form and (b) extensive form.

One can note that the extensive form in Figure 8b is irreversible, i.e., an observer cannot reconstruct the input states from the output states, if the game extensive form is given, since the payoff cell (1, 3) appears twice. Following the algorithm RDGT, the following steps are performed:

(1) Encode the players' inputs (decisions; choices). This can be done as follows:

$$A: R_1 \to "0", R_2 \to "1"$$
$$B: C_1 \to "0", C_2 \to "1"$$

(2) Encode the outputs (payoffs) in the payoff matrix. This can be done as follows:

$$A: 1 \to "0", 2 \to "1"$$
$$B: 3 \to "0", 1 \to "1"$$

(3) Obtain the map for steps 1 and 2:

A	B	u_A	u_B
0	0	**0**	**0**
0	1	**0**	**0**
1	0	0	1
1	1	1	0

The map is clearly irreversible.

(4) Apply RevBF algorithm:

A	B	u_A	u_B	u_x
0	0	0	0	**0**
0	1	0	0	**1**
1	0	0	1	**0**
1	1	1	0	**0**

The decision path is now reversible. The modified payoff matrix that corresponds to the reversible map in step (4) is now as follows:

		B	
		0	**1**
A	**0**	**000**	**001**
	1	010	100

and the reversible extensive form of the above payoff table is as follows:

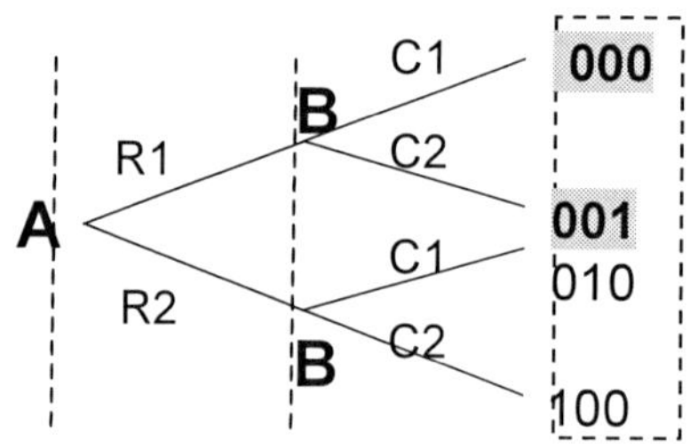

One can observe that the encodings in steps (1) and (2) are not unique, i.e., other encoding schemes can be obtained as well. *Note also that the new modified payoff matrix still reserves the dynamics of the original game, i.e., the dynamics do not change and the game's solution does not change.*

Since reversibility is a required property for quantum computing (QC) [4,44], the following section introduces the error correction in games to maintain both (1) correctness and (2) reversibility of the sent decisions, and Section 5 introduces the use of MVQC to perform GT dynamics' computations.

4. Reversible Error Correction of Games

While the previous section ensures game-theoretic reversibility explicitly over *time*, this section ensures game-theoretic reversibility over *space*. Also, while in subsection 2.2 error correction of communicated decisions was performed for the case of two-player games (i.e., single-input single-output (SISO) systems), this section introduces reversible error correction of communicated batch (parallel) of decisions in multi-player games (i.e., multiple-input multiple-output (MIMO) systems). Reversibility in parallel-based decision communication is directly observed since:

$$\vec{O}_1 = \vec{I}_2 \quad (14)$$

where $\vec{O}_1$ is the unique output (transmitted) decisions from playing team1 and $\vec{I}_2$ is the unique input (received) decisions to playing team2.

In multi-player games, the existence of noise will cause an error that may lead to irreversibility in decision communication (i.e., irreversibility in decision mapping) since $\vec{O}_1 \neq \vec{I}_2$. As will be introduced in this section, the implementation of reversible error correction can be performed (1) in software using reversible error correction algorithm and (2) in hardware using reversible error correction hardware. The following algorithm, called Reversible Viterbi (RV) Algorithm, introduces the implementation of reversible error correction in games.

Algorithm RV

---

1. Use the RevBF Algorithm to reversibly encode the communicated batch of decisions
2. Given a specific convolutional encoder circuit, determine the generator polynomials for all paths
3. **For** each communicated message within the batch, determine the encoded message sequence
4. **For** each received message, use the Viterbi Algorithm to decode the received erroneous message
5. Generate the total maximum-likelihood trellis resulting from the iterative application of
6. the Viterbi decoding algorithm
7. Generate the corrected communicated batch of decisions
8. **End**

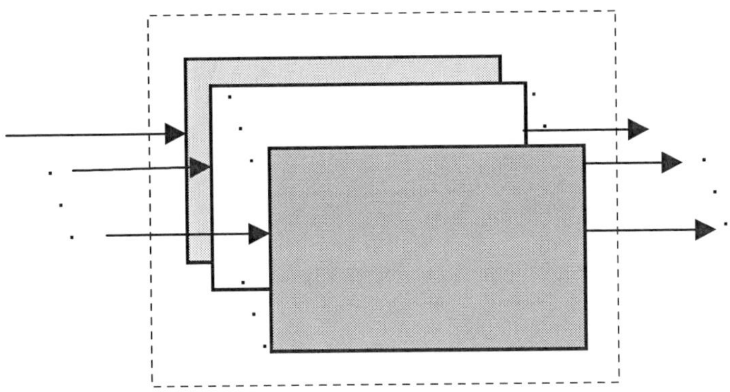

Figure 9. General encoder circuit for the parallel generation of convolutional codes where each box represents a convolutional encoder such as the one shown in Figure 3.

The convolutional encoding for the RV algorithm can be performed *serially* using a single convolutional encoder from Figure 3, or in *parallel* using the general parallel convolutional encoder circuit in Figure 9 in which several s convolutional encoders operate in parallel for encoding s number of simultaneously submitted decisions (messages) (i.e., a decision set of cardinality (size) equal to s).

Example 8. Suppose reversibility implementation (e.g., RevBF Algorithm) produces the following reversible set of message sequences:

$m_1 = (101)$
$m_2 = (001)$
$m_3 = (011)$

For the convolutional encoder in Figure 9, the following is the D-domain polynomial representations, respectively:

$m_1(D) = 1 \cdot D^0 + 0 \cdot D^1 + 1 \cdot D^2 = 1 + D^2$
$m_2(D) = 0 \cdot D^0 + 0 \cdot D^1 + 1 \cdot D^2 = D^2$
$m_3(D) = 0 \cdot D^0 + 1 \cdot D^1 + 1 \cdot D^2 = D + D^2$

The resulting encoded sequences are generated in parallel as follows, respectively:

$c_1 = (1110001011)$
$c_2 = (0000111011)$
$c_3 = (0011010111)$

Now suppose noise sources corrupt these sequences, and the noisy received sequences are as follows:

$c'_1 = (1111001001)$
$c'_2 = (0100101011)$
$c'_3 = (0010011111)$

Using the RV algorithm, Figure 10 shows the resulting survivor paths which generate the correct sent messages: $c_1 = (1110001011)$, $c_2 = (0000111011)$, $c_3 = (0011010111)$.

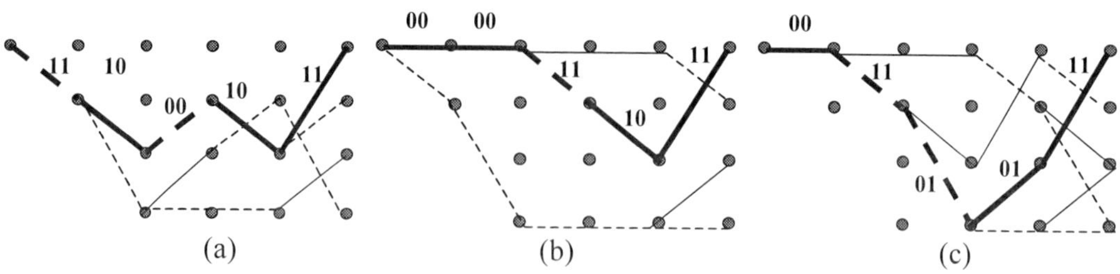

Figure 10. Continued on next page.

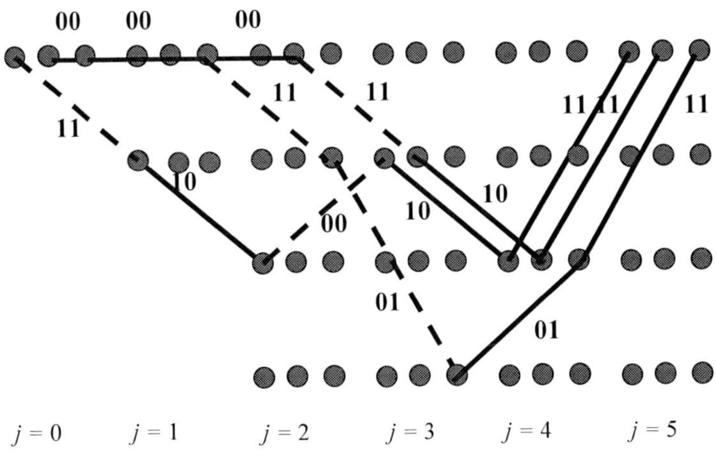

(d)

Figure 10. The resulting survivor paths of the RV algorithm when applied to Example 8.

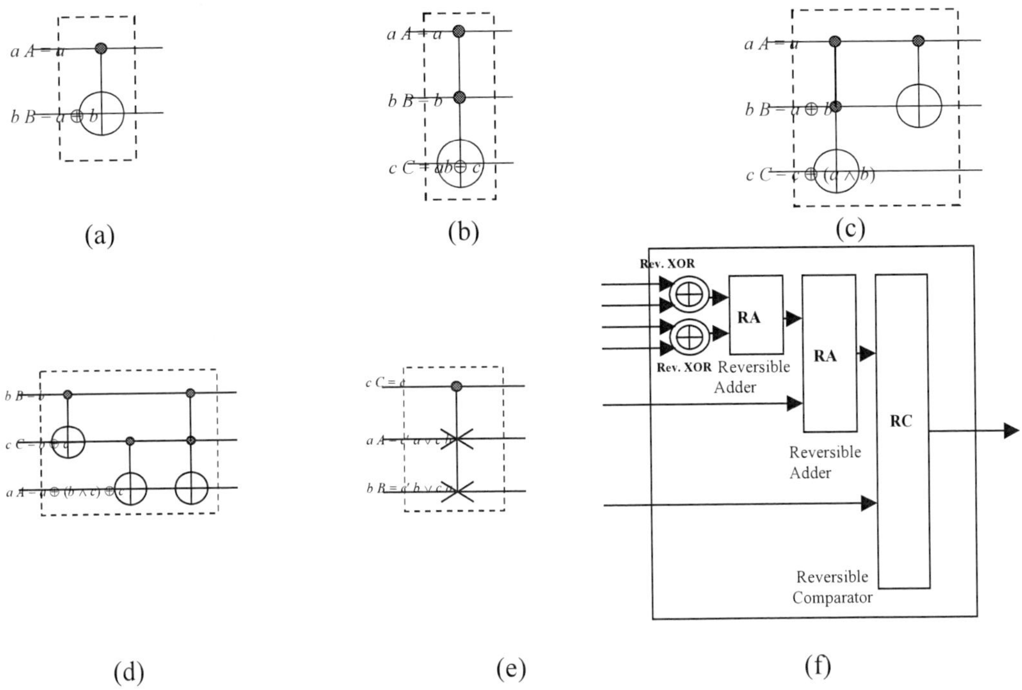

Figure 11. Reversible circuits: (a) reversible XOR gate (Feynman gate), (b) reversible Toffoli gate, (c) reversible half-adder, (d) reversible subtractor, (e) reversible multiplexer (Fredkin gate), and (f) basic reversible Viterbi cell. The symbol ⊕ is logic XOR (exclusive OR or omodulo-2 addition), ∧ is logic AND, ∨ is logic OR, and ′ is logic NOT.

The reversible hardware implementation for each convolutional encoder in the RV algorithm demands the following reversible components: reversible modulo-2 adder, reversible adder, reversible subtractor and reversible selector (multiplexer) to be used in the corresponding reversible comparator. Table 2 shows the truth tables of an irreversible half-adder (HA) and subtractor, and Figure 11 illustrates the needed reversible hardware components.

Table 2. Truth tables of an irreversible half-adder (HA) and subtractor.

Inputs		Half-Adder Outputs		Subtractor Outputs	
a	*b*	*a* + *b*	Carry	*a* - *b*	Borrow
0	0	0	0	0	0
0	1	1	0	1	1
1	0	1	0	1	0
1	1	0	1	0	0

5. Multiple-valued (*m*-valued) Quantum Computing of Game Theory

In addition to Prisoner's Dilemma (PD) classical solutions using: (1) iterated games, (2) non-myopic rationality , and (3) meta-games, recent development showed that the PD game can also be solved in the quantum domain by means of using a quantum computing (QC) method [8,16,28,29,39,46,49].

In this Section, the modeling of games' dynamics using MVQC methods is implemented. In doing so, the following convention is used: (1) game dynamics is represented as linear unitary transformations; (2) players' decisions are indices; (3) current payoffs (utilities) are inputs represented as input qudits; and (4) next payoffs (utilities) are outputs represented as output qudits. Figure 12 illustrates this notion.

		Player2	
		Decision1′: Index1′	**Decision2′: Index2′**
Player1	**Decision1: Index1**	Utilities1: qudits1	Utilities2: qudits2
	Decision2: Index2	Utilities3: qudits3	Utilities4: qudits4

Figure 12. 2-player quantum game: associating qudits to utilities.

In general, the games' maximin dynamics can be modeled (tracked) using either: (1) the cells' indices (players' decisions) or (2) the cells' values (utilities) as in Figure 12. The first is performed in the decision space **D** and the second is performed in the utility space **U**.

To perform MVQC in the utility space **U**, we describe the notion of associating a quantum state to a point in the utility (payoff) space of a game using a 2-player *m*-utility game (i.e., *two* types of utilities each can potentially take up to *m* distinct values) as an example (extension to an *N*-player games is straightforward).

- Assume each player's utility can take a finite (discrete) set of values.
- Form a 2-dimensional grid of all possible combinations of utility value pairs (2-tuples).

- Assign each of the grid points (i.e., each 2-tuple) to be a quantum basis state (in the quantum state space **Q**).
- If each of the two utilities (u_1 and u_2) can take m values, then there will be m^2 quantum basis states, each with dimension m^2 (*to yield an orthonormal basis set*).
- Let:

$$\vec{u}_{ij} = \begin{bmatrix} u_{1i} \\ u_{2j} \end{bmatrix}, \qquad i, j = 0,1,2,\ldots,m-1 \tag{15}$$

represent the m^2 points in the 2-dimensional utility space, where the (i, j) are position indices for the vector u_{ij} and the components of u_{ij} are utility values at the corresponding positions (i, j).

- Then define:

$$\vec{u}^{\,2,m} = \begin{bmatrix} \vec{u}_{00}^{\,T} \\ \vec{u}_{01}^{\,T} \\ \ldots \\ \vec{u}_{0,m-1}^{\,T} \\ \vec{u}_{10}^{\,T} \\ \ldots \\ \vec{u}_{m-1,m-1}^{\,T} \end{bmatrix} \tag{16}$$

where the superscript refers to 2 dimensions (2 players; 2 utility types), with m (discrete) values in each dimension. To reference a subset of all these possibilities, an appropriate subscript may be provided.

For example, by letting each utility take three values from the set $\{a, b, c\}$ where $\{a, b, c\}$ are any discrete real values (i.e., $m = 3$) then one would have nine grid points {00, 01, 02, 10, 11, 12, 20, 21, 22}, and Equation (16) becomes:

$$\vec{u}^{\,2,3} = \begin{bmatrix} u_{00}^{\ T} \\ u_{01}^{\ T} \\ u_{02}^{\ T} \\ u_{10}^{\ T} \\ u_{11}^{\ T} \\ u_{12}^{\ T} \\ u_{20}^{\ T} \\ u_{21}^{\ T} \\ u_{22}^{\ T} \end{bmatrix} = \begin{bmatrix} aa \\ ab \\ ac \\ ba \\ bb \\ bc \\ ca \\ cb \\ cc \end{bmatrix}$$

Based on Equation (11), the following nine ternary orthonormal computational basis states for the multiple-valued (MV) quantum space are obtained:

$$|00\rangle = |0\rangle \otimes |0\rangle = \begin{bmatrix}1\\0\\0\end{bmatrix} \otimes \begin{bmatrix}1\\0\\0\end{bmatrix} = [1\ 0\ 0\ 0\ 0\ 0\ 0\ 0\ 0]^T, |01\rangle = |0\rangle \otimes |1\rangle = \begin{bmatrix}1\\0\\0\end{bmatrix} \otimes \begin{bmatrix}0\\1\\0\end{bmatrix} = [0\ 1\ 0\ 0\ 0\ 0\ 0\ 0\ 0]^T,$$

$$|02\rangle = |0\rangle \otimes |2\rangle = \begin{bmatrix}1\\0\\0\end{bmatrix} \otimes \begin{bmatrix}0\\0\\1\end{bmatrix} = [0\ 0\ 1\ 0\ 0\ 0\ 0\ 0\ 0]^T, |10\rangle = |1\rangle \otimes |0\rangle = \begin{bmatrix}0\\1\\0\end{bmatrix} \otimes \begin{bmatrix}1\\0\\0\end{bmatrix} = [0\ 0\ 0\ 1\ 0\ 0\ 0\ 0\ 0]^T,$$

$$|11\rangle = |1\rangle \otimes |1\rangle = \begin{bmatrix}0\\1\\0\end{bmatrix} \otimes \begin{bmatrix}0\\1\\0\end{bmatrix} = [0\ 0\ 0\ 0\ 1\ 0\ 0\ 0\ 0]^T, |12\rangle = |1\rangle \otimes |2\rangle = \begin{bmatrix}0\\1\\0\end{bmatrix} \otimes \begin{bmatrix}0\\0\\1\end{bmatrix} = [0\ 0\ 0\ 0\ 0\ 1\ 0\ 0\ 0]^T,$$

$$|20\rangle = |2\rangle \otimes |0\rangle = \begin{bmatrix}0\\0\\1\end{bmatrix} \otimes \begin{bmatrix}1\\0\\0\end{bmatrix} = [0\ 0\ 0\ 0\ 0\ 0\ 1\ 0\ 0]^T, |21\rangle = |2\rangle \otimes |1\rangle = \begin{bmatrix}0\\0\\1\end{bmatrix} \otimes \begin{bmatrix}0\\1\\0\end{bmatrix} = [0\ 0\ 0\ 0\ 0\ 0\ 0\ 1\ 0]^T,$$

$$|22\rangle = |2\rangle \otimes |2\rangle = \begin{bmatrix}0\\0\\1\end{bmatrix} \otimes \begin{bmatrix}0\\0\\1\end{bmatrix} = [0\ 0\ 0\ 0\ 0\ 0\ 0\ 0\ 1]^T.$$

Note that this has resulted in an orthonormal basis set. One can perform MVQC by making the following assignments between the utility space **U** and the quantum space **Q**:

MV Utility Space U **MV Quantum Space Q**

$$\vec{u}^{2,3} = \begin{bmatrix} aa \\ ab \\ ac \\ ba \\ bb \\ bc \\ ca \\ cb \\ cc \end{bmatrix} \quad \rightleftarrows \quad \left|\vec{u}^{2,3}\right\rangle = \begin{bmatrix} |00\rangle \\ |01\rangle \\ |02\rangle \\ |10\rangle \\ |11\rangle \\ |12\rangle \\ |20\rangle \\ |21\rangle \\ |22\rangle \end{bmatrix} \tag{17}$$

Using the notation of Equation (8), the above may be written as follows:

$$\begin{aligned} |\psi_1\psi_2\rangle &= |\psi_1\rangle \otimes |\psi_2\rangle \\ &= (\alpha_1|0\rangle + \beta_1|1\rangle + \gamma_1|2\rangle) \otimes (\alpha_2|0\rangle + \beta_2|1\rangle + \gamma_2|2\rangle) \\ &= \alpha_1\alpha_2|00\rangle + \alpha_1\beta_2|01\rangle + \alpha_1\gamma_2|02\rangle + \beta_1\alpha_2|10\rangle + \\ &\quad \beta_1\beta_2|11\rangle + \beta_1\gamma_2|12\rangle + \gamma_1\alpha_2|20\rangle + \gamma_1\beta_2|21\rangle + \gamma_1\gamma_2|22\rangle \end{aligned} \tag{18}$$

Note that each component of the tensor product is associated with a product of two probabilities' amplitudes.

To perform MVQC to model games' maximin dynamics in the decision space **D**, one uses the same previous procedure (used for the utility space), with the only difference of using qudits to implement decisions (rather than using qudits to implement utilities as in the utility space method). This is achieved by modeling a 2-player *M*-decision game in a 2-D quantum space **Q** as $\vec{d}^{2,M}$ and, in general, by modeling an *N*-player *M*-decision game in an *N*-D quantum space **Q** as $\vec{d}^{N,M}$.

The coefficients (probabilities) of the quantum basis functions (cf. Equation (5)) are the system parameters, obtained by solving the wave equation with the specified potential function V applied. We note that different V's will (normally) result in different solutions (i.e., different probabilities) for each of the quantum basis states. Upon measurement of an observable variable in a physical quantum implementation, by definition, the highest probability state is the most likely one to occur.

A quantum operator ξ is a linear transformation, where the matrix of such transformation must be unitary since *matrix unitarity* (*orthogonality for matrices with real elements*) *leads to computational reversibility* [4,44]. Each transformation in the quantum domain corresponds to specific type of logic gate (primitive). Figure 13 shows examples of important quantum gates and their quantum representations using unitary transformations (i.e., unitary matrices; unitary operators) [4,44].

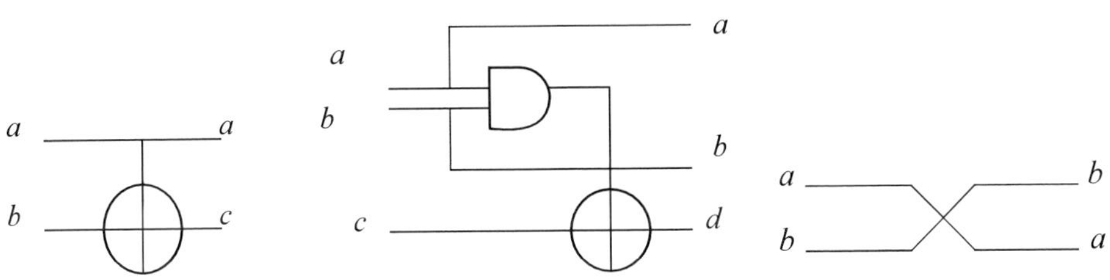

$$\begin{bmatrix} 1 & 0 & 0 & 0 \\ 0 & 1 & 0 & 0 \\ 0 & 0 & 0 & 1 \\ 0 & 0 & 1 & 0 \end{bmatrix} \quad \begin{bmatrix} 1 & 0 & 0 & 0 & 0 & 0 & 0 & 0 \\ 0 & 1 & 0 & 0 & 0 & 0 & 0 & 0 \\ 0 & 0 & 1 & 0 & 0 & 0 & 0 & 0 \\ 0 & 0 & 0 & 1 & 0 & 0 & 0 & 0 \\ 0 & 0 & 0 & 0 & 1 & 0 & 0 & 0 \\ 0 & 0 & 0 & 0 & 0 & 1 & 0 & 0 \\ 0 & 0 & 0 & 0 & 0 & 0 & 0 & 1 \\ 0 & 0 & 0 & 0 & 0 & 0 & 1 & 0 \end{bmatrix} \quad \begin{bmatrix} 1 & 0 & 0 & 0 \\ 0 & 0 & 1 & 0 \\ 0 & 1 & 0 & 0 \\ 0 & 0 & 0 & 1 \end{bmatrix}$$

(a) (b) (c)

Figure 13. Binary quantum gates: (a) (2, 2) Feynman gate which uses XOR, (b) (3, 3) Toffoli gate which uses AND and XOR, and (c) (2, 2) Swap gate which is two permuted wires.

One can use the quantum operators for modeling the maximin dynamics of a game. The following example shows how to model the dynamics of a 2-player game using MVQC.

Example 9. For the chicken game in Figure 1a, using the following encoding {W = "0", T = "1", S = "2", B = "3"}, the following is the 2-player interval-utility payoff matrix.

		B		
		S	**~S**	
A	**S**	po**2,2**	po**1,3**	**1**
	~S	po**3,1**	npo**0,0**	**0**
		1	**0**	

The payoffs in the above table are four valued, i.e., take values from the set {0, 1, 2, 3}. By using, in the 4-valued quantum logic, the following qudits:

$$|0\rangle = \begin{bmatrix}1\\0\\0\\0\end{bmatrix}, |1\rangle = \begin{bmatrix}0\\1\\0\\0\end{bmatrix}, |2\rangle = \begin{bmatrix}0\\0\\1\\0\end{bmatrix}, |3\rangle = \begin{bmatrix}0\\0\\0\\1\end{bmatrix}$$

then the combined utilities (in each cell of the payoff matrix) in the 4-valued quantum space take the following form:

$$|22\rangle = \big||2\rangle \otimes |2\rangle\big\rangle, |13\rangle = \big||1\rangle \otimes |3\rangle\big\rangle, |31\rangle = \big||3\rangle \otimes |1\rangle\big\rangle, |00\rangle = \big||0\rangle \otimes |0\rangle\big\rangle$$

then by using the following quantum linear unitary permutation operators ξ_1, ξ_2, ξ_3, and ξ_4, one models the dynamics in the 2-player interval-utility chicken game using 4-valued QC as follows:

$$\begin{aligned}
&[\xi_1]_{16x16}|00\rangle = |13\rangle, [\xi_2]_{16x16}|13\rangle = |22\rangle \\
&\rightarrow [[\xi_2][\xi_1]]_{16x16}|00\rangle = |22\rangle \\
&[\xi_3]_{16x16}|00\rangle = |31\rangle, [\xi_4]_{16x16}|31\rangle = |22\rangle \\
&\rightarrow [[\xi_4][\xi_3]]_{16x16}|00\rangle = |22\rangle \\
&\Rightarrow [[\xi_2][\xi_1]]_{16x16} = [[\xi_4][\xi_3]]_{16x16}
\end{aligned}$$

and by solving for the upper equations, one obtains the following set of *unitary* operators:

$$[\xi_1] = \begin{bmatrix} 0 & 1 & 0 & 0 & 0 & 0 & 0 & 0 & 0 & 0 & 0 & 0 & 0 & 0 & 0 & 0 \\ 0 & 0 & 1 & 0 & 0 & 0 & 0 & 0 & 0 & 0 & 0 & 0 & 0 & 0 & 0 & 0 \\ 0 & 0 & 0 & 1 & 0 & 0 & 0 & 0 & 0 & 0 & 0 & 0 & 0 & 0 & 0 & 0 \\ 0 & 0 & 0 & 0 & 1 & 0 & 0 & 0 & 0 & 0 & 0 & 0 & 0 & 0 & 0 & 0 \\ 0 & 0 & 0 & 0 & 0 & 1 & 0 & 0 & 0 & 0 & 0 & 0 & 0 & 0 & 0 & 0 \\ 0 & 0 & 0 & 0 & 0 & 0 & 1 & 0 & 0 & 0 & 0 & 0 & 0 & 0 & 0 & 0 \\ 0 & 0 & 0 & 0 & 0 & 0 & 0 & 1 & 0 & 0 & 0 & 0 & 0 & 0 & 0 & 0 \\ 1 & 0 & 0 & 0 & 0 & 0 & 0 & 0 & 0 & 0 & 0 & 0 & 0 & 0 & 0 & 0 \\ 0 & 0 & 0 & 0 & 0 & 0 & 0 & 0 & 1 & 0 & 0 & 0 & 0 & 0 & 0 & 0 \\ 0 & 0 & 0 & 0 & 0 & 0 & 0 & 0 & 0 & 1 & 0 & 0 & 0 & 0 & 0 & 0 \\ 0 & 0 & 0 & 0 & 0 & 0 & 0 & 0 & 0 & 0 & 1 & 0 & 0 & 0 & 0 & 0 \\ 0 & 0 & 0 & 0 & 0 & 0 & 0 & 0 & 0 & 0 & 0 & 1 & 0 & 0 & 0 & 0 \\ 0 & 0 & 0 & 0 & 0 & 0 & 0 & 0 & 0 & 0 & 0 & 0 & 1 & 0 & 0 & 0 \\ 0 & 0 & 0 & 0 & 0 & 0 & 0 & 0 & 0 & 0 & 0 & 0 & 0 & 1 & 0 & 0 \\ 0 & 0 & 0 & 0 & 0 & 0 & 0 & 0 & 0 & 0 & 0 & 0 & 0 & 0 & 1 & 0 \\ 0 & 0 & 0 & 0 & 0 & 0 & 0 & 0 & 0 & 0 & 0 & 0 & 0 & 0 & 0 & 1 \end{bmatrix}$$

$$[\xi_2] = \begin{bmatrix} 1 & 0 & 0 & 0 & 0 & 0 & 0 & 0 & 0 & 0 & 0 & 0 & 0 & 0 & 0 & 0 \\ 0 & 1 & 0 & 0 & 0 & 0 & 0 & 0 & 0 & 0 & 0 & 0 & 0 & 0 & 0 & 0 \\ 0 & 0 & 1 & 0 & 0 & 0 & 0 & 0 & 0 & 0 & 0 & 0 & 0 & 0 & 0 & 0 \\ 0 & 0 & 0 & 1 & 0 & 0 & 0 & 0 & 0 & 0 & 0 & 0 & 0 & 0 & 0 & 0 \\ 0 & 0 & 0 & 0 & 1 & 0 & 0 & 0 & 0 & 0 & 0 & 0 & 0 & 0 & 0 & 0 \\ 0 & 0 & 0 & 0 & 0 & 1 & 0 & 0 & 0 & 0 & 0 & 0 & 0 & 0 & 0 & 0 \\ 0 & 0 & 0 & 0 & 0 & 0 & 1 & 0 & 0 & 0 & 0 & 0 & 0 & 0 & 0 & 0 \\ 0 & 0 & 0 & 0 & 0 & 0 & 0 & 0 & 1 & 0 & 0 & 0 & 0 & 0 & 0 & 0 \\ 0 & 0 & 0 & 0 & 0 & 0 & 0 & 0 & 0 & 1 & 0 & 0 & 0 & 0 & 0 & 0 \\ 0 & 0 & 0 & 0 & 0 & 0 & 0 & 0 & 0 & 0 & 1 & 0 & 0 & 0 & 0 & 0 \\ 0 & 0 & 0 & 0 & 0 & 0 & 0 & 1 & 0 & 0 & 0 & 0 & 0 & 0 & 0 & 0 \\ 0 & 0 & 0 & 0 & 0 & 0 & 0 & 0 & 0 & 0 & 0 & 1 & 0 & 0 & 0 & 0 \\ 0 & 0 & 0 & 0 & 0 & 0 & 0 & 0 & 0 & 0 & 0 & 0 & 1 & 0 & 0 & 0 \\ 0 & 0 & 0 & 0 & 0 & 0 & 0 & 0 & 0 & 0 & 0 & 0 & 0 & 1 & 0 & 0 \\ 0 & 0 & 0 & 0 & 0 & 0 & 0 & 0 & 0 & 0 & 0 & 0 & 0 & 0 & 1 & 0 \\ 0 & 0 & 0 & 0 & 0 & 0 & 0 & 0 & 0 & 0 & 0 & 0 & 0 & 0 & 0 & 1 \end{bmatrix}$$

$$[\xi_3] = \begin{bmatrix}
0 & 1 & 0 & 0 & 0 & 0 & 0 & 0 & 0 & 0 & 0 & 0 & 0 & 0 & 0 & 0 \\
0 & 0 & 1 & 0 & 0 & 0 & 0 & 0 & 0 & 0 & 0 & 0 & 0 & 0 & 0 & 0 \\
0 & 0 & 0 & 1 & 0 & 0 & 0 & 0 & 0 & 0 & 0 & 0 & 0 & 0 & 0 & 0 \\
0 & 0 & 0 & 0 & 1 & 0 & 0 & 0 & 0 & 0 & 0 & 0 & 0 & 0 & 0 & 0 \\
0 & 0 & 0 & 0 & 0 & 1 & 0 & 0 & 0 & 0 & 0 & 0 & 0 & 0 & 0 & 0 \\
0 & 0 & 0 & 0 & 0 & 0 & 1 & 0 & 0 & 0 & 0 & 0 & 0 & 0 & 0 & 0 \\
0 & 0 & 0 & 0 & 0 & 0 & 0 & 1 & 0 & 0 & 0 & 0 & 0 & 0 & 0 & 0 \\
0 & 0 & 0 & 0 & 0 & 0 & 0 & 0 & 1 & 0 & 0 & 0 & 0 & 0 & 0 & 0 \\
0 & 0 & 0 & 0 & 0 & 0 & 0 & 0 & 0 & 1 & 0 & 0 & 0 & 0 & 0 & 0 \\
0 & 0 & 0 & 0 & 0 & 0 & 0 & 0 & 0 & 0 & 1 & 0 & 0 & 0 & 0 & 0 \\
0 & 0 & 0 & 0 & 0 & 0 & 0 & 0 & 0 & 0 & 0 & 1 & 0 & 0 & 0 & 0 \\
0 & 0 & 0 & 0 & 0 & 0 & 0 & 0 & 0 & 0 & 0 & 0 & 1 & 0 & 0 & 0 \\
0 & 0 & 0 & 0 & 0 & 0 & 0 & 0 & 0 & 0 & 0 & 0 & 0 & 1 & 0 & 0 \\
1 & 0 & 0 & 0 & 0 & 0 & 0 & 0 & 0 & 0 & 0 & 0 & 0 & 0 & 0 & 0 \\
0 & 0 & 0 & 0 & 0 & 0 & 0 & 0 & 0 & 0 & 0 & 0 & 0 & 0 & 1 & 0 \\
0 & 0 & 0 & 0 & 0 & 0 & 0 & 0 & 0 & 0 & 0 & 0 & 0 & 0 & 0 & 1
\end{bmatrix}$$

$$[\xi_4] = \begin{bmatrix}
1 & 0 & 0 & 0 & 0 & 0 & 0 & 0 & 0 & 0 & 0 & 0 & 0 & 0 & 0 & 0 \\
0 & 1 & 0 & 0 & 0 & 0 & 0 & 0 & 0 & 0 & 0 & 0 & 0 & 0 & 0 & 0 \\
0 & 0 & 1 & 0 & 0 & 0 & 0 & 0 & 0 & 0 & 0 & 0 & 0 & 0 & 0 & 0 \\
0 & 0 & 0 & 1 & 0 & 0 & 0 & 0 & 0 & 0 & 0 & 0 & 0 & 0 & 0 & 0 \\
0 & 0 & 0 & 0 & 1 & 0 & 0 & 0 & 0 & 0 & 0 & 0 & 0 & 0 & 0 & 0 \\
0 & 0 & 0 & 0 & 0 & 1 & 0 & 0 & 0 & 0 & 0 & 0 & 0 & 0 & 0 & 0 \\
0 & 0 & 0 & 0 & 0 & 0 & 1 & 0 & 0 & 0 & 0 & 0 & 0 & 0 & 0 & 0 \\
0 & 0 & 0 & 0 & 0 & 0 & 0 & 1 & 0 & 0 & 0 & 0 & 0 & 0 & 0 & 0 \\
0 & 0 & 0 & 0 & 0 & 0 & 0 & 0 & 1 & 0 & 0 & 0 & 0 & 0 & 0 & 0 \\
0 & 0 & 0 & 0 & 0 & 0 & 0 & 0 & 0 & 1 & 0 & 0 & 0 & 0 & 0 & 0 \\
0 & 0 & 0 & 0 & 0 & 0 & 0 & 0 & 0 & 0 & 0 & 0 & 0 & 1 & 0 & 0 \\
0 & 0 & 0 & 0 & 0 & 0 & 0 & 0 & 0 & 0 & 0 & 1 & 0 & 0 & 0 & 0 \\
0 & 0 & 0 & 0 & 0 & 0 & 0 & 0 & 0 & 0 & 0 & 0 & 1 & 0 & 0 & 0 \\
0 & 0 & 0 & 0 & 0 & 0 & 0 & 0 & 0 & 0 & 1 & 0 & 0 & 0 & 0 & 0 \\
0 & 0 & 0 & 0 & 0 & 0 & 0 & 0 & 0 & 0 & 0 & 0 & 0 & 0 & 1 & 0 \\
0 & 0 & 0 & 0 & 0 & 0 & 0 & 0 & 0 & 0 & 0 & 0 & 0 & 0 & 0 & 1
\end{bmatrix}$$

The implementation of the above quantum model using quantum circuits [4,44] requires the serial interconnect of the quantum logic primitives ξ_1 and ξ_2, and the serial interconnect of the quantum logic primitives ξ_3 and ξ_4, as the serial interconnect in QC is implemented formally using regular matrix multiplication.

One notes that QC is used in Example 9 to model a pure strategy game. The next example shows the use of QC to model a mixed-strategy game and show its quantum representations using quantum decision trees (QDTs) [4].

Example 10. Let us consider the case of a simple 2-player mixed strategy ZSG as in Figure 14, where p is the probability of player A to choose decision A_1, $(1 - p)$ is the probability of player A to choose decision A_2, q is the probability of player B to choose decision B_1, and $(1 - q)$ is the probability of player B to choose decision B_2.

		B		
		B_1	**B_2**	
A	**A_1**	α	β	p
	A_2	γ	δ	$1 - p$
		q	$1 - q$	

Figure 14. 2-player mixed strategy game.

Using the probability definition in Equation (7), one can use the following QC assignments:

$p = |\alpha_1|^2$, $1 - p = |\beta_1|^2$, $p + (1 - p) = |\alpha_1|^2 + |\beta_1|^2 = 1$,
$q = |\alpha_2|^2$, $1 - q = |\beta_2|^2$, and $q + (1 - q) = |\alpha_2|^2 + |\beta_2|^2 = 1$.

By encoding A_1 = "0", A_2 = "1", B_1 = "0" and B_2 = "1", then the modeling of each player's decision in the game can be performed using qubit representation as follows:

$$|\psi_A\rangle = \alpha_1|0\rangle + \beta_1|1\rangle \tag{19}$$

$$|\psi_B\rangle = \alpha_2|0\rangle + \beta_2|1\rangle \tag{20}$$

In terms of QC representation, solving the above game leads to the following equations:

$$\alpha|\alpha_1|^2 + \gamma|\beta_1|^2 = \beta|\alpha_1|^2 + \delta|\beta_1|^2 \tag{21}$$

$$-\alpha|\alpha_2|^2 + -\beta|\beta_2|^2 = -\gamma|\alpha_2|^2 + -\delta|\beta_2|^2 \tag{22}$$

Then the probability amplitudes $|\alpha_1|$, $|\alpha_2|$, $|\beta_1|$, and $|\beta_2|$ of having decisions A_1, A_2, B_1, and B_2 are, respectively:

$$|\alpha_1|^2 = \frac{\delta-\gamma}{\alpha-\beta-\gamma+\delta} \Rightarrow |\alpha_1| = \sqrt{\frac{\delta-\gamma}{\alpha-\beta-\gamma+\delta}} \tag{23}$$

$$|\alpha_2|^2 = \frac{\beta-\delta}{\beta+\gamma-\alpha-\delta} \Rightarrow |\alpha_2| = \sqrt{\frac{\beta-\delta}{\beta+\gamma-\alpha-\delta}} \tag{24}$$

$$|\beta_1|^2 = \frac{\alpha-\beta}{\alpha-\beta-\gamma+\delta} \Rightarrow |\beta_1| = \sqrt{\frac{\alpha-\beta}{\alpha-\beta-\gamma+\delta}} \tag{25}$$

$$|\beta_2|^2 = \frac{\gamma-\alpha}{\beta+\gamma-\alpha-\delta} \Rightarrow |\beta_2| = \sqrt{\frac{\gamma-\alpha}{\beta+\gamma-\alpha-\delta}} \tag{26}$$

and the value of the game is:

$$u^* = \frac{\alpha\delta-\beta\gamma}{\alpha-\beta-\gamma+\delta} \tag{27}$$

As noted earlier, the quantum analog for the game dynamics are the dynamics described by the SE. Each cell of a game is represented as a single point in the utility quantum grid space (cf. Figure 15b). An approach to implement a quantum game suggested here is as follows (cf. Figure 15): (1) specify maximin dynamics of game$_i$; (2) construct a separate wavefunction ψ_i in the MV quantum space for each cell such that its *highest probability* is at the *joint maximin* utility vector, and relatively low at all other utility vectors (cf. Figure 15b); and (3) substitute this ψ_i into the TISE and solve for V_i. Solutions of potential functions V_k for a collection of games G_k can be tabulated as a look-up-table (LUT) where for each game (input) the output is the corresponding V_k to solve for the corresponding joint maximin dynamics (cf. Figure 15c).

We call the quantum space that includes the wave function as an extra dimension (as in Figure 15b) by the augmented quantum space. The dimension of the augmented quantum space equals to (# players + 1) since the extra dimension accounts for the wavefunction ψ_i.

For example, for a 2-player game one needs a 2-D quantum space **Q** and 3-D augmented quantum space $\mathbf{Q}^+$ (cf. Figure 15b), for 3-player games one needs 3-D quantum space **Q** and 4-D augmented quantum space $\mathbf{Q}^+$, and in general for *N*-player games one needs an *N*-D quantum space **Q** and $(N+1)$-D augmented quantum space $\mathbf{Q}^+$.

Several physical models can be utilized to perform quantum games. This includes: (1) rigid box problem, (2) non-rigid box problem, (3) Hydrogen atom problem, and (4) simple harmonic oscillator (SHO) problem. The quantum formulation in Figure 15 can be extended to the more general case as follows: in a higher dimensional quantum space **Q**, one would like to design for a potential vector $\vec{V}$ that solves for maximin dynamics of *several* games, i.e., potential with several maxima with each maximum corresponds to a specific joint maximin game dynamics, rather than several independent potential fields as in Figure 15c. For example, one can use the following N-dimensional solution of the N-D rigid box problem in order to perform several maximin games' dynamics using QC as shown in Figure 16.

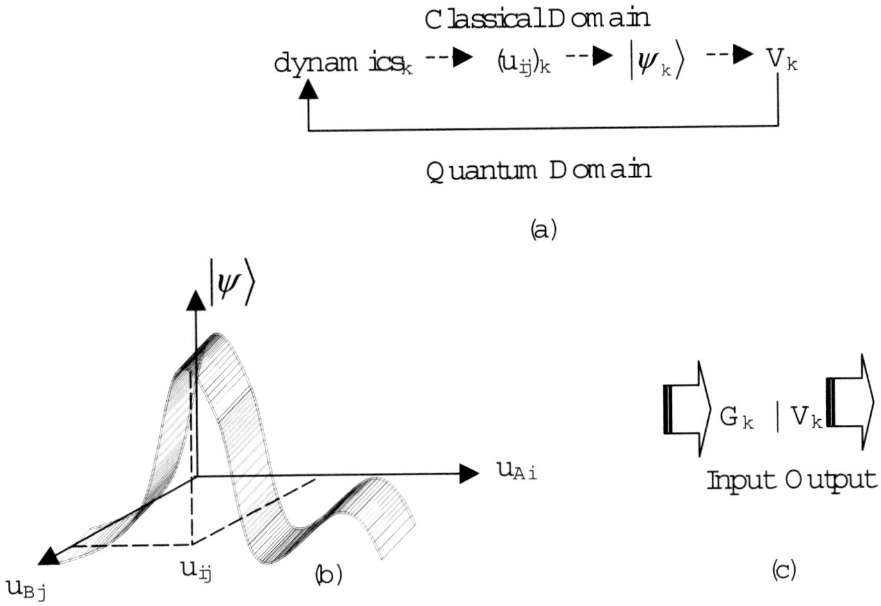

Figure 15. (a) Possible MVQC strategy to implement a game dynamics, (b) MV quantum-utility space to obtain $|\psi\rangle$, and (c) Look-Up-Table to implement joint maximin dynamics for many games.

$$\psi(\omega_0, \omega_1, ..., \omega_{N-1}) = A\prod_{k=0}^{N-1} \sin\left(\frac{n_k \pi}{a}\omega_k\right) \tag{28a}$$

$$|\psi(\omega_0, \omega_1, ..., \omega_{N-1})|^2 = A^2\prod_{k=0}^{N-1} \sin^2\left(\frac{n_k \pi}{a}\omega_k\right) \tag{28b}$$

Where $n_k = 1,2,3,..., \omega_k$, a is the width of the box which equals (in general) 2π, and A is a normalization constant that can be found using the following Equation:

$$\int_{-\infty}^{\infty}\int_{-\infty}^{\infty}...\int_{-\infty}^{\infty}|\psi(\omega_0, \omega_1, ..., \omega_{N-1})|^2 d\omega_0 d\omega_1 ... d\omega_{N-1} = 1 \tag{29}$$

In general, for an N-player game, with each player has M-decisions, the quantum representation for each player will be in the form of Equation (13) for $k = 0, \ldots, M-1$.

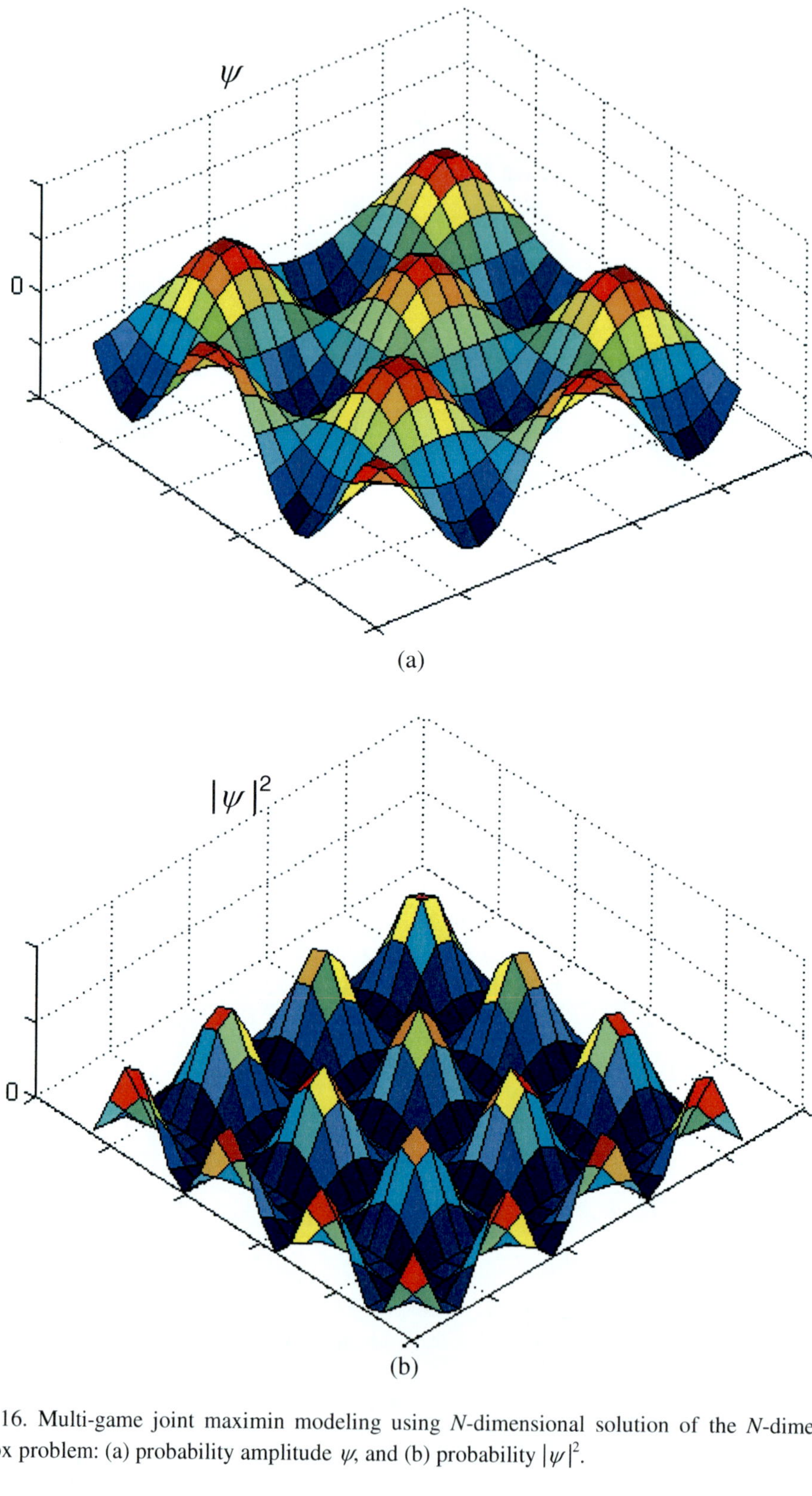

(b)

Figure 16. Multi-game joint maximin modeling using N-dimensional solution of the N-dimensional rigid box problem: (a) probability amplitude ψ, and (b) probability $|\psi|^2$.

For a game of two players, where each player has two decisions, each player decision affects the other. One can model *all* mutual (joint) decisions using QC, i.e., the *superposition* between the quantum states in Equations (19) and (20) ($|\psi_{AB}\rangle$) can be implemented using Equation (8).

For the mixed-strategy game in Example 10, one can use the quantum decision tree (QDT) [4] as a quantum data representation as follows:

$$|\psi_A\rangle = [|0\rangle \quad |1\rangle]\begin{bmatrix} 1 & 0 \\ 0 & 1 \end{bmatrix}\begin{bmatrix} \alpha_1 \\ \beta_1 \end{bmatrix} \tag{30}$$

$$|\psi_B\rangle = [|0\rangle \quad |1\rangle]\begin{bmatrix} 1 & 0 \\ 0 & 1 \end{bmatrix}\begin{bmatrix} \alpha_2 \\ \beta_2 \end{bmatrix} \tag{31}$$

Where $\begin{bmatrix} 1 & 0 \\ 0 & 1 \end{bmatrix}$ is the Buffer quantum operator [4]. Quantum decision tree representations, using the computational basis states $\{|0\rangle, |1\rangle\}$, for Equations (30), (31), and (8), are shown in Figure 17, where $|\alpha_1| = \sqrt{p}$, $|\beta_1| = \sqrt{1-p}$, $|\alpha_2| = \sqrt{q}$, and $|\beta_2| = \sqrt{1-q}$.

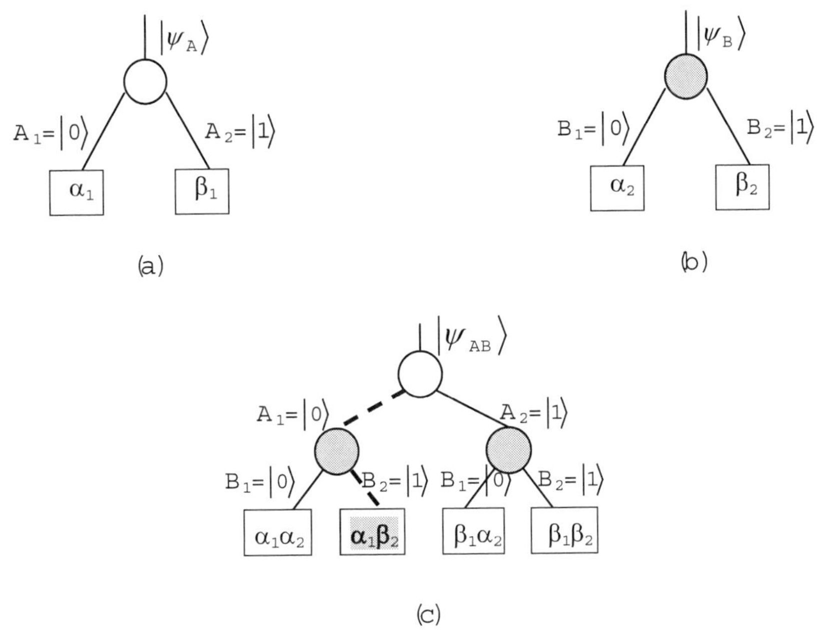

Figure 17. Orthonormal computational basis states QDT representations for: (a) Equation (30), (b) Equation (31), and (c) Equation (8).

As an example, Figure 17c shows the quantum decision path $|AB\rangle = |01\rangle$ in a dashed dark line that leads to the *highest* probability $\alpha_1\beta_2$ as a possible solution of the 2-player mixed strategy ZSG in Example 10.

The QDTs in Figure 17 use the quantum computational basis states to model a game's dynamics. Other quantum basis states [4,44] such as the 1-qubit quantum systems' orthonormal composite basis states $\left\{\frac{|0\rangle+|1\rangle}{\sqrt{2}}, \frac{|0\rangle-|1\rangle}{\sqrt{2}}\right\}$ and the 2-qubit quantum systems' Einstein-Podolsky-Rosen (EPR) basis states $\left\{\frac{|00\rangle+|11\rangle}{\sqrt{2}}, \frac{|00\rangle-|11\rangle}{\sqrt{2}}, \frac{|01\rangle+|10\rangle}{\sqrt{2}}, \frac{|01\rangle-|10\rangle}{\sqrt{2}}\right\}$ can be used [4] for the quantum representation of games, where various tree representations will lead to different computational optimizations in terms of (1) number of internal nodes used (i.e., the amount of memory used or *spatial complexity*) and (2) the speed of implementation operations using such representation (i.e., *temporal complexity*).

For instance, by using the quantum Walsh-Hadamard operator $\frac{1}{\sqrt{2}}\begin{bmatrix}1 & 1\\ 1 & -1\end{bmatrix}$ [4,44], Equations (32) and (33) are the equivalence of Equations (30) and (31) in terms of using the orthonormal *composite* basis states $\left\{\frac{|0\rangle+|1\rangle}{\sqrt{2}}, \frac{|0\rangle-|1\rangle}{\sqrt{2}}\right\}$.

$$\begin{aligned}|\Psi_A\rangle = \alpha_1|0\rangle + \beta_1|1\rangle &= \alpha_1\frac{|+\rangle+|-\rangle}{\sqrt{2}} + \beta_1\frac{|+\rangle-|-\rangle}{\sqrt{2}}\\ &= \frac{\alpha_1+\beta_1}{\sqrt{2}}|+\rangle + \frac{\alpha_1-\beta_1}{\sqrt{2}}|-\rangle\\ &= \lambda_1|+\rangle + \mu_1|-\rangle\end{aligned} \tag{32}$$

$$\begin{aligned}|\Psi_B\rangle = \alpha_2|0\rangle + \beta_2|1\rangle &= \alpha_2\frac{|+\rangle+|-\rangle}{\sqrt{2}} + \beta_2\frac{|+\rangle-|-\rangle}{\sqrt{2}}\\ &= \frac{\alpha_2+\beta_2}{\sqrt{2}}|+\rangle + \frac{\alpha_2-\beta_2}{\sqrt{2}}|-\rangle\\ &= \lambda_2|+\rangle + \mu_2|-\rangle\end{aligned} \tag{33}$$

where $\left\{|+\rangle = \frac{|0\rangle+|1\rangle}{\sqrt{2}}, |-\rangle = \frac{|0\rangle-|1\rangle}{\sqrt{2}}\right\}$ and $\left\{|0\rangle = \frac{|+\rangle+|-\rangle}{\sqrt{2}}, |1\rangle = \frac{|+\rangle-|-\rangle}{\sqrt{2}}\right\}$.

Consequently, measuring $|\Psi_A\rangle$ with respect to the new basis $\{|+\rangle, |-\rangle\}$ will result in the state (basis) $|+\rangle$ with probability $\frac{|\alpha_1+\beta_1|^2}{2}$ and the state (basis) $|-\rangle$ with probability $\frac{|\alpha_1-\beta_1|^2}{2}$. Similarly, measuring $|\Psi_B\rangle$ with respect to the new basis $\{|+\rangle, |-\rangle\}$ will result in the state (basis) $|+\rangle$ with probability $\frac{|\alpha_2+\beta_2|^2}{2}$ and the state (basis) $|-\rangle$ with probability $\frac{|\alpha_2-\beta_2|^2}{2}$. Figure 18 shows the corresponding QDTs using Equations (32) and (33) for the game in Example 10.

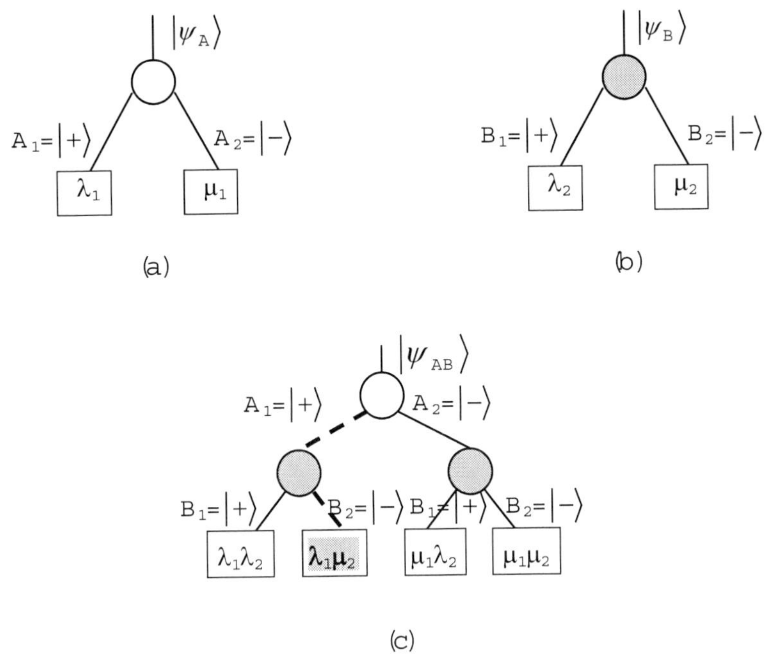

Figure 18. Orthonormal composite basis states QDT representation for: (a) Equation (32), (b) Equation (33), and (c) Equation (8).

As an example, Figure 18c shows the quantum decision path $|AB\rangle = |+-\rangle$ in a dashed dark line that leads to the *highest* probability $\lambda_1\mu_2$ as a possible solution of the 2-player mixed strategy ZSG in Example 10.

For the more general case, Figures 17 and 18 can be extended to several levels and several branches per level where: (1) the number of levels equals the number of players (i.e., an *N*-player game will have an *N*-level QDT), and (2) the number of branches per level is equal to the number of decisions per player (i.e., an *M*-decision player will have an *M*-branch QDT). Such generalized QDTs are called multiple-valued QDTs (MvQDTs) [4].

Figure 19 shows a QDT where d is the number of decisions per player, n is the number of players, and D_k is a level branch. Due to the fact that, in QC, *all* systems' states can occur at the same time (e.g., atomic spin-up and spin-down in 2-valued QC), all of the *tree paths* (from the root to the leaves) in Figure 19 (and Figures 17c and 18c as a special case) can occur *simultaneously* (*i.e., in parallel*), and *only* after measurement a *single* path will be observed (as the whole system's composite (superimposed; correlated) state will collapse into that single path (state) *after* measurement). As an example, the observed path is shown in Figure 19 as a dark line. (From computation point of view, each path in the QDT in Figure 19 is a single calculation (computation; processing), and thus a massive computational parallelism occurs with $d^{\,n}$ calculations performed simultaneously, and the QDT path superposition will collapse after measurement into a single path, where the path with the *highest* probability (leaf) α_j has the *highest* probability to be measured.)

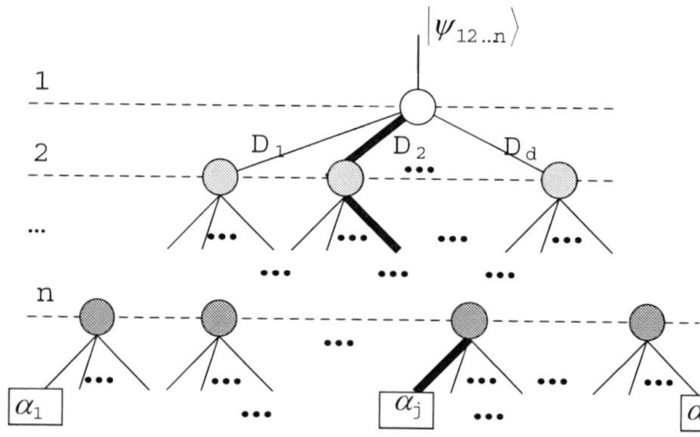

Figure 19. An n-player d-decision QDT.

In general, for d is the number of decisions per player (number of branches per tree level), n is the number of players (number of tree levels), and D_k is an edge (branch) in the quantum tree (per level) (cf. Figure 19), then the quantum state of *all paths* is expressed as:

$$|\psi_{12...n}\rangle = \sum_{i=1}^{d^n} \alpha_i |path_i\rangle \tag{34}$$

The game-theoretic quantum system represented in Equation (34) can be either: (1) *decomposable* into the tensor product of *all* levels as shown in Equation (35), or (2) *non-decomposable* (*entangled*; product of *some* levels) as shown in Equation (36).

$$|\psi_{12...n}\rangle = \prod_{p=1}^{n} \left(\sum_{k=1}^{d} \alpha_k |D_k\rangle \right)_p \tag{35}$$

$$|\psi_{12...n}\rangle \neq \prod_{p=1}^{n} \left(\sum_{k=1}^{d} \alpha_k |D_k\rangle \right)_p \tag{36}$$

If the state vectors (e.g., decisions in the GT context) of certain quantum systems (e.g., players in the GT context) were entangled with each other, then *if one changes the state vector of one system, the corresponding state vector of the other system is also changed, instantaneously and independently of the medium* (*channel*).

As a practical (realization) example, when two photons are created in the total spin zero state, and their spin conserved, as it must, one photon has a spin of +1 and a spin of -1. By measuring one of the state vectors of the photon, the state vector collapses into a knowable state. *Instantaneously* and *automatically*, the state vector of the other photon collapses into the other (opposite) knowable state, e.g., when one photon's spin is measured and found to be +1, the other photon's spin of -1 immediately becomes known too. (There are no forces (or communication) involved, and no physical explanation of the entanglement mechanism has been provided yet.)

An important quantum operator that is used for entanglement generation is the (1, 1) Walsh-Hadamard unitary operator (transform) $\frac{1}{\sqrt{2}}\begin{bmatrix} 1 & 1 \\ 1 & -1 \end{bmatrix}$ [4,44], that generates for a single qubit (with two possible states $|0\rangle$ or $|1\rangle$) the superimposed output state $\frac{1}{\sqrt{2}}|0\rangle + \frac{1}{\sqrt{2}}|1\rangle$. As an example, utilizing Equation (36), a two-decision photon-based entanglement can be expressed as follows (where each 2-qubit state occurs with 50% probability): $\left\{\frac{1}{\sqrt{2}}|01\rangle + \frac{1}{\sqrt{2}}|10\rangle\right\}$. This entanglement is shown to be produced, as a one possible output state within a set of other output states, using a serially interconnected (cascaded) 2-stage quantum circuit: the first stage is the (1, 1) Walsh-Hadamard in parallel with a (1, 1) buffer (wire), and the second stage is a (2, 2) Feynman gate [4,44].

6. Conclusion and Future Work

This research introduces the implementation of reversibility in playing games and error correction in multi-player games. Many-valued quantum computation (MVQC) of games' dynamics utilizing many-valued quantum discrete digits (qudits) is also introduced, and the associated GT quantum representations of: (1) the orthonormal computational basis states quantum decision trees (QDTs) and (2) the orthonormal composite basis states QDTs are also introduced as possible quantum representations of games' dynamics. The many-valued reversibility and QC approaches introduced in this research can be used for the modeling and processing of the (maximin) games' dynamics using low-power consuming quantum circuits (operators).

Future work will include items such as: (1) numerical evaluations of the computational efficiencies of using various QDTs as representations of various games' dynamics; (2) more extensive mathematical (formal) and algorithmic investigations of the potential use of MVQC in solving non-zero sum games; and (3) further statistical noise modeling in various different GT situations and the investigation of using advanced coding schemes to correct the corresponding corrupted multi-player communicated decisions.

References

[1] N. Abramson, *Information Theory and Coding*, McGraw-Hill, New York, 1963.
[2] J. J. Adamék, *Foundations of Coding*, Wiley, New York, 1991.
[3] Y. Akaiwa, *Introduction to Digital Mobile Communication*, Wiley, New York, 1997.
[4] A. N. Al-Rabadi, *Reversible Logic Synthesis: From Fundamentals to Quantum Computing*, Springer-Verlag, New York, 2004.
[5] J. B. Anderson and S. Mohan, *Source and Channel Coding: An Algorithmic Approach*, Kluwer Academic, Boston, Mass., 1991.
[6] R. Axelrod, *The Evolution of Cooperation*, Basic Books, New York, 1984.
[7] B. Barry and R. Hardin (Editors), *Rational Man and Irrational Society?* Sage Publications, Beverly Hills, 1982.
[8] S. C. Benjamin and P. M. Hayden, "Multiplayer Quantum Games," *Physical Review A*, **64**, 030301, 2001.
[9] C. Bennett, "Logical Reversibility of Computation," *IBM Journal of Research and Development,* **17**, pp. 525 - 532, 1973.
[10] E. R. Berlekamp, *Algebraic Coding Theory*, McGraw-Hill, New York, 1968.
[11] V. K. Bhargava, D. Haccoun, R. Matyas, and P. Nuspl, *Digital Communications by Satellite: Modulation, Multiple Access, and Coding*, Wiley, New York, 1981.
[12] G. C. Clark, Jr., and J.B. Cain, *Error-Correction Coding for Digital Communications*, Plenum Publishers, New York, 1981.
[13] T. M. Cover and J.A. Thomas, *Elements of Information Theory*, Wiley, New York, 1991.
[14] M. D. Davis, *Game Theory: A Nontechnical Introduction*, Dover Publications, Mineola, New York, 1997.
[15] A. De Vos, "Reversible Computing," *Progress in Quantum Electronics,* **23**, pp.1 - 49, 1999.
[16] J. Eisert, M. Wilkens, and M. Lewenstein, "Quantum Games and Quantum Strategies," *Phy. Rev. Letters*, 83, pp. 3077 - 3080, 1999.
[17] G. D. Forney, "The Viterbi Algorithm," *Proceedings of the IEEE*, **61**(3), pp. 268 - 278, 1973.
[18] R.G. Gallager, *Low-Density Parity-Check Codes*, MIT Press, Cambridge, Mass., 1963.
[19] R. D. Gitlin, J.F. Hayes, and S.B. Weinstein, *Data Communications Principles*, Plenum, New York, 1992.
[20] N. S. Glance and B. A. Huberman, "The Dynamics of Social Dilemmas," *Scientific American*, pp. 76 - 81, 1994.
[21] S. W. Golomb, *Shift Register Sequences*, Holden-Day, San Francisco, 1967.
[22] H. Hamburger, *Games as Models of Social Phenomena*, W. H. Freeman & Co., San Francisco, 1979.
[23] R. W. Hamming, *Coding and Information Theory*, Prentice-Hall, Englewood Cliffs, N. J., 1980.
[24] G. Hardin, "The Tragedy of the Commons," *Science*, Vol. 162, pp. 1243 - 1248, 1968.
[25] S. Haykin, *Communication Systems*, 4th Edition, John Wiley & Sons, 2001.
[26] C. Heegard and S.B. Wicker, *Turbo Coding*, Kluwer Academic Publishers, Boston, Mass., 1999.

[27] N. S. Jayant and P. Noll, *Digital Coding of Waveforms: Principles and Applications to Speech and Video*, Prentice-Hall, Englewood Cliffs, N. J., 1984.

[28] N. F. Johnson, "Playing a Quantum Game with a Corrupt Source," *Physical Review A*, **63**, 020302, 2001.

[29] R. Kay, S. C. Benjamin, and N. F. Johnson, "Evolutionary Quantum Game," *J. Phys. A*, **34**, L547, 2001.

[30] P. Lafrance, *Fundamental Concepts in Communication*, Prentice-Hall, Englewood Cliffs, N. J., 1990.

[31] R. Landauer, "Irreversibility and Heat Generation in the Computational Process," *IBM Journal of Research and Development*, **5**, pp.183 - 191, 1961.

[32] B. P. Lathi, *Modern Digital and Analog Communication Systems,* 2nd edition, Oxford University Press, 1995.

[33] E. A. Lee and D. G. Messerschmitt, *Digital Communication,* 2nd edition, Kluwer Academic, Boston, Mass., 1994.

[34] S. Lin and D.J. Costello, Jr., *Error Control Coding: Fundamentals and Applications*, Prentice-Hall, Englewood Cliffs, N. J., 1983.

[35] R. D. Luce and H. Raiffa, *Games and Decisions: Introduction and Critical Review*, Dover Publications, New York, 1989.

[36] R. W. Lucky, J. Salz, and E.J. Weldon, Jr., *Principles of Data Communication*, McGraw- Hill, New York, 1968.

[37] F. J. MacWilliams and N.J.A. Sloane, *The Theory of Error-Correcting Codes*, North-Holland, Amsterdam, 1977.

[38] R. J. McEliece, *The Theory of Information and Coding: A Mathematical Framework for Communication*, Addison-Wesley, Mass., 1977.

[39] D. Meyer, "Quantum Strategies," *Phy. Rev. Lett.*, **82**, pp. 1052 - 1055, 1999.

[40] A. M. Michelson and A.H. Levesque, *Error-Control Techniques for Digital Communication*, Wiley, New York, 1985.

[41] D. Middleton, *An Introduction to Statistical Communication Theory*, McGraw-Hill, New York, 1960.

[42] A. Muthukrishnan and C. R. Stroud, "Multivalued Logic Gates for Quantum Computation," *Phy. Rev. A*, V. 62, 052309, 2000.

[43] J. Neumann and O. Morgenstern, *Theory of Games and Economic Behavior*, Princeton University Press, Princeton, 1953.

[44] M. Nielsen and I. L. Chuang, *Quantum Computation and Quantum Information,* Cambridge University Press, 2000.

[45] A. Papoulis, *Probability, Random Variables, and Stochastic Processes,* 2nd edition, McGraw-Hill, New York, 1984.

[46] I. Peterson, "Quantum Games," *Science News*, Nov. 20, 1999.

[47] W. W. Peterson and E.J. Weldon, Jr., *Error Correcting Codes,* 2nd edition, M.I.T. Press, Cambridge, Mass., 1972.

[48] P. Picton, "Optoelectronic Multi-Valued Conservative Logic," *Int. J. of Optical Computing*, Vol. 2, pp.19 - 29, 1991.

[49] E. W. Piotrowski and J. Sladkowski, "Quantum Bargaining Games," *Physica A*, **308**, 391, 2002.

[50] L. R. Rabiner, "A Tutorial on Hidden Markov Models and Selected Applications in Speech Recognition," *Proceedings of the IEEE*, **77**(2), pp. 257 – 286, 1989.

[51] A. Rapoport, "Games," *Peace and Change*, Vol. XIII, pp. 18 - 43, 1988.
[52] A. Rapoport and M. Guyer, "A Taxonomy of 2x2 Games," *General Systems*, Vol. XXIII, pp. 125 - 136, 1978.
[53] T. S. Rappaport, *Wireless Communications: Principles and Practice*, IEEE Press, Piscataway, N. J., 1996.
[54] K. Roy and S. Prasad, *Low-Power CMOS VLSI Circuit Design*, John Wiley & Sons Inc., 2000.
[55] C. Schlegel, *Trellis Coding*, IEEE Press, Piscataway, N. J., 1997.
[56] C. E. Shannon and W. Weaver, *The Mathematical Theory of Communication*, University of Illinois Press, Urbana, 1949.
[57] Z. Shi and R. Lee, "Bit Permutation Instructions for Accelerating Software Cryptography," *Int. Conf. Application-Specific Systems, Architectures, and Processors*, pp. 138 - 148, July 2000.
[58] R. Steele, *Delta Modulation Systems*, Wiley, New York, 1975.
[59] A. S. Tanenbaum, *Computer Networks,* 2nd ed., Prentice-Hall, Englewood Cliffs, N. J., 1995.
[60] T. M. Thompson, *From Error-Correcting Codes Through Sphere Packings to Simple Groups*, The Mathematical Association of America, Washington D. C., 1983.
[61] P. E. Turner and L. Cho, "Prisoner's Dilemma in an RNA Virus," *Nature*, **398**, pp. 441 - 443, 1999.
[62] A. J. Viterbi, "Error Bounds for Convolutional Codes and an Asymptotically Optimum Decoding Algorithm," *IEEE Trans. on Information Theory,* Vol. IT-13, pp. 260 - 269, 1967.
[63] A. J. Viterbi and J.K. Omura, *Principles of Digital Communication and Coding*, McGraw-Hill, New York, 1979.
[64] A. J. Viterbi, "Wireless digital communication: A view based on three lessons learned," *IEEE Communications Magazine,* Vol. 29, No. 9, pp. 33 - 36, 1991.
[65] S. B. Wicker and V.K. Bhargava (editors), *Reed-Solomon Codes*, IEEE Press, Piscataway, N. J., 1994.
[66] S. G. Wilson, *Digital Modulation and Coding*, Prentice-Hall, Englewood Cliffs, N. J., 1996.
[67] R. Wright, *Nonzero*: *The Logic of Human Destiny*, Pantheon, New York, 2000.
[68] R. D. Yates and D.J. Goodman, *Probability and Stochastic Processes: A Friendly Introduction for Electrical and Computer Engineers*, Wiley, New York, 1999.
[69] J. H. Yuen (editor), *Deep Space Telecommunications Systems Engineering*, Plenum, New York, 1983.
[70] F. C. Zagare, *Game Theory* (*Concepts and Applications*), Monograph # 41 of Series: Quantitative Applications in the Social Sciences, Sage, Beverly Hills, 1984.
[71] R. E. Ziemer and W.H. Tranter, *Principles of Communications,* 3rd edition, Houghton Miflin, Boston, Mass., 1990.

In: Game Theory: Strategies, Equilibria and Theorems ISBN 978-1-60456-844-8
Editors: Ingrid N. Haugen et al, pp. 83-115

Chapter 3

QUANTUM GAMES AND THE RELATIONSHIPS BETWEEN QUANTUM MECHANICS AND GAME THEORY

Esteban Guevara Hidalgo[1,2]
[1]Center for Nonlinear and Complex Systems, Università degli Studi dell'Insubria, Via Valleggio 11, 22100 Como, Italy
[2]SIÓN, Autopista General Rumiñahui, Urbanización Edén del Valle, Sector 5, Calle 1 y Calle A # 79, Quito, Ecuador

Abstract

Quantum games have proposed a new point of view for the solution of the classical problems and dilemmas in game theory. It has been shown that are more efficient than classical games and provide a saturated upper bound for this efficiency. Certain quantization relationships can be proposed with the objective that a game can be generalized into a quantum domain where the linear superposition of actions is allowed. This quantization let us describe and solution problems originated by conflicting or cooperative behaviors among the members of a system from the point of view of quantum mechanical interactions. This leads us to interesting aspects which only can be observed through the quantization of a game like the possibility of the entanglement between players, the definition of a socioeconomical temperature in a system and the analysis of a game through elements of quantum information theory.

Although both systems analyzed are described through two theories apparently different (quantum mechanics and game theory) both are analogous and thus exactly equivalents. The quantum analogue of the replicator dynamics is the von Neumann equation. The classical equilibrium concepts in game theory can be also generalized through a maximum entropy approach in the so called Collective Welfare Principle. Nature is a game in where its players compete for the equilibrium of the system that they are members. They act as a whole besides individuals like they obey a rule in where they prefer to work for the collective besides the individual welfare. If it is maximized the welfare of the individual above the collective welfare the system gets unstable and eventually it collapses.

Quantum mechanics (and physics) could be used to explain more correctly biological and economical processes (econophysics). A special consequence of the relationships between quantum mechanics and game theory is analyzed. It is shown that the so called "globalization" process (i.e., the inexorable integration of markets, currencies, nation-states, technologies and the intensification of consciousness of the world as a whole) has a behavior exactly equivalent to a system that is tending to a maximum entropy state i.e., to its state of equilibrium. This let us predict the apparition of big common markets and strong common currencies that will reach the "equilibrium" by decreasing its number until they get a state characterized by only one common currency and only one big common community around the world.

1. Introduction

The present work offers an overview of game theory. Starting from the classical theory the basic concepts of game, strategy, equilibrium, the evolutionary theory, the replicator dynamics and evolutionary stable strategies and a review of the most important works in the recent field of quantum game theory, its importance, and how the quantization can improve the results in a game and solution the dilemmas of game theory. Finally, we analyze the relationships between game theory and quantum mechanics, its consequences and applications.

2. Game Theory: From Nash to the Replicator Dynamics

2.1. Classical Game Theory

Game theory [1–3] is the study of decision making of competing agents in some conflict situation. It tries to understand the birth and development of conflicting or cooperative behaviors among a group of individuals who behave rationally and strategically according to their personal interests. Each member in the group strive to maximize its welfare, state, utilities or payoffs by choosing the best courses of strategies from a cooperative or individual point of view. Game theory has been applied to solve many problems in economics, social sciences, biology, computer science, international relations, engineering and more recently, in physics.

2.1.1. Basic Definitions

A *game* $G = (N, S, E)$ consists of a set of players N, a set of strategies $S = \{S_1, ..., S_N\}$, where S_j is the set of strategies available to the jth player and a set of payoff functions $E = \{E_1, ..., E_N\}$, where E_j is the payoff function for the jth player. A *payoff function* E for a player is a mapping from the cross-product of player's strategy spaces to the player's set of payoffs. E assigns a real number to the pair (s_i, s_j), $E(s_i, s_j)$ is the payoff obtained by a player who plays the strategy s_i against an opponent who plays the strategy s_j. An *action* or a move is a choice available to a player. It could be taken by a player during some moment in a game.

2.1.2. Strategies

A *strategy* is a complete plan of action for every stage of the game, regardless of whether that stage actually arises in play. A *strategy space* for a player is the set of all strategies available to the player. A *pure strategy* is a strategy that specifies a unique move in a given game position, i.e., an action with probability 1. A *mixed strategy* x is a probability distribution over S which corresponds to how frequently each move is chosen. A *dominant strategy* is a strategy that does at least as well as any competing strategy against any possible moves by the other player(s).

2.1.3. Games

A *zero sum game* is a game in where the sum of all players payoffs is zero regardless of the strategies they choose. A player gains only at the expense of others. In a *constant-sum game*, the sum of all players' payoffs is the same for any outcome. A *cooperative game* is a game in which two or more players strive toward a unique objective and therefore win or lose as a group. In a *non-cooperative game* no outside authority assures that players stick to the same predetermined rules, and so binding agreements are not feasible. In these games players may cooperate but any cooperation must be self-enforcing. In a *game of perfect information* the knowledge about other players is available to all participants i.e., every player knows the payoff functions and the strategies available to other players. A *symmetric game* is a game in where all agents have the same set of strategies and identical payoffs functions, except for the interchange of roles of the players. A *symmetric two-person game* $G = (S, E)$ consists of a finite nonempty pure strategy set S and a payoff function E which assigns a real number to the pair (s_i, s_j). A $n_1 \times n_2 \times ... \times n_N$ game is a N player game where the jth player has available n_j strategies.

2.1.4. Equilibrium Notions

Let be $p, r \in S_i$ and $q \in S_j$. A *best reply* to q is a strategy p which maximizes $E(p, q)$. A *dominant strategy equilibrium* is a strategy profile in which each player plays best replies that do not depend on the strategies of other players. An *equilibrium point* is a pair (p, q) with the property that p and q are best replies to each other. A strategy r is a *strict best reply* to a strategy q if it is the only best reply to q. A strict best reply must be a pure strategy. An equilibrium point (p, q) is called *strict equilibrium point* if p and q are strict best replies to each other. A best reply to p which is different from p is called *alternative best reply*.

A *Nash equilibrium* (NE) [4, 5] is a set of strategies, one for each player, such that no player has an incentive to unilaterally change his action. Players are in equilibrium if a change in strategies by any one of them would lead that player to earn less than if he remained with his current strategy. A Nash equilibrium satisfies the following condition

$$E(p, p) \geq E(r, p). \tag{1}$$

A player cannot increase his payoff if he decides to play the strategy r instead of p. A *focal point* is one amongst several NE which for psychological reasons is particularly compelling.

In a zero-sum game between players A and B, player A should attempt to minimize player B's maximum payoff while player B attempts to maximize his own minimum payoff. When they do so the minimum of the maximum (minimax) payoffs equals the maximum of the minimum (maximin) payoffs. Neither player can improve his position, and so these strategies form an equilibrium of the game. The *minimax theorem* [1] states that for every two-person, zero-sum game, there always exists a mixed strategy for each player such that the expected payoff for one player is the same as the expected cost for the other. In other words, there is always a rational solution to a precisely defined conflict between two people whose interests are completely opposite. It is a rational solution in that both parties can convince themselves that they cannot expect to do any better, given the nature of the conflict.

A *Pareto optimal* (PO) [6] is a game result from which no player can improve their payoff without another player being worse off, that is, if $\forall k$, $\exists l$ such that

$$\begin{aligned} &E_k(s_1, ..., s'_k, s_l, ..., s_N) > E_k(s_1, ..., s_k, s_l, ..., s_N), \\ \text{then } &E_l(s_1, ..., s'_k, s_l, ..., s_N) < E_l(s_1, ..., s_k, s_l, ..., s_N). \end{aligned} \tag{2}$$

Then the unprimed strategy profile is Pareto optimal. An outcome of a game is Pareto optimal if there is no other outcome that makes every player at least as well off and at least one player strictly better off. That is, a Pareto optimal outcome cannot be improved upon without hurting at least one player.

2.2. Evolutionary Game Theory

Evolutionary game dynamics is the application of population dynamical methods to game theory. It has been introduced by evolutionary biologists, anticipated in part by classical game theorists [7] and first introduced under the name of *evolutionary game theory* by J. Smith and G. Price in the context of animal conflict [8]. Evolutionary game theory [7,9,10] does not rely on rational assumptions (like classical game theory) but on the idea that the Darwinian process of natural selection [11] drives organisms towards the optimization of reproductive success [12]. It combines the principles of game theory, evolution, non linear dynamics and dynamical systems to explain the distribution of different phenotypes in biological populations. Instead of working out the optimal strategy, the different phenotypes in a population are associated with the basic strategies that are shaped by trial and error by a process of natural selection or learning. *Strategies* are considered to be inherited programs for any conceivable situation which control the individual's behavior. The members of a population interact in game situations and the joint action of mutation and selection replaces strategies by others with a higher reproductive success. In this kind of games is less important to know which member plays which strategy within a population but it is important to know the relative frequency of actions (the probability of playing a strategy) [12]. *Payoffs* in biological games are in terms of fitness a measure of reproductive success.

In contrast with classical game theory its evolutionary version deals with entire populations of players, all "programmed" to use some strategy (or type of behavior). Strategies with high payoff will spread within the population (this can be achieved by learning, by copying or inheriting strategies, or even by infection). The payoffs depend on the actions of the coplayers and hence on the frequencies of the strategies within the population. Since these frequencies change according to the payoffs, this yields a *feedback loop* [7]. The dynamics of this feedback loop is the object of evolutionary game theory. The feedback dynamics depend strongly, of course, on the population structure, on the underlying game and on the way strategies spread. Thus there are many "game dynamics", which can be discrete or continuous, stochastic or deterministic [7, 10, 13, 14].

Many successful applications of evolutionary game theory appeared in mathematical biology to explain biological phenomena (e.g., to predict the behavior of bacteria and insects) but it can also be used to interpret classical games from a different perspective. Instead of directly calculating properties of a game, populations of players using different strategies are simulated and a process similar to natural selection is used to determine how the population evolves. This is made through the stability analysis of differential equations and the implications to the games [13].

2.2.1. Evolutionary Stable Strategies

The central equilibrium concept of evolutionary game theory is the notion of Evolutionary Stable Strategy (ESS) introduced by J. Smith and G. Price [8, 9]. An ESS is described as a strategy which has the property that if all the members of a population adopt it, no mutant strategy could invade the population under the influence of natural selection. ESS are interpreted as stable results of processes of natural selection. The natural selection process that determines how populations playing specific strategies evolve is known as the replicator dynamics [7, 10, 13, 14] whose stable fixed points are Nash equilibria [2].

Each agent in a n-player game (where the ith player has as strategy space S_i) is modeled by a population of players which have to be partitioned into groups. Individuals in the same group would all play the same strategy. Randomly, we make play the members of the subpopulations against each other. The subpopulations that perform the best will grow and those that do not will shrink and eventually will vanish. The process of natural selection assures survival of the best players at the expense of the others. A population equilibrium occurs when the population shares are such that the expected payoffs for all strategies are equal.

Consider a large population in which a two person game $G = (S, E)$ is played by randomly matched pairs of animals generation after generation. Let p be the strategy played by the vast majority of the population, and let r be the strategy of a mutant present in small frequency. Both p and r can be pure or mixed. An *evolutionary stable strategy* (ESS) p of a symmetric two-person game $G = (S, E)$ is a pure or mixed strategy for G which satisfies the following two conditions

$$\begin{gathered} E(p,p) > E(r,p), \\ \text{If } E(p,p) = E(r,p) \text{ then } E(p,r) > E(r,r). \end{gathered} \tag{3}$$

Since the stability condition only concerns to alternative best replies, p is always evolutionarily stable if (p, p) is an strict equilibrium point. An ESS is also a Nash equilibrium since it is the best reply to itself and the game is symmetric. The set of all the strategies that are ESS is a subset of the NE of the game. A population which plays an ESS can withstand an invasion by a small group of mutants playing a different strategy. It means that if a few individuals which play a different strategy are introduced into a population in an ESS, the selection process would eventually eliminate the invaders.

2.2.2. The Replicator Dynamics

The natural selection process that determines how populations playing specific strategies evolve is known as the *replicator dynamics*. It describes the evolution of a polymorphic state in a population represented by a mixed strategy x for G whose members are involved in a conflict described by a symmetric two-person game $G = (S, E)$. The probability assigned to a pure strategy s is denoted by $x(s)$. If $s_i, i = 1, ..., n \in S$ are the pure strategies available to a player, then the *player's strategy* will be denoted by the column vector x with $x_i \in [0, 1]$ and $\sum_{i=1}^{n} x_i = 1$. The ith component of x gives the probability of playing strategy s_i and also is interpreted as the relative frequency of individuals using strategy s_i. Playing a pure strategy s_j is represented by the vector x whose jth component is 1, and all the other components are 0. The *fitness function* $E = f_i(x)$, $i = 1, ..., n$ specifies how successful each subpopulation is and must be defined for each component of x. The fitness for x_i is the expected utility of playing strategy s_i against a player with a mixed strategy defined by the vector x. It is given by

$$f_i(x) = (Ax)_i = \sum_{j=1}^{n} a_{ij} x_j, \tag{4}$$

where A is the payoff matrix (a_{ij} are its elements) and the subscript $i = 1, ..., n$ in $(Ax)_i$ denotes the ith component of the matrix-vector product (Ax). The *average fitness of the population* $\langle f(x) \rangle = \sum_{i=1}^{n} x_i f_i(x)$ is

$$\langle f(x) \rangle = x^T A x = \sum_{k,l=1}^{n} a_{kl} x_k x_l, \tag{5}$$

where the superscript T denotes transpose.

The evolution of relative frequencies in a population is described by the replicator dynamics

$$\frac{dx_i(t)}{dt} = [f_i(x) - \langle f(x) \rangle] \, x_i(t) = \left[(Ax)_i - x^T A x \right] x_i(t), \tag{6}$$

or also

$$\frac{dx_i(t)}{dt} = \left[\sum_{j=1}^{n} a_{ij} x_j - \sum_{k,l=1}^{n} a_{kl} x_k x_l \right] x_i(t). \tag{7}$$

The stable fixed points of the replicator dynamics are Nash equilibria. It is important to note that the fixed points of a system do not change in the time. It means that if a population reaches a state which is a Nash equilibrium, it will remain there. The replicator dynamics rewards strategies that outperform the average by increasing their frequency, and penalizes poorly performing strategies by decreasing their frequency.

In a symmetric game payoff matrices and actions are identical for both agents. These games can be modeled by a single population of individuals playing against each other. When the game is *asymmetric*, a different population of players must be used to simulate each agent. The strategy vector for player one is represented by x and for player two is represented by y. Player one has n strategies $s_{1i} \in S_1, i = 1, ..., n$ and player two has m strategies $s_{2j} \in S_2, j = 1, ..., m$. Each player will have also a distinct payoff matrix $A \in \mathbb{R}^{n \times m}$ and $B \in \mathbb{R}^{m \times n}$, respectively. The fitness for a player who plays the strategy s_{1i} will be $f_{1i} = (Ay)_i$ and the average fitness of the first population will be $\langle f_1 \rangle = x^T A y$. Similarly, the fitness for a player who plays the strategy s_{2i} will be $f_{2i} = (Bx)_i$ and the average fitness of the second population will be $\langle f_2 \rangle = y^T Bx$. The evolution of this game would be described for the next equations system [7]

$$\frac{dx_i(t)}{dt} = \left[(Ay)_1 - x^T Ay\right] x_i(t),$$
$$\frac{dy_i(t)}{dt} = \left[(Bx)_1 - y^T Bx\right] y_i(t). \tag{8}$$

3. Quantum Mechanics Foundations

3.1. Physical System Representation

Lets represent a physical system through a Hilbert space. A state of that system is completely described through a state vector $|\Psi(t)\rangle$ (element of that Hilbert space) and it is postulated that contains all the information about that system. The possible configurations of a physical system are described by the state space which is spanned by a set of basis states B. The number of basis states equals the dimension of the state space which can be infinite and even uncountable. A basis state $i \in B$ is denoted by $|i\rangle$ and every state $|\Psi\rangle$ can be described with a linear combination on their basis states B

$$|\Psi(t)\rangle = \sum_{i \in B} \alpha_i |i\rangle \tag{9}$$

with $\alpha_i \in \mathbb{C}$. It can be shown that the inner product in the Hilbert space relates to the probability amplitudes in quantum mechanics and viceversa. By definition

$$|\Psi(t)\rangle = \sum_{i \in B} |i\rangle \langle i|\Psi\rangle . \tag{10}$$

The probability of measuring a state $|\Psi\rangle$ in the basis state i equals $|\langle i|\Psi\rangle|^2$ which satisfies a normalization condition $\sum_{i \in B} |\langle i|\Psi\rangle|^2 = 1$ for every $|\Psi\rangle$ in the Hilbert space.

In classical mechanics we can precisely specify the state of a system by one point in its phase space. Its trajectory through the phase space describes the time evolution of the system and this evolution follows Newton's laws or Hamilton equations. When the information about the system is incomplete the state of a system is not perfectly defined for which we have to describe our system in terms of probabilities.

An *ensemble* is a collection of identically prepared physical systems. When each member of the ensemble is characterized by the same state vector $|\Psi(t)\rangle$ it is called *pure ensemble*. If each member has a probability p_i of being in the state $|\Psi_i(t)\rangle$ we have a *mixed ensemble*. Each member of a *mixed ensemble* is a pure state and its evolution is given by Schrödinger equation. Probabilities for each state are constrained to satisfy the normalization condition $\sum_{i=1}^{n} p_i = 1$ and obviously $0 \leq p_1, p_2, ..., p_n \leq 1$.

An *observable* is a property of the system state that can be determined by some sequence of physical operations. Observables in our model are represented through Hermitian (or self adjoint) operators acting over the Hilbert space. An operator A is Hermitian when $A = A^{\dagger}$. Every operator can be represented in matrix form in some basis. In quantum physics we can know only the expectation value of an observable. Suppose we make a *measurement* on a mixed ensemble of some observable A. The *ensemble average* of A is defined by the average of the expected values measured in each member of the ensemble described by $|\Psi_i(t)\rangle$ and with probability p_i, $\langle A\rangle_\rho = p_1\langle A\rangle_1 + p_2\langle A\rangle_2 + ... + p_n\langle A\rangle_n$

$$\langle A\rangle_\rho = \sum_{i=1}^{n} p_i \langle\Psi_i(t)| A |\Psi_i(t)\rangle = \sum_{i,j,k=1}^{n} p_i a_{jk} c_j^{(i)*}(t) c_k^{(i)}(t), \tag{11}$$

where a_{jk} are the elements of the matrix that represents the observable A. The terms $c_k^{(i)}(t) = \langle k|\Psi_i(t)\rangle$ and $c_j^{(i)*}(t) = \langle\Psi_i(t)|j\rangle$ are the elements of certain *density operator* $\rho(t)$ defined as

$$\rho(t) = \sum_{i=1}^{n} p_i |\Psi_i(t)\rangle\langle\Psi_i(t)|. \tag{12}$$

For a mixed state ρ is $Hermitian$, $Tr\rho(t) = 1$, $\rho^2(t) \leq \rho(t)$ and $Tr\rho^2(t) \leq 1$.

To describe correctly a statistical mixture of states it is necessary the introduction of the density operator. It contains all the physically significant information we can obtain about the ensemble in question. Any two ensembles that produce the same density operator are physically indistinguishable. A pure state is specified by $p_i = 1$ for some $|\Psi_i(t)\rangle$, $i = 1, ..., n$ and its density operator $\rho(t)$ by a matrix with all its elements equal to zero except one 1 on the diagonal. The diagonal elements ρ_{nn} of the density operator $\rho(t)$ represents the average probability of finding the system in the state $|n\rangle$.

$$\rho_{nn} = \langle n|\rho(t)|n\rangle = \sum_{i=1}^{n} \langle n|\Psi_i(t)\rangle p_i \langle\Psi_i(t)|n\rangle = \sum_{i=1}^{n} p_i \left|c_n^{(i)}\right|^2, \tag{13}$$

where $c_n^{(i)} = \langle n|\Psi i(t)\rangle$ and $\left|c_n^{(i)}\right|^2 \in \mathbb{R}^+$. If the state of the system is $|\Psi_i(t)\rangle$, $\left|c_n^{(i)}\right|^2$ is the probability of finding, in a measurement, this system in the state $|n\rangle$. The diagonal elements ρ_{nn} are zero if and only if all $\left|c_n^{(i)}\right|^2$ are zero. The non-diagonal elements ρ_{np} expresses the interference effects between the states $|n\rangle$ and $|p\rangle$ which can appear when the state $|\Psi_i\rangle$ is a coherent linear superposition of these states.

$$\rho_{np} = \langle n|\rho(t)|p\rangle = \sum_{i=1}^{n} \langle n|\Psi_i(t)\rangle p_i \langle\Psi_i(t)|p\rangle = \sum_{i=1}^{n} p_i c_n^{(i)}(t) c_p^{(i)*}(t) \tag{14}$$

with $c_n^{(i)}(t) = \langle n|\Psi_i(t)\rangle$, $c_p^{(i)*}(t) = \langle\Psi_i(t)|p\rangle$ and $c_n^{(i)} c_p^{(i)*} \in \mathbb{C}$. If $\rho_{np} = 0$ it means that the average has cancelled out any interference effects between $|n\rangle$ and $|p\rangle$ but if it is different from zero subsists certain coherence between these states.

3.2. Evolution of a Physical System

3.2.1. Schrödinger & von Neumann Equations

Each pure state evolves following the Schrödinger equation but the evolution of the system as a statistical mixture of states described through a density operator is given by the von Neumann equation

$$i\hbar\frac{d\rho}{dt} = \left[\hat{H}, \rho\right], \tag{15}$$

$\hat{H}$ is the Hamiltonian of the physical system. The von Neumann equation is only a generalization (and/or a matrix/operator representation) of the Schrödinger equation and the quantum analogue of Liouville's theorem from classical statistical mechanics.

3.2.2. Unitary Operators

The evolution of an isolated quantum system is also described by unitary transformations (U is unitary if $U^\dagger = U^{-1}$). The states $|\Psi_1\rangle$ at time t_1 and $|\Psi_2\rangle$ at time t_2 are related by an unitary transformation U by

$$|\Psi_2\rangle = U\,|\Psi_1\rangle\,. \tag{16}$$

For a statistical mixture of states the ensemble evolves unitarily in time by

$$\rho(t_2) = U(t_2, t_1)\rho(t_1)U^\dagger(t_2, t_1). \tag{17}$$

3.3. Quantum Bits and Quantum Registers

A general *qubit* state in a two-dimensional Hilbert space whose orthonormal basis can be written as $\{|0\rangle\,, |1\rangle\}$ is

$$|\Psi\rangle = a\,|0\rangle + b\,|1\rangle \tag{18}$$

with $a, b \in \mathbb{C}$ and $|a|^2 + |b|^2 = 1$. In other words, $|\Psi\rangle$ is a unit vector in a two-dimensional complex vector space for which a particular basis has been fixed. If we expand this state space to that of a system whose basis set is described by $\{0, 1\}^n$ we get the definition of a n-qubit system. The possible configurations of such a *quantum register* are covered by

$$|\Psi\rangle = \sum_{i\in\{0,1\}^n} \alpha_i\,|i\rangle\,. \tag{19}$$

The state space of a n-qubit system equals the tensor product of n separate qubit systems $\mathcal{H}_{\{0,1\}^n} = \mathcal{H}_{\{0,1\}} \otimes \mathcal{H}_{\{0,1\}} \otimes ... \otimes \mathcal{H}_{\{0,1\}}$. Qubits and quantum registers are used to describe the memory of quantum computers.

3.4. Quantum Entanglement

Consider a system which has associated a Hilbert space $\mathcal{H}$ that can be divided into two subsystems. Assume $\mathcal{H}_A$ and $\mathcal{H}_B$ to be the Hilbert spaces corresponding to the subsystems A and B, respectively. The two state spaces A and B are spanned by $|i\rangle_A$ and $|j\rangle_B$. The state space is expanded into $\mathcal{H}$ through the tensor product $\otimes$ of $\mathcal{H}_A$ and $\mathcal{H}_B$ i.e., $\mathcal{H} = \mathcal{H}_A \otimes \mathcal{H}_B$. This space is spanned by the basis vectors $|k\rangle = |i\rangle_A \otimes |j\rangle_B$ sometimes denoted $|i\rangle_A\,|j\rangle_B$, $|i, j\rangle_{AB}$ or $|ij\rangle_{AB}$. Any state $|\Psi\rangle_{AB}$ of $\mathcal{H}$ is a linear combination of the basis states $|i\rangle_A\,|j\rangle_B$

$$|\Psi\rangle_{AB} = |\Psi\rangle_A \otimes |\Psi\rangle_B = (\sum_{i=1} \alpha_i\,|i\rangle_A) \otimes (\sum_{j=1} \beta_j\,|j\rangle_B) = \sum_{i,j} c_{ij}\,|i\rangle_A\,|j\rangle_B\,, \tag{20}$$

where c_{ij} are complex coeficients which satisfy a normalization condition $\sum_{i,j} |\ c_{ij}|^2 = 1$. The state $|\Psi\rangle_{AB}$ is called *direct product* (or separable) state if it is possible to factor it into two normalized states from the Hilbert spaces $\mathcal{H}_A$ and $\mathcal{H}_B$. A state $|\Psi\rangle_{AB}$ in $\mathcal{H} = \mathcal{H}_A \otimes \mathcal{H}_B$ is called *entangled* if it is not a direct product state i.e., it is entangled if it cannot be factored into two normalized states elements of the two subsystems that compose the system. Entanglement describes the situation when the state of whole cannot be written in terms of the states of its constituent parts. The basis for $\mathcal{H}_A \otimes \mathcal{H}_B$, where $\mathcal{H}_A$ and $\mathcal{H}_B$ are two-dimensional Hilbert spaces, is $\{|0\rangle_A \otimes |0\rangle_B, |0\rangle_A \otimes |1\rangle_B, |1\rangle_A \otimes |0\rangle_B, |1\rangle_A \otimes |1\rangle_B\}$. The most general state in the Hilbert space $\mathcal{H}_A \otimes \mathcal{H}_B$ is

$$|\Psi\rangle_{AB} = \sum_{i,j=0}^{n=1} c_{ij} |i\rangle_A |j\rangle_B = (c_0^A |0\rangle_A + c_1^A |1\rangle_A) \otimes (c_0^B |0\rangle_B + c_1^B |1\rangle_B), \tag{21}$$

where $|c_0^{(A)}|^2 + |c_1^{(A)}|^2 = 1$ and $|c_0^{(B)}|^2 + |c_1^{(B)}|^2 = 1$.

3.5. Von Neumann Entropy & Quantum Information Theory

Entropy [15, 16] is the central concept of information theory. In classical physics, information processing and communication is best described by Shannon information theory. The *Shannon entropy* expresses the average information we expect to gain on performing a probabilistic experiment of a random variable A which takes the value a_i with the respective probability p_i. It also can be seen as a measure of uncertainty before we learn the value of A. We define the Shannon entropy of a random variable A by

$$H(A) \equiv H(p_1, ..., p_n) \equiv -\sum_{i=1}^{n} p_i \log_2 p_i. \tag{22}$$

The entropy of a random variable is completely determined by the probabilities of the different possible values that the random variable takes. Due to the fact that $p = (p_1, ..., p_n)$ is a probability distribution, it must satisfy $\sum_{i=1}^{n} p_i = 1$ and $0 \leq p_1, ..., p_n \leq 1$. The Shannon entropy of the probability distribution associated with the source gives the minimal number of bits that are needed in order to store the information produced by a source, in the sense that the produced string can later be recovered. Suppose A and B are two random variables. The *joint entropy* $H(A, B)$ measures our total uncertainty about the pair (A, B). The *conditional entropy* $H(A \mid B)$ is a measure of how uncertain we are about the value of A, given that we know the value of B. The *mutual or correlation entropy* $H(A : B)$ measures how much information A and B have in common. The *relative entropy* $H(p \parallel q)$ measures the closeness of two probability distributions, p and q, defined over the same random variable A. The classical relative entropy of two probability distributions is related to the probability of distinguishing the two distributions after a large but finite number of independent samples (Sanov's theorem). Since Shannon [15], *information theory* or the mathematical theory of communication changed from an engineering discipline that dealed with communication channels and codes [17] to a physical theory [18] in where the introduction of the concepts of entropy and information were indispensable to our understanding of the physics of measurement. Classical information theory has two primary goals [19]: The first is the development of the fundamental theoretical limits on the achievable performance when communicating a given information source over a given communications channel using coding schemes from within a prescribed class. The second goal is the development of coding schemes that provide performance that is reasonably good in comparison with the optimal performance given by the theory.

The *von Neumann entropy* [20, 21] is the quantum analogue of the Shannon's entropy. It appeared 21 years before Shannon's and generalizes Boltzmann's expression. Von Neumann defined the entropy of a quantum state ρ by the formula

$$S(\rho) \equiv -Tr(\rho \ln \rho). \tag{23}$$

The entropy $S(\rho)$ is non-negative and takes its maximum value $\ln n$ when ρ is maximally mixed, and its minimum value zero if ρ is pure. If λ_i are the eigenvalues of ρ then von Neumann's definition can be expressed as $S(\rho) = -\sum_i \lambda_i \ln \lambda_i$. The von Neumann entropy reduces to a Shannon entropy if ρ is a mixed state composed of orthogonal quantum states [22]. By analogy with the Shannon entropies it is possible to define conditional, mutual and relative entropies. The negativity of the conditional entropy always indicates that two systems are entangled and indeed, how negative the conditional entropy is provides a lower bound on how entangled the two systems are. *Quantum information theory* [23, 24] may be defined as the study of the achievable limits to information processing possible within quantum mechanics. Thus, the field of quantum information has two tasks: It aims to determine limits on the class of information processing tasks which are possible in quantum mechanics and provide constructive means for achieving information processing tasks. Quantum information theory appears to be the basis for a proper understanding of the emerging fields of quantum computation [25, 26], quantum communication [27, 28], and quantum cryptography [29, 30]. Entropy in quantum information theory plays prominent roles in many contexts, e.g., in studies of the classical capacity of a quantum channel [31, 32] and the compressibility of a quantum source [33, 34].

4. Quantum Game Theory

For a quantum physicist it is legitimate to ask what happens if linear superpositions of the strategies in a game are allowed for, that is if games are generalized into the quantum domain [35]. Quantum games have demonstrated to propose a new point of view for the solution of the classical problems and dilemmas in game theory. It has been shown that quantum games are more efficient than classical games and provide a saturated upper bound for this efficiency [35–45].

In *Blaquiere*'s [46] *Wave mechanics as a two-player game* (1980) game-theoretical ideas are discussed in the context of quantum physics. Blaquiere analyzes the connection between dynamic programming, the theory of differential games, and wave mechanics. The author argues that wave mechanics is born of a dynamic programming equation which Louis de Broglie obtained in 1923. He then expresses the stationarity principle in the form of a minimax principle written in the form of sufficiency conditions for the optimality of strategies in a two-player zero-sum differential game. The saddle-point condition, on which optimality of strategies is based, is an extension of Hamilton's principle of least action. *Wiesner*'s [47] *Quantum money* (1983) is believed to have started the field of quantum cryptography. Cryptographic protocols can be written in the language of game theory. Wiesner suggested to use the uncertainty principle for creating means of transmitting two messages. In 1990 *Mermin* [48, 49] presented an n-player quantum game that can be won with certainty when it involves n spin half particles in a Greenberger-Horne-Zeilinger (GHZ) state; no classical strategy can win the game with a probability greater than $1/2 + 1/2^{n/2}$.

The actual firsts works on quantum games were the developed for *Meyer* and *Eisert et al.* Meyer [36] quantized a coin tossing game and found out that one player could increase his expected payoff and win with certainty by implementing a quantum strategy against his opponent's classical strategy. Eisert et al. [35] developed a general protocol for two player-two strategy quantum games with entanglement by quantizing prisoner's dilemma. They found a unique Nash equilibrium, which is different from the classical one, and the dilemma could be solved if the two players are allowed to use quantum strategies. This was extended later to multiplayer games [50]. *Marinatto and Weber* [37] extended the concept of a classical two-person static game to the quantum domain by giving a Hilbert structure to the space of classical strategies. They showed that the introduction of entangled strategies in battle of the sexes game leads to a unique solution of this game. *Du et al.* [39] implemented a game via nuclear magnetic resonance (NMR) system. It was demonstrated that neither of the two players would win the game if they play rationally, but if they adopt quantum strategies both of them would win. Quantum games have been used to explore unsolved problems of quantum information [23, 24] and in the production of algorithms for quantum computers [51]. Quantum communication can be considered as a game where the objective is maximize effective communication. Also distributed computing, cryptography, watermarking and information hiding tasks can be modeled as games [52–58].

Azhar Iqbal [44, 45, 59–62] introduced the concepts of replicator dynamics and evolutionary stable strategies from evolutionary game theory to the analysis of quantum games. *Guevara* [63–70] analyzed the relationships between game theory and quantum mechanics. There exists a correspondence between the replicator dynamics and the von Neumann and in general between quantum mechanics and game theory, their concepts and equilibrium definitions. *Piotrowski and Sladkowski* have modeled markets, auctions and bargaining assuming traders can use quantum protocols [71–73]. In the new quantum market games, transactions are described in terms of projective operations acting on Hilbert spaces of strategies of traders. A quantum strategy represents a superposition of trading actions that can achieve outcomes not realizable by classical means. Furthermore, quantum mechanics has features that can be used to model aspects of market behavior. For example, traders observe the actions of other players and adjust their actions and the maximal capital flow at a given price corresponds to entanglement between buyers and sellers. Nature may be playing quantum survival games at the molecular level [74, 75]. It could lead us to describe many of the life processes through quantum mechanics like *Gogonea and Merz* [76] on protein molecules. Game theory and quantum game theory offer interesting and powerful tools and their results will probably find their applications in computation, complex system analysis and cognition sciences [76–79].

4.1. Quantum Strategies

Meyer describes his quantum penny-flip game as follows [36]. The starship Enterprise faces some imminent catastrophe. Q appears on the bridge and offers P to rescue the ship if he can beat him in a penny-flip game. Q asks P to place the penny in a small box, head up. Then Q, P, and finally Q reaches into the box, without looking at the penny, and either flips it over or leaves it as it is. After Q's second turn they open the box and Q wins if the penny is head up. Classically, we can take (H, T) as the basis of a 2 dimensional vector space. The players moves can be represented by a sequence of 2 ×2 matrices $F = \begin{pmatrix} 0 & 1 \\ 1 & 0 \end{pmatrix}$ (flip) and $N = \begin{pmatrix} 1 & 0 \\ 0 & 1 \end{pmatrix}$ (not to flip) which act on a vector representing the state of the coin. A general mixed strategy is described by the matrix $P = \begin{pmatrix} 1-p & p \\ p & 1-p \end{pmatrix}$, where $p \in [0, 1]$ is the probability with which the player flips the coin. A sequence of mixed actions puts the state of the coin into a convex linear combination $aH + (1 - a)T$, with $a \in [0, 1]$. The coin is then in the state H with probability a. Q plays his move first, after P puts the coin in the H state. The question is what happens when Q make use of a quantum strategy, namely a sequence of unitary, rather than stochastic, matrices. The basis for V is written $\{|H\rangle, |T\rangle\}$. A pure quantum state for the penny is a linear combination, where $a, b \in \mathbb{C}$ and $|a|^2 + |b|^2 = 1$, which means that if the box is opened, the penny will be head up with probability $|a|^2$. Since the penny starts in the state $|H\rangle$, the unitary action $U(a, b)$ by Q puts the coin into the state $a|H\rangle + b|T\rangle$, where $U(a, b)$ is $U_1 = U(a, b) = \begin{pmatrix} a & b \\ b^* & -a^* \end{pmatrix}$. The initial state of the coin can be written as $\rho_0 = |H\rangle\langle H|$. Q's action $U(a, b)$ changes the initial state ρ_0 to

$$\rho_1 = U_1 \rho_0 U_1^\dagger = \begin{pmatrix} aa^* & ab^* \\ ba^* & bb^* \end{pmatrix}. \tag{24}$$

P's mixed action acts on this density matrix, not as a stochastic matrix on a probabilistic state, but as a convex linear combination of unitary (deterministic) transformations

$$\rho_2 = pF\rho_1 F^\dagger + (1-p)N\rho_1 N^\dagger = \begin{pmatrix} pbb^* + (1-p)aa^* & pba^* + (1-p)ab^* \\ pab^* + (1-p)ba^* & paa^* + (1-p)bb^* \end{pmatrix}. \tag{25}$$

The next move of Q, U_3 transforms the state of the penny by conjugation to $\rho_3 = U_3 \rho_2 U_3^\dagger$. If Q's strategy consists of $U_1 = U(1/\sqrt{2}, 1/\sqrt{2}) = U_3$, his first action puts the penny into a simultaneous eigenvalue 1 eigenstate of both F and N, which is therefore invariant under any mixed strategy $pF + (1-p)N$ of P; and his second action inverts his first to give $\rho_3 = |H\rangle\langle H|$ and wins the game. This is the optimal quantum strategy for Q. All the pairs $([pF + (1-p)N], [U(1/\sqrt{2}, 1/\sqrt{2}), U(1/\sqrt{2}, 1/\sqrt{2})])$ are mixed/quantum equilibria for PQ Penny Flip, with value -1 to P; this is why he loses every game. The following chart explain the dynamics of the game [45]

$$|H\rangle \rightarrow \underset{\widehat{H}}{Q} \rightarrow \frac{1}{\sqrt{2}}(|H\rangle + |T\rangle) \rightarrow \underset{\hat{\sigma_x}}{P} \text{ or } \underset{\hat{I}}{P} \rightarrow \frac{1}{\sqrt{2}}(|H\rangle + |T\rangle) \rightarrow \underset{\widehat{H}}{Q} \rightarrow |H\rangle.$$

$\widehat{H}$ is the Hadamard transformation, $|H\rangle$ is head, $|T\rangle$ is tail, $\hat{I}$ is the identity which means leaving the penny alone and $\hat{\sigma_x}$ flips the penny over. Q's quantum strategy of putting the penny into the equal superposition of head and tail, on his first turn, means that whether P flips the penny over or not, it remains in an equal superposition which Q can rotate back to head by applying the Hadamard transformation, so Q always wins when they open the box. Q's classical strategy consists of implementing $\hat{\sigma_x}$ or $\hat{I}$ on his turn. When Q is restricted to play only classically, flipping the penny over or not on each turn with equal probability becomes an optimal strategy for both the players. By adapting this classical strategy Q wins only with probability 1/2. By using a quantum strategy Q can win with probability 1.

4.2. Classical, Evolutionary & Quantum Prisoners Dilemma

In this section we will analyze the famous Prisoners Dilemma from 3 points of view. It is shown how its quantization provides a solution to the problem.

4.2.1. Prisoners Dilemma

Two suspects, A and B, are arrested by the police but the police have insufficient evidence for a conviction. Both prisoners are separated. The police visit each of them to offer the same deal: if one testifies against the other (and the other remains silent), the betrayer goes free and the silent accomplice receives the full 10-year sentence. If both remain silent, both prisoners are sentenced to only six months in jail for a minor charge. If each betrays the other, each receives a five-year sentence. Each prisoner must take the decision of betray the other or to remain silent. But neither prisoner knows for sure what choice the other prisoner will make. The numerical values for the payoff matrix for the prisoners dilemma are chosen as in [75], $P_{A,B} = \begin{pmatrix} (3,3) & (0,5) \\ (5,0) & (1,1) \end{pmatrix}$. The first entry in the parenthesis denotes the payoff of A and the second is B's payoff. This choice corresponds for $r = 3$ (reward), $p = 1$ (punishment), $t = 5$ (temptation), and $s = 0$ (sucker's pay-off). The dilemma arises when one assumes that both prisoners only care about minimizing their own punishment. Each prisoner has only two options: to *cooperate* C with his accomplice and stay quiet, or to *defect* D from their implied pact and betray his accomplice in return for a lighter sentence. The outcome of each choice depends on the choice of the accomplice, but each prisoner must choose without knowing what his accomplice has chosen. The catch of the dilemma is that D is the dominant strategy, i.e., rational reasoning forces each player to defect, and thereby doing substantially worse than if they would both decide to cooperate. Mutual defection in prisoners dilemma is a Nash equilibrium: contemplating on the move DD in retrospect, each of the players comes to the conclusion that he or she could not have done better by unilaterally changing his or her own strategy.

4.2.2. Evolutionary Prisoners Dilemma

Due to the fact that the payoff matrix for each player is the same and there are only two pure strategies, a population with two groups can be constructed. We can denote the frequency of cooperators by $x_1 = x$, and obviously the frequency of the defectors by $x_2 = (1 - x)$, so the frequency vector would be $\left(\frac{x}{1-x}\right)$. To simplify the analysis consider the concrete example with $A = \begin{pmatrix} -1 & -20 \\ 0 & -10 \end{pmatrix}$. This choice is analytically convenient but not exceptional, the same results will hold for any r, p, t, and s as long as the ordering does not change. It is sufficient to study the evolution of the cooperators frequency, since the defectors frequency falls immediately from it. The average fitness of the population is $\langle f(x)\rangle = \sum_{k,l=1}^{n=2} a_{kl}x_k x_l = a_{11}x_1x_1 + a_{12}x_1x_2 + a_{21}x_2x_1 + a_{22}x_2x_2 = 9x^2 - 10$. The fitness function for the cooperators is $f_1(x) = \sum_{j=1}^{n=2} a_{ij}x_j = a_{11}x_1 + a_{12}x_2 = 19x - 20$. The evolution for the cooperators is given by the replicator dynamics

$$\dot{x} = [f_1 - \langle f(x)\rangle]x = -(9x - 10)(x - 1)x. \tag{26}$$

Note that $\forall x \in [0, 1]$, $x < 0$. That is, the frequency of cooperators is strictly decreasing. Under the replicator dynamics the frequency of cooperators will converge to 0, leaving a population purely composed of defectors. This indicates that a population of purely defecting players is a fixed point of the system, and hence D is a Nash equilibrium. The convergence to a stable point is due to the fact that the pure strategy D is an ESS. Under the given dynamic, the introduction of any number of cooperators to the population will result in the extinction of those cooperators and return to the stable state.

4.2.3. Quantum Prisoners Dilemma

After Meyer's work, *Eisert, Wilkens and Lewenstein* [35] formulated a quantization scheme for the Prisoners Dilemma and showed that the players can escape the dilemma if they both resort to quantum strategies. Also there exists a particular pair of quantum strategies (NE) which always gives reward. There exists a particular quantum strategy which always gives at least reward if played against any classical strategy.

The physical model consists of (i) a source of two bits, one bit for each player, (ii) a set of physical instruments which enable the player to manipulate his or her own bit in a strategic manner, and (iii) a physical measurement device which determines the players' pay-off from the state of the two bits. All three ingredients, the source, the players' physical instruments, and the pay-off physical measurement device are assumed to be perfectly known to both players. In this quantum formulation the classical strategies C and D are associated with two basis vectors $|C\rangle$ and $|D\rangle$ in the Hilbert space of two state system, i.e., a qubit. At each instance, the state of the game is described by a vector element of the tensor product space $\mathcal{H}_A \otimes \mathcal{H}_B$ which is spanned by the basis $|CC\rangle$, $|CD\rangle$, $|DC\rangle$, $|DD\rangle$, the first and second entry refers to A and B's qubit, respectively. The initial state of the game is $|\Psi_0\rangle = \hat{\mathcal{J}}\,|CC\rangle$, where $\hat{\mathcal{J}}$ is a unitary operator which is known for both players. Strategic moves of A and B are associated with unitary operators $\hat{U}_A$ and $\hat{U}_B$, respectively, which are chosen from a strategic space S. $\hat{U}_A$ and $\hat{U}_B$ operate exclusively on the qubits of A and B, respectively. Having executed their moves, which leaves the game in a state $(\hat{U}_A \otimes \hat{U}_B)\hat{\mathcal{J}}\,|CC\rangle$, A and B forward their qubits for the final measurement which determines their payoff. The measurement device consists of a reversible two-bit gate $\hat{\mathcal{J}}$ which is followed by a pair of Stern Gerlach type detectors. The final state of the game prior to detection is given by

$$|\Psi_f\rangle = \hat{\mathcal{J}}^\dagger(\hat{U}_A \otimes \hat{U}_B)\hat{\mathcal{J}}\,|CC\rangle\,. \tag{27}$$

The players' expected payoffs can then be written as the projections of the state $|\Psi_f\rangle$ onto the basis vectors of the tensor-product space $\mathcal{H}_A \otimes \mathcal{H}_B$. A expected payoff is given by $P_A = rP_{CC} + pP_{DD} + tP_{DC} + sP_{CD}$,

$$P_A = r|\langle CC\,|\Psi_f\rangle|^2 + p|\langle DD\,|\Psi_f\rangle|^2 + t|\langle DC\,|\Psi_f\rangle|^2 + s|\langle CD\,|\Psi_f\rangle|^2. \tag{28}$$

It is sufficient to restrict the strategic space to the 2-parameter set of unitary 2×2 matrices

$$\hat{U}(\theta,\phi) = \begin{pmatrix} e^{i\phi}\cos\theta/2 & \sin\theta/2 \\ -\sin\theta/2 & e^{-i\phi}\cos\theta/2 \end{pmatrix} \tag{29}$$

with $0 \leq \theta \leq \pi$ and $0 \leq \phi \leq \pi/2$. Specificaly, $\hat{C} \equiv \hat{U}(0,0) = \begin{pmatrix} 1 & 0 \\ 0 & 1 \end{pmatrix}$ and $\hat{D} \equiv \hat{U}(\pi,0) = \begin{pmatrix} 0 & 1 \\ -1 & 0 \end{pmatrix}$. $\hat{C}$ and $\hat{D}$ are the operators corresponding to the strategies of cooperation and defection respectively. To ensure that the ordinary PD is faithfully represented in its quantum version, Eisert et al. imposed additional conditions on $\hat{\mathcal{J}}$

$$\left[\hat{\mathcal{J}}, \hat{D}\otimes\hat{D}\right] = 0, \ \left[\hat{\mathcal{J}}, \hat{D}\otimes\hat{C}\right] = 0, \ \left[\hat{\mathcal{J}}, \hat{C}\otimes\hat{D}\right] = 0, \tag{30}$$

so the operator $\hat{\mathcal{J}}$ is

$$\hat{\mathcal{J}} = \exp\{i\gamma\hat{D}\otimes\hat{D}/2\}, \tag{31}$$

where $\gamma \in [0,\pi/2]$ is a real parameter. In fact, γ is a measure for the entanglement of the game. At $\gamma = 0$ the game reduces to its classical form. For a maximally entangled game $\gamma = \pi/2$ the classical NE, $\hat{D}\otimes\hat{D}$ is replaced by a different unique equilibrium $\hat{Q}\otimes\hat{Q}$ with $\hat{Q} \sim \hat{U}(0,\pi/2)$. The new equilibrium is also found to be Pareto optimal, that is, a player cannot increase his/her payoff by deviating from this pair of strategies without reducing the other player's payoff. Classically (C,C) is Pareto optimal, but is not an equilibrium. Eisert et al. claimed that in the quantum version the dilemma in PD disappears and quantum strategies give a superior performance if entanglement is present.

4.3. Marinatto & Weber's Quantum Approach

In *Quantum Approach to Static Games of Complete Information* [37] *Marinatto & Weber's* introduced a new scheme for quantizing bi-matrix games by presenting a quantum version of the Battle of Sexes. In this scheme a state in a $2 \otimes 2$ dimensional Hilbert space is referred to as a strategy. At the start of the game the players are supplied with this strategy. Then players manipulate the strategy. The state is finally measured and the payoffs are rewarded depending on the results of the measurement. A player can do actions within a two-dimensional subspace. Tactics are therefore local actions on a player's qubit. The final measurement, made independently on each qubit, takes into consideration the local nature of players' manipulations.

Suppose ρ_0 is the density operator of the initial strategy which the players A and B receive at the start of the game. A acts with the identity $\hat{I}$ with probability p and with $\hat{\sigma_x}$ with probability $(1-p)$. Similarly with B. It means that each player can modify his own strategy by applying to his reduced part of the total density matrix ρ_0

$$\rho_f^{A(B)} = [p\hat{I}\rho_0^{A(B)}\hat{I}^\dagger + (1-p)\hat{\sigma}_x\rho_0^{A(B)}\hat{\sigma}_x^\dagger]. \tag{32}$$

After the actions of the players the state changes to

$$\begin{aligned}\rho_f = pq\hat{I}_A\otimes\hat{I}_B\rho_0\hat{I}_A^\dagger\otimes\hat{I}_B^\dagger + p(1-q)\hat{I}_A\otimes\hat{\sigma}_{xB}\rho_0\hat{I}_A^\dagger\otimes\hat{\sigma}_{xB}^\dagger \\ +q(1-p)\hat{\sigma}_{xA}\otimes\hat{I}_B\rho_0\hat{\sigma}_{xA}^\dagger\otimes\hat{I}_B^\dagger + (1-p)(1-q)\hat{\sigma}_{xA}\otimes\hat{\sigma}_{xB}\rho_0\hat{\sigma}_{xA}^\dagger\otimes\hat{\sigma}_{xB}^\dagger.\end{aligned} \tag{33}$$

In order to calculate the payoff functions $P_A = Tr\left\{\hat{P}_A \rho_f\right\}$ and $P_B = Tr\left\{\hat{P}_B \rho_f\right\}$ the following payoff operators were defined

$$\hat{P}_A = \alpha_A |00\rangle \langle 00| + \beta_A |01\rangle \langle 01| + \gamma_A |10\rangle \langle 10| + \delta_A |11\rangle \langle 11|,$$
$$\hat{P}_B = \alpha_B |00\rangle \langle 00| + \beta_B |01\rangle \langle 01| + \gamma_B |10\rangle \langle 10| + \delta_B |11\rangle \langle 11|. \quad (34)$$

The scheme presented by Marinatto and Weber differs from the scheme of Eisert et al. due to the absence of the reverse gate $\hat{\mathcal{J}}$ which makes that the classical game remains as a subset of its quantum version. Also in the Marinatto and Weber scheme it is possible to get the same results one obtains in the classical version of our game. Starting from an initial state $|\Psi_0\rangle = |00\rangle$ and allowing the two players to manipulate their own strategy with unitary and unimodular operators or, equivalently, through a particular transformation involving two hermitian operators, one interchanging states $\hat{\sigma_x}$ and the other leaving them unvaried $\hat{I}$. By the other hand, asuming that A and B have at their disposal the entangled state $|\Psi_0\rangle = a|00\rangle + b|11\rangle$ it is shown that both players have the same expected payoff functions making possible to choose a unique Nash equilibrium by discarding the ones which give the players the lesser reward. The entangled strategy can therefore be termed the unique solution of the quantum version of the Battle of the Sexes game.

4.4. Quantum Evolutionary Game Theory

Using Eisert et al. and Marinatto and Weber's quantization schemes, *Iqbal and Toor* [59] investigated the concept of evolutionary stable strategies within the context of quantum games and considered situations where quantization changes ESSs without affecting the corresponding Nash equilibria. In *Evolutionary Stable Strategies in Quantum Games* they investigated the consequences when a small group of mutants using quantum strategies try to invade a classical ESS in a population engaged in a symmetric bimatrix game of prisoners dilemma. The classical pure strategies C and D are realized as $C \sim \hat{U}(0)$, $D \sim \hat{U}(\pi)$ respectively for one-parameter strategies and $C \sim \hat{U}(0,0)$, $D \sim \hat{U}(\pi,0)$ respectively for two-parameter strategies. And considered three cases: (i). A small group of mutants appear equipped with one-parameter quantum strategy $\hat{U}(\theta)$ when D exists as a classical ESS. (ii). The mutants are equipped with two-parameter quantum strategy $\hat{U}(\theta,\phi)$ against classical ESS. (iii). The mutants have successfully invaded and a two-parameter quantum strategy $\hat{Q} \sim \hat{U}(0,\pi/2)$ has established itself as a new quantum ESS. Again another small group of mutants appear using some other two-parameter quantum strategy and try to invade the quantum ESS $\hat{Q}$. The results for these three cases were: (i). The fitness of a one-parameter quantum strategy, which also corresponds to the case of mixed (randomized) classical strategies, cannot be greater than that of a classical ESS. A one-parameter quantum strategy, therefore, cannot succeed to invade a classical ESS. (ii). D is an ESS if $\phi < \arcsin(1/\sqrt{5})$ otherwise the strategy $\hat{U}(\theta,\phi)$ will be in position to invade D. If most of the members of the population play $D \sim \hat{U}(\pi,0)$, then the fitness $W(D)$ will remain greater than the fitness $W[\hat{U}(\theta,\phi)]$ if $\phi < \arcsin(1/\sqrt{5})$. For $\phi > \arcsin(1/\sqrt{5})$ the strategy $\hat{U}(\theta,\phi)$ can invade the strategy D which is an ESS. The possession of a richer strategy by the mutants in this case leads to an invasion of D when $\phi > \arcsin(1/\sqrt{5})$. Mutants having access to richer strategies may seem non-judicious but even in classical setting an advantage by the mutants leading to invasion may be seen in similar context. (iii). A two parameter quantum strategy $\hat{U}(\theta, \phi)$ cannot invade the quantum ESS i.e., the strategy $\hat{Q} \sim \hat{U}(0, \pi/2)$ for this particular game. The mutants having access to richer strategy space remains an advantage not any more now. For the population as well as the mutants $\hat{Q}$ is the unique NE and ESS of the game. The invasion of the mutants in case (ii) does not seem so unusual given the richer structure of strategy space they exploit and they are unable to invade when it does not remain an advantage and most of the population have access to it. For an asymmetric quantum game between two players they showed that a strategy pair can be made an ESS for either classical (using unentangled $|\Psi_0\rangle$) or quantum (using entangled $|\Psi_0\rangle$) version of the game even when the strategy pair remains a Nash equilibrium in both the versions. They showed that in certain types of games entanglement can be used to make appear or disappear ESSs while retaining corresponding Nash equilibria.

In *Quantum mechanics gives stability to a Nash equilibrium* [61], Iqbal and Toor explored evolutionary stability in a modified Rock-Scissors-Paper quantum game. They showed that a mixed strategy NE which is not an ESS in the classical version of the game can be made an ESS when the two players play instead a quantum game by using a selected form of initial entangled state on which they apply the unitary operators in their possession. Quantum mechanics, thus, gives stability to a classical mixed NE against invasion by mutants. Stability against mutants for a mixed classical NE can be made to disappear in certain types of three player symmetric games when players decide to resort to quantum strategies. Stability against mutants in pair-wise contests coming as a result of quantum strategies have been shown a possibility for only pure strategies in certain type of symmetric games. Their results imply the selected method of quantization can bring stability against mutants to a classical mixed NE in pair-wise symmetric contests when the classically available number of pure strategies to a player is increased to three from two. A different behavior is also observed of mixed NE from pure NE in relation to quantization.

Using again Marinatto and Weber's scheme Iqbal et al. [62] analyzed the equilibria of replicator dynamics in quantum games. A well known theorem in evolutionary game theory says that an ESS is an attractor of replicator dynamics but not every attractor is an ESS. The quantization of matrix games can give or take away evolutionary stability to attractors of replicator dynamics when it is the underlying process of the game. They considered the effects of quantization on a saddle or a center of the dynamics. The quantization can be responsible for changing the evolutionary stability of an attractor of the dynamics.

In evolutionary game theory the Bishop-Cannings theorem does not permit pure ESSs when a mixed ESS exists in a bi-matrix game. However, evolutionary stability of a mixed symmetric NE cannot be changed with such a control. Following the approach developed for the quantum version of the rock-scissor-paper game, Iqbal et al. [60] allowed the game to be played with a general form of a two-qubit initial quantum state. It becomes possible to change evolutionary stability of a mixed NE. For a bi-matrix game we worked out a symmetric mixed NE that remains intact in both the classical and quantum versions of the game. For this mixed NE they found conditions making it possible that evolutionary stability of a mixed symmetric NE changes with a switch-over of the game between its two forms, one classical and the other quantum.

In the next sections the present chapter is concluded with the analysis of the relationships between quantum mechanics and game theory, its consequences and applications. We

Table 1.

Quantum Mechanics	Game Theory
n system members	n players
Quantum States	Strategies
Density Operator	Relative Frequencies Vector
Von Neumann Equation	Replicator Dynamics
Von Neumann Entropy	Shannon Entropy
System Equilibrium	Payoff
Maximum Entropy	Maximum Payoff

will propose quantization relationships for a classical game, and the quantum analogues of the replicator dynamics and the Nash equilibrium.

5. Relationships between Quantum Mechanics & Game Theory

Lets analyze some characteristic aspects of quantum mechanics and game theory.

Physics is a mathematical model which describes nature which usually is represented like a system composed by n members. In quantum mechanics we represent and describe the state of each member through quantum states and the state of the ensemble through a density operator. The system evolves following the von Neumann equation. A measure of its order or disorder is the von Neumann entropy which is also a measure of its entanglement. The objective is the equilibrium of the system.

Game theory describes conflictive situations in populations composed by n players (not necessarily people). The state of each player is given by its strategy or its relative frequencies vector which evolves in time following the replicator dynamics. The purpose of the game is a payoff and each member struggle to maximize it.

The clear resemblances and apparent differences between both theories were a motivation to try to find an actual relationship between both systems. Due to our interests, it is important to try to analyze deeply both systems starting from its constituent elements. To start with this analysis it is important to note that the replicator dynamics is a vectorial differential equation while von Neumann equation describes the evolution of an operator (or matrix). So, if we would like to compare both systems the first we would have to do is to try to compare their evolution equations by trying to find a matrix representation of the replicator dynamics. So, the procedure we followed was the next.

5.1. Lax Form of the Replicator Dynamics & Its Properties

As we saw the replicator dynamics is a differential equation where x is a column vector. Obviously, the matrix $U = (Ax)_i - x^T Ax$ has to be diagonal and its elements are given by

$u_{ii} = \sum_{j=1}^{n} a_{ij}x_j - \sum_{k,l=1}^{n} a_{kl}x_k x_l$. The replicator dynamics can be expressed as

$$\frac{dx}{dt} = \left[(Ax)_i - x^T Ax\right] x_i(t) = Ux. \tag{35}$$

By multiplying each element of the vector x by its corresponding $(x_i)^{-1/2}$ in both parts of equation (35) we can get

$$v = U\hat{x}, \tag{36}$$

where v and $\hat{x}$ are column vectors with $v_i = \frac{1}{(x_i)^{1/2}}\frac{dx_i}{dt}$ and $\hat{x}_i = (x_i)^{1/2}$ respectively. Lets multiply the last equation by $\hat{x}^T$ and lets define the matrix

$$G = \frac{1}{2}v\hat{x}^T = \frac{1}{2}U\hat{x}\hat{x}^T, \tag{37}$$

where $g_{ij} = \frac{1}{2}\frac{(x_j)^{1/2}}{(x_i)^{1/2}}\frac{dx_i}{dt} = \left[\sum_{k=1}^{n} a_{ik}x_k - \sum_{k,l=1}^{n} a_{kl}x_k x_l\right](x_i x_j)^{1/2}$. The transpose of the matrix G is equal to $G^T = \frac{1}{2}\hat{x}v^T = \frac{1}{2}\hat{x}\hat{x}^T U^T$, where $g_{ij}^T = \frac{1}{2}\frac{(x_i)^{1/2}}{(x_j)^{1/2}}\frac{dx_j}{dt} = (x_j x_i)^{1/2}\left[\sum_{k=1}^{n} a_{jk}x_k - \sum_{k,l=1}^{n} a_{kl}x_k x_l\right]$ are the elements of G^T. We can define certain matrix X as

$$\frac{dX}{dt} = G + G^T = \frac{1}{2}U\hat{x}\hat{x}^T + \frac{1}{2}\hat{x}\hat{x}^T U^T \tag{38}$$

so X has as elements $x_{ij} = (x_i x_j)^{1/2}$. It can be shown that the elements of $G + G^T$ are given by

$$\left(G + G^T\right)_{ij} = \frac{1}{2}\sum_{k=1}^{n} a_{ik}x_k (x_i x_j)^{1/2} + \frac{1}{2}\sum_{k=1}^{n} a_{jk}x_k (x_j x_i)^{1/2} - \sum_{k,l=1}^{n} a_{kl}x_k x_l (x_i x_j)^{1/2}. \tag{39}$$

Lets call $(G_1)_{ij} = \frac{1}{2}\sum_{k=1}^{n} a_{ik}x_k(x_i x_j)^{1/2}$, $(G_2)_{ij} = \frac{1}{2}\sum_{k=1}^{n} a_{jk}x_k(x_j x_i)^{1/2}$, and $(G_3)_{ij} = \sum_{k,l=1}^{n} a_{kl}x_k x_l (x_i x_j)^{1/2}$ the elements of the matrixes G_1, G_2 and G_3 that compose by adding the matrix $\left(G + G^T\right)$. The matrix G_3 can also be represented as $(G_3)_{ij} = \sum_{l=1}^{n}(x_i x_l)^{1/2}\sum_{k=1}^{n} a_{kl}x_k (x_l x_j)^{1/2}$. G_1, G_2 and G_3 can be factored in function of the matrix X and the diagonal matrix Q, $q_{ii} = \frac{1}{2}\sum_{k=1}^{n} a_{ik}x_k$ so that $G_1 = QX$, $G_2 = XQ$ and $G_3 = 2XQX$. It is easy to show that $X^2 = X$. We can write the equation (38) like

$$\frac{dX}{dt} = QXX + XXQ - 2XQX = \left[[Q, X], X\right] = [\Lambda, X], \tag{40}$$

where $\Lambda = [Q, X]$ and $(\Lambda)_{ij} = \frac{1}{2}\left[\left(\sum_{k=1}^{n} a_{ik}x_k\right)(x_i x_j)^{1/2} - (x_j x_i)^{1/2}\left(\sum_{k=1}^{n} a_{jk}x_k\right)\right]$.

Matrix X has the following properties: $Tr(X) = 1$, $X^2 = X$ and $X^T = X$. Each component of this matrix will evolve following the replicator dynamics so that we could call equation (40) the matrix form of the replicator dynamics.

It is easy to realize that the matrix commutative form of the replicator dynamics (40) follows the same dynamic than the von Neumann equation (15). As will be shown, the properties of their correspondent elements (matrixes) are similar, being the properties corresponding to our quantum system more general than the classical system.

Table 2.

Quantum Statistical Mechanics	Evolutionary Game Theory
n system members	n population members
Each member in the state $\|\Psi_k\rangle$	Each member plays strategy s_i
$\|\Psi_k\rangle$ with $p_k \rightarrow \rho_{ij}$	$s_i \rightarrow x_i$
$\rho, \sum_i \rho_{ii} = 1$	$X, \sum_i x_i = 1$
$i\hbar \frac{d\rho}{dt} = \left[\hat{H}, \rho\right]$	$\frac{dX}{dt} = [\Lambda, X]$
$S = -Tr\{\rho \ln \rho\}$	$H = -\sum_i x_i \ln x_i$

Table 3.

Density Operator	Relative freq. Matrix
ρ is Hermitian	X is Hermitian
$Tr\rho(t) = 1$	$TrX = 1$
$\rho^2(t) \leq \rho(t)$	$X^2 = X$
$Tr\rho^2(t) \leq 1$	$TrX^2(t) = 1$

5.2. Actual Relationships between Quantum Mechanics & Game Theory

The following table shows some specific resemblances between quantum statistical mechanics and evolutionary game theory.

A physical system is modeled mathematically through quantum mechanics while a socioeconomical is modeled through game theory. However it is evident that both systems seem to have a similar behavior. Both are composed by n members (particles, subsystems, players, states, etc.). Each member is described by a state or a strategy which has assigned a determined probability. The quantum mechanical system is described by a density operator ρ whose elements represent the system average probability of being in a determined state. The socioeconomical system is described through a relative frequencies matrix X whose elements represent the frequency of players playing a determined strategy. The evolution equation of the relative frequencies matrix X (which describes our socioeconomical system) is given by the Lax form of the replicator dynamics which follows the same dynamic than the evolution equation of the density operator (i.e., the von Neumann equation). The following table shows how the properties of the matrix that describe the quantum system are more general that the properties of the matrix that describe the classical one.

Although both systems analyzed are described through two theories apparently different both are analogous and thus exactly equivalents.

6. Direct Consequences of the Relationships between Quantum Mechanics & Game Theory

6.1. Quantization Relationships

We can propose the next "quantization relationships"

$$x_i \to \sum_{k=1}^{n} \langle i | \Psi_k \rangle p_k \langle \Psi_k | i \rangle = \rho_{ii},$$
$$(x_i x_j)^{1/2} \to \sum_{k=1}^{n} \langle i | \Psi_k \rangle p_k \langle \Psi_k | j \rangle = \rho_{ij}. \tag{41}$$

A population is represented by a quantum system in which each subpopulation playing strategy s_i is represented by a pure ensemble in the state $|\Psi_k(t)\rangle$ and with probability p_k. The probability x_i of playing strategy s_i or the relative frequency of the individuals using strategy s_i in that population will be represented as the probability ρ_{ii} of finding each pure ensemble in the state $|i\rangle$ [63].

6.2. Quantum Replicator Dynamics

Through the last quantization relationships the replicator dynamics (in matrix commutative form) takes the form of the equation of evolution of mixed states. The von Neumann equation is the quantum analogue of the replicator dynamics. Also $X \to \rho$, $\Lambda \to -\frac{i}{\hbar}\hat{H}$, and $H(x) \to S(\rho)$ [63,65].

6.3. Games Analysis from Quantum Information Theory

If we define an entropy over a random variable S^A (player's A strategic space) [65] which can take the values s_i^A with the respective probabilities x_i^A i.e., $H(A) \equiv -\sum_{i=1}^{n} x_i \log_2 x_i$, we could interpret the entropy of our game as a measure of uncertainty before we learn what strategy player A is going to use. If we do not know what strategy a player is going to use every strategy becomes equally probable and our uncertainty becomes maximum and it is greater while greater is the number of strategies.

If a player B decides to play strategy s_j^B against player A (who plays strategy s_i^A) our total uncertainty about the pair (A, B) can be measured by an external "referee" through the joint entropy of the system $H(A, B) \equiv -\sum_{i,j} x_{ij} \log_2 x_{ij}$, x_{ij} is the joint probability to find A in state s_i and B in state s_j. This is smaller or at least equal than the sum of the uncertainty about A and the uncertainty about B, $H(A, B) \leq H(A) + H(B)$. The interaction and the correlation between A and B reduces the uncertainty due to the sharing of information. There can be more predictability in the whole than in the sum of the parts. The uncertainty decreases while more systems interact jointly creating a new only system.

We can measure how much information A and B share and have an idea of how their strategies or states are correlated by their mutual or correlation entropy $H(A:B) \equiv -\sum_{i,j} x_{ij} \log_2 x_{i:j}$, with $x_{i:j} = \frac{\sum_i x_{ij} \sum_j x_{ij}}{x_{ij}}$. It can be seen easily as $H(A:B) \equiv H(A) + H(B) - H(A, B)$. The joint entropy would equal the sum of each of A's and

B's entropies only in the case that there are no correlations between A's and B's states. In that case, the mutual entropy vanishes and we could not make any predictions about A just from knowing something about B.

If we know that B decides to play strategy s_j^B we can determinate the uncertainty about A through the conditional entropy $H(A \mid B) \equiv H(A, B) - H(B) = -\sum_{i,j} x_{ij} \log_2 x_{i|j}$ with $x_{i|j} = \frac{x_{ij}}{\sum_i x_{ij}}$. If this uncertainty is bigger or equal to zero then the uncertainty about the whole is smaller or at least equal than the uncertainty about A, i.e., $H(A : B) \leq H(A)$. Our uncertainty about the decisions of player A knowing how B and C plays is smaller or at least equal than our uncertainty about the decisions of A knowing only how B plays $H(A \mid B, C) \leq H(A \mid B)$ i.e., conditioning reduces entropy. If the behavior of the players of a game follows a Markov chain i.e., $A \rightarrow B \rightarrow C$ then $H(A) \geq H(A : B) \geq H(A : C)$ i.e., the information can only reduces in time. Also any information C shares with A must be information which C also shares with B, $H(C : B) \geq H(C : A)$.

Two external observers of the same game can measure the difference in their perceptions about certain strategy space of a same player A by its relative entropy. Each of them could define a relative frequency vector, x and y, and the relative entropy over these two probability distributions is a measure of its closeness $H(x \parallel y) \equiv \sum_i x_i \log_2 x_i - \sum_i x_i \log_2 y_i$. We could also suppose that A could be in two possible states i.e., we know that A can play of two specific but different ways and each way has its probability distribution (again x and y that also are known). Suppose that this situation is repeated exactly N times or by N people. We can made certain "measure", experiment or "trick" to determine which the state of the player is. The probability that these two states can be confused is given by the classical or the quantum Sanov's theorem.

6.3.1. Quantum Games Entropy

Classically, the entropy of our system is given by $H = -Tr\left\{X \ln X\right\}$. When the non diagonal elements of matrix X are equal to zero it turns to the Shannon entropy over the elements of the relative frequency vector x, i.e., $H = -\sum_{i=1}^{n} x_i \ln x_i$. By supposing that the vector of relative frequencies $x(t)$ evolves in time following the replicator dynamics the evolution of the entropy of our system would be given by

$$\frac{dH}{dt} = Tr\left\{U(\tilde{H} - X)\right\}, \tag{42}$$

where $U_i = [f_i(x) - \langle f(x) \rangle]$, and $\tilde{H}$ comes from $H = Tr\tilde{H}$. For a quantum system the entropy is given by the von Neumann entropy which in a far from equilibrium system also vary in time until it reaches its maximum value. When the dynamics is chaotic the variation with time of the physical entropy goes through three successive, roughly separated stages [25]. In the first one, $S(t)$ is dependent on the details of the dynamical system and of the initial distribution, and no generic statement can be made. In the second stage, $S(t)$ is a linear increasing function of time ($\frac{dS}{dt} = const.$). In the third stage, $S(t)$ tends asymptotically towards the constant value which characterizes equilibrium ($\frac{dS}{dt} = 0$). With the purpose of calculating the time evolution of entropy we approximated the logarithm of

ρ by series i.e., $\ln \rho = (\rho - I) - \frac{1}{2}(\rho - I)^2 + \frac{1}{3}(\rho - I)^3$... and [65]

$$\frac{dS(t)}{dt} = \frac{11}{6}\sum_i \frac{d\rho_{ii}}{dt} - 6\sum_{i,j} \rho_{ij}\frac{d\rho_{ji}}{dt} + \frac{9}{2}\sum_{i,j,k} \rho_{ij}\rho_{jk}\frac{d\rho_{ki}}{dt} - \frac{4}{3}\sum_{i,j,k,l} \rho_{ij}\rho_{jk}\rho_{kl}\frac{d\rho_{li}}{dt} + \zeta. \quad (43)$$

6.4. Thermodynamical Temperature of a Socioeconomical System

In statistical mechanics, entropy can be regarded as a quantitative measure of disorder. It takes its maximum possible value $\ln n$ in a completely random ensemble in which all quantum mechanical states are equally likely and is equal to zero if ρ is pure i.e., when all its members are characterized by the same quantum mechanical state ket.

Entropy can be maximized subject to different constraints. Generally, the result is a probability distribution function. If we maximize $S(\rho)$ subject to the constraints $\delta Tr(\rho) = 0$ and $\delta \langle E \rangle = 0$ then

$$\rho_{ii} = \frac{e^{-\beta E_i}}{\sum_k e^{-\beta E_k}} \quad (44)$$

which is the condition that the density operator must satisfy to our system tends to maximize its entropy S. Without the internal energy constraint $\delta \langle E \rangle = 0$, $\rho_{ii} = \frac{1}{N}$ which is the $\beta \to 0$ limit ("high-temperature limit") in equation (44) in where a canonical ensemble becomes a completely random ensemble in which all energy eigenstates are equally populated. In the opposite low-temperature limit $\beta \to \infty$ tell us that a canonical ensemble becomes a pure ensemble where only the ground state is populated. The parameter β is related inversely to the "temperature" τ of the system, $\beta = \frac{1}{\tau}$. We can rewrite entropy in terms of the partition function $Z = \sum_k e^{-\beta E_k}$, β and $\langle E \rangle$ via $S = \ln Z + \beta \langle E \rangle$. From the partition function we can know some parameters that define the system like $\langle E \rangle$ and $\langle \Delta E^2 \rangle$. We can also analyze the variation of entropy with respect to the average energy of the system

$$\frac{\partial S}{\partial \langle E \rangle} = \frac{1}{\tau}, \quad (45)$$

$$\frac{\partial^2 S}{\partial \langle E \rangle^2} = -\frac{1}{\tau^2}\frac{\partial \tau}{\partial \langle E \rangle} \quad (46)$$

and with respect to the parameter β

$$\frac{\partial S}{\partial \beta} = -\beta \langle \Delta E^2 \rangle, \quad (47)$$

$$\frac{\partial^2 S}{\partial \beta^2} = \frac{\partial \langle E \rangle}{\partial \beta} + \beta \frac{\partial^2 \langle E \rangle}{\partial \beta^2}. \quad (48)$$

6.5. Econophysics: from Physics to Economics

Why has it been possible to apply some methods of physics to economics and biology? It is a good reason to say that physics is a model which tries to describe phenomena and behaviors and if this model fits and describes almost exactly the observed and the measured even in the economic world then there is no problem or impediment to apply physics to solve problems in economics and biology. But, could economics, biology and physics be correlated? Could it have a relationship between quantum mechanics and game theory? Could quantum mechanics even enclose theories like games and the evolutionary dynamics?

Based in the analysis done in this paper we can conclude that although both systems analyzed (a physical and a socioeconomical) are described through two theories apparently different (quantum mechanics and game theory) both are analogous and thus exactly equivalents. So, we could make use of some of the concepts, laws and definitions in physics for the best understanding of the behavior of economics and biology. Quantum mechanics could be a much more general theory that we had thought. From this point of view many of the equations, concepts and its properties defined quantically must be more general that its classical analogues.

It is important to note that we are dealing with very general and unspecific terms, definitions, and concepts like state, game and system. Look that the state of a system can be its strategy, and the game its behavior. Due to this, the theories that have been developed around these terms (like quantum mechanics, statistical physics, information and game theories) enjoy of this generality quality and could be applicable to model any system depending on what we want to mean for game, state, or system. Once we have defined what system is in our model, we could try to understand what kind of "game" is developing between its members and how they accommodate their "states" in order to get their objectives and also understand the game in terms of temperature, energy, entropy, the properties and laws like if it were a physical system [66–69].

6.6. The Collective Welfare Principle

If our systems are analogous and thus exactly equivalents, our physical equilibrium should be also exactly equivalent to our socioeconomical equilibrium. Morerover, the natural trend of a socioeconomical system should be is to a maximum entropy state.

Based specially on the analogous behavior between quantum mechanical systems and game theoretical systems, it is suggested the following (quantum) understanding of our (classical) system: If in an isolated system each of its accessible states do not have the same probability, the system is not in equilibrium. The system will vary and will evolve in time until it reaches the equilibrium state in where the probability of finding the system in each of the accessible states is the same. The system will find its more probable configuration in which the number of accessible states is maximum and equally probable. The whole system will vary and rearrange its state and the states of its ensembles with the purpose of maximize its entropy and reach its equilibrium state. We could say that the purpose and maximum payoff of a physical system is its maximum entropy state. The system and its members will vary and rearrange themselves to reach the best possible state for each of them which is also the best possible state for the whole system.

This can be seen like a microscopical cooperation between quantum objects to improve their states with the purpose of reaching or maintaining the equilibrium of the system. All the members of our quantum system will play a game in which its maximum payoff is the equilibrium of the system. The members of the system act as a whole besides individuals like they obey a rule in where they prefer the welfare of the collective over the welfare of the individual. This equilibrium is represented in the maximum entropy of the system in where the system resources are fairly distributed over its members. The system is stable only if it maximizes the welfare of the collective above the welfare of the individual. If it is maximized the welfare of the individual above the welfare of the collective the system gets unstable and eventually it collapses *(Collective Welfare Principle)* [63–70].

6.7. The Equilibrium Process Called Globalization

Lets discuss how the world process that it is called "globalization" (i.e., the inexorable integration of markets, currencies, nation-states, technologies and the intensification of consciousness of the world as a whole) has a behavior exactly equivalent to a system that is tending to a maximum entropy state.

6.7.1. Globalization: Big Communities & Strong Currencies

Globalization represents the inexorable integration of markets, nation-states, currencies, technologies [80] and the intensification of consciousness of the world as a whole [81]. This refers to an increasing global connectivity, integration and interdependence in the economic, social, technological, cultural, political, and ecological spheres [82].

Economic globalization can be measured around the four main economic flows that characterize globalization such as goods and services (e.g., exports plus imports as a proportion of national income or per capita of population), labor/people (e.g., net migration rates; inward or outward migration flows, weighted by population), capital (e.g., inward or outward direct investment as a proportion of national income or per head of population), and technology.

Maybe the firsts of these so called big communities were the Unites States of America and the USSR (now the Russian Federation). Both consists in a set or group of different nations or states under the same basic laws or principles (constitution), objectives and an economy characterized by a same economy characterized by a same currency. Although each state or nation is a part of a big community each of them can take its own decisions and its own way of government, policies, laws and punishments (e.g., death penalty) but subject to a constitution (which is no more than a common agreement) and also subject to the decisions of a congress of the community which regulates the whole and the decisions of the parts.

The European Union stands as an example that the world should emulate by its sharing rights, responsibilities, and values, including the obligation to help the less fortunate. The most fundamental of these values is democracy, understood to entail not merely periodic elections, but also active and meaningful participation in decision making, which requires an engaged civil society, strong freedom of information norms, and a vibrant and diversified media that are not controlled by the state or a few oligarchs. The second value is social justice. An economic and political system is to be judged by the extent to which individuals are able to flourish and realize their potential. As individuals, they are part of an ever-widening circle of communities, and they can realize their potential only if they live in harmony with each other. This, in turn, requires a sense of responsibility and solidarity [83].

The meeting of 16 national leaders at the second East Asia Summit (EAS) on the Philippine island of Cebu in January 2007 offered the promise of the politically fractious but economically powerful Asian mega-region one day coalescing into a single meaningful unit [84].

Seth Kaplan has offered the innovative idea of a West African Union (the 15 west african countries stretching from Senegal to Nigeria) to help solve West Africa's deep-rooted problems [85].

In South America has been proposed the creation of a Latin-American Community which is an offer for the integration, the struggle against the poverty and the social exclusion of the countries of Latin America. It is based on the creation of mechanisms to establish cooperative advantages between countries. This would balance the asymmetries between the countries of the hemisphere and the cooperation of funds to correct the inequalities of the weak countries against the powerful nations. The economy ministers of Paraguay, Brazil, Argentina, Ecuador, Venezuela and Bolivia agreed in the "Declaración de Asunción" to create the Bank of the South and invite the rest of countries to add to this project. The Brazilian economy minister Mantega stand out that the new bank is going to consolidate the economic, social and politic block that is appearing in South America and now they have to point to the creation of a common currency. Recently, Uruguay and Colombia have also accepted this offer and is expected the addition of more countries [86].

6.7.2. The Equilibrium Process called Globalization

After analyzing our systems we concluded that a socioeconomical system has a behavior exactly equivalent that a physical system. Both systems evolve in analogous ways and to analogous states. A system where its members are in Nash Equilibrium (or ESS) is exactly equivalent to a system in a maximum entropy state. The stability of the system is based on the maximization of the welfare of the collective above the welfare of the individual. The natural trend of a physical system is to a maximum entropy state, should not a socioeconomical system trend be also to a maximum entropy state which would have to be its state of equilibrium? Has a socioeconomical system something like a "natural trend"?

From our analysis a population can be represented by a quantum system in which each subpopulation playing strategy s_i is represented by a pure ensemble in the state $|\Psi_k(t)\rangle$ and with probability p_k. The probability x_i of playing strategy s_i or the relative frequency of the individuals using strategy s_i in that population is represented as the probability ρ_{ii} of finding each pure ensemble in the state $|i\rangle$. Through these quantization relationships the replicator dynamics takes the form of the equation of evolution of mixed states. The von Neumann equation is the quantum analogue of the replicator dynamics.

Our now "quantum statistical system" composed by quantum objects represented by quantum states which represent the strategies with which "players" interact is characterized by certain interesting physical parameters like temperature, entropy and energy.

In this statistical mixture of ensembles (each ensemble characterized by a state and each state with a determined probability) its natural trend is to its maximum entropy state. If each of its accessible states do not have the same probability, the system will vary and will evolve in time until it reaches the equilibrium state in where the probability of finding the system in each of the accessible states is the same and its number is maximum. In this equilibrium state or maximum entropy state the system "resources" are fairly distributed over its members. Each ensemble is equally probable, is characterized by a same temperature and by a stable state.

Socioeconomically and based on our analysis, our world could be understood as a statistical mixture of "ensembles" (countries for example). Each of these ensembles are characterized by a determined state and a determined probability. But more important, each "country" is characterized by a specific "temperature" which is a measure of the socioeconomical activity of that ensemble. That temperature is related with the interactions between the members of the ensemble. The system will evolve naturally to a maximum entropy state. Each pure ensemble of this statistical mixture will vary and accommodate its state until get the "thermal equilibrium" . First with its nearest neighbors creating new big ensembles characterized each of them by a same temperature. Then through the time, these new big ensembles will seek its "thermal equilibrium" between themselves and with its nearest neighbors creating new bigger ensembles. The system will continue evolving naturally in time until the whole system get an only state characterized by a same "temperature".

This behavior is very similar to what has been called globalization. The process of equilibrium that is absolutely equivalent to a system that is tending to a maximum entropy state is the actual globalization. This analysis predicts the apparition of big common "markets" or (economical, political, social, etc.) communities of countries (European Union, Asian Union, Latin-American Community, African Union, Mideast Community, Russia and USA) and strong common currencies (dollar, euro, yen, sol, etc.). The little and poor economies eventually will be unavoidably absorbed by these "markets" and these currencies. If this process continues these markets or communities will find its "equilibrium" by decreasing its number until reach a state in where there exists only one big common community (or market) and only one common currency around the world [70].

7. Conclusion

We have reviewed the evolution of games analysis from classical through evolutionary and quantum game theory. From the basic concepts and definitions for 2 players we have extended them to a population and its dynamics. We analyzed the equilibria, Nash equilibrium, the evolutionary stable strategies and the collective welfare principle. And some new concepts in game theory like entanglement, entropy and temperature.

Moreover, the relationships between quantum mechanics and game theory and its direct consequences and applications. The correspondence between the replicator dynamics and the von Neumann equation and between the NE and the Collective Welfare Principle. And how quantum mechanics (and physics) could be used to explain more correctly biological and economical processes (econophysics).

Finally, we presented an interesting result consequence from our analysis which proves that the so called "globalization" process (i.e., the inexorable integration of markets, currencies, nation-states, technologies and the intensification of consciousness of the world as a whole) has a behavior exactly equivalent to a system that is tending to a maximum entropy state. This let us predict the apparition of big common markets and strong common currencies that will reach the "equilibrium" by decreasing its number until they get a state characterized by only one common currency and only one big common community around the world.

Acknowledgments

The author (Esteban Guevara) undertook this work with the support of the Abdus Salam International Centre for Theoretical Physics ICTP at Trieste, Italy and its Programme for Training and Research in Italian Laboratories ICTP-TRIL which let him continue with his research at the Center for Nonlinear and Complex Systems at Università degli Studi dell'Insubria in Como.

References

[1] von Neumann, J. & Morgenstern, O. *The Theory of Games and Economic Behavior* (Princeton University Press, Princeton, 1947).

[2] Myerson, R. B. *Game Theory: An Analysis of Conflict* (MIT Press, Cambridge, 1991).

[3] Nowak, M. A. & Sigmund, K. *Nature* **398**, 367 (1999).

[4] Nash, J. Equilibrium Points in N-Person Games, *Proc. Nat. Acad. Sci.* **36**, 48 (1950).

[5] Nash, J. Non-cooperative games, *Annals of Mathematics* **54**, 286 (1951).

[6] Rasmusen, E. *Games and Information* (Blackwell, Cambridge MA, 1989).

[7] Hofbauer, J. & Sigmund, K. *Evolutionary Games and Replicator Dynamics* (Cambridge University Press, Cambridge, UK, 1998).

[8] Smith, J. M. & Price, G. R. The logic of animal conflict, *Nature* **246**, 15 (1973).

[9] Smith, J. M. *Evolution and the Theory of Games* (Cambridge University Press, Cambridge, UK, 1982).

[10] Weibul, J. *Evolutionary Game Theory* (MIT Press, Cambridge, MA, 1995).

[11] Fisher, R. A. *The Genetic Theory of Natural Selection* (Oxford, Clarendon Press, 1930).

[12] Hammerstein, P. & Selten, R. *Game Theory and Evolutionary Biology* (Handbook of Game Theory. Vol 2. Elsevier B.V., 1994).

[13] Cressman, R. *The Stability Concept of Evolutionary Game Theory: A Dynamic Approach* (Springer-Verlag, New York, 1992).

[14] Taylor, P. D. & Jonker, L.B. Evolutionary stable strategies and game dynamics, *Mathematical Biosciences* **40**, 145-156 (1978).

[15] Shannon, C. A mathematical theory of communication. *Bell System Tech. Jour.* **27**, 379-423 (1948).

[16] Wehrl, A. General properties of entropy, *Rev. Mod. Phys.* **50**, 221-260 (1978).

[17] Cover, T. M. & Thomas, J.A. *Elements of Information Theory* (Wiley, New York, 1991).

[18] Landauer, R. Information is physical. *Phys. Today 44*, 23-29 (1991).

[19] Gray, R. *Entropy and Information Theory* (Springer-Verlag, New York, 1990).

[20] von Neumann, J. Thermodynamik quantummechanischer Gesamheiten, *Gött. Nach.* **1** 273-291(1927).

[21] von Neumann, J. *Mathematische Grundlagen der Quantenmechanik* (Springer, Berlin, 1932).

[22] Cerf N. J. & Adami, C. Negative entropy and information in quantum mechanics, *Phys. Rev. Lett.* **79**, 5194-5197(1997).

[23] Nielsen, M. A. & Chuang, I. L. *Quantum Computation and Quantum Information* (Cambridge University Press, Cambridge, 2000).

[24] Benenti, G., Casati G. & Strini G., *Principles of Quantum Computation and Information, Volume 1: Basic Concepts* (Wold Scientific, Singapore, 2004).

[25] Bennett, C. H. & DiVincenzo, D. P. *Nature* **377**, 389 (1995).

[26] DiVincenzo, D. P. *Science* **270**, 255 (1995).

[27] Bennett, C. H. & Wiesner, S. J. *Phys. Rev. Lett.* **69**, 2881 (1992).

[28] Bennett, C. H. et al. *Phys. Rev. Lett.* **70**, 1895 (1993).

[29] Ekert, A. *Nature* **358**, 14 (1992).

[30] Bennett, C. H., Brassard, G. & Mermin, N. D. *Phys. Rev. Lett.* **68**, 557 (1992).

[31] Schumacher, B. & Westmoreland, M. D. *Phys. Rev. A* **56**, 131 (1997).

[32] Holevo, A. S. *IEEE Trans. Inf. Theory* **44**, 269 (1998).

[33] Schumacher, B. *Phys. Rev. A* **51**, 2738 (1995).

[34] Jozsa, R. & Schumacher, B. *J. Mod. Opt.* **41**, 2343 (1994).

[35] Eisert, J., Wilkens & Lewenstein, *M. Phys. Rev. Lett.* **83**, 3077 (1999).

[36] Meyer, D. A. *Phys. Rev. Lett.* **82**, 1052-1055 (1999).

[37] Marinatto, L. & Weber, T. *Phys. Lett. A* **272**, 291 (2000).

[38] Lee, C. F. & Johnson, N. F. *Phys. Rev. A* **67**, 022311 (2003).

[39] Du, J. et al. *Phys. Rev. Lett.* **88**, 137902 (2002).

[40] Flitney, A. P. & Abbott, D. *Proc. R. Soc. (London) A* **459**, 2463-74 (2003).

[41] Shimamura, J. Ozdemir, S. K., Morikoshi F. & Imoto, N. *Int. J. of Quant. Inf.* **2/1**, **79** (2004).

[42] Ozdemir, S. K., Shimamura, J., Morikoshi F. & Imoto, N. *Phys. Lett. A* **333**, 218 (2004).

[43] Piotrowski, E. W. & Sladkowski, J. *Int. J. Theor. Phys.* **42**, 1089 (2003).

[44] Iqbal, A. *Studies in the Theory of Quantum Games*, (PhD thesis, Quaid-i-Azam University, 2004), quant-ph/0503176.

[45] Iqbal, A. *Investigations in quantum games using EPR-type set-ups*, (PhD thesis, University of Hull, 2006), quant-ph/0604188.

[46] Blaquiere, A. Wave mechanics as a two-player game, *Dynamical Systems and Microphysics* **33**, (1980).

[47] Wiesner, S. Conjugate coding, *SIGACT News* **15/1**, 78 (1983).

[48] Mermin, N. D. Quantum mysteries revisited, *Am. J. Phys.* **58**, 731 (1990).

[49] Mermin, N. D. Extreme quantum entanglement in a superposition of macroscopically distinct states, *Phys. Rev. Lett.* **65**, 1838 (1990).

[50] Benjamin S. C. & Hayden, P. M. *Phys. Rev. A* **64**, 030301 (2001).

[51] Lee C. F. & Johnson, N. F. *Parrondo Games and Quantum Algorithms*, quant-ph/0204043.

[52] Moulin, P. & Ivanovic, A. *Proc. of Int. Conf. on Image Process.* **3**, 975-978 (2001).

[53] Cohen, A. S. & Lapidoth, A. *IEEE Trans. on Inf. Theory* **48**, 1639-1667 (2002).

[54] Conway J. & Sloane, N. *IEEE Trans. on Inf. Theory* **32**, 337-348 (1986).

[55] Shen X. M. & Deng, L. *IEEE Trans. on Signal Process.* **45**, 1092-1095 (1997).

[56] Ettinger, J. M. 2nd Inf. Hiding Workshop, Portland, OR, Apr 15-17 (1998).

[57] Pateux S. & Le Guelvouit, G. *Signal Process. Image Comm.* **18**, 283-296 (2003).

[58] DaSilva, L. A. & Srivastava, V. *The First Workshop on Games and Emergent Behaviors in Distributed Computing Environments*, Birmingham, UK, September (2004).

[59] Iqbal, A. & Toor, A. H. Evolutionarily stable strategies in quantum games. *Physics Letters A* **280/5-6**, 249-256 (2001).

[60] Iqbal, A. & Toor, A. H. Entanglement and dynamic stability of Nash equilibria in a symmetric quantum game. *Physics Letters A* **286/4**, 245-250 (2001).

[61] Iqbal, A. & Toor, A. H. Quantum mechanics gives stability to a Nash equilibrium. *Physical Review A* **65**, 022306 (2002).

[62] Iqbal, A. & Toor, A. H. *Equilibria of Replicator Dynamics in Quantum Games*, quant-ph/0106135.

[63] Guevara, E. Quantum Replicator Dynamics, *Physica A* **369/2**, 393-407 (2006).

[64] Guevara, E. *Introduction to the study of Entropy in Quantum Games*, quant-ph/0604170.

[65] Guevara, E. Quantum Games Entropy, *Physica A* **383/2**, 797-804 (2007).

[66] Guevara, E. *The Why of the applicability of Statistical Physics to Economics*, physics/0609088.

[67] Guevara, E. *Quantum Econophysics*, in Proceedings of Quantum Interaction 2007, AAAI Spring Symposia Series, Stanford University, Palo Alto, published by the American Association of Artificial Intelligence, AAAI PRESS Technical Report **SS-07-08**, 158-165 (2007).

[68] Guevara, E. *EGT through Quantum Mechanics & from Statistical Physics to Economics*, arXiv:0705.0029v1.

[69] Guevara, E. *Maximum Entropy, the Collective Welfare Principle and the Globalization Process*, arXiv:0707.1897v1

[70] Guevara, E. *Common Markets, Strong Currencies & the Collective Welfare*, arXiv:0710.1307v1

[71] Piotrowski, E. W. & Sladkowski, J. Quantum Market Games, *Physica A* **312**, 208-216 (2002).

[72] Piotrowski, E. W. & Sladkowski, J. Quantum-like Approach to Financial Risk: Quantum Anthropic Principle, *Acta Phys. Pol. B* **32**, 3873-3879 (2001).

[73] Piotrowski, E. W. & Sladkowski, J. *Interference of quantum market strategies*, Physica A **318**, 516-518 (2003).

[74] Dawkins, R. *The Selfish Gene* (Oxford University Press, Oxford, 1976).

[75] Axelrod, R. *The Evolution of Cooperation* (Basic Books, New York, 1984).

[76] Gogonea, V. & Merz, K. M. Fully quantum mechanical description of proteins in solution - combining linear scaling quantum mechanical methodologies with the Poisson-Boltzmann equation, *J. Phys. Chem. A* **103**, 5171-5188 (1999).

[77] Patel, A. Quantum algorithms and the genetic code, *Pramana* **56**, 367-381 (2001).

[78] Patel, A. Testing Quantum Dynamics in Genetic Information Processing, *Journal of Genetics* **80**, 39 (2001).

[79] Home, D. & Chattopadhayaya, R. *Determination of When an Outcome is Actualised in a Quantum Measurement using DNA - Photolyase System*, quant-ph/9903036.

[80] Friedman, T. L. *The Lexus and the Olive Tree*, **7-8** (1999).

[81] Robertson, R. *Globalization*, **8** (1992).

[82] Wikipedia.

[83] Stiglitz, J. *The European Union's Global Mission*, Joseph Stiglitz talks of Europe's achievements and future challenges. Date Posted on Global Envision: May 14, 2007.

[84] Richardson, B. *An East Asian Community? Not So Fast*. Date Posted on Global Envision: May 02, 2007.

[85] Kaplan, S. *A West African Union*. Date Posted on Global Envision: May 09, 2007.

[86] http://www.alternativabolivariana.org

In: Game Theory: Strategies, Equilibria and Theorems
Editors: Ingrid N. Haugen et al, pp. 117-160
ISBN 978-1-60456-844-8

Chapter 4

ITERATIVE SOLUTION METHODS FOR MIXED EQUILIBRIUM PROBLEMS AND VARIATIONAL INEQUALITIES WITH NON-SMOOTH FUNCTIONS*

Igor V. Konnov†
Department of System Analysis and Information Technologies,
Kazan University, ul. Kremlevskaya, 18,
konn-igor@yandex.ru

Abstract

In this paper, a general class of mixed equilibrium problems involving non - differentiable functions is considered. We describe some recent advances in the theory and solution methods for such mixed equilibrium and variational inequality problems under monotonicity type assumptions. The methods are based on descent and combined relaxation approaches and involve different linesearch procedures. As a result, they do not require a priori information on the problem. Their numerical implementation is further simplified by making use of specific features of the problems, in particular, their decomposable structure. Then methods' parts can be executed in parallel, making them suitable for rather large problems. Some additional examples of applications are also presented.

Keywords. Equilibrium problem; non - differentiable functions; mixed variational inequalities; Nash equilibrium problem; iterative solution methods; descent methods; combined relaxation methods.

1. Introduction

A general equilibrium problem (EP) gives rather a common and suitable format for a lot of problems arising in Economics, Mathematical Physics and Operations Research. The search of an equilibrium state of a system is an actual problem in these fields. Besides, EP

*In this work, the author was supported by RFBR Grant No. 07-01-00499 and by joint RFBR–NNSF Grant No. 07-01-92101.

†E-mail address: konn-igor@yandex.ru

is closely related with other general problems in Nonlinear Analysis, such as fixed point, saddle point, complementarity, and variational inequality problems. Especially, EP can be treated as a normalized form of the Nash equilibrium problem (NEP). NEP represents a way of determining the optimal strategies in various mathematical models of conflict situations where participants (players) behave in a non-cooperative manner. In a natural way, this leads to necessity of finding the corresponding equilibrium points. Since the problem is essentially nonlinear and we do not suppose that utility functions of players are differentiable, its solution can be find via a suitable iterative method.

It is known that iterative solution methods are very powerful and efficient in solving nonlinear optimization problems. Hence, one can try to apply analogues of the known optimization methods to finding equilibrium points. However, most of EPs are not equivalent to optimization problems in the sense that they involve mappings which are non-integrable (i.e. non gradientmaps) in general. For this reason, convergence of such methods is guaranteed under rather restrictive additional assumptions such as smoothness and stability of the set of solutions. Besides, iterative methods requiring a priori information on the problem under solution, say, for computation of stepsizes, may appear non-efficient since these constants are usually known only approximately. Moreover, there exist few approaches for constructing convergent methods for such EPs without differentiability of utility functions. Even non - differentiable optimization problems require special linesearch and direction finding procedures for ensuring convergence of solution methods. On the one hand, most simple methods, such as averaging and iterative regularization methods have low convergence due to the divergent series stepsize rules. On the other hand, more complicated methods such as proximal and level ones involve auxiliary procedures which are very hard to implement and adjust to the possible decomposable structure of NEPs. The fact that most applied problems are typically of large dimensions has stimulated the development of methods which not only do not suffer from these drawbacks, but are also simpler in computational respects.

Thus, in order to construct an effective solution method for non - differentiable EPs one has to take into account all the particular properties of the problem under examination. For this reason, various splitting schemes seem rather suitable for such problems. In the paper, we first describe some recent advances in the theory of mixed equilibrium and variational inequality problems under monotonicity type assumptions. Next, we develop iterative solution methods for solving EPs and related problems, which are based on descent and combined relaxation approaches and involve splitting type direction finding procedures and different linesearch strategies. Hence, they do not require a priori information on the problem. These methods are applied either to the initial problem or to an equivalent mixed variational inequality problem. Convergence of the methods are based on monotonicity type assumptions. Their numerical implementation is further simplified by making use of the specific features of the problems. Owing to the decomposable structure of the problem, methods' parts can be executed in parallel, making them suitable for rather large problems. Some examples of applications are also presented.

A few words about our notation. As usual, we denote by $\mathbb{R}^n$ the real n-dimensional Euclidean space, all elements of such spaces being column vectors represented by a lower case Roman alphabet in boldface, e.g. $\mathbf{x}$. We use superscripts to denote different vectors, the superscript ^{T} denotes transpose. Subscripts are used to denote different scalars or components of vectors. For any vectors $\mathbf{x}$ and $\mathbf{y}$ of $\mathbb{R}^n$, we denote by $\langle \mathbf{x}, \mathbf{y} \rangle$ their scalar product, i.e.,

$$\langle \mathbf{x}, \mathbf{y} \rangle = \mathbf{x}^T \mathbf{y} = \sum_{i=1}^{n} x_i y_i,$$

and by $\|\mathbf{x}\|$ the Euclidean norm of $\mathbf{x}$, i.e., $\|\mathbf{x}\| = \sqrt{\langle \mathbf{x}, \mathbf{x} \rangle}$. The symbol $\square$ is used to denote the end of proofs.

Part I

Equilibrium Problems and Their Properties

In this part, we describe some recent advances in the theory of mixed equilibrium and variational inequality problems under monotonicity type assumptions and their applications to decomposable problems.

2. Mixed Equilibrium and Variational Inequality Problems

Let X be a nonempty, convex, and closed set in the n-dimensional Euclidean space $\mathbb{R}^n$ and let $\Phi : X \times X \to \mathbb{R}$ be an equilibrium bifunction, i.e.,

$$\Phi(\mathbf{x}, \mathbf{x}) = 0 \quad \text{for every } \mathbf{x} \in X.$$

Then one can define the *general equilibrium problem* (EP for short), which is to find an element $\mathbf{x}^* \in X$ such that

$$\Phi(\mathbf{x}^*, \mathbf{y}) \geq 0 \quad \forall \mathbf{y} \in X. \tag{1}$$

Clearly, EP (1) gives rather general and suitable format for many basic problems of Nonlinear Analysis, such as optimization, saddle point, Nash equilibrium, fixed point, and variational inequality ones, which can be viewed as particular cases of EP (1); see e.g. [2,4,24] and references therein. Significant recent achievements in both the theory and solution methods of EPs are described, e.g., in [4,16,24]. Nevertheless, it is well known that extracting certain generic properties of a general problem under consideration and investigation of peculiarities of its subclasses may strengthen essentially the results obtained for general problems. In this work, we consider the following mixed equilibrium problem (MEP for short): Find $\mathbf{x} \in X$ such that

$$f(\mathbf{x}^*, \mathbf{y}) + h(\mathbf{y}) - h(\mathbf{x}^*) \geq 0 \quad \forall \mathbf{y} \in X, \tag{2}$$

where $f : X \times X \to \mathbb{R}$ is a differentiable equilibrium bifunction and $h : X \to \mathbb{R}$ is a convex, but not necessarily differentiable, function. Clearly, we can always represent the bifunction Φ in (1) as follows:

$$\Phi(\mathbf{x}, \mathbf{y}) = \Phi_1(\mathbf{x}, \mathbf{y}) + \Phi_2(\mathbf{x}, \mathbf{y}), \tag{3}$$

where Φ_1 is differentiable and Φ_2 is non differentiable. Hence, MEP (2) can be treated as a particular case of EP (1), (3), where

$$\Phi_1(\mathbf{x}, \mathbf{y}) = f(\mathbf{x}, \mathbf{y}) \text{ and } \Phi_2(\mathbf{x}, \mathbf{y}) = h(\mathbf{y}) - h(\mathbf{x}),$$

i.e., the non-differentiable part has here the special form. In the sequel, we intend to develop solution methods which are adjusted for this special form of problem (2). In order to motivate our work we first reveal relationships of just MEP (2) with some other basic problems and outline possible fields of applications.

Let us consider the mixed variational inequality problem (MVI for short): Find $\mathbf{x}^* \in X$ such that

$$\langle \mathbf{G}(\mathbf{x}^*), \mathbf{y} - \mathbf{x}^* \rangle + h(\mathbf{y}) - h(\mathbf{x}^*) \geq 0 \quad \forall \mathbf{y} \in X, \tag{4}$$

where $\mathbf{G} : X \to \mathbb{R}^n$ is a continuous mapping, $\langle \cdot, \cdot \rangle$ denotes the scalar (inner) product of vectors. This problem was first proposed in [6, 50] and further developed by many authors; see e.g. [10, 24, 55, 59]. We now give an equivalence result between MEP (2) and MVI (4).

Proposition 2.1 *Suppose that $f(\mathbf{x}, \cdot)$ is convex and differentiable for each $\mathbf{x} \in X$. Then MEP (2) is equivalent to MVI (4), where*

$$\mathbf{G}(\mathbf{x}) = f'_{\mathbf{y}}(\mathbf{x}, \mathbf{y})|_{\mathbf{y}=\mathbf{x}}. \tag{5}$$

Proof. If $\mathbf{x}^*$ solves MVI (4), (5), then, by convexity, we have

$$\begin{aligned} f(\mathbf{x}^*, \mathbf{y}) + h(\mathbf{y}) - h(\mathbf{x}^*) &= f(\mathbf{x}^*, \mathbf{y}) - f(\mathbf{x}^*, \mathbf{x}^*) + h(\mathbf{y}) - h(\mathbf{x}^*) \\ &\geq \langle \mathbf{G}(\mathbf{x}^*), \mathbf{y} - \mathbf{x}^* \rangle + h(\mathbf{y}) - h(\mathbf{x}^*) \geq 0 \end{aligned}$$

for every $\mathbf{y} \in X$, i.e. $\mathbf{x}^*$ solves MEP (2). Conversely, if $\mathbf{x}^*$ solves MEP (2), then it solves the optimization problem

$$\min_{\mathbf{y} \in X} \to \{\varphi(\mathbf{y}) + h(\mathbf{y})\}, \tag{6}$$

where

$$\varphi(\mathbf{y}) = f(\mathbf{x}^*, \mathbf{y}).$$

Since the objective function in (6) is convex and φ is differentiable, we can apply Proposition 2.2.2 in [55] and obtain that $\mathbf{x}^*$ solves the problem

$$\langle \varphi'(\mathbf{x}^*), \mathbf{y} - \mathbf{x}^* \rangle + h(\mathbf{y}) - h(\mathbf{x}^*) \geq 0 \quad \forall \mathbf{y} \in X;$$

hence $\mathbf{x}^*$ is a solution to MVI (4). □

We observe that mixed problems of forms (2) and (4) arise after application of the exact (non-smooth) penalty approach to the usual EPs and VIs, having nonlinear constraints; see [25, 46]. Some other applications are given in Section 3..

We now recall some monotonicity concepts for mappings and bifunctions.

Definition 2.1 A mapping $\mathbf{F} : X \to \mathbb{R}^n$ is said to be

(a) *monotone* if, for each pair of points $\mathbf{x}, \mathbf{y} \in X$, it holds that

$$\langle \mathbf{F}(\mathbf{x}) - \mathbf{F}(\mathbf{y}), \mathbf{x} - \mathbf{y} \rangle \geq 0;$$

(b) *strictly monotone*, if, for each pair of points $\mathbf{x}, \mathbf{y} \in X, \mathbf{x} \neq \mathbf{y}$, it holds that

$$\langle \mathbf{F}(\mathbf{x}) - \mathbf{F}(\mathbf{y}), \mathbf{x} - \mathbf{y} \rangle > 0;$$

(c) *strongly monotone* with constant $\tau > 0$ if, for each pair of points $\mathbf{x}, \mathbf{y} \in X$, it holds that

$$\langle \mathbf{F}(\mathbf{x}) - \mathbf{F}(\mathbf{y}), \mathbf{x} - \mathbf{y} \rangle \geq \tau \|\mathbf{x} - \mathbf{y}\|^2;$$

(d) *weakly monotone* with constant $\sigma > 0$ if, for each pair of points $\mathbf{x}, \mathbf{y} \in X$, it holds that

$$\langle \mathbf{F}(\mathbf{x}) - \mathbf{F}(\mathbf{y}), \mathbf{x} - \mathbf{y} \rangle \geq -\sigma \|\mathbf{x} - \mathbf{y}\|^2;$$

(e) *co-coercive* (or *inverse strongly monotone*) with constant $\kappa > 0$ if, for each pair of points $\mathbf{x}, \mathbf{y} \in X$, it holds that

$$\langle \mathbf{F}(\mathbf{x}) - \mathbf{F}(\mathbf{y}), \mathbf{x} - \mathbf{y} \rangle \geq \kappa \|\mathbf{x} - \mathbf{y}\|^2.$$

Clearly, we have $(c) \Longrightarrow (b) \Longrightarrow (a) \Longrightarrow (d)$ and $(c) \Longrightarrow (e) \Longrightarrow (a)$, but the reverse implications do not hold in general. Observe that co-coercivity of $\mathbf{F}$ implies the strong monotonicity of the inverse mapping $\mathbf{F}^{-1}$ and the Lipschitz continuity of $\mathbf{F}$ with constant $1/\kappa$. Also, each Lipschitz continuous mapping with constant L is weakly monotone with constant $1/L$.

Definition 2.2 An equilibrium bifunction $\Psi : X \times X \to \mathbb{R}$ is said to be

(a) *monotone* if, for each pair of points, it holds that

$$\Psi(\mathbf{x}, \mathbf{y}) + \Psi(\mathbf{y}, \mathbf{x}) \leq 0;$$

(b) *strictly monotone*, if, for each pair of points $\mathbf{x}, \mathbf{y} \in X, \mathbf{x} \neq \mathbf{y}$, it holds that

$$\Psi(\mathbf{x}, \mathbf{y}) + \Psi(\mathbf{y}, \mathbf{x}) < 0;$$

(c) *strongly monotone* with constant $\tau > 0$ if, for each pair of points $\mathbf{x}, \mathbf{y} \in X$, it holds that

$$\Psi(\mathbf{x}, \mathbf{y}) + \Psi(\mathbf{y}, \mathbf{x}) \leq -\tau \|\mathbf{x} - \mathbf{y}\|^2;$$

(d) *weakly monotone* with constant $\sigma > 0$ if, for each pair of points $\mathbf{x}, \mathbf{y} \in X$, it holds that

$$\Psi(\mathbf{x}, \mathbf{y}) + \Psi(\mathbf{y}, \mathbf{x}) \leq \sigma \|\mathbf{x} - \mathbf{y}\|^2.$$

Again, we have $(c) \Longrightarrow (b) \Longrightarrow (a) \Longrightarrow (d)$, but the reverse implications are not true in general. Moreover, if we set

$$\Psi(\mathbf{x}, \mathbf{y}) = \langle \mathbf{F}(\mathbf{x}), \mathbf{y} - \mathbf{x} \rangle + \mu(\mathbf{y}) - \mu(\mathbf{x}),$$

where $\mu : X \to \mathbb{R}$ is an arbitrary function, in Definition 2.2, then these properties become equivalent to (a)–(d) in Definition 2.1, respectively.

We also recall some convexity and continuity concepts for functions.

Definition 2.3 A function $\mu : X \to \mathbb{R}$ is said to be

(a) *convex* if, for each pair of points $\mathbf{x}, \mathbf{y} \in X$ and for all $\alpha \in [0, 1]$, it holds that

$$\mu(\alpha \mathbf{x} + (1 - \alpha)\mathbf{y}) \leq \alpha \mu(\mathbf{x}) + (1 - \alpha)\mu(\mathbf{y});$$

(b) *strictly convex*, if, for each pair of points $\mathbf{x}, \mathbf{y} \in X, \mathbf{x} \neq \mathbf{y}$ and for all $\alpha \in (0,1)$, it holds that

$$\mu(\alpha \mathbf{x} + (1-\alpha)\mathbf{y}) < \alpha\mu(\mathbf{x}) + (1-\alpha)\mu(\mathbf{y});$$

(c) *strongly convex* with constant $\tau > 0$ if, for each pair of points $\mathbf{x}, \mathbf{y} \in X$ and for all $\alpha \in [0,1]$, it holds that

$$\mu(\alpha \mathbf{x} + (1-\alpha)\mathbf{y}) \leq \alpha\mu(\mathbf{x}) + (1-\alpha)\mu(\mathbf{y}) - 0.5\tau\alpha(1-\alpha)\|\mathbf{x}-\mathbf{y}\|^2;$$

(d) *weakly convex* with constant $\sigma > 0$ if, for each pair of points $\mathbf{x}, \mathbf{y} \in X$ and for all $\alpha \in [0,1]$, it holds that

$$\mu(\alpha \mathbf{x} + (1-\alpha)\mathbf{y}) \leq \alpha\mu(\mathbf{x}) + (1-\alpha)\mu(\mathbf{y}) + 0.5\sigma\alpha(1-\alpha)\|\mathbf{x}-\mathbf{y}\|^2;$$

(e) *lower* (respectively, *upper*) *semicontinuous*, if for each sequence $\{\mathbf{x}^k\} \to \bar{\mathbf{x}}$, it holds that

$$\liminf_{k\to\infty} \mu(\mathbf{x}^k) \geq \mu(\bar{\mathbf{x}}).$$

(respectively, $\limsup_{k\to\infty} \mu(\mathbf{x}^k) \leq \mu(\bar{\mathbf{x}})$).

The implications $(c) \Longrightarrow (b) \Longrightarrow (a) \Longrightarrow (d)$ are obvious, but the reverse ones are not true again. Additionally, the function μ is said to be *concave* (strictly concave, strongly concave, weakly concave) if the function $-\mu$ is convex (strictly convex, strongly convex, weakly convex).

We now recall a somewhat simplified version of the Minty lemma taken from [2, Section 10.1].

Proposition 2.2 *Suppose that the equilibrium bifunction $\Phi : X \times X \to \mathbb{R}$ is monotone, $\Phi(\cdot, \mathbf{y})$ is upper semicontinuous for each $\mathbf{y} \in X$, $\Phi(\mathbf{x}, \cdot)$ is convex and lower semicontinuous for each $\mathbf{x} \in X$. Then EP (1) is equivalent to the following: Find $\mathbf{y}^* \in X$ such that*

$$\Phi(\mathbf{x}, \mathbf{y}^*) \leq 0 \quad \forall \mathbf{x} \in X, \tag{7}$$

and its solution set is convex and closed.

Observe that problem (7) can be treated as a saddle point (minimax) dual of EP (1); see [32, Section 2.2].

We can apply this assertion to MEP (2) and MVI (4) and obtain similar properties of their solution sets.

Corollary 2.1 *Suppose that the equilibrium bifunction $f : X \times X \to \mathbb{R}$ is monotone and differentiable and that $h : X \to \mathbb{R}$ is a lower semicontinuous convex function. Then MEP (2) is equivalent to the following problem: Find $\mathbf{y}^* \in X$ such that*

$$f(\mathbf{x}, \mathbf{y}^*) + h(\mathbf{y}^*) - h(\mathbf{x}) \leq 0 \quad \forall \mathbf{x} \in X, \tag{8}$$

and its solution set is convex and closed.

Corollary 2.2 *Suppose that the mapping* $\mathbf{G} : X \to \mathbb{R}^n$ *is continuous and monotone and that* $h : X \to \mathbb{R}$ *is a lower semicontinuous convex function. Then MVI (4) is equivalent to the following problem: Find* $\mathbf{y}^* \in X$ *such that*

$$\langle G(\mathbf{x}), \mathbf{x} - \mathbf{y}^* \rangle + h(\mathbf{x}) - h(\mathbf{y}^*) \geq 0 \quad \forall \mathbf{x} \in X, \tag{9}$$

and its solution set is convex and closed.

Also, problems (8) and (9) can be treated as dual ones of MEP (2) and MVI (4), respectively. To prove the above corollaries it suffices to set

$$\Phi(\mathbf{x}, \mathbf{y}) = f(\mathbf{x}, \mathbf{y}) + h(\mathbf{y}) - h(\mathbf{x})$$

and

$$\Phi(\mathbf{x}, \mathbf{y}) = \langle G(\mathbf{x}), \mathbf{y} - \mathbf{x} \rangle + h(\mathbf{y}) - h(\mathbf{x}),$$

respectively, and verify the conditions of Proposition 2.2. We now turn to the relations between monotonicity and convexity properties.

Lemma 2.1 *(see e.g. [61, Chapter 1, Section 1.4]) Let* Y *be an open convex subset of* X*. A differentiable function* $\varphi : X \to \mathbb{R}$ *is strongly convex with constant* τ *(respectively, strictly convex, convex) on* Y*, if and only if its gradient map* $\varphi' : X \to \mathbb{R}^n$ *is strongly monotone with constant* τ *(respectively, strictly monotone, monotone) on* Y*.*

This property remains true for weakly convex functions.

Lemma 2.2 *Let* Y *be an open convex subset of* X*. A differentiable function* $\varphi : X \to \mathbb{R}$ *is weakly convex with constant* σ *on* Y *if and only it its gradient map* $\varphi' : X \to \mathbb{R}^n$ *is weakly monotone with constant* σ *on* Y*.*

Proof. In fact, weak convexity of φ with constant σ is equivalent to convexity of the function

$$\mu(\mathbf{x}) = \varphi(\mathbf{x}) + 0.5\sigma \|\mathbf{x}\|^2$$

since

$$\begin{aligned} \|\alpha\mathbf{x} + (1-\alpha)\mathbf{y}\|^2 &= \alpha^2\|\mathbf{x}\|^2 + 2\alpha(1-\alpha)\langle \mathbf{x}, \mathbf{y}\rangle + (1-\alpha)\|\mathbf{y}\|^2 \\ &= \alpha\|\mathbf{x}\|^2 + (1-\alpha)\|\mathbf{y}\|^2 \\ &\quad -\alpha(1-\alpha)\left\{\|\mathbf{x}\|^2 - 2\langle \mathbf{x}, \mathbf{y}\rangle + \|\mathbf{y}\|^2\right\} \\ &= \alpha\|\mathbf{x}\|^2 + (1-\alpha)\|\mathbf{y}\|^2 - \alpha(1-\alpha)\|\mathbf{x} - \mathbf{y}\|^2. \end{aligned}$$

By Lemma 2.1, it is equivalent to monotonicity of the gradient $\mu' = \varphi' + \sigma I$, where I is the identity map, i.e.

$$\langle \varphi'(\mathbf{x}) - \varphi'(\mathbf{y}), \mathbf{x} - \mathbf{y} \rangle \geq -\sigma\|\mathbf{x} - \mathbf{y}\|^2,$$

and the result follows. □

We now recall another basic property of convex differentiable functions.

Lemma 2.3 *(see e.g. [61, Chapter 1, Section 1.4, Lemma 3]) Let Y be an open convex subset of X. A differentiable function $\varphi : X \to \mathbb{R}$ is convex on Y if and only if, for each point $\mathbf{x} \in X$, it holds that*

$$\varphi(\mathbf{y}) \geq \varphi(\mathbf{x}) + \langle \varphi'(\mathbf{x}), \mathbf{y} - \mathbf{x} \rangle \quad \forall \mathbf{y} \in Y.$$

By using the technique from the proof of Lemma 2.2, we obtain similar properties for weakly and strongly convex functions.

Corollary 2.3 *Let Y be an open convex subset of X. A differentiable function $\varphi : X \to \mathbb{R}$ is strongly convex with constant $\tau > 0$ (respectively, weakly convex with constant $\sigma > 0$) on Y if and only if, for each point $\mathbf{x} \in X$, it holds that*

$$\varphi(\mathbf{y}) \geq \varphi(\mathbf{x}) + \langle \varphi'(\mathbf{x}), \mathbf{y} - \mathbf{x} \rangle + 0.5\tau \|\mathbf{x} - \mathbf{y}\|^2 \quad \forall \mathbf{y} \in Y$$

(respectively,

$$\varphi(\mathbf{y}) \geq \varphi(\mathbf{x}) + \langle \varphi'(\mathbf{x}), \mathbf{y} - \mathbf{x} \rangle - 0.5\sigma \|\mathbf{x} - \mathbf{y}\|^2 \quad \forall \mathbf{y} \in Y).$$

Proof. In fact, φ is strongly convex with constant τ if and only if

$$\mu(\mathbf{x}) = \varphi(\mathbf{x}) - 0.5\tau \|\mathbf{x}\|^2$$

is convex and $\mu' = \varphi' - \tau I$. Applying Lemma 2.3 we see that

$$\begin{aligned} \varphi(\mathbf{y}) - \varphi(\mathbf{x}) &\geq \langle \varphi'(\mathbf{x}), \mathbf{y} - \mathbf{x} \rangle + 0.5\tau \left(\|\mathbf{y}\|^2 - \|\mathbf{x}\|^2 - 2\langle \mathbf{x}, \mathbf{y} - \mathbf{x} \rangle \right) \\ &= \langle \varphi'(\mathbf{x}), \mathbf{y} - \mathbf{x} \rangle + 0.5\tau \|\mathbf{x} - \mathbf{y}\|^2. \end{aligned}$$

Also, φ is weakly convex with constant σ if and only if

$$\eta(\mathbf{x}) = \varphi(\mathbf{x}) + 0.5\sigma \|\mathbf{x}\|^2$$

is convex and the result again follows from Lemma 2.3. □

The monotonicity properties play an important role in both the theory and substantiation of solution methods of MEP's and MVI's. We now discuss the relationship between these properties of the bifunction f and the mapping $\mathbf{G}$ in (5).

Proposition 2.3 *Suppose that $\Phi : X \times X \to \mathbb{R}$ is an equilibrium bifunction, such that $\Phi(\mathbf{x}, \cdot)$ is convex and differentiable for each $\mathbf{x} \in X$. If Φ is monotone (respectively, strictly monotone, strongly monotone with constant τ, weakly monotone with constant σ), then so is $\mathbf{G}$ in (5).*

Proof. The cases when Φ is (strictly, strongly) monotone were substantiated in [24, Proposition 2.1.17]. Clearly, Φ is weakly monotone with constant σ if and only if

$$\tilde{\Phi}(\mathbf{x}, \mathbf{y}) = \Phi(\mathbf{x}, \mathbf{y}) + \sigma \langle \mathbf{x}, \mathbf{y} - \mathbf{x} \rangle$$

is monotone. Hence, by the above, so is

$$\tilde{\mathbf{G}}(\mathbf{x}) = \tilde{\Phi}'_{\mathbf{y}}(\mathbf{x}, \mathbf{y})|_{\mathbf{y}=\mathbf{x}},$$

but

$$\tilde{\mathbf{G}}(\mathbf{x}) = \Phi'_{\mathbf{y}}(\mathbf{x},\mathbf{y})|_{\mathbf{y}=\mathbf{x}} + \sigma\mathbf{x},$$

and we see that

$$\mathbf{G}(\mathbf{x}) = \Phi'_{\mathbf{y}}(\mathbf{x},\mathbf{y})|_{\mathbf{y}=\mathbf{x}} = \tilde{\mathbf{G}}(\mathbf{x}) - \sigma\mathbf{x}$$

must be weakly monotone with constant σ, as desired. □

Observe that the reverse assertions are not true in general as the following simple example, which is a modification of that in [24, Example 3.2.7], illustrates. We recall that if $\mathbf{G}$ is differentiable, then there exists the Jacobian of $\mathbf{G}$ at $\mathbf{x}$, denoted by $\nabla\mathbf{G}(\mathbf{x})$, whose entries are partial derivatives $\frac{\partial \mathbf{G}_i(\mathbf{x})}{\partial x_j}$ for $i, j = 1, \dots, n$.

Example 2.1 Let us consider the so-called normalized equilibrium problem from the known Cournot oligopoly; see Example 3.1, Section 3. below for details. Namely, let

$$\Phi(\mathbf{x},\mathbf{y}) = f_1(x_1, x_2) + f_2(x_1, x_2) - f_1(y_1, x_2) - f_2(x_1, y_2),$$

where $\mathbf{x} = (x_1, x_2)^T, \mathbf{y} = (y_1, y_2)^T$,

$$f_i(x_1, x_2) = x_i\left[\alpha - \beta(x_1 + x_2)\right] - \gamma x_i - \delta, \quad i = 1, 2;\ \alpha, \beta, \gamma, \delta > 0$$

(cf. (19)). Then

$$\mathbf{G}(\mathbf{x}) = (\gamma - \alpha)\mathbf{e} + \beta(x_1 + x_2)\mathbf{e} + \beta\mathbf{x}$$

where $\mathbf{e} = (1, 1)^T$, and the Jacobian

$$\nabla\mathbf{G}(\mathbf{x}) = \beta\begin{pmatrix} 2 & 1 \\ 1 & 2 \end{pmatrix},$$

hence $\mathbf{G}$ is the gradient map of the strongly convex function

$$\mu(\mathbf{x}) = \beta(x_1^2 + x_2^2 + x_1x_2),$$

i.e. $\mathbf{G}$ is integrable and strongly monotone. However,

$$\begin{aligned} \Phi(\mathbf{x},\mathbf{y}) + \Phi(\mathbf{y},\mathbf{x}) &= 2\beta(x_1y_2 + y_1x_2 - x_1x_2 - y_1y_2) \\ &= 2\beta(x_1 - y_1)(y_2 - x_2) > 0 \end{aligned}$$

if $x_1 > y_1$ and $x_2 < y_2$. For instance, if we take $\mathbf{x} = (2, 1)^T$ and $\mathbf{y} = (1, 2)^T$, then $\Phi(\mathbf{x},\mathbf{y}) + \Phi(\mathbf{y},\mathbf{x}) = 2\beta$. If means that Φ is not even monotone.

So, monotonicity properties of mappings allow for covering broader classes of problems and consideration of MVI (4) instead of MEP (2) may give certain advantages.

MVI (4) can be in principle replaced with the equivalent VI with the multi-valued cost mapping $\mathbf{G}(\mathbf{x}) + \partial h(\mathbf{x})$, where

$$\partial h(\mathbf{x}) = \{g \in \mathbb{R}^n \mid h(\mathbf{y}) - h(\mathbf{x}) \geq \langle g, \mathbf{y} - \mathbf{x}\rangle \quad \forall \mathbf{y} \in \mathbb{R}^n\}$$

is the subdifferential of h at $\mathbf{x}$. In fact, let us consider the problem of finding a point $\mathbf{x}^* \in X$ such that

$$\exists \mathbf{t}^* \in \partial h(\mathbf{x}^*), \langle \mathbf{G}(\mathbf{x}^*) + \mathbf{t}^*, \mathbf{y} - \mathbf{x}^*\rangle \geq 0 \quad \forall \mathbf{y} \in X. \tag{10}$$

Proposition 2.4 *(see [59, Proposition 1.3] and [24, Proposition 2.1.18]) Suppose that $h : V \to \mathbb{R}$ is a convex function and V is an open convex set containing X. Then problems (4) and (10) are equivalent.*

Observe that Lemma 2.1 remains true for the case where φ is nondifferentiable, then the gradient map should be replaced by the subdifferential $\partial\varphi$. Hence convexity of h and monotonicity of $\mathbf{G}$ yield the monotonicity of the cost mapping in (10). If $\mathbf{G}$ is defined by (5), then (10) becomes equivalent to MEP (2). However, MVI (4) seems more preferable since its formulation does not contain multi-valued mappings explicitly. Besides, there exist continuous convex functions whose subdifferentials can be empty at some points; e.g. $\varphi(x) = -\sqrt{1-x^2}$ on $X = [-1, 1]$ (see [64, pp.95, 96, and 104]). For this reason, we will consider mainly formulations (2) and (4).

In addition, we give the known monotonicity criteria for continuously differentiable mappings; see e.g. [11, Proposition 2.3.2].

Lemma 2.4 *Let Y be an open convex subset of X and let $\mathbf{G} : X \to \mathbb{R}^n$ be continuously differentiable on Y. Then:*

(a) $\mathbf{G}$ is monotone on Y if and only if $\nabla\mathbf{G}$ is positive semidefinite on Y;

(b) $\mathbf{G}$ is strictly monotone on Y if $\nabla\mathbf{G}$ is positive definite on Y;

(c) $\mathbf{G}$ is strongly monotone on Y with constant τ if and only if it holds that

$$\langle \mathbf{p}, \nabla\mathbf{G}(\mathbf{x})\mathbf{p}\rangle \geq \tau\|\mathbf{p}\|^2 \qquad \forall \mathbf{p} \in \mathbb{R}^n, \mathbf{x} \in Y.$$

Note that the Jacobian of a differentiable strictly monotone mapping need not be positive definite.

We close this section by a useful equivalence result for EP (1).

Lemma 2.5 *Suppose that $\Phi : X \times X \to \mathbb{R}$ is an equilibrium bifunction such that $\Phi(\mathbf{x}, \cdot)$ is convex for each $\mathbf{x} \in X$, $\Gamma : X \times X \to \mathbb{R}$ is an equilibrium bifunction such that $\Gamma(\mathbf{x}, \mathbf{y}) \geq 0$ for all $(\mathbf{x}, \mathbf{y}) \in X \times X$, $\Gamma(\mathbf{x}, \cdot)$ is convex and differentiable and $\Gamma'_{\mathbf{y}}(\mathbf{x}, \mathbf{y})|_{\mathbf{y}=\mathbf{x}} = \mathbf{0}$ for each $\mathbf{x} \in X$. Then EP (1) is equivalent to the perturbed problem: Find $\mathbf{x}^* \in X$ such that*

$$\Phi(\mathbf{x}^*, \mathbf{y}) + \Gamma(\mathbf{x}^*, \mathbf{y}) \geq 0 \quad \forall \mathbf{y} \in X. \tag{11}$$

Proof. Since $\Gamma(\mathbf{x}, \mathbf{y})$ is always non-negative, the implication (1) $\Longrightarrow$ (11) is obvious. Conversely, let $\mathbf{x}^*$ be a solution to (11). Then it solves the optimization problem

$$\min_{\mathbf{y} \in X} \to \{\varphi_1(\mathbf{y}) + \varphi_2(\mathbf{y})\},$$

where $\varphi_1(\mathbf{y}) = \Phi(\mathbf{x}^*, \mathbf{y})$, $\varphi_2(\mathbf{y}) = \Gamma(\mathbf{x}^*, \mathbf{y})$. Since φ_1 and φ_2 are convex and φ_2 is differentiable, we can apply Proposition 2.2.2 in [55] and obtain that $\mathbf{x}^*$ solves the problem

$$\varphi_1(\mathbf{y}) - \varphi_1(\mathbf{x}^*) + \langle \varphi_2'(\mathbf{x}^*), \mathbf{y} - \mathbf{x}^*\rangle \geq 0 \quad \forall \mathbf{y} \in X,$$

but it is clearly equivalent to (1). □

3. Decomposable Mixed Equilibrium Problems

We now consider some applications of equilibrium problems which have a decomposable structure. We start our considerations from non-cooperative games.

We recall that such a game model contains m players so that the i-th player has the strategy set X_i and the utility function $\tilde{\varphi}_i : X \to \mathbb{R}$, where

$$X = X_1 \times \ldots \times X_m, \tag{12}$$

for $i = 1, \ldots, m$. We suppose that $X_i \subseteq \mathbb{R}^{n_i}, i = 1, \ldots, m$ and $n = \sum\limits_{i=1}^{m} n_i$. Hence we can consider the corresponding partition of any vector $\mathbf{x} = (\mathbf{x}_1, \ldots, \mathbf{x}_m)$ with $\mathbf{x}_i \in \mathbb{R}^{n_i}$ for $i = 1, \ldots, m$. For brevity, set

$$(\mathbf{x}_{-i}, \mathbf{y}_i) = (\mathbf{x}_1, \ldots, \mathbf{x}_{i-1}, \mathbf{y}_i, \mathbf{x}_{i+1}, \ldots, \mathbf{x}_m).$$

We recall that a point $\mathbf{x}^* = (\mathbf{x}_1^*, \ldots, \mathbf{x}_m^*) \in X$ is said to be a *Nash equilibrium point* (NEP) of this game if

$$\tilde{\varphi}_i(\mathbf{x}_{-i}^*, \mathbf{y}_i) \le \tilde{\varphi}_i(\mathbf{x}^*) \quad \forall \mathbf{y}_i \in X_i, i = 1, \ldots, m. \tag{13}$$

By using the approach from [53], we can convert this problem into EP of form (1). In fact, let us define the normalized bifunction

$$\tilde{\Phi}(\mathbf{x}, \mathbf{y}) = \tilde{\Psi}(\mathbf{x}, \mathbf{y}) - \tilde{\Psi}(\mathbf{x}, \mathbf{x}), \tag{14}$$

where

$$\tilde{\Psi}(\mathbf{x}, \mathbf{y}) = -\sum_{i=1}^{m} \tilde{\varphi}_i(\mathbf{x}_{-i}, \mathbf{y}_i). \tag{15}$$

We then obtain that problem (13) becomes equivalent to the problem: Find $\mathbf{x}^* \in X$ defined in (12) such that

$$\tilde{\Phi}(\mathbf{x}^*, \mathbf{y}) \ge 0 \quad \forall \mathbf{y} \in X \tag{16}$$

(cf. (1)). If each utility function $\tilde{\varphi}_i$ is concave in $\mathbf{x}_i$, then $\tilde{\Phi}(\mathbf{x}, \cdot)$ is clearly convex. Next, let us consider a class of non-cooperative games with the following utility functions

$$\tilde{\varphi}_i(\mathbf{x}) = \varphi_i(\mathbf{x}) + \psi_i(\mathbf{x}_i), \tag{17}$$

where $\varphi_i : X \to \mathbb{R}$ is concave and differentiable in $\mathbf{x}_i$ and $\psi_i : X_i \to \mathbb{R}$ is concave; see e.g. [59, Example 1.6]. Then problem (12)–(16) becomes equivalent to MEP (2), where

$$\begin{aligned} f(\mathbf{x}, \mathbf{y}) &= \sum_{i=1}^{m} f_i(\mathbf{x}, \mathbf{y}_i), \ f_i(\mathbf{x}, \mathbf{y}_i) = \varphi_i(\mathbf{x}) - \varphi_i(\mathbf{x}_{-i}, \mathbf{y}_i), i = 1, \ldots, m; \\ h(\mathbf{x}) &= \sum_{i=1}^{m} h_i(\mathbf{x}_i), \ h_i(\mathbf{x}_i) = -\psi_i(\mathbf{x}_i), i = 1, \ldots, m. \end{aligned} \tag{18}$$

Non-cooperative games with similar utility functions are typical for oligopolistic competition problems; see e.g. [48, 54].

Example 3.1 Let us consider an oligopolistic market in which n firms supply a homogeneous commodity. Let $p(\sigma)$ denote the inverse demand function, i.e., it is the price at which consumers will purchase a quantity σ. If each i-th firm supplies x_i units of the product, then the total supply is defined by

$$\sigma_{\mathbf{x}} = \sum_{i=1}^{n} x_i.$$

Next, let $h_i(x_i)$ be the total cost of supplying x_i units of the commodity for the i-th firm. Suppose that each h_i is a convex, but not necessarily differentiable function. This means that firms can utilize different technologies of production. Then we obtain the non-cooperative game of n players where each i-th player has the strategy set $\mathbb{R}_+ = \{\alpha \mid \alpha \geq 0\}$ and the utility function

$$\tilde{\varphi}_i(\mathbf{x}) = x_i p(\sigma_{\mathbf{x}}) - h_i(x_i), \quad \text{for} \quad i = 1, \ldots, n. \tag{19}$$

Hence, we have a particular case of MEP (2), (18) where $m = n, \varphi_i(\mathbf{x}) = x_i p(\sigma_{\mathbf{x}})$ for $i = 1, \ldots, n$.

Now we can write the decomposable form of MEP (2): Find $\mathbf{x}^* \in X = X_1 \times \ldots \times X_m$ such that

$$\sum_{i=1}^{m} f_i(\mathbf{x}^*, \mathbf{y}_i) + \sum_{i=1}^{m} [h_i(\mathbf{y}_i) - h_i(\mathbf{x}_i^*)] \geq 0 \quad \forall \mathbf{y}_i \in X_i, \quad i = 1, \ldots, m. \tag{20}$$

Therefore,

$$f(\mathbf{x}, \mathbf{y}) = \sum_{i=1}^{m} f_i(\mathbf{x}, \mathbf{y}_i)$$

is an equilibrium bifunction which is convex and differentiable in $\mathbf{y}$. It is easy to see that MEP (20) is equivalent to the system of partial MEP's: Find $\mathbf{x}^* \in X = X_1 \times \ldots \times X_m$ such that

$$f_i(\mathbf{x}^*, \mathbf{y}_i) + h_i(\mathbf{y}_i) - h_i(\mathbf{x}_i^*) \geq 0 \quad \forall \mathbf{y}_i \in X_i, \quad \text{for } i = 1, \ldots, m. \tag{21}$$

The decomposable form of MVI (4) can be defined similarly: Find $\mathbf{x}^* \in X = X_1 \times \ldots \times X_m$ such that

$$\begin{gathered}\sum_{i=1}^{m} \langle \mathbf{G}_i(\mathbf{x}^*), \mathbf{y}_i - \mathbf{x}_i^* \rangle + \sum_{i=1}^{m} [h_i(\mathbf{y}_i) - h_i(\mathbf{x}_i^*)] \geq 0 \\ \forall \mathbf{y}_i \in X_i, \quad \text{for } i = 1, \ldots, m;\end{gathered} \tag{22}$$

where

$$\mathbf{G}(\mathbf{x}) = (\mathbf{G}_1(\mathbf{x}), \ldots, \mathbf{G}_m(\mathbf{x})), \ \mathbf{G}_i : X \to \mathbb{R}^{n_i}, \quad i = 1, \ldots, m.$$

Also, MVI (22) is equivalent to the system of partial MVI's: Find $\mathbf{x}^* \in X = X_1 \times \ldots \times X_m$ such that

$$\begin{gathered}\langle \mathbf{G}_i(\mathbf{x}^*), \mathbf{y}_i - \mathbf{x}_i^* \rangle + h_i(\mathbf{y}_i) - h_i(\mathbf{x}_i^*) \geq 0 \\ \forall \mathbf{y}_i \in X_i, \quad \text{for } i = 1, \ldots, m.\end{gathered} \tag{23}$$

The equivalence result between MEP (20) and MVI (22) follows from Proposition 2.1.

Proposition 3.1 *Suppose that the functions $f_i(\mathbf{x}, \cdot), i = 1, \dots, m$ are convex and differentiable for each $\mathbf{x} \in X$, and that the functions $h_i : \mathbb{R}^{n_i} \to \mathbb{R}, i = 1, \dots, m$ are convex. Then MEP (20) is equivalent to MVI (22), where*

$$\mathbf{G}_i(\mathbf{x}) = \frac{\partial f_i(\mathbf{x}, \mathbf{y}_i)}{\partial \mathbf{y}_i} \quad \text{for } i = 1, \dots, m. \tag{24}$$

Clearly, then the systems (21) and (23) also become equivalent.

Let us consider the particular case of problem (13), (17), when $m = 2, \varphi_1(\mathbf{x}) = -\varphi_2(\mathbf{x})$. It is easy to see that we then obtain the saddle point problem: Find a pair $\mathbf{x}^* = (\mathbf{x}_1^*, \mathbf{x}_2^*) \in X_1 \times X_2$ such that

$$M(\mathbf{x}_1, \mathbf{x}_2^*) \leq M(\mathbf{x}_1^*, \mathbf{x}_2^*) \leq M(\mathbf{x}_1^*, \mathbf{x}_2) \quad \forall \mathbf{x}_1 \in X_1, \forall \mathbf{x}_2 \in X_2, \tag{25}$$

where

$$M(\mathbf{x}_1, \mathbf{x}_2) = \varphi_1(\mathbf{x}) + \psi_1(\mathbf{x}_1) - \psi_2(\mathbf{x}_2). \tag{26}$$

Note that the game is not antagonistic, i.e.

$$\tilde{\varphi}_1(\mathbf{x}) \neq -\tilde{\varphi}_2(\mathbf{x}).$$

If the bifunction $\varphi_1 : X_1 \times X_2 \to \mathbb{R}$ is concave in $\mathbf{x}_1$ and convex in $\mathbf{x}_2$, and the functions h_1 and h_2 are concave, then the bifunction $M : X_1 \times X_2 \to \mathbb{R}$ becomes also concave-convex. Suppose in addition that the bifunction $\varphi_1 : X_1 \times X_2 \to \mathbb{R}$ is differentiable. By setting

$$\begin{array}{l} f(\mathbf{x}, \mathbf{y}) = f_1(\mathbf{y}_1, \mathbf{x}_2) + f_2(\mathbf{x}_1, \mathbf{y}_2), \\ f_1(\mathbf{x}) = -\varphi_1(\mathbf{x}), \ f_2(\mathbf{x}) = \varphi_1(\mathbf{x}), \\ h_1(\mathbf{x}_1) = -\psi_1(\mathbf{x}_1), \ h_2(\mathbf{x}_2) = -\psi_2(\mathbf{x}_2) \end{array} \tag{27}$$

(cf. (18)), problem (25)–(26) then reduces to MEP (20), where $m = 2$, the equilibrium bifunction $f : X \times X \to \mathbb{R}$ is differentiable and concave-convex and the functions $h_1 : X_1 \to \mathbb{R}$ and $h_2 : X_2 \to \mathbb{R}$ are convex under the above assumptions, hence, by Proposition 3.1, hence problem (25)–(27) becomes equivalent to MVI (22), (24) with $m = 2$, where the mapping G defined by

$$\mathbf{G}(\mathbf{x}) = \left(-\frac{\partial \varphi_1(\mathbf{x})}{\partial \mathbf{x}_1}, \frac{\partial \varphi_2(\mathbf{x})}{\partial \mathbf{x}_2} \right)$$

is monotone. Moreover, if φ_1 is strictly (respectively, strongly) concave-convex, then $\mathbf{G}$ is strictly (respectively, strongly) monotone (see [62, Chapter VII] and [65, Chapter 5]). Observe that in the general case where $m > 2$ we can not guarantee the concavity-convexity of the bifunction $\tilde{\Phi}$ defined in (14)–(15), but taking into account Proposition 2.3, we can deduce the monotonicity properties of $\mathbf{G}$ in (24) from similar ones of f in (18) if $f(\mathbf{x}, \cdot)$ is still convex.

We now give two additional examples of saddle point problems of form (25)–(26).

First we consider the constrained convex optimization problem:

$$\min_{\mathbf{u} \in D} \to f_0(\mathbf{u}), \tag{28}$$

where

$$D = \{\mathbf{u} \in U \mid f_i(\mathbf{u}) \leq 0 \quad i = 1, \dots, m\}, \tag{29}$$

U is a convex set in $\mathbb{R}^l$, $f_0 : U \to \mathbb{R}$ is a convex function, and $f_i : U \to \mathbb{R}, i = 1, \dots, m$, are convex differentiable functions. Then we can consider the saddle point problem: Find a pair $(\mathbf{u}^*, \mathbf{v}^*) \in U \times \mathbb{R}^m_+$ such that

$$L(\mathbf{u}^*, \mathbf{v}) \le L(\mathbf{u}^*, \mathbf{v}^*) \le L(\mathbf{u}, \mathbf{v}^*) \quad \forall \mathbf{u} \in U, \forall \mathbf{v} \in \mathbb{R}^m_+, \tag{30}$$

where

$$L(\mathbf{u}, \mathbf{v}) = f_0(\mathbf{u}) + \langle \mathbf{v}, \mathbf{f}(\mathbf{u}) \rangle \tag{31}$$

is the Lagrange function for problem (28)–(29), $\mathbf{f}(\mathbf{u}) = (f_1(\mathbf{u}), \dots, f_m(\mathbf{u}))^T$. It is well-known (see e.g. [61]) that each saddle point $(\mathbf{u}^*, \mathbf{v}^*)$ in (30)–(31) yields the solution $\mathbf{u}^*$ of problem (28)–(29). Conversely, if $\mathbf{u}^*$ solve (28)–(29) and certain constraint qualification holds, then there exists $\mathbf{v}^* \in \mathbb{R}^m_+$ such that $\mathbf{u}^*, \mathbf{v}^*$ constitutes a saddle point in (30)–(31). Thus, the saddle point problem gives a suitable way of eliminating nonlinear constraints. Note that the bifunction L in (31) is convex-concave and that f_0 clearly a particular case of problem (25)–(26) with $\mathbf{x}_1 = \mathbf{u}, \mathbf{x}_2 = \mathbf{v}$, $M(\mathbf{x}_1, \mathbf{x}_2) = -L(\mathbf{u}, \mathbf{v})$, $\psi_1(\mathbf{x}_1) = -f_0(\mathbf{u})$, $\psi_2(\mathbf{x}_2) \equiv 0$, $\varphi_1(\mathbf{x}_1, \mathbf{x}_2) = -\langle \mathbf{v}, \mathbf{f}(\mathbf{u}) \rangle$.

Next, let us consider again the saddle point problem (30) where

$$L(\mathbf{u}, \mathbf{v}) = f_0(\mathbf{u}) + \langle \mathbf{v}, \mathbf{f}(\mathbf{u}) \rangle - \mu(\mathbf{v}), \tag{32}$$

where $f_i : U \to \mathbb{R}, i = 1, \dots, m$, were defined above, $\mu : \mathbb{R}^m \to \mathbb{R}$ is a convex function. It is known that many equilibrium problems in Economics, Transportation and Operations Research can be formulated as problem (30), (32); see e.g. [26, 29, 31]. Clearly, it reduces to (30)–(31) if $\mu \equiv 0$. At the same time, if $(\mathbf{u}^*, \mathbf{v}^*)$ is a solution to problem (30), (32), then $\mathbf{u}^*$ solves the implicit optimization problem:

$$\min_{\mathbf{u} \in \tilde{D}} \to f_0(\mathbf{u}),$$

where

$$\tilde{D} = \{\mathbf{u} \in U \mid f_i(\mathbf{u}) \le \mathbf{b}^*, i = 1, \dots, m, \quad \mathbf{b}^* \in \partial \mu(\mathbf{v}^*)\}.$$

Clearly, (30), (32) is also a particular case of problem (25)–(26), where $\mathbf{x}_1 = \mathbf{u}$, $\mathbf{x}_2 = \mathbf{v}$, $M(\mathbf{x}_1, \mathbf{x}_2) = -L(\mathbf{u}, \mathbf{v})$, $\psi_1(\mathbf{x}_1) = -f_0(\mathbf{u})$, $\psi_2(\mathbf{x}_2) = -\mu(\mathbf{v})$, $\varphi_1(\mathbf{x}_1, \mathbf{x}_2) = -\langle \mathbf{v}, \mathbf{f}(\mathbf{u}) \rangle$ with M being concave-convex and ψ_i being non-differentiable.

4. Existence and Uniqueness Results

A great number of works are traditionally devoted to the theory of equilibrium problems and variational inequalities; see e.g. [2, 4, 16] and references therein. We first recall the existence result for EP (1), which was obtained by Ky Fan [12] and extended by H. Brézis, L. Nirenberg, and G. Stampacchia [5].

Proposition 4.1 *Suppose that X is a nonempty, convex, and closed subset in $\mathbb{R}^n$, $\Phi : X \times X \to \mathbb{R}$ is an equilibrium bifunction such that $\Phi(\cdot, \mathbf{y})$ is upper semicontinuous for each $\mathbf{y} \in X$, $\Phi(\mathbf{x}, \cdot)$ is convex for each $\mathbf{x} \in X$. Suppose also that at least one of the following assumptions holds:*

(a) X is bounded;

(b) there exists a nonempty bounded subset $\tilde{X}$ of X such that for every $\mathbf{x} \in X \tilde{X}$ there is $\tilde{\mathbf{x}} \in \tilde{X}$ with $\Phi(\mathbf{x}, \tilde{\mathbf{x}}) < 0$.

Then EP (1) has a solution.

Observe that there exist more general coercivity conditions in comparison with that in (b), which are utilized for ensuring the existence of a solution on unbounded feasible sets; see e.g. [4,38].

Now we turn to uniqueness results.

Proposition 4.2 *Suppose that X is a convex set in $\mathbb{R}^n$, $\Phi : X \times X \to \mathbb{R}$ is an equilibrium bifunction. Suppose also that at least one of the following assumptions holds:*

(a) Φ is strictly monotone;

(b) Φ is monotone and $\Phi(\mathbf{x}, \cdot)$ is strictly convex for each $\mathbf{x} \in X$.

Then EP (1) has at most one solution.

Proof. Suppose that $\mathbf{x}'$ and $\mathbf{x}''$ are solutions of EP (1) and $\mathbf{x}' \neq \mathbf{x}''$. In case (a) we have

$$\Phi(\mathbf{x}', \mathbf{x}'') \geq 0 \quad \text{and} \quad \Phi(\mathbf{x}'', \mathbf{x}') \geq 0,$$

which contradicts the strict monotonicity of Φ.

In case (b), set $\tilde{\mathbf{x}} = 0.5(\mathbf{x}' + \mathbf{x}'')$. Then $0 \leq \Phi(\mathbf{x}'', \tilde{\mathbf{x}}) < 0.5\Phi(\mathbf{x}', \mathbf{x}') + 0.5\Phi(\mathbf{x}', \mathbf{x}'') = 0.5\Phi(\mathbf{x}', \mathbf{x}'')$. However, $\Phi(\mathbf{x}'', \mathbf{x}') \geq 0$, and, by monotonicity, $\Phi(\mathbf{x}', \mathbf{x}'') \leq 0$, which is a contradiction. □

Now we give a joint existence and uniqueness result, which requires strengthened monotonicity/convexity properties.

Proposition 4.3 *Suppose that X is a nonempty, convex, and closed subset in $\mathbb{R}^n$, $\Phi : X \times X \to \mathbb{R}$ is an equilibrium bifunction such that $\Phi(\cdot, \mathbf{y})$ is upper semicontinuous for each $\mathbf{y} \in X$, $\Phi(\mathbf{x}, \cdot)$ is convex and lower semicontinuous for each $\mathbf{x} \in X$. Suppose also that at least one of the following assumptions holds:*

(a) Φ is strongly monotone;

(b) Φ is monotone and $\Phi(\mathbf{x}, \cdot)$ is strongly convex for each $\mathbf{x} \in X$.

Then EP (1) has a unique solution.

Proof. In case (a), the proof is given e.g. in [24, Proposition 2.1.16]. In case (b), if $\Phi(\mathbf{x}, \cdot)$ is strongly convex with constant $\tau > 0$, then the bifunction

$$\tilde{\Phi}(\mathbf{x}, \mathbf{y}) = \Phi(\mathbf{x}, \mathbf{y}) - 0.5\tau\|\mathbf{x} - \mathbf{y}\|^2$$

is clearly an equilibrium bifunction, which is strongly monotone with constant $\tau > 0$. Moreover, $\tilde{\Phi}(\mathbf{x}, \cdot)$ is still convex and lower semicontinuous for each $\mathbf{x} \in X$, and $\tilde{\Phi}(\cdot, \mathbf{y})$ is upper semicontinuous for each $\mathbf{y} \in X$. Hence, due to (a), there exists a unique point $\mathbf{x}^* \in X$ such that

$$\tilde{\Phi}(\mathbf{x}^*, \mathbf{y}) \geq 0 \quad \forall \mathbf{y} \in X.$$

However, on account of Lemma 2.5, this EP is equivalent to EP (1) and the result follows.
□

Now we specialize the above results for MEP (2). Let us consider the following basic assumptions.

(A1) *X is a nonempty, convex and closed subset in $\mathbb{R}^n$.*

(A2) *$f : X \times X \to \mathbb{R}$ is an equilibrium bifunction such that $f(\cdot, \mathbf{y})$ is upper semicontinuous for each $\mathbf{y} \in X$, $f(\mathbf{x}, \cdot)$ is convex and lower semicontinuous for each $\mathbf{x} \in X$.*

(A3) *$h : X \to \mathbb{R}$ is a lower semicontinuous convex function.*

Corollary 4.1 *Let (A1)–(A3) hold. If either*

(a) X is bounded,

or

(b) there exists a nonempty bounded subset $\tilde{X}$ of X such that for each every $\mathbf{y} \in X \setminus \tilde{X}$ there is $\tilde{\mathbf{x}} \in \tilde{X}$ with $f(\mathbf{x}, \tilde{\mathbf{x}}) + h(\mathbf{x}) < 0$,

then MEP (2) has a solution.

Corollary 4.2 *Let (A1) and (A2) hold. Suppose that at least one of the following assumptions holds:*

(a) f is strictly monotone;

(b) (A3) is fulfilled, f is monotone, $f(\mathbf{x}, \cdot)$ is strictly convex for each $\mathbf{x} \in X$;

(c) (A3) is fulfilled, f is monotone, h is strictly convex.

Then MEP (2) has at most one solution.

Corollary 4.3 *Let (A1)–(A3) hold. Suppose also that at least one of the following assumptions hold:*

(a) f is strongly monotone;

(b) f is monotone, $f(\mathbf{x}, \cdot)$ is strongly convex.

Then MEP (2) has a unique solution.

The results can be adjusted easily for MVI (4) as well.

Part II

Iterative Methods for Mixed Variational Inequalities

In this part we consider iterative solution methods for MVI (4). First of all, we give a brief survey of results related to the classical forward-backward splitting method, which is intermediate between the projection and proximal ones, and to its combinations with descent schemes with respect to artificial gap functions. Next, we propose a descent method based on the primal gap function and utilizing an inexact linesearch procedure. In order to attain convergence without strengthened convexity/ monotonicity assumptions we utilize combined descent and regularization (or proximal point) methods. Also, we present an alternative to these approaches. In fact, a combined relaxation method, which makes use of a forward-backward splitting iteration to finding a hyperplane which separates strictly the current iterate and solution set, converges to a solution under even weakened assumptions. All these methods can be adjusted to the decomposable structure of the problem under solution.

5. Descent Methods for Mixed Variational Inequalities

5.1. Splitting Type Methods: Traditional Approaches

It was mentioned in Section 2. that MEP (2) can be replaced not only by MVI (4), but also by the multi-valued VI (10). However, construction of efficient solution methods for (10) can meet certain difficulties. In fact, projection type methods have rather low convergence due to their "off-line" divergent series stepsize rules [7, 15, 65], whereas more complicated methods such as proximal and level (bundle) ones [15, 49, 63] involve auxiliary procedures which are very hard to be implemented. So, if we turn to MVI (4), then it seems natural to suppose that the optimization problem with the cost function h, despite of its possible non-smoothness, can be solved rather easily, at least in comparison with the initial problem (4). Indeed, it is true for decomposable problems described in Section 3.. For this reason, the *forward-backward splitting method*, which is intermediate between the projection and proximal ones, may lead to more efficient computational schemes. It consists in generating a sequence $\{\mathbf{x}^k\}$ in conformity with the rule:

$$\begin{array}{c}\langle \mathbf{G}(\mathbf{x}^k) + \alpha^{-1}(\mathbf{x}^{k+1} - \mathbf{x}^k), \mathbf{y} - \mathbf{x}^{k+1}\rangle \\ +h(\mathbf{y}) - h(\mathbf{x}^{k+1}) \geq 0 \quad \forall \mathbf{y} \in X,\end{array} \tag{33}$$

where $\alpha > 0$ is a stepsize parameter. This method was proposed first by P.L.Lions and B.Mercier [51] and substantiated by D.Gabay [14]. In addition to (A1), we introduce the following basic assumptions.

(B1) $\mathbf{G} : X \to \mathbb{R}^n$ *is a continuous mapping.*

(B2) $h : V \to \mathbb{R}$ *is a convex function where* V *is an open convex set such that* $V \supset X$.

Then, due to Corollary 2.3, problem (33) is well-defined and has a unique solution. Then convergence of a sequence $\{\mathbf{x}^k\}$ defined in (33) will require the additional co-coercivity of $\mathbf{G}$, α in (33) depending on the co-coercivity constant. In [56], a combined averaging and splitting method, which requires for $\mathbf{G}$ to be only monotone, was proposed. However, its convergence rate is rather low due to the same divergent series stepsize rule. These facts force one to look for some other ways of enhancing convergence properties of the splitting type methods.

In the case when $\mathbf{G}$ is integrable, i.e. it is the gradientmap, some descent methods were proposed in [13, 52]; see also [58]. They do not use a priori information about the initial problem. However, construction of such methods for general MVI (4) requires an analog of the cost function, which is not given explicitly.

5.2. Regularized Gap Function Approach

We now consider one of the most popular approaches to solve MVI (4), which consists in its converting into an equivalent constrained optimization problem with the help of some artificial gap (or merit) function. This approach allows for construction of descent splitting based methods; see [59]. The simplest regularized gap function is defined as follows:

$$\varphi_\alpha(\mathbf{x}) = \max_{\mathbf{y}\in X} \Phi_\alpha(\mathbf{x}, \mathbf{y}), \tag{34}$$

where

$$\Phi_\alpha(\mathbf{x},\mathbf{y}) = \langle \mathbf{G}(\mathbf{x}), \mathbf{x}-\mathbf{y}\rangle - 0.5\alpha\|\mathbf{x}-\mathbf{y}\|^2 + h(\mathbf{x}) - h(\mathbf{y}), \alpha > 0. \tag{35}$$

Under assumptions (A1) and (B2), $\Phi_\alpha(\mathbf{x},\cdot)$ is strongly concave and continuous on X, hence there exists the unique element $\mathbf{y}_\alpha(\mathbf{x}) \in X$ such that

$$\varphi_\alpha(\mathbf{x}) = \Phi_\alpha(\mathbf{x}, \mathbf{y}_\alpha(\mathbf{x})).$$

It is easy to see that computation of $\mathbf{y}_\alpha(\mathbf{x})$ is equivalent to the splitting iteration in (33) with $\mathbf{x}^k = \mathbf{x}$. Next, from Lemma 2.5 it follows that MVI (4) is equivalent to the following perturbed problem: Find $\tilde{\mathbf{x}} \in X$ such that

$$\langle \mathbf{G}(\tilde{\mathbf{x}}), \mathbf{y}-\tilde{\mathbf{x}}\rangle + h(\mathbf{y}) - h(\tilde{\mathbf{x}}) + 0.5\alpha\|\mathbf{y}-\tilde{\mathbf{x}}\|^2 \geq 0 \quad \forall \mathbf{y} \in X. \tag{36}$$

In fact, it suffices to set

$$\begin{aligned}\Phi(\mathbf{x},\mathbf{y}) &= \langle \mathbf{G}(\mathbf{x}), \mathbf{y}-\mathbf{x}\rangle + h(\mathbf{y}) - h(\mathbf{x}),\\ \Gamma(\mathbf{x},\mathbf{y}) &= 0.5\alpha\|\mathbf{y}-\mathbf{x}\|^2\end{aligned}$$

in Lemma 2.5; cf. (35) and (36). From the above properties we obtain immediately the *non-negativity*

(a) $\varphi_\alpha(\mathbf{x}) \geq 0$ for every $\mathbf{x} \in X$,

and the *basic equivalence result*:

(b)

$$\varphi_\alpha(\mathbf{x}) = 0 \text{ and } \mathbf{x} \in X \Longleftrightarrow \mathbf{x} = \mathbf{y}_\alpha(\mathbf{x}) \Longleftrightarrow \mathbf{x} \in X^*, \tag{37}$$

where X^* denotes the solution set of MVI (4). Therefore, MVI (4) is equivalent to the optimization problem:

$$\min_{\mathbf{x}\in X} \to \varphi_\alpha(\mathbf{x}), \tag{38}$$

note that φ_α is non-convex and non-differentiable in general. Nevertheless, it is possible to utilize φ_α for constructing descent splitting based method. Following [57], we describe such a method with exact linesearch.

(DSE) Choose a point $\mathbf{x}^0 \in X$ and a number $\alpha > 0$.

At the k-th iteration, $k = 0, 1, \ldots$, we have a point $\mathbf{x}^k \in X$, compute $\mathbf{y}_\alpha(\mathbf{x}^k)$ and set $\mathbf{d}^k = \mathbf{y}_\alpha(\mathbf{x}^k) - \mathbf{x}^k$. If $\mathbf{d}^k = 0$, stop. Otherwise, find $\lambda_k \in [0, 1]$ such that

$$\varphi_\alpha(\mathbf{x}^k + \lambda_k \mathbf{d}^k) = \min_{\lambda\in[0,1]} \varphi_\alpha(\mathbf{x}^k + \lambda\mathbf{d}^k),$$

set $\mathbf{x}^{k+1} = \mathbf{x}^k + \lambda_k \mathbf{d}^k$ and go to the next iteration.

Proposition 5.1 *(see [59, Theorem 6.1]) Suppose that assumptions (A1) and (B2) are fulfilled and that* $\mathbf{G} : X \to \mathbb{R}^n$ *is continuously differentiable and strongly monotone with constant* τ *on* X*. Then method (DSE) either terminates with a unique solution of MVI (4) or generates an infinite sequence* $\{\mathbf{x}^k\}$ *which converges to this unique solution of MVI (4).*

Observe that existence and uniqueness of a solution of MVI (4) follows from Corollary 4.3 (a), the strong monotonicity guarantees that each stationary point of problem (38) is a solution of MVI (4), thus ensuring the global convergence. The main part of the proof is based on Zangwill's convergence theorems [66]. At the same time, an inexact linesearch procedure seems more suitable for implementation. We describe such a method from [39].

(DSI) Choose a point $\mathbf{x}^0 \in X$ and numbers $\alpha > 0, \beta \in (0,1)$, and $\gamma \in (0,1)$.

At the k-th iteration, $k = 0, 1, \ldots$, we have a point $\mathbf{x}^k \in X$, compute $\mathbf{y}_\alpha(\mathbf{x}^k)$ and set $\mathbf{d}^k = \mathbf{y}_\alpha(\mathbf{x}^k) - \mathbf{x}^k$. If $\mathbf{d}^k = 0$, stop. Otherwise, find m as the smallest non-negative integer such that

$$\varphi_\alpha(\mathbf{x}^k + \gamma^m \mathbf{d}^k) \leq \varphi_\alpha(\mathbf{x}^k) - \beta\gamma^m \|\mathbf{d}^k\|^2,$$

set $\lambda_k = \gamma^m$, $\mathbf{x}^{k+1} = \mathbf{x}^k + \lambda_k \mathbf{d}^k$ and go to the next iteration.

Convergence of this method is based on somewhat different techniques.

Proposition 5.2 *(see [39, Theorem 4.1]) Suppose that all the assumptions of Proposition 5.1 are fulfilled. Then method (DSI) with $\beta < \tau$ either terminates with a unique solution of MVI (4) or generates an infinite sequence $\{\mathbf{x}^k\}$ which converges to this unique solution of MVI (4).*

5.3. D-gap Function Approach

Let us consider the difference of two gap functions from (34):

$$\psi_{\alpha\beta}(\mathbf{x}) = \varphi_\alpha(\mathbf{x}) - \varphi_\beta(\mathbf{x}), \tag{39}$$

where $0 < \alpha < \beta$. This D-gap function was proposed by J.-M. Peng [60] for the usual VI when $h \equiv 0$. I.V. Konnov [19] proposed to apply this gap function for MVI (4). If $\mathbf{G}$ and h are defined on the whole space $\mathbb{R}^n$, then

$$\|\mathbf{x} - \mathbf{y}_\beta(\mathbf{x})\|^2 \leq 2\psi_{\alpha\beta}(\mathbf{x})/(\beta - \alpha) \leq \|\mathbf{x} - \mathbf{y}_\alpha(\mathbf{x})\|^2 \quad \forall \mathbf{x} \in \mathbb{R}^n,$$

hence MVI (4) is equivalent to the unconstrained optimization problem

$$\min_{\mathbf{x} \in \mathbb{R}^n} \to \psi_{\alpha\beta}(\mathbf{x});$$

cf. (38). Moreover, it was noticed in [19], that $\psi_{\alpha\beta}$ in (39) is continuously differentiable if $\mathbf{G}$ is so and that

$$\nabla\psi_{\alpha\beta}(\mathbf{x}) = \nabla\mathbf{G}(\mathbf{x})^T[\mathbf{y}_\beta(\mathbf{x}) - \mathbf{y}_\alpha(\mathbf{x})] + \beta(\mathbf{x} - \mathbf{y}_\beta(\mathbf{x})) - \alpha(\mathbf{x} - \mathbf{y}_\alpha(\mathbf{x}));$$

note that φ_α and φ_β need not be differentiable. If $\nabla\mathbf{G}(\mathbf{x})$ is positive definite on $\mathbb{R}^n$, then MVI (4) becomes equivalent to the system of equations:

$$\nabla\psi_{\alpha\beta}(\mathbf{x}) = 0$$

and we can apply efficient iterative methods to find a solution. In [19], the following inexact descent method without computation of derivatives was proposed.

(DIDG) Choose a point $\mathbf{x}^0 \in \mathbb{R}^n$ and numbers $\beta > \alpha > 0, \mu > 0, \gamma \in (0,1), \theta > 0$.

At the k-th iteration, $k = 0, 1, \ldots$, we have a point $\mathbf{x}^k$, compute $\mathbf{y}_\alpha(\mathbf{x}^k)$ and $\mathbf{y}_\beta(\mathbf{x}^k)$, set

$$\mathbf{r}(\mathbf{x}^k) = \mathbf{y}_\alpha(\mathbf{x}^k) - \mathbf{y}_\beta(\mathbf{x}^k), \mathbf{s}(\mathbf{x}^k) = \alpha(\mathbf{x}^k - \mathbf{y}_\alpha(\mathbf{x}^k)) - \beta(\mathbf{x}^k - \mathbf{y}_\beta(\mathbf{x}^k))$$

and $\mathbf{d}^k = \mathbf{r}(\mathbf{x}^k) + \mu \mathbf{s}(\mathbf{x}^k)$. If $\mathbf{d}^k = 0$, stop. Otherwise find m as the smallest nonnegative integer such that

$$\psi_{\alpha\beta}(\mathbf{x}^k + \gamma^m \mathbf{d}^k) \leq \psi_{\alpha\beta}(\mathbf{x}^k) - \gamma^m \theta \left(\|\mathbf{r}(\mathbf{x}^k)\| + \mu \|\mathbf{s}(\mathbf{x}^k)\| \right)^2,$$

set $\lambda_k = \gamma^m$, $\mathbf{x}^{k+1} = \mathbf{x}^k + \lambda_k \mathbf{d}^k$ and go to the next iteration.

Under the additional strong monotonicity assumption on $\mathbf{G}$ the convergence result similar to that in Proposition 5.2, was established in [19, Theorem 2].

In the case when $\mathbf{G}$ is only monotone, all the above descent methods can be combined with either regularization or proximal point methods such that their auxiliary problems are solved approximately; see [23, 30, 40, 42, 44].

5.4. Primal Gap Function Approach

The primal gap function is defined by setting $\alpha = 0$ in the formulas for the regularized gap function from (34), (35):

$$\varphi_0(\mathbf{x}) = \max_{\mathbf{y} \in X} \Phi_0(\mathbf{x}, \mathbf{y}), \tag{40}$$

where

$$\Phi_0(\mathbf{x}, \mathbf{y}) = \langle \mathbf{G}(\mathbf{x}), \mathbf{x} - \mathbf{y} \rangle + h(\mathbf{x}) - h(\mathbf{y}). \tag{41}$$

If (A1), (B1), and (B2) are fulfilled, then $\Phi_0(\mathbf{x}, \cdot)$ is concave, but the function φ_0 in (40) may be nondifferentiable and undefined at some points, in particular, if X is not bounded; see e.g. [59]. For this reason, we replace (B2) with the following.

(B2′) *$h : V \to \mathbb{R}$ is a strongly convex function with constant $\tau > 0$ where V is an open convex set such that $V \supset X$.*

Then $\Phi_0(\mathbf{x}, \cdot)$ in (41) is strongly concave and continuous on X for each $\mathbf{x} \in X$, hence φ_0 in (40) becomes well-defined. More precisely, for each $\mathbf{x} \in X$, there exists the unique point $\mathbf{y}(\mathbf{x}) \in X$ such that

$$\varphi_0(\mathbf{x}) = \Phi_0(\mathbf{x}, \mathbf{y}(\mathbf{x})).$$

We now show that φ_0 is a gap function for MVI (4); cf. (37).

Lemma 5.1 *Let assumptions (A1), (B1), and (B2′) be fulfilled. Then:*
(a) $\varphi_0(\mathbf{x}) \geq 0 \quad \forall \mathbf{x} \in X$;
(b) $\tilde{\mathbf{x}} = \mathbf{y}(\tilde{\mathbf{x}}) \Longleftrightarrow \tilde{\mathbf{x}} \in X$ and $\varphi_0(\tilde{\mathbf{x}}) = 0 \Longleftrightarrow \tilde{\mathbf{x}} \in X^*$.

Proof. Assertion (a) is obvious since $\varphi_0(\mathbf{x}) \geq \Phi_0(\mathbf{x}, \mathbf{x}) = 0$. Next, from the definition of $\mathbf{y}(\mathbf{x})$ we have

$$\langle \mathbf{G}(\tilde{\mathbf{x}}), \mathbf{z} - \mathbf{y}(\tilde{\mathbf{x}}) \rangle + h(\mathbf{z}) - h(\mathbf{y}(\tilde{\mathbf{x}})) \geq 0 \quad \forall \mathbf{z} \in X.$$

If $\tilde{\mathbf{x}} = \mathbf{y}(\tilde{\mathbf{x}})$, then the above inequality clearly gives $\tilde{\mathbf{x}} \in X^*$. In turn, $\tilde{\mathbf{x}} \in X^*$ means that

$$\langle \mathbf{G}(\tilde{\mathbf{x}}), \tilde{\mathbf{x}} - \mathbf{y} \rangle + h(\tilde{\mathbf{x}}) - h(\mathbf{y}) \leq 0 \quad \forall \mathbf{y} \in X,$$

hence $\varphi_0(\tilde{\mathbf{x}}) \leq 0$ and $\varphi_0(\tilde{\mathbf{x}}) = 0$ due to (a). Next, $\tilde{\mathbf{x}} \in X$ and $\varphi_0(\tilde{\mathbf{x}}) = 0$ yield

$$0 = \max_{\mathbf{y} \in X} \{ \langle \mathbf{G}(\tilde{\mathbf{x}}), \tilde{\mathbf{x}} - \mathbf{y} \rangle + h(\tilde{\mathbf{x}}) - h(\mathbf{y}) \} = \Phi_0(\tilde{\mathbf{x}}, \tilde{\mathbf{x}}),$$

and $\tilde{\mathbf{x}} = \mathbf{y}(\tilde{\mathbf{x}})$, as desired. □

Therefore, MVI (4) is equivalent to the optimization problem:

$$\min_{\mathbf{x} \in X} \to \varphi_0(\mathbf{x}). \tag{42}$$

However, φ_0 is nonconvex and nonsmooth in general and we need some additional properties and assumptions in order to create an efficient solution method.

Lemma 5.2 *Let assumptions (A1) and (B2′) be fulfilled. Then:*

$$\varphi_0(\mathbf{x}) - \Phi_0(\mathbf{x}, \mathbf{y}) \geq 0.5\tau \|\mathbf{y}(\mathbf{x}) - \mathbf{y}\|^2 \quad \forall \mathbf{y} \in X. \tag{43}$$

Proof. Take arbitrary $\mathbf{x}, \mathbf{y} \in X$ and set $\mathbf{x}(\alpha) = \alpha \mathbf{y} + (1 - \alpha)\mathbf{y}(\mathbf{x})$. Then, by strong convexity of h we have

$$\begin{aligned}
\varphi_0(\mathbf{x}) &\geq \Phi_0(\mathbf{x}, \mathbf{x}(\alpha)) = \langle \mathbf{G}(\mathbf{x}), \mathbf{x} - \mathbf{x}(\alpha) \rangle + h(\mathbf{x}) - h(\mathbf{x}(\alpha)) \\
&\geq (1 - \alpha) [\langle \mathbf{G}(\mathbf{x}), \mathbf{x} - \mathbf{y}(\mathbf{x}) \rangle + h(\mathbf{x}) - h(\mathbf{y}(\mathbf{x}))] \\
&\quad + \alpha [\langle \mathbf{G}(\mathbf{x}), \mathbf{x} - \mathbf{y} \rangle + h(\mathbf{x}) - h(\mathbf{y})] + 0.5\tau\alpha(1 - \alpha)\|\mathbf{y} - \mathbf{y}(\mathbf{x})\|^2,
\end{aligned}$$

hence

$$\varphi_0(\mathbf{x}) - \Phi_0(\mathbf{x}, \mathbf{y}) \geq 0.5\tau(1 - \alpha)\|\mathbf{y} - \mathbf{y}(\mathbf{x})\|^2 \quad \forall \alpha \in [0, 1].$$

Taking the limit $\alpha \to 0$ in this inequality, we obtain (43). □

Proposition 5.3 *Let (A1), (B1), and (B2′) be fulfilled. Then:*
(a) the mapping $\mathbf{x} \mapsto \mathbf{y}(\mathbf{x})$ *is continuous on* X*;*
(b) it holds that

$$\varphi_0(\mathbf{x}) \geq 0.5\tau \|\mathbf{x} - \mathbf{y}(\mathbf{x})\|^2 \quad \forall \mathbf{x} \in X.$$

Proof. Fix $\mathbf{x}', \mathbf{x}'' \in X$ and set $\mathbf{y}' = \mathbf{y}(\mathbf{x}')$ and $\mathbf{y}'' = \mathbf{y}(\mathbf{x}'')$ for brevity. Then from (43) it follows that

$$\varphi_0(\mathbf{x}') - \Phi_0(\mathbf{x}', \mathbf{y}'') \geq 0.5\tau \|\mathbf{y}' - \mathbf{y}''\|^2$$

and

$$\varphi_0(\mathbf{x}'') - \Phi_0(\mathbf{x}'', \mathbf{y}') \geq 0.5\tau \|\mathbf{y}' - \mathbf{y}''\|^2.$$

Adding these inequalities gives

$$\langle \mathbf{G}(\mathbf{x}'), \mathbf{y}'' - \mathbf{y}' \rangle + \langle \mathbf{G}(\mathbf{x}''), \mathbf{y}' - \mathbf{y}'' \rangle \geq \tau \|\mathbf{y}' - \mathbf{y}''\|^2,$$

hence by the Cauchy-Schwarz inequality

$$\|\mathbf{G}(\mathbf{x}') - \mathbf{G}(\mathbf{x}'')\| \geq \tau\|\mathbf{y}' - \mathbf{y}''\|,$$

and we see that assertion (a) is true. Assertion (b) follows from (43) with $\mathbf{y} = \mathbf{x}$. □

We now fix our assumptions on $\mathbf{G}$.

(B1′) $\mathbf{G} : X \to \mathbb{R}^n$ *is continuously differentiable and monotone on* X.

Under assumptions (B1′) and (B2′), φ_0 is a composition of smooth and convex functions and hence subdifferentiable and regular in the sense of Clarke. Utilizing the formula for the subdifferential of composite functions we can determine explicitly the Clarke subdifferential and directional derivative of the function φ_0 (e.g. see [59, Proposition 4.19]).

Lemma 5.3 *Let assumptions (A1), (B1′) and (B2′) be fulfilled. Then, at any point* $\mathbf{x} \in X$, *there exist the Clarke subdifferential of the function* φ_0, *defined by the formula:*

$$\partial\varphi_0(\mathbf{x}) = \mathbf{G}(\mathbf{x}) - \nabla\mathbf{G}(\mathbf{x})^T[\mathbf{y}(\mathbf{x}) - \mathbf{x}] + \partial h(\mathbf{x}),$$

and the directional derivative with respect to any vector d:

$$\partial\varphi_0'(\mathbf{x}; \mathbf{d}) = \langle \mathbf{G}(\mathbf{x}) - \nabla\mathbf{G}(\mathbf{x})^T[\mathbf{y}(\mathbf{x}) - \mathbf{x}], \mathbf{d}\rangle + h'(\mathbf{x}; \mathbf{d}). \tag{44}$$

Observe that MVI (4) (hence (42)) has a unique solution under assumptions (A1), (B1′) and (B2′); see Corollary 4.3 (b). We give additional properties of this solution.

Proposition 5.4 *Let assumptions (A1), (B1), and (B2′) be fulfilled, and let* $\mathbf{G}$ *be monotone. Then, for any* $\mathbf{x} \in X$, *it holds that:*

(a)

$$\langle \mathbf{G}(\mathbf{x}^*), \mathbf{x} - \mathbf{x}^*\rangle + h(\mathbf{x}) - h(\mathbf{x}^*) \geq 0.5\tau\|\mathbf{x} - \mathbf{x}^*\|^2$$

and that

(b)

$$\varphi_0(\mathbf{x}) \geq 0.5\tau\|\mathbf{x} - \mathbf{x}^*\|^2,$$

where $\mathbf{x}^*$ *is a unique solution of MVI (4).*

Proof. Take any $\mathbf{x} \in X$ and set $\mathbf{x}(\alpha) = \alpha\mathbf{x} + (1 - \alpha)\mathbf{x}^*$ with $\alpha \in [0, 1]$. Then $\mathbf{x}(\alpha) \in X$ and, by definition,

$$\langle \mathbf{G}(\mathbf{x}^*), \mathbf{x}(\alpha) - \mathbf{x}^*\rangle + h(\mathbf{x}(\alpha)) - h(\mathbf{x}^*) \geq 0,$$

hence, by strong convexity of h,

$$\alpha\langle \mathbf{G}(\mathbf{x}^*), \mathbf{x} - \mathbf{x}^*\rangle + \alpha h(\mathbf{x}) + (1 - \alpha)h(\mathbf{x}^*) - 0.5\tau\alpha(1 - \alpha)\|\mathbf{x} - \mathbf{x}^*\|^2 - h(\mathbf{x}^*) \geq 0.$$

It follows that

$$\langle \mathbf{G}(\mathbf{x}^*), \mathbf{x} - \mathbf{x}^*\rangle + h(\mathbf{x}) - h(\mathbf{x}^*) \geq 0.5\tau(1 - \alpha)\|\mathbf{x} - \mathbf{x}^*\|^2.$$

Taking the limit $\alpha \to 0$ in this inequality, we see that (a) is true.

Next, by monotonicity of $\mathbf{G}$, we obtain

$$\begin{aligned}\varphi_0(\mathbf{x}) &\geq \Phi_0(\mathbf{x}, \mathbf{x}^*) = \langle \mathbf{G}(\mathbf{x}), \mathbf{x} - \mathbf{x}^* \rangle + h(\mathbf{x}) - h(\mathbf{x}^*) \\ &\geq \langle \mathbf{G}(\mathbf{x}^*), \mathbf{x} - \mathbf{x}^* \rangle + h(\mathbf{x}) - h(\mathbf{x}^*) \\ &\geq 0.5\tau \|\mathbf{x} - \mathbf{x}^*\|^2,\end{aligned}$$

and (b) is also true. □

Observe that the differentiability of $\mathbf{G}$ has not been used in Proposition 5.4. We now show that $\mathbf{y}(\mathbf{x}) - \mathbf{x}$ is a sufficient descent direction for φ_0 at any point $\mathbf{x}$, thus obtaining a basis for creating an iteration solution method.

Lemma 5.4 *Let assumptions (A1), (B1′) and (B2′) be fulfilled. Then for any point* $\mathbf{x} \in X$, *it holds that*

$$\varphi_0'(\mathbf{x}; \mathbf{y}(\mathbf{x}) - \mathbf{x}) \leq -\varphi_0(\mathbf{x}).$$

Proof. Using Lemma 5.3, we have

$$\begin{aligned}\varphi_0'(\mathbf{x}; \mathbf{y}(\mathbf{x}) - \mathbf{x}) &= \langle \mathbf{G}(\mathbf{x}) + \nabla \mathbf{G}(\mathbf{x})^T [\mathbf{y}(\mathbf{x}) - \mathbf{x}], \mathbf{y}(\mathbf{x}) - \mathbf{x} \rangle \\ &\quad + h'(\mathbf{x}; \mathbf{y}(\mathbf{x}) - \mathbf{x}) \\ &\leq \langle \mathbf{G}(\mathbf{x}), \mathbf{y}(\mathbf{x}) - \mathbf{x} \rangle + h(\mathbf{y}(\mathbf{x})) - h(\mathbf{x}) = -\varphi_0(\mathbf{x})\end{aligned}$$

since $\mathbf{G}$ is monotone and h is convex. □

In order to construct a descent method for MVI (4) we will utilize an inexact Armijo-type linesearch procedure (cf. (DSI)).

(DPI) Choose a point $\mathbf{x}^0 \in X$ and numbers $\beta \in (0, 1)$ and $\gamma \in (0, 1)$.

At the k-th iteration, $k = 0, 1, \ldots$, we have a point $\mathbf{x}^k \in X$, compute $\mathbf{y}(\mathbf{x}^k)$, and set $\mathbf{d}^k = \mathbf{y}(\mathbf{x}^k) - \mathbf{x}^k$. If $\mathbf{d}^k = 0$, stop. Otherwise, find m as the smallest non-negative integer such that

$$\varphi_0(\mathbf{x}^k + \gamma^m \mathbf{d}^k) \leq \varphi_0(\mathbf{x}^k) - \beta \gamma^m \varphi_0(\mathbf{x}^k), \tag{45}$$

set $\lambda_k = \gamma^m$, $\mathbf{x}^{k+1} = \mathbf{x}^k + \lambda_k \mathbf{d}^k$ and go to the next iteration.

First we note that the termination of (DPI) yields $\mathbf{x}^k \in X^*$ due to Lemma 5.1. For this reason, in what follows we consider only the case when (DPI) generates an infinite sequence.

Theorem 5.1 *Let assumptions (A1), (B1′) and (B2′) be fulfilled. Then method (DPI) generates an infinite sequence* $\{\mathbf{x}^k\}$ *which converges to a unique solution of MVI (4).*

Proof. We first show that the linesearch procedure with criterion (45) is always finite. In fact, otherwise, using Lemma 5.3, we have

$$\varphi_0'(\mathbf{x}^k; \mathbf{d}^k) = \lim_{m \to \infty} \gamma^{-m} \left[\varphi_0(\mathbf{x}^k + \gamma^m \mathbf{d}^k) - \varphi_0(\mathbf{x}^k) \right] \geq -\beta \varphi_0(\mathbf{x}^k).$$

On the other hand, by Lemma 5.4 and Proposition 5.3 (b), we obtain

$$\varphi_0'(\mathbf{x}^k;\mathbf{d}^k) \le -\varphi_0(\mathbf{x}^k).$$

Combining both the inequalities gives

$$(1-\beta)\varphi_0(\mathbf{x}^k) \le 0,$$

but $\varphi_0(\mathbf{x}^k) \ge 0$, hence $\varphi_0(\mathbf{x}^k) = 0$, which is a contradiction in view of Lemma 5.1 (b). So, (DPI) is well-defined and $\lambda_k > 0$.

Due to (45) and Lemma 5.1, the sequence $\{\varphi_0(\mathbf{x}^k)\}$ is nonincreasing and bounded below, hence there exists a number $\mu \ge 0$ such that

$$\lim_{k\to\infty} \varphi_0(\mathbf{x}^k) = \mu.$$

From Proposition 5.4 (b) it follows that the sequence $\{\mathbf{x}^k\}$ is bounded and so is $\{\mathbf{d}^k\}$ due to Proposition 5.3 (a), hence they have limit points. Suppose that $\mu > 0$, then there exists a subsequence $\{\varphi_0(\mathbf{x}^{k_s})\} \to \mu$, but (45) now gives $\lim\limits_{k_s\to\infty} \lambda_{k_s} = 0$. It follows that

$$\varphi_0\left(\mathbf{x}^{k_s} + (\lambda_{k_s}/\gamma)\mathbf{d}^{k_s}\right) - \varphi_0(\mathbf{x}^{k_s}) > -\beta(\lambda_{k_s}/\gamma)\varphi_0(\mathbf{x}^{k_s}).$$

Using the mean value theorem (e.g. see [9, Ch.1, Theorem 3.1]) gives

$$\varphi_0'\left(\mathbf{x}^{k_s} + \theta_{k_s}(\lambda_{k_s}/\gamma)\mathbf{d}^{k_s};\mathbf{d}^{k_s}\right) > -\beta\varphi_0(\mathbf{x}^{k_s})$$

for some $\theta_{k_s} \in [0,1]$, but, due to (44), we have

$$\begin{aligned}&\langle G\left(\mathbf{x}^{k_s} + \theta_{k_s}(\lambda_{k_s}/\gamma)\mathbf{d}^{k_s}\right) - \nabla G\left(\mathbf{x}^{k_s} + \theta_{k_s}(\lambda_{k_s}/\gamma)\mathbf{d}^{k_s}\right)^T \mathbf{d}^{k_s} + \mathbf{g}^{k_s}, \mathbf{d}^{k_s}\rangle\\ &> -\beta\varphi_0(\mathbf{x}^{k_s})\end{aligned}$$

for some $\mathbf{g}^{k_s} \in \partial h\left(\mathbf{x}^{k_s} + \theta_{k_s}(\lambda_{k_s}/\gamma)\mathbf{d}^{k_s}\right)$. Since $\mathbf{G}$ and $\nabla\mathbf{G}$ are continuous and ∂h is upper semicontinuous, taking the limit $k_s \to \infty$ and a subsequence, if necessary, we obtain

$$\langle \mathbf{G}(\tilde{\mathbf{x}}) - \nabla\mathbf{G}(\tilde{\mathbf{x}})^T\tilde{\mathbf{d}} + \tilde{\mathbf{g}}, \tilde{\mathbf{d}}\rangle \ge -\beta\varphi_0(\tilde{\mathbf{x}}), \tag{46}$$

where $\tilde{\mathbf{x}}$, $\tilde{\mathbf{g}}$, and $\tilde{\mathbf{d}}$ are the corresponding limit points for $\{\mathbf{x}^{k_s}\}$, $\{\mathbf{g}^{k_s}\}$, and $\{\mathbf{d}^{k_s}\}$, respectively. We also see that $\tilde{\mathbf{g}} \in \partial h(\tilde{\mathbf{x}})$. But Lemmas 5.3 and 5.4 give

$$-\varphi_0(\tilde{\mathbf{x}}) \ge \varphi_0'(\tilde{\mathbf{x}};\tilde{\mathbf{d}}) \ge \langle \mathbf{G}(\tilde{\mathbf{x}}) - \nabla\mathbf{G}(\tilde{\mathbf{x}})^T\tilde{\mathbf{d}} + \tilde{\mathbf{g}}, \tilde{\mathbf{d}}\rangle,$$

hence combining this inequality with (46) yields

$$0 \ge (1-\beta)\varphi_0(\tilde{\mathbf{x}}) \ge 0$$

and $\varphi_0(\tilde{\mathbf{x}}) = 0$, which is a contradiction.

Therefore, $\mu = 0$, and applying now Proposition 5.4 (b), we obtain

$$\lim_{k\to\infty} \|\mathbf{x}^k - \mathbf{x}^*\| = 0,$$

where $\mathbf{x}^*$ is a unique solution of MVI (4), as desired. □

Observe that method (DPI) requires neither evaluation of the strong convexity modulus of h as in (DSI) nor computation of the Jacobian of $\mathbf{G}$ (see [40,59]). Again, we can replace the inexact linesearch in (45) with the exact one of the form:

$$\varphi_0(\mathbf{x}^k + \lambda_k \mathbf{d}^k) = \min_{\lambda \in [0,1]} \varphi_0(\mathbf{x}^k + \lambda \mathbf{d}^k)$$

and then obtain the similar convergence result as in Theorem 5.1 by using the above properties of the function φ_0 and Zangwill's convergence theorems [66] (cf. Proposition 5.1). However, the inexact linesearch seems more suitable for implementation.

5.5. Combined Descent and Regularization Methods

The above descent method (DPI) requires the strengthened convexity properties of h, whereas (DSE) and (DSI) require the strong monotonicity of G. In this section we consider two-level methods which allow us to solve MVI (4) without strengthened convexity / monotonicity assumptions. Such an approach was first proposed in [23] where gap functions were utilized for evaluating error bounds in solving auxiliary subproblems, thus making the whole method implementable. Further development of this approach for regularized gap function and different problems can be found in [40,42–44].

We now give a modification of convergence results of the Tikhonov-Browder regularization method, which was obtained in [44, Theorem 2.1], and [40, Theorem 2.1] for general equilibrium problems in Banach space, for MVI (4).

Proposition 5.5 *Suppose that assumptions (A1)–(A3) are fulfilled,* $\mathbf{G}$ *is monotone on* X *and*

(H) $\mu : V \to \mathbb{R}$ *is a continuous and strongly convex with constant* $\kappa > 0$ *on an open set* $V \supset X$.

Then, for each $\varepsilon > 0$*, there exists a unique point* $\tilde{\mathbf{x}}^\varepsilon \in X$ *such that*

$$\langle \mathbf{G}(\tilde{\mathbf{x}}^\varepsilon), \mathbf{x} - \tilde{\mathbf{x}}^\varepsilon \rangle + h(\mathbf{x}) - h(\tilde{\mathbf{x}}^\varepsilon) + \varepsilon\,[\mu(\mathbf{x}) - \mu(\tilde{\mathbf{x}}^\varepsilon)] \geq 0 \quad \forall \mathbf{x} \in X. \tag{47}$$

If MVI (4) is solvable and $\{\varepsilon_l\} \searrow 0$*, then the corresponding sequence* $\{\tilde{\mathbf{x}}^{\varepsilon_l}\}$ *converges to the point* $\mathbf{x}^*_n \in X^*$ *such that*

$$\mu(\mathbf{x}^*_n) = \min_{\mathbf{x} \in X^*} \mu(\mathbf{x}).$$

Proof. First we note that the perturbed problem (47) has the unique solution $\tilde{\mathbf{x}}^\varepsilon$ due to Corollary 4.3 (b). For brevity, set

$$\Psi(\mathbf{x}, \mathbf{y}) = \langle \mathbf{G}(\mathbf{x}), \mathbf{y} - \mathbf{x} \rangle + h(\mathbf{y}) - h(\mathbf{x}).$$

Let $\mathbf{x}^*$ be a solution of MVI (4). Then

$$\Psi(\mathbf{x}^*, \tilde{\mathbf{x}}^\varepsilon) \geq 0,$$

but

$$\Psi(\tilde{\mathbf{x}}^\varepsilon, \mathbf{x}^*) + \varepsilon\,[\mu(\mathbf{x}^*) - \mu(\tilde{\mathbf{x}}^\varepsilon)] \geq \varepsilon\kappa \|\mathbf{x}^* - \tilde{\mathbf{x}}^\varepsilon\|^2$$

due to Proposition 5.4 (a). Adding these inequalities and taking into account the monotonicity of Ψ, we have

$$\varepsilon\left[\mu(\mathbf{x}^*)-\mu(\tilde{\mathbf{x}}^\varepsilon)\right] \geq \Psi(\mathbf{x}^*,\tilde{\mathbf{x}}^\varepsilon)+\Psi(\tilde{\mathbf{x}}^\varepsilon,\mathbf{x}^*)+\varepsilon\left[\mu(\mathbf{x}^*)-\mu(\tilde{\mathbf{x}}^\varepsilon)\right] \geq \varepsilon\kappa\|\mathbf{x}^*-\tilde{\mathbf{x}}^\varepsilon\|^2,$$

hence

$$\mu(\mathbf{x}^*)-\mu(\tilde{\mathbf{x}}^\varepsilon) \geq \kappa\|\mathbf{x}^*-\tilde{\mathbf{x}}^\varepsilon\|^2. \tag{48}$$

Since μ is strongly monotone, $\|\mathbf{x}\| \to \infty$ implies $\mu(\mathbf{x}) \to +\infty$ (see e.g. [36, Proposition 1.15]). This means that any sequence $\{\tilde{\mathbf{x}}^\varepsilon\}$ is bounded and must have limit points. Utilizing Corollary 2.2, we have

$$0 \geq \Psi(\mathbf{x},\tilde{\mathbf{x}}^\varepsilon)+\varepsilon\left[\mu(\mathbf{x}^\varepsilon)-\mu(\tilde{\mathbf{x}})\right] \quad \forall \mathbf{x} \in X.$$

If $\mathbf{x}'$ is a limit point of $\{\tilde{\mathbf{x}}^\varepsilon\}$ with $\varepsilon \searrow 0$, then the above inequality gives

$$0 \geq \Psi(\mathbf{x},\mathbf{x}') \quad \forall \mathbf{x} \in X,$$

i.e. $\mathbf{x}' \in X^*$. So, all the limit points of $\{\tilde{\mathbf{x}}^\varepsilon\}$ belong to X^*, moreover, X^* is convex and closed. Hence, there exists the point $\mathbf{x}^*_n$. Setting $\mathbf{x}^* = \mathbf{x}^*_n$ in (48) and taking the limit $\varepsilon \to 0$ and a subsequence if necessary, we obtain

$$0 \geq \mu(\mathbf{x}^*_n)-\mu(\mathbf{x}') \geq \kappa\|\mathbf{x}^*_n-\mathbf{x}'\|$$

where $\mathbf{x}'$ is an arbitrary limit point of $\{\tilde{\mathbf{x}}^\varepsilon\}$. It follows that

$$\lim_{\varepsilon_l \searrow 0} \tilde{\mathbf{x}}^{\varepsilon_l} = \mathbf{x}^*_n,$$

and the proof is complete. □

The combined regularization and descent method for MVI (4) is described as follows.

(CRDPI) Choose a point $\mathbf{u}^0 \in X$, a number $\delta > 0$, a positive sequence $\{\varepsilon_l\} \searrow 0$, and a function μ satisfying (H). For each $l = 0, 1, \ldots$, we have a point $\mathbf{u}^{l-1} \in X$, apply (DPI) to the perturbed MVI (47) with $\mathbf{x}^0 = \mathbf{u}^{l-1}$ and $\varepsilon = \varepsilon_l$, i.e. construct a sequence $\{\mathbf{x}^k\}$ until

$$\varphi_0^{(\varepsilon)}(\mathbf{x}^k) \leq \varepsilon^{1+\delta}, \tag{49}$$

where

$$\varphi_0^{(\varepsilon)}(\mathbf{x}) = \max_{\mathbf{y} \in X}\left\{\langle \mathbf{G}(\mathbf{x}),\mathbf{x}-\mathbf{y}\rangle + h(\mathbf{x}) - h(\mathbf{y}) + \varepsilon[\mu(\mathbf{x})-\mu(\mathbf{y})]\right\}. \tag{50}$$

Then set $\mathbf{u}^l = \mathbf{x}^k$ and increase $l = l+1$.

Note that $\varphi_0^{(\varepsilon)}$ in (50) is the primal gap function for MVI (47) and that we solve each problem approximately within the accuracy in (49). Instead of (B2′) we utilize the relaxed assumption.

(B2″) *$h : V \to \mathbb{R}$ is a convex function where V is an open convex set such that $V \supset X$.*

Now we can state the basic convergence result.

Theorem 5.2 *Suppose that assumptions (A1), (B1′) and (B2″) are fulfilled and that MVI (4) is solvable. Then*

(a) the number of iterations of (DPI) for each l is finite;

(b) the sequence $\{\mathbf{u}^l\}$ converges to the point $\mathbf{x}_n^*$.

Proof. Since the function $h + \varepsilon\mu$ satisfies (B2′) with $\tau = \varepsilon\kappa$ we see that, for each $\varepsilon > 0$, the sequence $\{\mathbf{x}^k\}$ generated by (DPI) converges to the unique solution $\tilde{\mathbf{x}}^\varepsilon$ of MVI (47) because of Theorem 5.1, hence

$$\lim_{k\to\infty} \varphi_0^{(\varepsilon)}(\mathbf{x}^k) = 0,$$

and assertion (a) is true. Next, using Proposition 5.4 (b), we have

$$\varphi_0^{(\varepsilon)}(\mathbf{x}) \geq 0.5\varepsilon\kappa\|\mathbf{x} - \tilde{\mathbf{x}}^\varepsilon\|^2.$$

Combining this inequality with (49) gives

$$\|\mathbf{x}^k - \tilde{\mathbf{x}}^\varepsilon\|^2 \leq 2\varepsilon^\delta\kappa^{-1},$$

hence

$$\|\mathbf{u}^l - \tilde{\mathbf{x}}^{\varepsilon_l}\|^2 \leq 2\varepsilon_l^\delta\kappa^{-1}.$$

Since

$$\begin{aligned} \|\mathbf{x}_n^* - \mathbf{u}^l\| &\leq \|\mathbf{x}_n^* - \tilde{\mathbf{x}}^{\varepsilon_l}\| + \|\tilde{\mathbf{x}}^{\varepsilon_l} - \mathbf{u}^l\| \\ &\leq \|\mathbf{x}_n^* - \tilde{\mathbf{x}}^{\varepsilon_l}\| + 2\varepsilon_l^\delta\kappa^{-1} \to 0 \end{aligned}$$

as $\{\varepsilon_l\} \searrow 0$ on account of Proposition 5.5, we conclude that assertion (b) is also true. □

5.6. Combined Descent and Proximal Point Methods

Similarly, we can combine descent and proximal point methods. To this end, we first adjust a convergence result for a usual inexact proximal point method from [30, Theorem 2.1] (see also [34, Theorem 1]) to monotone MVI (4).

Proposition 5.6 *Suppose that assumptions (A1)–(A3) are fulfilled,* $\mathbf{G}$ *is monotone on* X*, and that a sequence* $\{\mathbf{v}^l\}$ *is constructed in conformity with the rules:*

$$\begin{array}{c} \mathbf{v}^{l+1} \in X, \quad \|\mathbf{v}^{l+1} - \tilde{\mathbf{v}}^{l+1}\| \leq \varepsilon_{l+1}, \tilde{\mathbf{v}}^{l+1} \in X; \\ \langle \mathbf{G}(\tilde{\mathbf{v}}^{l+1}), \mathbf{x} - \tilde{\mathbf{v}}^{l+1}\rangle + 0.5\theta\left[\|\mathbf{x} - \mathbf{v}^l\|^2 - \|\tilde{\mathbf{v}}^{l+1} - \mathbf{v}^l\|^2\right] \\ +h(\mathbf{x}) - h(\tilde{\mathbf{v}}^{l+1}) \geq 0 \quad \forall \mathbf{x} \in X; \\ \varepsilon_{l+1} > 0, \quad \sum\limits_{l=0}^{\infty} \varepsilon_l < \infty, \quad \theta > 0, \quad \mathbf{v}^0 \in X. \end{array} \tag{51}$$

Then,

$$\lim_{l\to\infty} \mathbf{v}^l = \mathbf{x}^* \in X^*.$$

Observe that $\{\tilde{\mathbf{v}}^{l+1}\}$ is nothing but the unique solution of the MVI in (51), which exists due to Corollary 4.3 (b). Hence $\{\mathbf{v}^l\}$ is a sequence of approximate solutions and we can describe the combined method as follows.

(CPDPI) Choose a point $\mathbf{v}^0 \in X$, a number $\theta > 0$, and a positive sequence $\{\tilde{\varepsilon}_l\}$ such that

$$\sum_{l=0}^{\infty} \tilde{\varepsilon}_l < \infty.$$

For each $l = 0, 1, \ldots$, we have a point $\mathbf{v}^l \in X$, apply (DPI) to the perturbed MVI in (51) with $\mathbf{x}^0 = \mathbf{v}^l$, i.e. construct a sequence $\{\mathbf{x}^k\}$ until

$$\psi_0^{(l)}(\mathbf{x}^k) \leq \tilde{\varepsilon}_l^2, \tag{52}$$

where

$$\begin{aligned} \psi_0^{(l)}(\mathbf{x}) = \max_{\mathbf{y} \in X} \{ \langle \mathbf{G}(\mathbf{x}), \mathbf{x} - \mathbf{y} \rangle \\ + h(\mathbf{x}) - h(\mathbf{y}) + 0.5\theta \left[\|\mathbf{x} - \mathbf{v}^l\|^2 - \|\mathbf{y} - \mathbf{v}^l\|^2 \right] \}. \end{aligned} \tag{53}$$

Then set $\mathbf{v}^{l+1} = \mathbf{x}^k$ and increase $l = l + 1$.

Again, $\psi_0^{(l)}$ in (53) is the primal gap function applied to the perturbed MVI in (51). The convergence result of (CPDPI) is similar to that in Theorem 5.2.

Theorem 5.3 *Suppose that assumptions (A1), (B1′) and (B2″) are fulfilled and that MVI (4) is solvable. Then*

(a) the number of iterations of (DPI) for each l is finite;

(b) the sequence $\{\mathbf{v}^l\}$ converges to a solution of MVI (4).

Proof. Since the function $h(\mathbf{x}) + 0.5\theta\|\mathbf{x} - \mathbf{v}^l\|^2$ satisfies (B2′) with $\tau = \theta$ we see that, for each l, the sequence $\{\mathbf{x}^k\}$ generated by (DPI) converges to the unique solution $\tilde{\mathbf{v}}^{l+1}$ of MVI in (51) because of Theorem 5.1, hence

$$\lim_{k \to \infty} \psi_0^{(l)}(\mathbf{x}^k) = 0,$$

and assertion (a) is true. Next, using Proposition 5.4 (b), we have

$$\psi_0^{(l)}(\mathbf{x}) \geq 0.5\theta\|\mathbf{x} - \tilde{\mathbf{v}}^{l+1}\|^2 \quad \forall \mathbf{x} \in X.$$

Combining this inequality with (52) gives

$$\|\mathbf{v}^{l+1} - \tilde{\mathbf{v}}^{l+1}\|^2 \leq 2\tilde{\varepsilon}_l^2/\theta,$$

hence the relations in (51) hold with $\varepsilon_l = \sqrt{2\theta^{-1}}\tilde{\varepsilon}_l$, and, by Proposition 5.6, assertion (b) is also true. □

By using the results from [30, 35], we can construct a combined descent and proximal point method even without basic monotonicity and convexity assumptions. We begin our considerations from the general EP (1).

Proposition 5.7 *Suppose that X is a nonempty, convex, and closed subset in $\mathbb{R}^n$, $\Phi : X \times X \to \mathbb{R}$ is an equilibrium bifunction, which is weakly monotone with constant $\sigma' > 0$, $\Phi(\cdot, \mathbf{y})$ is upper semicontinuous for each $\mathbf{y} \in X$, $\Phi(\mathbf{x}, \cdot)$ is weakly convex with constant $\sigma'' > 0$ and lower semicontinuous for each $\mathbf{x} \in X$. Suppose also that problem (7) is solvable, a sequence $\{\mathbf{v}^l\}$ is constructed in conformity with the rules:*

$$\begin{array}{l} \mathbf{v}^{l+1} \in X, \quad \|\mathbf{v}^{l+1} - \tilde{\mathbf{v}}^{l+1}\| \leq \varepsilon_{l+1}, \tilde{\mathbf{v}}^{l+1} \in X; \\ \Phi(\tilde{\mathbf{v}}^{l+1}, \mathbf{x}) + 0.5\theta\left[\|\mathbf{x} - \mathbf{v}^l\|^2 - \|\tilde{\mathbf{v}}^{l+1} - \mathbf{v}^l\|^2\right] \geq 0 \quad \forall \mathbf{x} \in X; \\ \varepsilon_{l+1} > 0, \quad \sum\limits_{l=0}^{\infty} \varepsilon_l < \infty, \quad \theta > 0, \quad \mathbf{v}^0 \in X. \end{array} \tag{54}$$

Then, there exists a number $\theta' > 0$ such that, for any $\theta > \theta'$, $\{\mathbf{v}^l\}$ has limit points and all these points are solutions of EP (1). If, in addition, solution sets of problems (1) and (7) coincide, then $\{\mathbf{v}^l\}$ converges to a solution of EP (1).

Proof. Set $\theta = \sigma' + \sigma'' + \tau$, where $\tau > 0$, and consider the following EP: Find a point $\mathbf{x}^\theta \in X$ such that

$$\tilde{\Phi}_\theta(\mathbf{x}^\theta, \mathbf{y}) \geq 0 \quad \forall \mathbf{y} \in X, \tag{55}$$

where

$$\begin{array}{rcl} \tilde{\Phi}_\theta(\mathbf{x}, \mathbf{y}) & = & \Phi(\mathbf{x}, \mathbf{y}) + \sigma'\langle \mathbf{x} - \mathbf{v}^l, \mathbf{y} - \mathbf{x}\rangle \\ & & +0.5(\sigma'' + \tau)\left[\|\mathbf{y} - \mathbf{v}^l\|^2 - \|\mathbf{x} - \mathbf{v}^l\|^2\right]. \end{array}$$

Clearly, $\tilde{\Phi}_\theta : X \times X \to \mathbb{R}$ is a monotone equilibrium bifunction such that $\tilde{\Phi}_\theta(\cdot, \mathbf{y})$ is upper semicontinuous for each $\mathbf{y} \in X$, $\tilde{\Phi}_\theta(\mathbf{x}, \cdot)$ is strongly convex with constant $\tau > 0$ and lower semicontinuous for each $\mathbf{x} \in X$. On account of Proposition 4.3 (b), we conclude that EP (55) has always the unique solution $\mathbf{x}^\theta$. However, by Lemma 2.5, EP (55) is equivalent to EP (53), and the latter also has the unique solution $\mathbf{x}^\theta$. In fact, we can choose $\Gamma(\mathbf{x}, \mathbf{y}) = 0.5\sigma'\|\mathbf{y} - \mathbf{x}\|^2$, then

$$\tilde{\Phi}_\theta(\mathbf{x}, \mathbf{y}) + \Gamma(\mathbf{x}, \mathbf{y}) = \Phi(\mathbf{x}, \mathbf{y}) + 0.5\theta\left[\|\mathbf{y} - \mathbf{v}^l\|^2 - \|\mathbf{x} - \mathbf{v}^l\|^2\right].$$

The proof of convergence of the sequence $\{\mathbf{v}^l\}$ follows from [30, Theorem 2.1] or [35, Corollary 3.3] since their proofs do not utilize in fact the convexity assumption on $\Phi(\mathbf{x}, \cdot)$. □

We adjust this assertion to MEP (2) and MVI (4) and obtain similar convergence properties.

Corollary 5.1 *Suppose that X is a nonempty, convex, and closed subset in $\mathbb{R}^n$, $f : X \times X \to \mathbb{R}$ is an equilibrium bifunction, which is weakly monotone with constant $\sigma' > 0$, $f(\cdot, \mathbf{y})$ is upper semicontinuous for each $\mathbf{y} \in X$, $f(\mathbf{x}, \cdot)$ is weakly convex with constant $\rho' > 0$ and lower semicontinuous for each $\mathbf{x} \in X$, and that $h : X \to \mathbb{R}$ weakly convex with constant $\rho'' > 0$ and lower semicontinuous. Suppose also that problem (8) is solvable, a* *sequence* $\{\mathbf{v}^l\}$ *is constructed in conformity with the rules:*

$$\begin{array}{l} \mathbf{v}^{l+1} \in X, \quad \|\mathbf{v}^{l+1} - \tilde{\mathbf{v}}^{l+1}\| \leq \varepsilon_{l+1}, \tilde{\mathbf{v}}^{l+1} \in X; \\ f(\tilde{\mathbf{v}}^{l+1}, \mathbf{x}) + 0.5\theta \left[\|\mathbf{x} - \mathbf{v}^l\|^2 - \|\tilde{\mathbf{v}}^{l+1} - \mathbf{v}^l\|^2\right] \\ +h(\mathbf{x}) - h(\tilde{\mathbf{v}}^{l+1}) \geq 0 \quad \forall \mathbf{x} \in X; \\ \varepsilon_{l+1} > 0, \quad \sum\limits_{l=0}^{\infty} \varepsilon_l < \infty, \quad \theta > 0, \quad \mathbf{v}^0 \in X. \end{array}$$

Then, there exists a number $\theta' > 0$ *such that, for any* $\theta > \theta'$, $\{\mathbf{v}^l\}$ *has limit points and all these points are solutions of MEP (2). If, in addition, solution sets of problems (2) and (8) coincide, then* $\{\mathbf{v}^l\}$ *converges to a solution of MEP (2).*

To prove this assertion it suffices to set

$$\Phi(\mathbf{x}, \mathbf{y}) = f(\mathbf{x}, \mathbf{y}) + h(\mathbf{y}) - h(\mathbf{x})$$

and notice that all the assumptions of Proposition 5.7 are fulfilled with $\sigma' = \rho' + \rho''$.

Corollary 5.2 *Suppose that assumptions (A1) and (B1) are fulfilled,* $\mathbf{G} : \mathbb{R}^n \to \mathbb{R}^n$ *is weakly monotone with constant* $\sigma' > 0$, *and that* $h : X \to \mathbb{R}$ *weakly convex with constant* $\sigma'' > 0$ *and lower semicontinuous. Suppose also that problem (9) is solvable, a sequence* $\{\mathbf{v}^l\}$ *is constructed in conformity with the rules in (51). Then, there exists a number* $\theta' > 0$ *such that, for any* $\theta > \theta'$, $\{\mathbf{v}^l\}$ *has limit points and all these points are solutions of MVI (4). If, in addition, solution sets of problems (4) and (9) coincide, then* $\{\mathbf{v}^l\}$ *converges to a solution of MVI (4).*

Again, to prove the above corollary it suffices to set

$$\Phi(\mathbf{x}, \mathbf{y}) = \langle \mathbf{G}(\mathbf{x}), \mathbf{y} - \mathbf{x} \rangle + h(\mathbf{y}) - h(\mathbf{x}),$$

and notice that all the assumptions of Proposition 5.7 are fulfilled.

In order to implement the above methods, we have to provide an approximate solution to the auxiliary problems within the prescribed accuracy. As in (CPDPI), we can then utilize gap functions; see [30] for more details. This approach is illustrated for the case of MVI (4).

(CPDPI-NM) Choose a point $\mathbf{v}^0 \in X$, numbers $\eta' > 0$, $\eta'' > 0$, and a positive sequence $\{\tilde{\varepsilon}_l\}$ such that

$$\sum_{l=0}^{\infty} \tilde{\varepsilon}_l < \infty.$$

For each $l = 0, 1, \ldots$, we have a point $\mathbf{v}^l \in X$, apply (DPI) to the perturbed MVI: Find $\tilde{\mathbf{v}}^{l+1} \in X$ such that

$$\begin{array}{c} \langle \mathbf{G}(\tilde{\mathbf{v}}^{l+1}) + \eta'(\tilde{\mathbf{v}}^{l+1} - \mathbf{v}^l), \mathbf{x} - \tilde{\mathbf{v}}^{l+1} \rangle \\ +0.5\eta'' \left[\|\mathbf{x} - \mathbf{v}^l\|^2 - \|\tilde{\mathbf{v}}^{l+1} - \mathbf{v}^l\|^2\right] + h(\mathbf{x}) - h(\tilde{\mathbf{v}}^{l+1}) \geq 0 \quad \forall \mathbf{x} \in X; \end{array} \tag{56}$$

with $\mathbf{x}^0 = \mathbf{v}^l$, i.e. construct a sequence $\{\mathbf{x}^k\}$ until

$$\tilde{\psi}_0^{(l)}(\mathbf{x}^k) \leq \tilde{\varepsilon}_l^2, \tag{57}$$

where

$$\begin{array}{rl}\tilde{\psi}_0^{(l)}(\mathbf{x}) & = \max\limits_{\mathbf{y}\in X}\{\langle \mathbf{G}(\mathbf{x})+\eta'(\mathbf{x}-\mathbf{v}^l), \mathbf{x}-\mathbf{y}\rangle \\ & +h(\mathbf{x})-h(\mathbf{y})+0.5\eta''\left[\|\mathbf{x}-\mathbf{v}^l\|^2-\|\mathbf{y}-\mathbf{v}^l\|^2\right]\}.\end{array} \tag{58}$$

Then set $\mathbf{v}^{l+1} = \mathbf{x}^k$ and increase $l = l + 1$.

Theorem 5.4 *Suppose that assumption (A1) is fulfilled,* $\mathbf{G} : \mathbb{R}^n \to \mathbb{R}^n$ *is continuously differentiable and weakly monotone with constant* $\sigma' > 0$ *on* X*, and that* $h : X \to \mathbb{R}$ *is weakly convex with constant* $\sigma'' > 0$ *on* X *and continuous on an open convex set* V *such that* $V \supset X$*. Suppose also that problem (9) is solvable, a sequence* $\{\mathbf{v}^l\}$ *is constructed by (CPDPI-NM). Then, there exists a number* $\theta' > 0$ *such that, for any* $\theta > \theta'$*:*

(a) the number of iterations of (DPI) for each l is finite;

(b) the sequence $\{\mathbf{v}^l\}$ *has limit points and all these points are solutions of MVI (4);*

(c) if, in addition, solution sets of problems (2) and (9) coincide, then $\{\mathbf{v}^l\}$ *converges to a solution of MVI (4).*

Proof. Choose $\eta' \geq \sigma'$, $\eta'' \geq \sigma'' + \tau$ and set $\theta = \eta' + \eta''$, where $\tau > 0$. Then the mapping $\mathbf{G}(\cdot) + \eta'(\cdot - \mathbf{v}^l)$ satisfies (B1′), whereas the function $h(\cdot) + 0.5\eta''\|\cdot - \mathbf{v}^l\|^2$ satisfies (B2′). This means that, for each l, the sequence $\{\mathbf{x}^k\}$ generated by (DPI) converges to the unique solution $\tilde{\mathbf{v}}^{l+1}$ of MVI (56) because of Theorem 5.1, hence

$$\lim_{k\to\infty} \tilde{\psi}_0^{(l)}(\mathbf{x}^k) = 0,$$

and assertion (a) is true. Next, if we set

$$\begin{array}{rcl}\Phi_\theta(\mathbf{x}, \mathbf{y}) & = & \langle \mathbf{G}(\mathbf{x}) + \eta'(\mathbf{x} - \mathbf{v}^l), \mathbf{y} - \mathbf{x}\rangle \\ & & +h(\mathbf{y}) - h(\mathbf{x}) + 0.5\eta''\left[\|\mathbf{y} - \mathbf{v}^l\|^2 - \|\mathbf{x} - \mathbf{v}^l\|^2\right],\end{array}$$

then MVI (56) coincides with the following EP: Find a point $\mathbf{x}^\theta \in X$ such that

$$\Phi_\theta(\mathbf{x}^\theta, \mathbf{y}) \geq 0 \quad \forall \mathbf{y} \in X. \tag{59}$$

Clearly, $\Phi_\theta : X \times X \to \mathbb{R}$ is a monotone equilibrium bifunction such that $\Phi_\theta(\cdot, \mathbf{y})$ is upper semicontinuous for each $\mathbf{y} \in X$, $\Phi_\theta(\mathbf{x}, \cdot)$ is strongly convex with constant $\tau > 0$ and lower semicontinuous for each $\mathbf{x} \in X$. On account of Proposition 4.3 (b), we conclude that EP (59) has always the unique solution $\mathbf{x}^\theta$. However, by Lemma 2.5, EP (59) is equivalent to EP (51), and the latter also has the unique solution $\mathbf{x}^\theta = \tilde{\mathbf{v}}^{l+1}$. In fact, by setting $\Gamma(\mathbf{x}, \mathbf{y}) = 0.5\eta'\|\mathbf{y} - \mathbf{x}\|^2$, we obtain

$$\begin{array}{rcl}\Phi_\theta(\mathbf{x}, \mathbf{y}) + \Gamma(\mathbf{x}, \mathbf{y}) & = & \langle \mathbf{G}(\mathbf{x}), \mathbf{y} - \mathbf{x}\rangle + h(\mathbf{y}) - h(\mathbf{x}) \\ & & +0.5\theta\left[\|\mathbf{y} - \mathbf{v}^l\|^2 - \|\mathbf{x} - \mathbf{v}^l\|^2\right].\end{array}$$

Next, applying Proposition 5.4 (b) to MVI (56), we have

$$\tilde{\psi}_0^{(l)}(\mathbf{x}) \geq 0.5\tau\|\mathbf{x} - \tilde{\mathbf{v}}^{l+1}\|^2 \quad \forall \mathbf{x} \in X.$$

Combining this inequality with (57) gives

$$\|\mathbf{v}^{l+1} - \tilde{\mathbf{v}}^{l+1}\|^2 \leq 2\tilde{\varepsilon}_l^2/\tau,$$

hence the relations in (51) hold with $\varepsilon_l = \sqrt{2\tau^{-1}}\tilde{\varepsilon}_l$. Therefore, all the assumptions of Corollary 5.2 are fulfilled and assertions (b) and (c) are also true. □

It should be noted that the idea of utilizing regularization type methods with adding nonlinear regularization terms and applying the primal gap function for auxiliary subproblems was proposed first in [23] and called "nonlinear smoothing". The combined methods can be created for more general versions of regularization and proximal point methods; see [8, 17, 34, 44]. Moreover, the regularization and proximal point methods are applicable for some other classes of nonmonotone problems; see [1, 30, 35, 38, 41] and references therein.

5.7. Modifications and Applications

In this section, we describe some applications of the above descent methods to MEP (2) and to the decomposable MVI's and MEP's. First we consider MEP (2). By using the equivalence result from Proposition 2.1 we see that all the above methods applied to MVI (4) where $\mathbf{G}$ is defined in (5) can solve MEP (2) if the equilibrium bifunction f is convex and differentiable in the second variable and the corresponding conditions on $\mathbf{G}$ from the convergence results (see Propositions 5.1 and 5.2, Theorems 5.1–5.3) are fulfilled. In particular, these additional assumptions for the descent methods (DSE), (DSI), and (DPE) involve either strong monotonicity of $\mathbf{G}$ in (5) and convexity of h or monotonicity of $\mathbf{G}$ and strong convexity of h. On account of Proposition 2.3, (strong) monotonicity of $\mathbf{G}$ follows from (strong) monotonicity of f. At the same time, we can construct descent methods directly for MEP (2) by using analogues of the gap function φ_α. In particular, we can replace φ_α by

$$\tilde{\varphi}_\alpha(\mathbf{x}) = \max_{\mathbf{y}\in X} \tilde{\Phi}_\alpha(\mathbf{x}, \mathbf{y}), \tag{60}$$

where

$$\begin{aligned}\tilde{\Phi}_\alpha(\mathbf{x}, \mathbf{y}) = & -f(\mathbf{x}, \mathbf{y}) - 0.5\alpha\|\mathbf{x} - \mathbf{y}\|^2 \\ & -h(\mathbf{y}) + h(\mathbf{x})\end{aligned}$$

(cf. (34), (35)). Then the problem in (60) will have a unique solution, denoted by $\tilde{\mathbf{y}}_\alpha(\mathbf{x})$, if $f(\mathbf{x}, \cdot)$ and h are convex. For instance, the analog of (DSE) from [45] is described as follows.

(DSEME) Choose a point $\mathbf{x}^0 \in X$ and a number $\alpha > 0$. At the k-th iteration, $k = 0, 1, \ldots$, we have a point $\mathbf{x}^k \in X$, compute $\tilde{\mathbf{y}}_\alpha(\mathbf{x}^k)$ and set $\mathbf{d}^k = \tilde{\mathbf{y}}_\alpha(\mathbf{x}) - \mathbf{x}^k$. If $\mathbf{d}^k = 0$, stop. Otherwise, find $\lambda_k \in [0, 1]$ such that

$$\tilde{\varphi}_\alpha(\mathbf{x}^k + \lambda_k\mathbf{d}^k) = \min_{\lambda\in[0,1]} \tilde{\varphi}_\alpha(\mathbf{x}^k + \lambda\mathbf{d}^k),$$

set $\mathbf{x}^{k+1} = \mathbf{x}^k + \lambda_k\mathbf{d}^k$ and go to the next iteration.

Observe that (DSEME), unlike (DSE), does not use any derivatives of f, i.e. it is a zero order method with respect to MEP (2). However, the differentiability of f is still necessary for convergence of (DSEME). At the same time, (DSEME) requires strengthened monotonicity properties of f. More precisely (see [45, Theorem 4.2] and also [40]), together with the strong monotonicity of $\mathbf{G}$ in (5), the following property is necessary for convergence:

$$\langle f'_{\mathbf{x}}(\mathbf{x},\mathbf{y}) + f'_{\mathbf{y}}(\mathbf{x},\mathbf{y}), \mathbf{y}-\mathbf{x}\rangle \geq \kappa\|\mathbf{x}-\mathbf{y}\|^2 \quad \forall \mathbf{x},\mathbf{y} \in X. \tag{61}$$

In case

$$f(\mathbf{x},\mathbf{y}) = \langle \mathbf{C}(\mathbf{x}), \mathbf{y}-\mathbf{x}\rangle,$$

where $\mathbf{C}$ is a continuously differentiable mapping, (61) becomes equivalent to the following inequality

$$\langle \nabla \mathbf{C}(\mathbf{x})^T(\mathbf{y}-\mathbf{x}), \mathbf{y}-\mathbf{x}\rangle \geq \kappa\|\mathbf{y}-\mathbf{x}\|^2 \quad \forall \mathbf{x},\mathbf{y} \in X,$$

and corresponds to the strong monotonicity of $\mathbf{C}$. In the general nonlinear case (61) seems rather restrictive.

Example 5.1 Take

$$f(x,y) = \frac{x^2}{y} - y + \alpha(xy - x^2)$$

on $[0.5,+\infty) \times [0.5,+\infty)$, $\alpha > 0$. Then f is clearly an equilibrium bifunction, $f(x,\cdot)$ is convex for each $x \in X$. Next,

$$G(x) = f'_y(x,y)|_{y=x} = -\frac{x^2}{x^2} - 1 + \alpha x = \alpha x - 2$$

is a strongly monotone mapping. At the same time,

$$\left[f'_x(x,y) + f'_y(x,y)\right](y-x) = \left(2\frac{x}{y} + \alpha y - 2\alpha x - \frac{x^2}{y^2} - 1 + \alpha x\right)(y-x)$$

$$= -\left(1 - \frac{x}{y}\right)^2 (y-x) + \alpha(y-x)^2 = \left[\alpha - \frac{1}{y^2}(y-x)\right](y-x)^2.$$

Setting $x = 1, y = 2, \alpha = 1/8$ in the above expression, we have

$$[f'_x(x,y) + f'_y(x,y)](y-x) = \left(\frac{1}{8} - \frac{1}{4}\right) = -\frac{1}{8} < 0,$$

hence (61) does not hold.

Therefore the descent methods applied to MVI (4)–(5) instead of MEP (2) have certain advantages. Nevertheless, we can adjust the methods based on the primal gap function for MEP (2) if necessary. As to the methods (DSE), (DSI), and (DIDG), they can utilize more general non-quadratic auxiliary functions; see e.g. [8, 40, 59].

Next, we applied the regularized gap function approach to MVI (4) under the assumptions that $\mathbf{G}$ is continuously differentiable and strongly monotone with constant $\tau > 0$ and that h is convex. However, by Proposition 2.1, MVI (4) is then equivalent to the problem of finding $\mathbf{x}^* \in X$ such that

$$\begin{array}{c}\langle \mathbf{G}(\mathbf{x}^*) - \alpha\mathbf{x}^*, \mathbf{y} - \mathbf{x}^*\rangle + \left[h(\mathbf{y}) + 0.5\alpha\|\mathbf{y}\|^2\right] \\ - \left[h(\mathbf{x}^*) + 0.5\alpha\|\mathbf{x}^*\|^2\right] \geq 0 \quad \forall \mathbf{y} \in X,\end{array} \tag{62}$$

where $\alpha \in (0, \tau)$. In fact, we can consider (62) as MEP (2) where

$$f(\mathbf{x}, \mathbf{y}) = \langle \mathbf{G}(\mathbf{x}) - \alpha\mathbf{x}, \mathbf{y} - \mathbf{x}\rangle + 0.5\alpha(\|\mathbf{y}\|^2 - \|\mathbf{x}\|^2)$$

and

$$f'_{\mathbf{y}}(\mathbf{x}, \mathbf{y})|_{\mathbf{y}=\mathbf{x}} = \mathbf{G}(\mathbf{x}) - \alpha\mathbf{x} + \alpha\mathbf{x} = \mathbf{G}(\mathbf{x}).$$

In fact, $h(\cdot) + 0.5\alpha\|\cdot\|^2$ is strongly convex with constant $\alpha > 0$ and $\mathbf{G} - \alpha I$ is monotone. Therefore, we can apply (DPI) to (62) in order to solve MVI (4). Note that this approach requires an evaluation of the strong monotonicity constant of $\mathbf{G}$.

Let us consider applications of splitting based methods to decomposable problems. Again, instead of the decomposable MEP (20) we can consider the decomposable MVI (22), (24) if all the assumptions of Proposition 3.1 are fulfilled. Note that (strong) convexity of the function

$$h(\mathbf{x}) = \sum_{i=1}^{m} h_i(\mathbf{x}_i)$$

becomes equivalent to (strong) convexity of each function h_i and that convexity and closedness of X in (12) become clearly equivalent to convexity and closedness of each X_i. The basic part of the splitting method (33) and the descent methods (DSE), (DSI), and (DIDG) consists in solving the auxiliary optimization problem:

$$\min_{\mathbf{y}\in X} \to \langle \mathbf{G}(\mathbf{x}), \mathbf{y}\rangle + 0.5\alpha\|\mathbf{x} - \mathbf{y}\|^2 + h(\mathbf{y}) \tag{63}$$

which gives the unique solution $\mathbf{y}_\alpha(\mathbf{x})$; see (34), (35) and Proposition 2.1. However, now (63) decomposes into m independent optimization problems:

$$\min_{\mathbf{y}_i\in X_i} \to \langle \mathbf{G}_i(\mathbf{x}), \mathbf{y}_i\rangle + 0.5\alpha\|\mathbf{x}_i - \mathbf{y}_i\|^2 + h_i(\mathbf{y}_i) \tag{64}$$

for $i = 1, \dots, m$; each of them has rather small dimensionality. Therefore, all the above methods can be implemented efficiently for decomposable problems. As to (DPI), computation of $\mathbf{y}(\mathbf{x})$ requires a solution of the problem:

$$\min_{\mathbf{y}\in X} \to \langle \mathbf{G}(\mathbf{x}), \mathbf{y}\rangle + h(\mathbf{y})$$

which is in turn replaced by the m decomposable problems:

$$\min_{\mathbf{y}_i\in X_i} \to \langle \mathbf{G}_i(\mathbf{x}), \mathbf{y}_i\rangle + h_i(\mathbf{y}_i), \quad i = 1, \dots, m;$$

each of them has a unique solution since h_i is strongly convex. Clearly, applying regularization or proximal point methods (see (CRDPI) and (CPDPI)) does not destroy the decomposable structure of basic subproblems which also decompose into a series of independent problems of form (64). Thus, splitting based methods become very suitable for decomposable problems and various parallelization technique can be applied; see also [59, Chapter 8] and references therein.

6. Combined Relaxation Method for Mixed Variational Inequalities

In this section, we consider some other approach to MVI (4) without strengthened convexity/ monotonicity assumptions. This approach was proposed first in [18] for usual VI's and called *combined relaxation* since it can be treated as a development of relaxation or Fejer-monotone methods for linear and convex inequalities; see [24] and [37] for more details. At each iteration of a combined relaxation method (CRM), an auxiliary procedure is used in order to find a hyperplane which separates strictly the current iterate and solution set. The main iteration involves the projection onto this hyperplane and onto the feasible set, if necessary. This approach appeared very flexible and enabled us to essentially weaken the basic assumptions on the problem together with rather rapid convergence; see [24]. For MVI (4) the first CR methods were proposed in [20] and [24, Section 2.2]. Now we consider the CR method from [24, Section 2.2] under somewhat weakened assumptions. Namely, together with (A1) we will utilize the following.

(C1) $\mathbf{G}:\mathbb{R}^n\to\mathbb{R}^n$ *is a continuous monotone mapping.*

(C2) $h:\mathbb{R}^n\to\mathbb{R}$ *is a convex function.*

The method is described as follows.

(CRM) *Step 0*: Choose a point $\mathbf{x}^0\in\mathbb{R}^n$, a sequence of functions $\psi_k:\mathbb{R}^n\to\mathbb{R}, k=0,1,\ldots$, such that

(a) ψ_k is differentiable and its gradient ψ' is Lipschitz continuous with constant $\tau''<\infty$;

(b) ψ_k is strongly convex with constant $\tau'>0$.

Also, choose numbers $\alpha\in(0,1),\beta\in(0,1),\gamma\in(0,2)$, set $k=0$.

Step 1 (Auxiliary procedure): Find m as the smallest non-negative integer such that

$$\begin{array}{l}\langle\mathbf{G}(\mathbf{x}^k)-\mathbf{G}(\mathbf{z}^{k,m}),\mathbf{x}^k-\mathbf{z}^{k,m}\rangle\\ \leq(1-\alpha)\beta^{-m}\langle\psi'_k(\mathbf{z}^{k,m})-\psi'_k(\mathbf{x}^k),\mathbf{z}^{k,m}-\mathbf{x}^k\rangle,\end{array}\tag{65}$$

where $\mathbf{z}^{k,m}$ is a solution of the problem:

$$\min_{y\in X}\to\langle\mathbf{G}(\mathbf{x}^k)-\beta^{-m}\psi'_k(\mathbf{x}^k),\mathbf{y}\rangle+\beta^{-m}\psi_k(\mathbf{y})+h(\mathbf{y}).\tag{66}$$

Set $\theta_k=\beta^{-m},\mathbf{y}^k=\mathbf{z}^{k,m}$. If $\mathbf{y}^k=\mathbf{x}^k$, stop.

Step 2 (Main iteration): Set

$$\begin{array}{l}\mathbf{g}^k=\mathbf{G}(\mathbf{y}^k)-\mathbf{G}(\mathbf{x}^k)-\theta_k^{-1}\left[\psi'_k(\mathbf{y}^k)-\psi'_k(\mathbf{x}^k)\right],\\ \omega_k=\langle\mathbf{g}^k,\mathbf{x}^k-\mathbf{y}^k\rangle,\\ \mathbf{x}^{k+1}=\mathbf{x}^k-\gamma(\omega_k/\|\mathbf{g}^k\|^2)\mathbf{g}^k,\end{array}\tag{67}$$

$k=k+1$ and go to Step 1.

First we note that, under assumptions (A1) and (C2), problem (66) has always the unique solution $\mathbf{z}^{k,m}$ since its cost function is strongly monotone and continuous, besides, by Proposition 2.1, it is equivalent to the following MVI: Find $\mathbf{z}^{k,m}\in X$ such that

$$\begin{array}{c}\langle\mathbf{G}(\mathbf{x}^k)+\beta^{-m}\left[\psi'_k(\mathbf{z}^{k,m})-\psi'_k(\mathbf{x}^k)\right],\mathbf{y}-\mathbf{z}^{k,m}\rangle\\ +h(\mathbf{y})-h(\mathbf{z}^{k,m})\geq 0\quad\forall\mathbf{y}\in X.\end{array}\tag{68}$$

In case

$$\psi_k(\mathbf{x}^k) = 0.5\alpha\|\mathbf{x}\|^2, \alpha > 0; \tag{69}$$

computation of $\mathbf{z}^{k,m}$ is equivalent to one step of the usual forward-backward splitting method (33); see also (34), (35), and (63). For a number $\theta \in (0,1)$ we define $\mathbf{z}^{(k,\theta)} \in X$ such that

$$\begin{array}{c} \langle \mathbf{G}(\mathbf{x}^k) + \theta^{-1}\left[\psi_k'(\mathbf{z}^{(k,\theta)}) - \psi_k'(\mathbf{x}^k)\right], \mathbf{y} - \mathbf{z}^{(k,\theta)}\rangle \\ +h(\mathbf{y}) - h(\mathbf{z}^{(k,\theta)}) \geq 0 \quad \forall \mathbf{y} \in X. \end{array} \tag{70}$$

We establish several useful properties of this auxiliary problem.

Lemma 6.1 *Let assumptions (A1) and (C2) be fulfilled. Fix $\theta \in (0,1)$. Then*
(a) $\tau'\|\mathbf{z}^{(k,\theta)} - \tilde{\mathbf{x}}\| \leq \theta\left(\|\mathbf{G}(\mathbf{x}^k)\| + \|\tilde{\mathbf{g}}\|\right) + \tau''\|\mathbf{x}^k - \tilde{\mathbf{x}}\|, \quad \tilde{\mathbf{x}} \in X, \tilde{\mathbf{g}} \in \partial h(\tilde{\mathbf{x}})$;
(b) $\mathbf{x}^k = \mathbf{z}^{(k,\theta)} \Longleftrightarrow \mathbf{x}^k \in X^*$;
(c) $\tau''(\theta^{-1} - 1)\|\mathbf{z}^{(k,\theta)} - \mathbf{x}^k\| \geq \tau'\|\mathbf{z}^{(k,\theta)} - \mathbf{z}^{(k,1)}\|$.

Proof. Fix $\tilde{\mathbf{x}} \in X$ and take any subgradient $\tilde{\mathbf{g}}$ of h at $\tilde{\mathbf{x}}$, then from (70) it follows that

$$\begin{array}{l} \theta\left(\|\mathbf{G}(\mathbf{x}^k)\| + \|\tilde{\mathbf{g}}\|\right)\|\tilde{\mathbf{x}} - \mathbf{z}^{(k,\theta)}\| \geq \theta\left(\langle \mathbf{G}(\mathbf{x}^k), \tilde{\mathbf{x}} - \mathbf{z}^{(k,\theta)}\rangle + \langle \tilde{\mathbf{g}}, \tilde{\mathbf{x}} - \mathbf{z}^{(k,\theta)}\rangle\right) \\ \geq \theta\left[\langle \mathbf{G}(\mathbf{x}^k), \tilde{\mathbf{x}} - \mathbf{z}^{(k,\theta)}\rangle + h(\tilde{\mathbf{x}}) - h(\mathbf{z}^{(k,\theta)})\right] \\ \geq \langle \psi_k'(\mathbf{z}^{(k,\theta)}) - \psi_k'(\mathbf{x}^k), \mathbf{z}^{(k,\theta)} - \tilde{\mathbf{x}}\rangle \\ = \langle \psi_k'(\mathbf{z}^{(k,\theta)}) - \psi_k'(\tilde{\mathbf{x}}), \mathbf{z}^{(k,\theta)} - \tilde{\mathbf{x}}\rangle + \langle \psi_k'(\tilde{\mathbf{x}}) - \psi_k'(\mathbf{x}^k), \mathbf{z}^{(k,\theta)} - \tilde{\mathbf{x}}\rangle \\ \geq \tau'\|\mathbf{z}^{(k,\theta)} - \tilde{\mathbf{x}}\|^2 - \tau''\|\tilde{\mathbf{x}} - \mathbf{x}^k\|\|\mathbf{z}^{(k,\theta)} - \tilde{\mathbf{x}}\|, \end{array}$$

hence

$$\theta\left(\|\mathbf{G}(\mathbf{x}^k)\| + \|\tilde{\mathbf{g}}\|\right) + \tau''\|\tilde{\mathbf{x}} - \mathbf{x}^k\| \geq \tau'\|\mathbf{z}^{(k,\theta)} - \tilde{\mathbf{x}}\|,$$

and assertion (a) is true.

If $\mathbf{z}^{(k,\theta)} = \mathbf{x}^k$, then (70) implies (4). Conversely, if $\mathbf{x}^k \in X^*$, but $\mathbf{z}^{(k,\theta)} \neq \mathbf{x}^k$, setting $\mathbf{x}^k = \mathbf{y}$ in (70) gives

$$\begin{array}{l} \langle \mathbf{G}(\mathbf{x}^k), \mathbf{z}^{(k,\theta)} - \mathbf{x}^k\rangle + h(\mathbf{z}^{(k,\theta)}) - h(\mathbf{x}^k) \\ \leq \theta^{-1}\langle \psi_k'(\mathbf{z}^{(k,\theta)}) - \psi_k'(\mathbf{x}^k), \mathbf{x}^k - \mathbf{z}^{(k,\theta)}\rangle \\ \leq -\theta^{-1}\tau'\|\mathbf{z}^{(k,\theta)} - \mathbf{x}^k\|^2 < 0, \end{array}$$

a contradiction. Hence, part (b) is true.

Next, adding (70) with $\theta = 1, \mathbf{y} = \mathbf{z}^{(k,\theta)}$ and (70) with $\mathbf{y} = \mathbf{z}^{(k,1)}$ gives

$$\begin{array}{l} \langle \psi_k'(\mathbf{z}^{(k,1)}) - \psi_k'(\mathbf{x}^k), \mathbf{z}^{(k,\theta)} - \mathbf{z}^{(k,1)}\rangle \\ +\theta^{-1}\langle \psi_k'(\mathbf{z}^{(k,\theta)}) - \psi_k'(\mathbf{x}^k), \mathbf{z}^{(k,1)} - \mathbf{z}^{(k,\theta)}\rangle \geq 0. \end{array}$$

It follows that

$$\begin{array}{l} (\theta^{-1} - 1)\langle \psi_k'(\mathbf{z}^{(k,\theta)}) - \psi_k'(\mathbf{x}^k), \mathbf{z}^{(k,1)} - \mathbf{z}^{(k,\theta)}\rangle \\ \geq \langle \psi_k'(\mathbf{z}^{(k,\theta)}) - \psi_k'(\mathbf{z}^{(k,1)}), \mathbf{z}^{(k,\theta)} - \mathbf{z}^{(k,1)}\rangle \\ \geq \tau'\|\mathbf{z}^{(k,\theta)} - \mathbf{z}^{(k,1)}\|^2, \end{array}$$

hence

$$\tau''(\theta^{-1} - 1)\|\mathbf{z}^{(k,\theta)} - \mathbf{x}^k\| \ge \tau'\|\mathbf{z}^{(k,\theta)} - \mathbf{z}^{(k,1)}\|,$$

and part (c) is also true. □

Note that the termination of (CRM), by part (b), means that $\mathbf{x}^k \in X^*$. For this reason, in what follows we consider only the case when (CRM) generates an infinite sequence.

Lemma 6.2 *Let assumptions (A1), (C1), and (C2) be fulfilled. Then the linesearch procedure is always finite, i.e.* $\theta_k > 0$.

Proof. If we suppose that (65) does not hold for any $m \to +\infty$, then

$$\begin{aligned} &\beta^m\langle \mathbf{G}(\mathbf{x}^k) - \mathbf{G}(\mathbf{z}^{k,m}), \mathbf{x}^k - \mathbf{z}^{k,m}\rangle \\ &> (1-\alpha)\langle \psi_k'(\mathbf{z}^{k,m}) - \psi_k'(\mathbf{x}^k), \mathbf{z}^{k,m} - \mathbf{x}^k\rangle \\ &\ge (1-\alpha)\tau'\|\mathbf{z}^{k,m} - \mathbf{x}^k\|^2, \end{aligned}$$

hence

$$\beta^m \left\|\mathbf{G}(\mathbf{x}^k) - \mathbf{G}(\mathbf{z}^{k,m})\right\| \ge (1-\alpha)\tau'\|\mathbf{z}^{k,m} - \mathbf{x}^k\|.$$

The sequence $\{\mathbf{z}^{k,m}\}$ is bounded because of Lemma 6.1 (a), hence

$$\|\mathbf{z}^{k,m} - \mathbf{x}^k\| \to 0 \quad \text{as} \quad m \to +\infty.$$

Since $\mathbf{x}^k \notin X^*$, by Lemma 6.1 (b), $\mathbf{x}^k \ne \mathbf{z}^{(k,1)}$. However, using Lemma 6.1 (c) gives

$$\begin{aligned} \left\|\mathbf{G}(\mathbf{x}^k) - \mathbf{G}(\mathbf{z}^{k,m})\right\| &\ge \frac{(1-\alpha)\tau'}{\beta^m}\|\mathbf{z}^{k,m} - \mathbf{x}^k\| \\ &\ge \frac{(\tau')^2(1-\alpha)}{\tau''(1-\beta^m)}\|\mathbf{z}^{k,m} - \mathbf{z}^{(k,1)}\| \ge \mu' > 0, \end{aligned}$$

which is a contradiction. □

We now show that our method follows the general CR approach.

Lemma 6.3 *Let assumptions (A1), (C1), and (C2) be fulfilled and let MVI (4) be solvable. If* $x^* \in X^*$, *then*

(a)

$$\begin{aligned} &\langle \mathbf{g}^k, \mathbf{x}^k - \mathbf{x}^*\rangle \ge \omega_k \ge (\alpha/\theta_k)\langle \psi_k'(\mathbf{y}^k) - \psi_k'(\mathbf{x}^k), \mathbf{y}^k - \mathbf{x}^k\rangle \\ &\ge (\alpha\tau'/\theta_k)\|\mathbf{y}^k - \mathbf{x}^k\|^2, \end{aligned}$$

and

(b)

$$\|\mathbf{x}^{k+1} - \mathbf{x}^*\|^2 \le \|\mathbf{x}^k - \mathbf{x}^*\|^2 - \gamma(2-\gamma)(\omega_k/\|\mathbf{g}^k\|)^2). \tag{71}$$

Proof. By definition and the monotonicity of $\mathbf{G}$, we have

$$\begin{aligned}
\langle \mathbf{g}^k, \mathbf{y}^k - \mathbf{x}^* \rangle &= \langle \mathbf{G}(\mathbf{y}^k), \mathbf{y}^k - \mathbf{x}^* \rangle \\
&\quad + \langle \mathbf{G}(\mathbf{x}^k) + \theta_k^{-1}\left(\psi'_k(\mathbf{y}^k) - \psi'_k(\mathbf{x}^k)\right), \mathbf{x}^* - \mathbf{y}^k \rangle \\
&\geq \left[\langle \mathbf{G}(\mathbf{y}^k), \mathbf{y}^k - \mathbf{x}^* \rangle + h(\mathbf{y}^k) - h(\mathbf{x}^*)\right] \\
&\quad + \left[\langle \mathbf{G}(\mathbf{x}^k) + \theta_k^{-1}\left(\psi'_k(\mathbf{y}^k) - \psi'_k(\mathbf{x}^k)\right), \mathbf{x}^* - \mathbf{y}^k \rangle \right.\\
&\quad \left. + h(\mathbf{x}^*) - h(\mathbf{y}^k)\right] \geq 0,
\end{aligned}$$

hence, using (65), we obtain

$$\begin{aligned}
\langle \mathbf{g}^k, \mathbf{x}^k - \mathbf{x}^* \rangle &= \langle \mathbf{g}^k, \mathbf{x}^k - \mathbf{y}^k \rangle + \langle \mathbf{g}^k, \mathbf{y}^k - \mathbf{x}^* \rangle \geq \omega_k \\
&= \langle \mathbf{g}^k, \mathbf{x}^k - \mathbf{y}^k \rangle = \langle \mathbf{G}(\mathbf{y}^k) - \mathbf{G}(\mathbf{x}^k), \mathbf{x}^k - \mathbf{y}^k \rangle \\
&\quad + \theta_k^{-1} \langle \psi'_k(\mathbf{y}^k) - \psi'_k(\mathbf{x}^k), \mathbf{y}^k - \mathbf{x}^k \rangle \\
&\geq \left(-(1-\alpha)\theta_k^{-1} + \theta_k^{-1}\right) \langle \psi'_k(\mathbf{y}^k) - \psi'_k(\mathbf{x}^k), \mathbf{y}^k - \mathbf{x}^k \rangle \\
&= (\alpha/\theta_k) \langle \psi'_k(\mathbf{y}^k) - \psi'_k(\mathbf{x}^k), \mathbf{y}^k - \mathbf{x}^k \rangle \geq \left(\frac{\alpha\tau'}{\theta_k}\right) \|\mathbf{y}^k - \mathbf{x}^k\|^2,
\end{aligned}$$

and part (a) is true.

Next, using (a) and (67), we see that

$$\begin{aligned}
\|\mathbf{x}^{k+1} - \mathbf{x}^*\|^2 &= \left\| \mathbf{x}^k - \gamma\left(\frac{\omega_k}{\|\mathbf{g}^k\|^2}\right)\mathbf{g}^k - \mathbf{x}^* \right\|^2 \\
&= \|\mathbf{x}^k - \mathbf{x}^*\|^2 - 2\gamma \frac{\omega_k}{\|\mathbf{g}^k\|^2} \langle \mathbf{g}^k, \mathbf{x}^k - \mathbf{x}^* \rangle + \left(\gamma \frac{\omega_k}{\|\mathbf{g}^k\|}\right)^2 \\
&\leq \|\mathbf{x}^k - \mathbf{x}^*\|^2 - 2\gamma\left(\frac{\omega_k}{\|\mathbf{g}^k\|}\right)^2 + \left(\gamma\frac{\omega_k}{\|\mathbf{g}^k\|}\right)^2 \\
&= \|\mathbf{x}^k - \mathbf{x}^*\|^2 - \gamma(2-\gamma)\left(\frac{\omega_k}{\|\mathbf{g}^k\|}\right)^2,
\end{aligned}$$

and part (b) also true. □

From part (a) of Lemma 6.3 we conclude that $\mathbf{g}^k \neq 0$ at Step 2 since $\mathbf{y}^k \neq \mathbf{x}^k$, i.e. (CRM) is well-defined. Next, the hyperplane

$$H_k = \left\{ \mathbf{y} \in \mathbb{R}^n \mid \langle \mathbf{g}^k, \mathbf{x}^k - \mathbf{y} \rangle = \omega_k \right\}$$

separates strictly the current iterate $\mathbf{x}^k$ and the solution set X^*, the next iterate $\mathbf{x}^{k+1}$ is nothing but the projection of $\mathbf{x}^k$ onto the shifted hyperplane

$$H_k(\gamma) = \left\{ \mathbf{y} \in \mathbb{R}^n \mid \langle \mathbf{g}^k, \mathbf{x}^k - \mathbf{y} \rangle = \gamma\omega_k \right\},$$

but (71) confirms the basic relaxation property. Therefore, the method follows the general CR approach.

We are now ready to establish convergence of (CRM).

Theorem 6.1 *Let assumptions (A1), (C1), and (C2) be fulfilled and let MVI (4) be solvable. Then (CRM) generates a sequence* $\{\mathbf{x}^k\}$ *which converges to a solution of MVI (4).*

Proof. First we see that, by (71), the sequence $\{\mathbf{x}^k\}$ is bounded, hence it has limit points. Again, due to (71) we have to only prove that at least one of these limit points belongs to X^*. Let us consider two possible cases.

Case 1: $\limsup\limits_{k\to\infty} \theta_k \geq \theta' > 0$.

Then there exists a subsequence $\{\theta_{k_s}\} \to \theta' > 0$ and (68) gives

$$\langle G(\mathbf{x}^{k_s}) + \theta_{k_s}^{-1}\left[\psi'_{k_s}(\mathbf{y}^{k_s}) - \psi'_{k_s}(\mathbf{x}^{k_s})\right], \mathbf{y} - \mathbf{y}^{k_s}\rangle + h(\mathbf{y}) - h(\mathbf{y}^{k_s}) \geq 0 \quad \forall \mathbf{y} \in X.$$

However, the sequence $\{\mathbf{y}^k\}$ is also bounded due to Lemma 6.1 (a), besides, Lemma 6.3 gives

$$\lim_{k_s\to\infty} \|\mathbf{y}^{k_s} - \mathbf{x}^{k_s}\| = 0.$$

Hence, taking the limit $k_s \to \infty$ in the above inequality we see that $\mathbf{x}' \in X^*$ if $\mathbf{x}'$ is the corresponding limit point of $\{\mathbf{x}^k\}$.

Case 2: $\lim \theta_k = 0$.

For each k then there exists m_k such that $\mathbf{z}^{k,m_k} = \mathbf{y}^k$. Set $\tilde{\mathbf{y}}^k = \mathbf{z}^{k,m_k-1}$, then from (65) we have

$$\begin{aligned}&\langle \mathbf{G}(\mathbf{x}^k) - \mathbf{G}(\tilde{\mathbf{y}}^k), \mathbf{x}^k - \tilde{\mathbf{y}}^k\rangle > (1-\alpha)(\theta_k/\beta)^{-1}\langle \psi'_k(\tilde{\mathbf{y}}^k) - \psi'_k(\mathbf{x}^k), \tilde{\mathbf{y}}^k - \mathbf{x}^k\rangle \\ &\geq (1-\alpha)\tau'(\theta_k/\beta)^{-1}\|\tilde{\mathbf{y}}^k - \mathbf{x}^k\|^2.\end{aligned}$$

Again, by Lemma 6.1 (a), the sequence $\{\tilde{\mathbf{y}}^k\}$ is bounded, hence

$$\lim_{k\to\infty} \|\tilde{\mathbf{y}}^k - \mathbf{x}^k\| = 0,$$

besides, Lemma 6.3 gives

$$\lim_{k\to\infty} \|\mathbf{y}^k - \mathbf{x}^k\| = 0.$$

By using Lemma 6.1 (c), we now have

$$\left\|\mathbf{G}(\mathbf{x}^k) - \mathbf{G}(\tilde{\mathbf{y}}^k)\right\| \geq \frac{(1-\alpha)\tau'}{(\theta_k/\beta)}\|\mathbf{x}^k - \tilde{\mathbf{y}}^k\| \geq \frac{(1-\alpha)(\tau')^2}{\tau''(1-(\theta_k/\beta))}\left\|\tilde{\mathbf{y}}^k - \mathbf{z}^{(k,1)}\right\|,$$

hence

$$\lim_{k\to\infty} \left\|\mathbf{x}^k - \mathbf{z}^{(k,1)}\right\| = 0.$$

By definition,

$$\langle G(\mathbf{x}^k) + \left[\psi'_k(\mathbf{z}^{(k,1)}) - \psi'_k(\mathbf{x}^k)\right], \mathbf{y} - \mathbf{z}^{(k,1)}\rangle + h(\mathbf{y}) - h(\mathbf{z}^{(k,1)}) \geq 0 \quad \forall \mathbf{y} \in X.$$

Taking the limit $k \to \infty$ in this inequality, we conclude that $\mathbf{x}' \in X^*$ where $\mathbf{x}'$ is any limit point of $\{\mathbf{x}^k\}$. The proof is complete. □

We see that (CRM) requires the same computational expenses per step as the descent methods (DSI) and (DPI), but converges under weaker assumptions. It was shown in [24,

Theorem 2.2.3] that its convergence rate is linear if $\mathbf{G}$ is strongly monotone. Moreover, the above convergence result remains true if we replace the monotonicity of $\mathbf{G}$ with somewhat weaker assumption: Each point $x^* \in X^*$ is also a solution of the dual problem (9); see Corollary 2.2. Also, (CRM) is also suitable for decomposable problems. When solving MVI (22) we should choose the separable auxiliary function

$$\psi_k(\mathbf{x}) = \sum_{i=1}^{m} \psi_{ki}(\mathbf{x}_i),$$

then (66) decomposes into m independent optimization problems. In case (69), they correspond to (64). Observe that the described method can be extended for more general classes of problems; see e.g. [28, 33, 47]. Some other techniques for creating decomposable CR methods can be found in [21, 22, 27]. For this reason, (CRM) can be adjusted to peculiarities of each special problem under solution of forms (2) and (4) and to creation of efficient computational procedures.

References

[1] Allevi E., Gnudi A., Konnov I.V. The proximal point method for nonmonotone variational inequalities, *Mathem. Meth. Oper. Res.*, 2006, Vol.63, pp. 553–565.

[2] Baiocchi, C., Capelo, A. *Variational and Quasivariational Inequalities: Applications to Free Boundary Problems*, John Wiley and Sons, New York, NY, 1984.

[3] Bertsekas, D.P. Necessary and sufficient conditions for a penalty method to be exact, *Mathem. Progr.*, 1975, Vol.9, pp. 87–99.

[4] Blum, E., Oettli, W. From optimization and variational inequalities to equilibrium problems, *The Mathem. Student*, 1994, Vol.63, pp. 123–145.

[5] Brézis, H., Nirenberg, L., Stampacchia, G. A remark on Ky Fan's minimax principle, *Boll. Unione Mat. Ital.*, 1972, Vol.6, pp.293–300.

[6] Browder F.E. On the unification of the calculus of variations and the theory of monotone nonlinear operators in Banach spaces, *Proc. Nat. Acad. Sci. USA*, 1966, Vol.56, pp.419–425.

[7] Bruck, R. On weak convergence of an ergodic iteration for the solution of variational inequalities for monotone operators in Hilbert space, *J. Mathem. Anal. Appl.*, Vol.61, pp.159–164, 1977.

[8] Chadli, O., Konnov, I.V., Yao, J.-C. Descent methods for equilibrium problems in a Banach space, *Comput. Mathem. Appl.*, 2004, Vol.48, pp.609–616.

[9] Dem'yanov, V.F., Rubinov, A.M. Principles of Nonsmooth Analysis and Quasidifferential Calculus, *Nauka*, Moscow, 1990.

[10] Duvaut, G., Lions, J.-L. *Les Inéquations en Mechanique et Physique*, Dunod, Paris, 1972.

[11] Facchinei, F., Pang, J.-S. *Finite-Dimensional Variational Inequalities and Complementarity Problems*, Springer-Verlag, Berlin, 2003.

[12] Fan Ky, A minimax inequality and applications, in: *Inequalities*. III, Edited by O. Shisha, Academic Press, New York, 1972, pp.103–113.

[13] Fukushima, M., Mine, H. A generalized proximal point algorithm for certain nonconvex minimization problems, *Intern. J. Syst. Sci.,* 1981, Vol.12, pp. 989–1000.

[14] Gabay, D. Application of the method of multipliers to variational inequalities, in: *Augmented Lagrangian Methods: Application to the Numerical Solution of Boundary-Value Problems,* Edited by M. Fortin and R. Glowinski, North-Holland, Amsterdam, The Netherlands, 1983, pp. 299–331.

[15] Gol'shtein, E.G., Tret'yakov, N.V. Modified Lagrange Functions, Nauka, Moscow 1989; English translation: *Modified Lagrangians and Monotone Maps in Optimization*, John Wiley and Sons, New York, 1996.

[16] Hadjisavvas, N., Komlósi, S., Schaible, S., Editors. *Handbook of Generalized Convexity and Generalized Monotonicity*, "Nonconvex Optimization and Applications", Vol.76, Springer, New York, NY, 2005.

[17] Kaplan, A., Tichatschke, R. Some results about proximal-like methods, in: Recent Advances in Optimization, Edited by A. Seeger, *Lecture Notes in Economics and Mathematical Systems,* Vol.563, Springer, New York, 2006, pp.61–86.

[18] Konnov, I.V. Combined relaxation methods for finding equilibrium points and solving related problems, *Russ. Math.* (Iz. VUZ), 1993, Vol. 37, no.2, pp.44–51.

[19] Konnov, I.V. On a class of D-gap functions for mixed variational inequalities, *Russ. Math.* (Iz. VUZ), 1999, Vol. 43, no. 12, pp.60–64.

[20] Konnov, I.V. Combined relaxation method for solving variational inequalities with monotone operators, *Comput. Maths. Math. Phys.,* 1999, Vol. 39, pp.1051–1056.

[21] Konnov, I.V. Combined relaxation methods for variational inequality problems over product sets, *Lobachevskii J. of Mathematics,* 1999, Vol.2, pp.3–9.

[22] Konnov, I.V. Combined relaxation method for decomposable variational inequalities, *Optimiz. Methods and Software*, 1999, Vol.10, pp.711–728.

[23] Konnov, I.V. Approximate methods for primal-dual mixed variational inequalities, *Russ. Math. (Iz. VUZ)*, 2000, Vol. 44, no. 12, pp.55–66.

[24] Konnov, I.V. Combined Relaxation Methods for Variational Inequalities, *Lecture Notes in Economics and Mathematical Systems*, Vol.495, Springer-Verlag, Berlin-Heidelberg- New York, 2001.

[25] Konnov, I.V. Mixed variational inequalities and exact penalty functions, in: Numerical Methods for Continuous Casting and Related Problems, Edited by A. Lapin and E. Laitinen, *Proceedings of Lobachevskii's Mathematical Center*, DAS, Kazan, 2001, pp.59–63.

[26] Konnov, I.V. Dual approach to one class of mixed variational inequalities, *Comput. Maths. Math. Phys.*, 2002, Vol.42, pp.1276–1288.

[27] Konnov, I.V. A class of combined relaxation methods for decomposable variational inequalities, *Optimization*, 2002, Vol.51, pp.109–125.

[28] Konnov, I.V. A combined relaxation method for nonlinear variational inequalities, *Optimiz. Meth. Software,* 2002, Vol.17, pp.271–292.

[29] Konnov, I.V. The splitting method with linear search for primal-dual variational inequalities, *Comput. Maths. Math. Phys.*, 2003, Vol.43, pp.494–507.

[30] Konnov, I.V. Application of the proximal point method to nonmonotone equilibrium problems, *J. Optim. Theory Appl.*, 2003, Vol.119, pp.317–333.

[31] Konnov, I.V. Dual approach for a class of implicit convex optimization problems, Math. Methods Oper. Res., 2004, Vol.60, pp.87–99.

[32] Konnov I.V. Generalized monotone equilibrium problems and variational inequalities, in: Handbook of Generalized Convexity and Generalized Monotonicity, Edited by N. Hadjisavvas, S. Komlósi, and S. Schaible, "Nonconvex Optimization and Applications", Vol.76. - New York: Springer, 2005, pp.559–618.

[33] Konnov I.V. Splitting-type method for systems of variational inequalities, *Comput. Oper. Res.*, 2006, Vol.33, pp.520–534.

[34] Konnov I.V. Application of the proximal point method to a system of extended primal-dual equilibrium problems, in: Recent Advances in Optimization, Edited by A. Seeger, *Lecture Notes in Economics and Mathematical Systems,* Vol.563, Springer, New York, 2006, pp.87–102.

[35] Konnov I.V. Partial proximal point method for nonmonotone equilibrium problems, *Optimiz. Meth. Software,* 2006, Vol.21, pp.373–384.

[36] Konnov, I.V. *Equilibrium Models and Variational Inequalities*, Elsevier, Amsterdam, 2007.

[37] Konnov, I.V. Combined relaxation methods for generalized monotone variational inequalities, in: Generalized Convexity and Related Topics, Edited by I.V. Konnov, D.T. Luc, and A.M. Rubinov, *Lecture Notes in Economics and Mathematical Systems,* Vol.583, Springer, Berlin-Heidelberg-New York, 2007, pp.3–31.

[38] Konnov, I.V. Regularization method for nonmonotone equilibrium problems, *J. Nonlin. Conv. Anal.* (submitted).

[39] Konnov, I.V. Descent method with inexact linesearch for mixed variational inequalities, *Russ. Math. (Iz. VUZ)*, to appear.

[40] Konnov, I.V., Ali, M.S.S. Descent methods for monotone equilibrium problems in Banach spaces, *J. Comput. Appl. Mathem.*, Vol. 188, pp. 165–179, 2006.

[41] Konnov I.V., Ali M.S.S., Mazurkevich E.O. Regularization of nonmonotone variational inequalities, *Appl. Mathem. Optim.*, 2006, Vol.53, pp.311–330.

[42] Konnov, I.V., Kum, S. Descent methods for mixed variational inequalities in a Hilbert space, *Nonlin. Anal.: Theory, Methods and Appl.,* 2001, Vol.47, pp.561–572.

[43] Konnov, I.V., Kum, S., Lee, G.M. On convergence of descent methods for variational inequalities in a Hilbert space, *Math. Methods Oper. Res.*, 2002, Vol.55, pp.371–382.

[44] Konnov I.V., Pinyagina O.V. D-gap functions and descent methods for a class of monotone equilibrium problems, *Lobachevskii J. Mathem.*, 2003, Vol.13, pp.57–65.

[45] Konnov I.V., Pinyagina O.V. Descent method with respect to the gap function for nonsmooth equilibrium problems, *Russ. Math. (Iz. VUZ),* 2003, Vol.47, no.12, pp.67–73.

[46] Konnov I.V., Pinyagina O.V. D-gap functions for a class of equilibrium problems in Banach spaces, *Comput. Methods Appl. Mathem.,* 2003, Vol.3, pp.274–286.

[47] Konnov I.V., Schaible, S., Yao, J.C. Combined relaxation method for equilibrium problems, *J. Optim. Theory Appl.,* 2005, Vol.126, pp.309–322.

[48] Konnov, I.V., Volotskaya, E.O. Mixed variational inequalities and economic equilibrium problems, *J. Appl. Mathem.*, Vol.2, pp.289–314, 2002.

[49] Lemaréchal, C., Nemirovskii, A., Nesterov, Y. New variants of bundle methods, *Mathem. Progr.*, 1995, Vol. 69, pp.111–147.

[50] Lescarret C. Cas d'addition des applications monotones maximales dan un espace de Hilbert, *Compt. Rend. Acad. Sci. (Paris),* 1965, Vol. 261, pp.1160–1163.

[51] Lions, P.L., Mercier, B. Splitting algorithms for the sum of two monotone operators, *SIAM J. Num. Anal.*, 1979, Vol.16, pp.964–979.

[52] Mine, H., Fukushima, M. A minimization method for the sum of a convex function and a continuously differentiable function, *J. Optim. Theory Appl.*, 1981, Vol.33, pp. 9–23.

[53] Nikaido, H., Isoda, K. Note on noncooperative convex games, *Pacific J. Mathem.,* 1955, Vol.5, pp.807–815

[54] Okuguchi, K., Szidarovszky, F. The Theory of Oligopoly with Multi-Product Firms, *Lecture Notes in Economics and Mathematical Systems*, Springer-Verlag, 1990.

[55] Panagiotopoulos, P.D. *Inequality Problems in Mechanics and Their Applications,* Birkhauser, Boston, 1985.

[56] Passty, G.B. Ergodic convergence to zero of the sum of two monotone operators in Hilbert space, *J. Math. Anal. Appl.,* 1979, Vol.72, pp.383–390.

[57] Patriksson M. Merit functions and descent algorithms for a class of variational inequality problems, *Optimization*, 1997, Vol.41, pp.37–55.

[58] Patriksson M. Cost approximation: a unified framework of descent algorithms for nonlinear programs, *SIAM J. Optim.*, 1998, Vol.8, pp. 561–582.

[59] Patriksson M. *Nonlinear Programming and Variational Inequality Problems: A Unified Approach*, Kluwer Academic Publishers, Dordrecht, 1999.

[60] Peng J.-M. Equivalence of variational inequality problems to unconstrained minimization, *Mathem. Progr.*, 1997, Vol.78, pp.347–355.

[61] Polyak, B.T. Introduction to Optimization, Nauka, Moscow, 1983 (Engl. transl. in *Optimization Software,* New York, 1987).

[62] Rockafellar, R.T. *Convex Analysis,* Princeton University Press, Princeton, 1970.

[63] Rockafellar, R.T. Monotone operators and the proximal point algorithm, *SIAM J. Contr. Optimiz.*, 1976, Vol. 14, pp.877–898

[64] Sukharev, A.G., Timokhov, A.V., Fedorov, V.V. A Course in Optimization Methods, *Nauka*, Moscow, 1986 (in Russian).

[65] Uryas'ev, S.P. Adaptive Algorithms of Stochastic Optimization and Game Theory. *Nauka*, Moscow, 1990 (in Russian).

[66] Zangwill, W.I. *Nonlinear Programming: A Unified Approach,* Prentice-Hall, Englewood Cliffs, NJ, 1969.

In: Game Theory: Strategies, Equilibria and Theorems
Editors: Ingrid N. Haugen et al, pp. 161-182
ISBN: 978-1-60456-844-8

Chapter 5

DISCRETE MODELS OF POLITICAL COMPETITION: STRATEGIES OF VICTORY AND STUDY OF EQUILIBRIUM

***Mª Dolores López*[a,1] *and Javier Rodrigo*[b,2]**

[a]Department of Applied Mathematical and Computer Science, Higher Technical School of Civil Engineering, Polytechnic University of Madrid. Profesor Aranguren s/n. 28040 Madrid, Spain

[b]Department of Applied Mathematics, Higher Technical School of Engineering, University Pontificia Comillas of Madrid. Alberto Aguilera, 23. 28015 Madrid, Spain

Abstract

A discrete two-dimensional political competition model has been proposed and addressed with geometric strategies that find the equilibrium positions, if they exist, and ensure their uniqueness. To adapt the problem to various political landscapes one simply assumes that the distribution of voters is not uniform. This complexity can be represented by assigning an appropriate weight to each position of a voter in the policy plane (we assume a finite number of voters). The case in which all the voters have the same weight, and the general case, when they have different weights is studied. It has been proved that, in both cases, except for the particular case where all voters are aligned along a single line of the plane, an equilibrium, if it exists, is attained only when both parties choose to offer the same policy to their voters. That is to say, the two parties will converge to essentially the same political program in order to maximise the number of voters. To avoid the uniqueness of the equilibrium position a definition of equilibrium weaker than the classical one is provided. As a result, a "region of equilibrium" appears. In this area, the parties can operate in a situation of "almost" equilibrium, in which they are not necessarily required to adopt the same policy. Finally, the maximum number of positions of weak equilibrium when the two parties choose positions of the voters is stated, and some examples where this maximum number is attained are presented.

Keywords: Game theory, Equilibrium, Strategies, Computational geometry, Political competition.

[1] E-mail address: ma08@caminos.upm.es.
[2] E-mail address: jrodrigo@upcomillas.es.

1. Introduction

The Nash equilibrium has been studied in general models of competition. It was first stated by John Forbes Nash in his dissertation, *Non-cooperative games* (Nash, 1951), as a way to obtain an optimum strategy for games with two or more players. Plott (1997), Kramer (1973), McKelvey (1976) and others have demonstrated that pure-strategy Nash equilibria generally do not exist when the competition takes place in a space of more than one dimension.

Various approaches to search for a resolution to this situation have been reported in the literature, among them mixed-strategy equilibria, uncovered sets, probabilistic voting, and valence criteria. See for example Laver and Shepsle, 1996; McKelvey, 1976; McKelvey, 1986; Enelow and Hinich, 1982; Londregan and Roemer, 1993; Ansolabehere and Snyder, 2000; Banks and Sundaram, 1993; Hinich and Mueger, 1995.

In this work, a discrete model of two-party competition is presented. It is expected to fit the political reality of a large number of countries. The study of the equilibrium positions in this game is performed.

In this chapter, we present a game related to Political Economy that is a discrete version of the well-known game of Voronoi (Fekete and Meijer, 2003; Ahn, et al., 2004) and it is also an equivalent model to a discrete version of the Downs game (Downs, 1957; Roemer, 2001). Strategies of victory are developed and the condition of equilibrium is studied by means of geometric tools.

The game we propose, is the following one: Two players choose their position on a plane in which n points are located given by their two coordinates. We consider that each player captures those points that are closer to him than to the other one. To count the points each player gets, we trace the perpendicular bisector of the two positions of the players. Then, each one will get the points located in their half-plane. The winner will be the player that gets more points (Serra and Revelle, 1994; Smid, 1997; Aurenhammer and Klein, 2000; Okabe et al., 2000; Abellanas et al., 2006).

This game can be translated in terms of Political Economy: Let us consider the two players as two political parties p and q whose positions are given by the politics they offer, t_1 and t_2, of the two-dimensional space of politics $T=R^2$. We can consider the set of points $v_i=(v_{i1}, v_{i2})$ with $i=1,\dots,n$, as the corresponding positions of the voters of a certain population, with v_i belonging to the set of types $H=\{v_1, \dots, v_n\} \subset R^2$ (Roemer, 2001).

In section 2 we assume that the set of types is equally distributed, that is to say, the types are supported by the same number of voters.

In order to adapt this model to a particular political reality, in section 3 we stipulate that the types positions v_i are not evenly distributed. That is to say, certain positions in the policy space will be supported by more voters. As an example, extreme positions (with respect to the majority of political actions) usually have fewer supporters than moderate positions. It thus seems more reasonable to consider a weighted distribution of voters.

It can be proven that, in both cases, except for the particular case where all voters are aligned along a single line of the plane, an equilibrium, if it exists, is attained only when both parties choose to offer the same policy to their voters. That is to say, the two parties will converge to essentially the same political program in order to maximise the number of voters. To avoid the uniqueness of the equilibrium position, in section 4 a definition of equilibrium weaker than the classical one is provided. As a result, a "region of equilibrium" that can be defined by geometric methods appears. In this area, the parties can operate in a situation of "almost" equilibrium, in which they are not necessarily required to adopt the same policy. Finally, we look for the maximum number of positions of weak equilibrium when the two parties choose positions of the voters.

2. Geometric Study of the Equilibrium: Types Equally Distributed

Along the paper we assume that voter preferences over the issue space are Euclidean, so the payoff functions in the presented game are given by:

$$\Pi^1(t_1,t_2) = \text{number of points } v_i \text{ such that } d(v_i,t_1) \le d(v_i,t_2)$$
$$\Pi^2(t_1,t_2) = \text{number of points } v_i \text{ such that } d(v_i,t_1) > d(v_i,t_2) = n - \Pi^1(t_1,t_2), \quad (1)$$
$$\text{if } t_1 \neq t_2$$

where $d(t,v_i)$ is the Euclidean distance between policy t and position v_i.

In the case where $t_1 = t_2$, Equation 1 becomes: $\Pi^1(t_1,t_2) = \Pi^2(t_1,t_2) = \frac{n}{2}$, and each party takes half of the voters.

2.1. Strategies to Win

We present strategies for the party p to choose a position so that it improves its gain when it knows the position of the other party q (Wendell and McKelvery, 1981).

Proposition 1:
If n is even, then there is a strategy for p that allows it to tie q whatever it is the position of q.

Proof:
We draw parallel straight lines such that none of them contains more than one point of the set.

When we get a line that leaves exactly $\frac{n}{2}$ points of the set in each open half-plane it determines, we locate p in the symmetrical of q about this line. Then, p will obtain exactly $\frac{n}{2}$ points of the set. (Figure 1) #

Remarks:
1) The strategy we have chosen is different to the trivial one, that is to say to locate p in the position of q.
2) For even n, there are situations in those it is impossible for p to win, see figure 2. In this case, it is impossible to locate p so that it captures three points of that set, this is because the half-plane that contains those three points will contain its convex hull (de Berg et al.,

1997) but q belongs to this convex hull so, they will be points captured by q. Therefore, p can only choose a situation in order to tie.

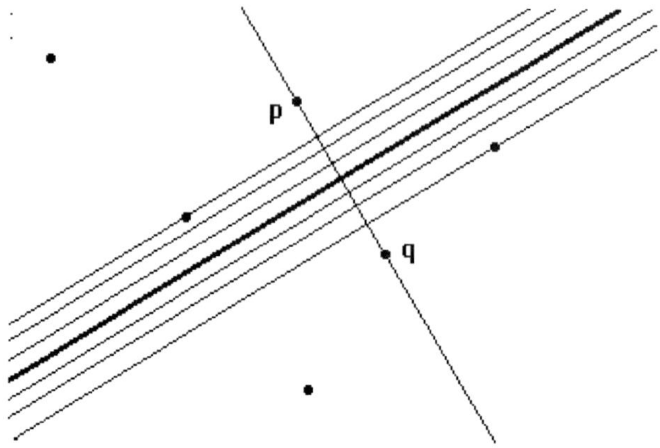

Figure 1. For even n, p can tie q.

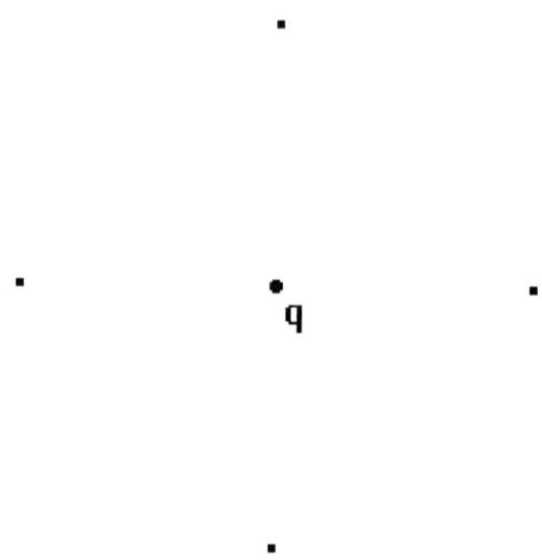

Figure 2. In this situation, p cannot win.

Proposition 2:

If n is odd, then there is a strategy to locate p to get $\left[\frac{n}{2}\right]+1$ voters and then win q, whenever q is not located in the position of some voter.

Proof:

We consider a family of parallel lines with slopes different to the straight line that join two points of the set or lines that join a point of the set and q. When we get a line of this family such that one of its half-plane contains $\left[\frac{n}{2}\right]+1$ points of the set and the other one contains the rest of the points and q, we locate p in the symmetrical of q with respect to this line. (Figure 3) #

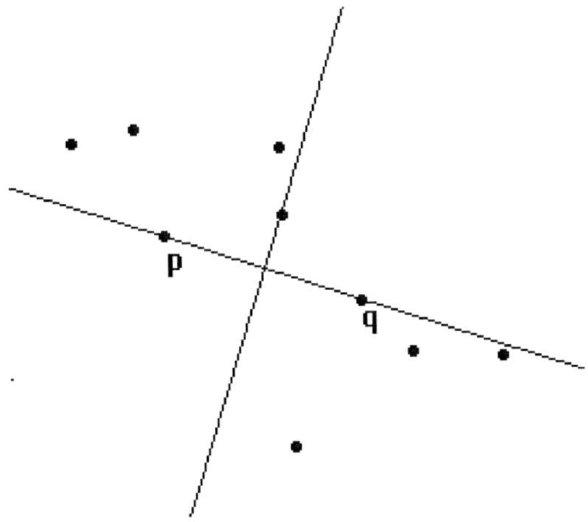

Figure 3. When n is odd, p can always win q if q is not located in the position of a voter.

2.2. Existence of Equilibrium

Necessary and sufficient conditions will now be developed for the existence of equilibrium in the game.

Proposition 3:

Consider $\left[\frac{n}{2}\right]+1$ points of the set of n points and a position of the first party, t. Then there exists a location of the second party, t', in which it captures those $\left[\frac{n}{2}\right]+1$ points, if and only if t does not belong to the convex hull of the $\left[\frac{n}{2}\right]+1$ points.

Proof:

First, it can be seen that, given a convex hull of $\left[\frac{n}{2}\right]+1$ points of the set and an outer point p, there exists a straight line that separates the point and the convex hull.

Consider the part of the hull visible from p. The convex hull is a compact set and then the visible part has two support points: p_1 and p_2. Let p_1 be the one with highest order, proceeding angularly clockwise from p. The straight line that goes through p and p_1 leaves the convex hull in a half-plane. Rotating this straight line "infinitesimally" clockwise, with its center at p, and translating it "infinitesimally" in the direction of the convex hull, a straight line can be found that has p on one side and the convex hull on the other (Figure 4).

In this situation, if q is located in the symmetrical point of p with respect to that line, then it will capture the $\left[\frac{n}{2}\right]+1$ voters (Figure 5).

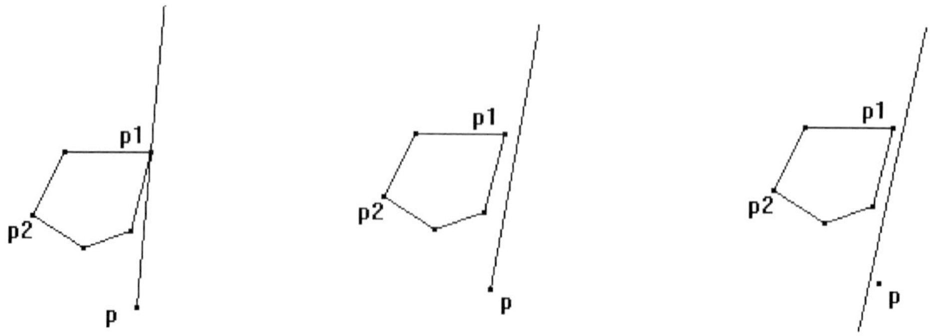

Figure 4. Rotating and then translating infinitesimally the straight line that goes by p and p_1, we separate p and the convex hull.

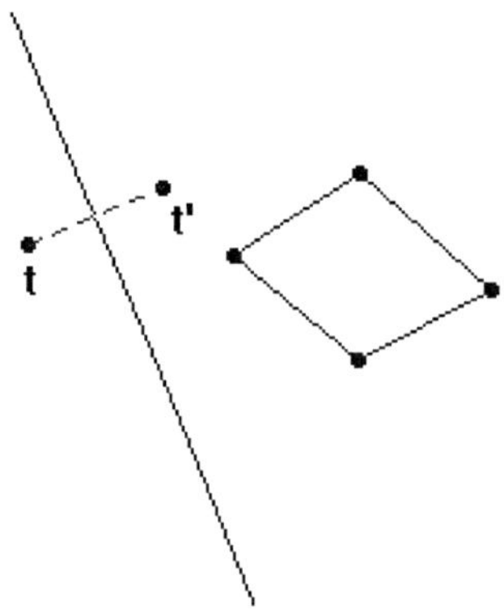

Figure 5. We locate q in t' that is the symmetrical of p located in t with respect to the line.

If t is in the convex hull of the $\left[\frac{n}{2}\right]+1$ points, then, there is no position t' for the party q that will allow it to capture these $\left[\frac{n}{2}\right]+1$ points, because any half-plane containing the points will also contain their convex hull, so it will contain t. Thus, the half-plane contains the voters of t and the points are assigned to t rather than to t'. #

Definition 1: Let $v_1,\ldots,v_n$ be n positions, and let $\left[\frac{n}{2}\right]+i$ of these n positions be considered in all possible ways, and define $C_{n,i}$ as the intersection of the convex hulls of those points.

Proposition 4:

There exist locations for a party in which the other party cannot remove from him $\left[\frac{n}{2}\right]+1$ points if and only if $C_{n,1}$ is not empty. Any point in $C_{n,1}$ will be one of those locations.

Proof:

If $C_{n,1}$ is not empty, then any location of p in a point of $C_{n,1}$ ensures that the other party cannot obtain any selection of $\left[\frac{n}{2}\right]+1$ voters, since p is included in the convex hull of these $\left[\frac{n}{2}\right]+1$ voters (Proposition 3).

If $C_{n,1}$ is empty, then for each position of p, $\left[\frac{n}{2}\right]+1$ points of the set can be found such that p is not in their convex hull, so by Proposition 3, a strategy for q to obtain them exists. #

Now, Proposition 4 can be applied to find the equilibrium positions in the proposed game.

Proposition 5:

In the game as presented, equilibrium positions exist if and only if $C_{n,1}$ is not empty. In this case, the only equilibrium positions will be any (t_1, t_2) such that t_1 and t_2 are in this set. Therefore, equilibrium situations of the (t,t) form are included.

Proof:

If $C_{n,1}$ is not empty, then any position (t_1, t_2) such that t_1 and t_2 belong to $C_{n,1}$ is an equilibrium situation.

If p is located in t_1, then it is known (from Proposition 4) that q cannot obtain more than $\frac{n}{2}$ voters wherever it is located, so $\Pi^2(t_1,t)\leq\frac{n}{2}$ for every t. The same reasoning applies for the first party when q is in t_2.

On the other hand, in the position (t_1, t_2) each party obtains $\frac{n}{2}$ voters, since $\Pi^1(t_1,t_2)\leq\frac{n}{2}$, $\Pi^2(t_1,t_2)\leq\frac{n}{2}$, and the gains are complementary.

Therefore, $\Pi^2(t_1,t)\leq\frac{n}{2}=\Pi^2(t_1,t_2)$, and the same pattern exists for p, so (t_1,t_2) is an equilibrium position.

These are the only equilibrium positions; if (t_1,t_2) is an equilibrium position, as the payoff are complementary, $\Pi^1(t_1,t_2)=\Pi^2(t_1,t_2)=\frac{n}{2}$. If, say, t_1 does not belong to $C_{n,1}$, then there is a strategy for t_2 to obtain $\left[\frac{n}{2}\right]+1$ points of the set in a position t (Proposition 3), so $\Pi^2(t_1,t)=\left[\frac{n}{2}\right]+1>\Pi^2(t_1,t_2)$, which is a contradiction since (t_1,t_2) is an equilibrium position.

If $C_{n,1}$ is empty and there is an equilibrium position (t_1, t_2), then one of the payoffs, say Π^2, satisfies the condition that $\Pi^2(t_1,t_2)\leq\frac{n}{2}$, since the gains are complementary. However, applying Proposition 3, since $C_{n,l}$ is empty, for the position t_1 of the party p, there exists a position t of the party q that obtains $\left[\frac{n}{2}\right]+1$ points of the set, resulting in: $\Pi^2(t_1,t)=\left[\frac{n}{2}\right]+1>\frac{n}{2}\geq\Pi^2(t_1,t_2)$, so the party can change its position profitably. That is a contradiction, because (t_1,t_2) is an equilibrium situation. #

2.3. Uniqueness of Equilibrium

The previous subsection presented a geometrical argument demonstrating the equilibrium positions for the game, if they exist, and where they are located. This section presents cases in which these positions are unique.

a) Case of odd *n*

Proposition 6:

$C_{n,l}$, n being odd, is empty or consists of points of the set. Because the intersection of convex sets is itself a convex set, in this last case, the intersection consists of only one point of the set.

Proof:

If there is a point in $C_{n,l}$ not belonging to the set, and p is located in this point, then q cannot gain $\left[\frac{n}{2}\right]+1$ points from p by Proposition 3, but for odd n, there is a strategy for a party to obtain $\left[\frac{n}{2}\right]+1$ points of a set of n points if the other is not situated in a point of the set (proposition 2). This yields a contradiction. #

b) Case of even *n*

A similar result can be presented for even n. A preliminary proposition is required:

Proposition 7:

Let v_l be a point of the set in the boundary of the convex hull of the n points of the set. There exists a halving line containing v_l (Erdõs et al., 1973).

Proof:

For any point in the boundary of the convex hull, a line can be found containing this point that leaves the convex hull in a half-plane. It can be assumed without loss of generality that the convex hull is in the half-plane below the line (or at the right side of the line if it is a vertical line).

Taking this approach, v_l can be designated the highest point of the set in the line, and the other points of the set can be arranged by their angles from v_l. Since all the points of the set are in the same half-plane, it can be concluded that the line that joins the middle point in this arrangement with v_l, leaves the same number of points on each side (Figure 6). #

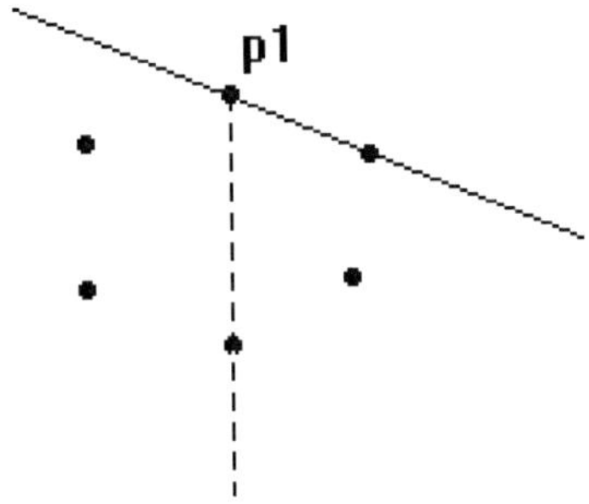

Figure 6. Angular arrangement and election of the point that, together with p_1, leaves half of the number of the points of the set to each side.

Proposition 8:

For even n, if the n points of the set are not aligned, then $C_{n,1}$ is a point or the empty set.

Proof:

Select a point of the set in the boundary of the convex hull of the n points. It is known by Proposition 7 that there is another point of the set such that the line connecting the two points leaves $\frac{n}{2}-1$ points on each side. Then, the intersection of the convex hull of $\frac{n}{2}-1$ points on one side plus the two points in the line, and the convex hull of $\frac{n}{2}-1$ points on the other side plus the two points in the line, are both segments contained in the line. As the points of the set are not aligned, there is a point in the boundary of the convex hull not in the line containing the segment. Applying Proposition 7, another point of the set can be obtained such that the line containing the two points leaves $\frac{n}{2}-1$ points on each side. Following the same reasoning as before, two convex hulls of $\frac{n}{2}+1$ points of the set can be found, whose intersection is another segment contained in the previous line. Then the intersection of the two segments is contained in the intersection of the two lines, so the intersection of the two segments is contained in a set of one point. Therefore, the intersection of the four convex hulls under consideration is contained in a point, and $C_{n,1}$ is empty or a unique point. #

Remark: In the case of even n and aligned points, there is an infinite number of equilibrium situations, since $C_{n,1}$ will be the segment determined by the two intermediate points (Figure 7).

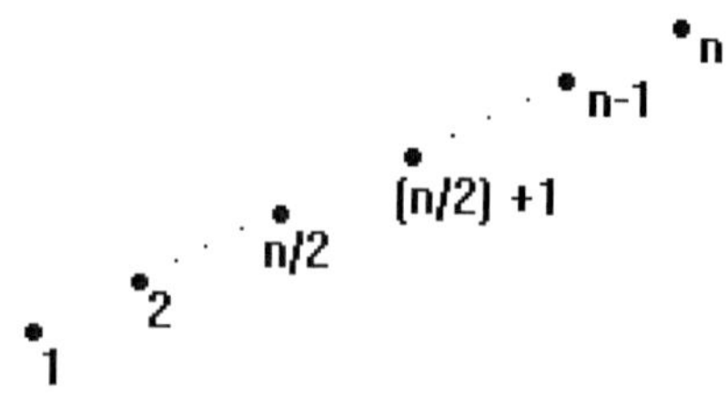

Figure 7. For even n, it is possible to find cases of infinite equilibrium positions.

c) General result

Taking into account the results demonstrated for cases a) and b), it is possible to state the following general result:

Proposition 9:

$C_{n,1}$ is a point or the empty set, unless the points are aligned with even n. Therefore, if there is equilibrium in the game as presented, then it is unique and is of the format (t, t), unless the points are aligned with even n.

3. Geometric Study of the Equilibrium: The Weighted Case

In this section we consider a distribution of the different types according to a measure of probability of the form $F(\{v_i\})=k_i$, where $k_1+k_2+...+k_n=1$ and all $k_i>0$. Keeping these considerations in mind, the players' payoff functions Π can be defined as

$$\left.\begin{array}{l}\Pi^1(t^1,t^2)=n\sum_{j=1}^{n_{t1}}k_{i_j}\\ \Pi^2(t^1,t^2)=n-\Pi^1(t_1,t_2)\end{array}\right\}\text{ if } t^1\neq t^2,\ \Pi^1(t^1,t^2)=\Pi^2(t^1,t^2)=\frac{n}{2}\quad\text{if } t^1=t^2 \quad (3)$$

If we define the weight of position v_i as $weight(v_i)=n\,k_i$, then the payoff of policy t^1 will be the sum of the weights of all positions located in the same half-plane as t^1, including the points on the bisector. Here we assume that the policies are distinct ($t^1\neq t^2$). The second policy (t^2) follows the same pattern, except for positions on the bisector.

We note that the total payoff is equal to the number of voters:

$$\sum_{i=1}^{n} weight(v_i)=\sum_{i=1}^{n} nk_i=n$$

3.1. Existence of Equilibrium

Let us see in the next two subsections the following conditions of existence and uniqueness of equilibrium, the proofs are analogous to those given in section 2, hence they are ommited. We need two previous definitions

Definition 2: We define the weight of a set $\{ v_{i_1}, \ldots, v_{i_k} \}$ as $\sum_{j=1}^{k} weight\ (v_{i_j})$

Definition 3: A *minimal subset* of the set $\{ v_1, \ldots, v_n \}$ is a subset of points whose weight is greater than $\frac{n}{2}$, and that itself contains no other subset with weight greater than $\frac{n}{2}$.

Proposition 10: Consider all the possible minimal subsets of $\{ v_1, \ldots, v_n \}$. There exist Nash equilibria in the game if and only if the intersection of the convex hulls of these subsets is not the empty set. Furthermore, the equilibria positions are the positions (t^1, t^2) such that t^1, t^2 are in the intersection.

3.2. Uniqueness of Equilibrium

We will now show that the intersection of the convex hulls of minimal subsets described in the previous section contains at most a single point, unless the n voter positions all lie on a single line. We will also look at a particular case where the intersection must belong to the set of positions $\{ v_1, \ldots, v_n \}$.

Proposition 11: If there is no combination of points from the set $\{ v_1, \ldots, v_n \}$ with weight $\frac{n}{2}$, then the intersection of the convex hulls of minimal subsets is at most in one point of the set: $\{ v_i \}$.

Now we can come to a more general result:

Proposition 12: If the n voter positions are not contained in a single line, then the convex hulls of minimal subsets intersect at most in one point.

Remark: In the degenerate case that all the points are in a single line and there exists a combination of points with weight $\frac{n}{2}$, the intersection of the convex hulls may be an infinite set.

This analysis leads to the following conclusion:

Proposition 13: The equilibrium in the present game, if it exists, is the unique point (t, t) for some $t \in R^2$. In other words, both parties will choose to offer the same policy, except in cases where the voter positions lie along a single line.

Remark: Propositions 10, 11, 12, 13 are the generalizations of propositions 5, 6, 8 and 9 respectively.

4. Weak Equilibrium

Sections 2 and 3 state that the equilibrium in the cases presented, if it exists, is attained only when both parties choose to offer the same policy to their voters. This result is the same as the obtained in the continuous version of this game where the equilibrium, if any, is obtained in the position of the ideal policy of the median voter (Roemer, 2001; Person and Tabellini, 1999). To avoid this uniqueness, we recall the definition of Nash equilibrium by means of:

Definition 4: A position $(t^0{}_1, t^0{}_2)$ is a weak equilibrium if: $\Pi^1(t_1, t_2^0) \leq \Pi^1(t^0{}_1, t^0{}_2) + 1, \quad \Pi^2(t_1^0, t_2) \leq \Pi^2(t^0{}_1, t^0{}_2) + 1 \quad \forall t_1, t_2 \in T$.

In this section, a geometric analysis is developed that extends that presented in the section 2, to search for equilibrium positions, if they exist, according to the new definition.

4.1. Existence of Equilibrium

Proposition 14

In a position of weak equilibrium $(t^0{}_1, t^0{}_2)$, it necessarily holds that

$$\Pi^1(t_1^0, t_2^0) \geq \frac{n}{2} - 1, \quad \Pi^2(t_1^0, t_2^0) \geq \frac{n}{2} - 1.$$

Proof:

Let $(t^0{}_1, t^0{}_2)$ be a position of weak equilibrium. Assume that, say, $\Pi^1(t_1^0, t_2^0) < \frac{n}{2} - 1$.

Then $\Pi^1(t_2^0, t_2^0) = \frac{n}{2} > \Pi^1(t_1^0, t_2^0) + 1$, which is a contradiction, since $(t^0{}_1, t^0{}_2)$ is a position of weak equilibrium. #

Remark: As a consequence of the last proposition, in the case of complementary payoffs, in a position of weak equilibrium with even n, the payoffs must be $\frac{n}{2} - 1$, $\frac{n}{2} + 1$, or $\frac{n}{2}$ for both. In a position of weak equilibrium with odd n, the payoffs must be $\left[\frac{n}{2}\right] = \frac{n-1}{2}$, $\left[\frac{n}{2}\right] + 1 = \frac{n+1}{2}$, or both equal to $\frac{n}{2}$ (in this last case, the two parties choose the same policy).

The next step is to search for necessary and sufficient conditions for a position to be a position of weak equilibrium.

Proposition 15

In the game as presented, there exist positions of weak equilibrium if and only if $C_{n,2}$ is not empty (n>2).

Proof:

Let t be a point belonging to $C_{n,2}$. It will be shown that the position (t,t) is a weak equilibrium:

$\Pi^1(t,t) = \Pi^2(t,t) = \dfrac{n}{2}$. If, say, $\Pi^1(t_1,t) \geq \left[\dfrac{n}{2}\right] + 2$ for some position t_1 of the first party, then there exists a straight line that separates at least $\left[\dfrac{n}{2}\right] + 2$ points of the set from t, so t would not belong to the convex hull of these $\left[\dfrac{n}{2}\right] + 2$ points. This is a contradiction with the initial assumption. #

Remark: The proof of the other implication of the proposition is analogous to that performed in the case studied in the section 2 (Proposition 5).

The next step is to characterize the positions of weak equilibrium according to the parity of n.

a) Case of odd n

Positions of weak equilibrium are (t_1,t_2) with $\Pi^1(t_1,t_2) = \left[\dfrac{n}{2}\right]$, $\Pi^2(t_1,t_2) = \left[\dfrac{n}{2}\right] + 1$, t_1 in $C_{n,3}$, and t_2 in $C_{n,2}$. This way, t_2 cannot earn $\left[\dfrac{n}{2}\right] + 3$ points to increase its payoff by two in any position, and in the same way, t_1 cannot obtain $\left[\dfrac{n}{2}\right] + 2$ points of the set. These are the only possible positions of weak equilibrium, with these payoffs: if one of the parties is not in the set referred to, the other one can separate it from $\left[\dfrac{n}{2}\right] + 3$ (respectively $\left[\dfrac{n}{2}\right] + 2$) points of the set by changing its position, thus increasing its gain by two units.

These positions make sense when $\left[\dfrac{n}{2}\right] + 3 \leq n$, that is to say, $n \geq 5$.

The other positions of weak equilibrium are (t, t) with t belonging to $C_{n,2}$ (n>1).

There are no more positions of weak equilibrium with other payoffs, since they would not satisfy the necessary condition (Proposition 14).

b) Case of even n

The positions of weak equilibrium will be:

1. Those in which one of the parties has gain $\frac{n}{2}-1$ and is located in $C_{n,3}$, and the other has gain $\frac{n}{2}+1$ and is located in $C_{n,1}$, assuming that this last set is not empty and $n>4$. In these positions, neither of the parties can increase its score by two by moving. These are the only possible positions of weak equilibrium with these gains.
2. The other positions of weak equilibrium are those in which each of the two parties has half of the payoff and is in $C_{n,2}$.

Remark: In the case $n=4$, 1. must be sustituted by: The positions with one of the parties located in $C_{4,1}$, and the other one located in $C_{4,2}$.

4.2. Examples

For odd n, with $n>1$, every position (t_1, t_2) such that t_1, t_2 are in $C_{n,2}$ is a position of weak equilibrium, since these positions with $t_1 \neq t_2$ ensure that the payoffs of the parties are $\left[\frac{n}{2}\right]$, $\left[\frac{n}{2}\right]+1$, which is case a) described above. However, these are not the only positions of weak equilibrium.

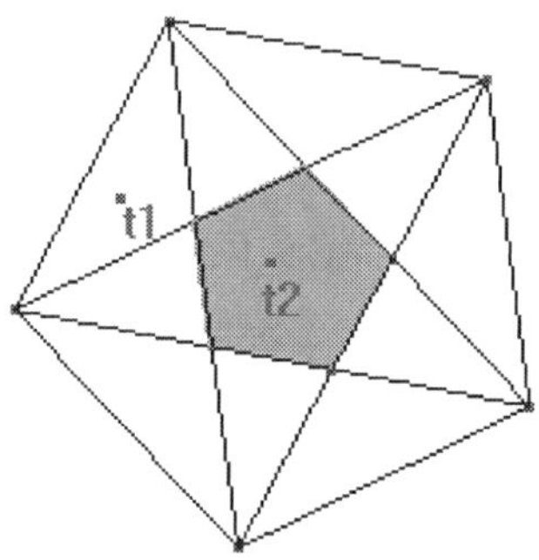

Figure 8. Points in position of weak equilibrium for odd n.

As an example, for $n=5$, with the set of points in Figure 8, $C_{5,2}$ is the shaded polygon. If t_1, t_2 are the positions labeled in the figure, then:

$$\Pi^1(t_1,t_2)=2, \quad \Pi^2(t_1,t_2)=3.$$

Since t_2 belongs to $C_{5,2}$, the first party cannot change its position to win four voters, so it cannot improve its payoff by two units. In the same way, because t_1 is inside the polygon, then the second party cannot win five voters in any position, so it is a position of weak equilibrium, but t_1 does not belong to $C_{5,2}$.

It should be remarked that in this set there is no position of Nash equilibrium, because the intersection of convex hulls of subsets of three points is empty (Proposition 5). Therefore, the search for a position of weak equilibrium in this example appears interesting.

For even n, $n > 2$, it should be noted that the positions (t_1, t_2) with t_1, t_2 in $C_{n,2}$ are not necessarily weak equilibrium positions, because these positions do not ensure the payoffs given in case b). As in the previous case, there are also positions of weak equilibrium in which one of the parties is not in $C_{n,2}$. As an example, for $n = 6$, in the situation shown in Figure 9, the gains are $\Pi^1(t_1,t_2) = 2$, $\Pi^2(t_1,t_2) = 4$ with t_2 in $C_{6,1}$, t_1 in $C_{6,3}$, so this is a position of weak equilibrium as it is stated in case b), but t_1 is not in $C_{6,2}$.

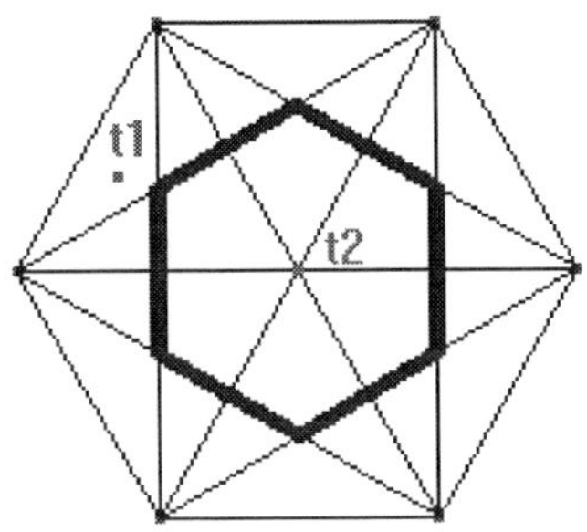

Figure 9. Points in position of weak equilibrium for even n.

This example yields a single position of Nash equilibrium, (t_2,t_2), in which the two parties choose the same policy, but it has an infinite number of positions of weak equilibrium.

4.3. Maximum Number of Positions of Weak Equilibrium with Parties Choosing Positions of Voters

In this subsection, we study the problem of determine the maximum number of positions of weak equilibrium (t_1,t_2), with t_1,t_2 belonging to the set of types H .We consider as the same position of weak equilibrium the positions (t_1,t_2), (t_2,t_1), and we assume along this subsection that the points of H are in general position, that is to say, there is no three points of H in the same line. We find examples of sets where this maximum is attained. As we have seen in subsection 4.1., in positions (t_1,t_2) of weak equilibrium, the points t_1,t_2 have to belong to the sets $C_{n,1}$, $C_{n,2}$ or $C_{n,3}$, so we have to maximize the number of points of

H in these sets in order to solve the problem. To do this, we maximize the number of points of the set that can be in $C_{n,s}$ for $\left\lceil \frac{n}{2} \right\rceil + s < n$, $s > 0$:

Definition 5: We define I_k as the intersection of the convex hulls of the possible subsets of n-k points that can be formed in a set of n points.

Proposition 16

If n is an odd number and $1 < s < \frac{n+1}{2}$, then the maximum number of points of the set that can belong to $C_{n,s}$ is $2s-2$.

If n is an even number and $s < \frac{n}{2}$, then the maximum number of points of the set that can belong to $C_{n,s}$ is $2s-1$.

Proof:

According to a result of Erdõs et al., 1973, the minimum number of points that can take part in some k-set in a set of n points in general position, is $2k+1$ for $0 < k < \frac{n}{2}$.

This implies that the maximum number of points of the set that can belong to I_k is $n-(2k+1)$, since these point can not take part in any k-set. Therefore, if n is an odd number, let s be a number such that $1 < s < \frac{n+1}{2}$, then taking $k = \frac{n+1}{2} - s$, we have that $0 < k < \frac{n-1}{2}$ and then the maximum number of points of the set that can belong to I_k, that is $C_{n,s}$, is $n-(n+1-2s+1) = 2s-2$. If n is an even number and we consider s with $s < \frac{n}{2}$, then taking $k = \frac{n}{2} - s$, we have that the maximum number of points of the set in I_k, that is $C_{n,s}$, is $n-(n-2s+1) = 2s-1$. #

Remark: Consequently, if $n \geq 5$ is an odd number and we have a set of n points, then the maximum number of points of the set belonging to $C_{n,2}$ is 2, and the maximum number of points of the set belonging to $C_{n,3}$ is 4 if $n > 5$, 5 if $n = 5$, if n is an even number greater or equal than 6 and we have a set of n points no all in a line, then the maximum number of points of the set belonging to $C_{n,1}$ is 1, the maximum number of points of the set belonging to $C_{n,2}$ is 3, and the maximum number of points of the set belonging to $C_{n,3}$ is 5 if $n > 6$, 6 if $n = 6$. If $n = 4$, the four points belong to $C_{n,2}$.

4.3.1. Examples

a) We are going to see an example of n points, n an even number, in which the maximum number of points of the set in $C_{n,1}$, $C_{n,2}$ is attained. Moreover, in all the possible positions of weak equilibrium (t_1, t_2), with t_1, t_2 belonging to the set, the conditions on the payoffs established in subsection 4.1. b) are fulfilled, so it is an example with the maximum number of positions of weak equilibrium (t_1, t_2), with t_1, t_2 belonging to the set.

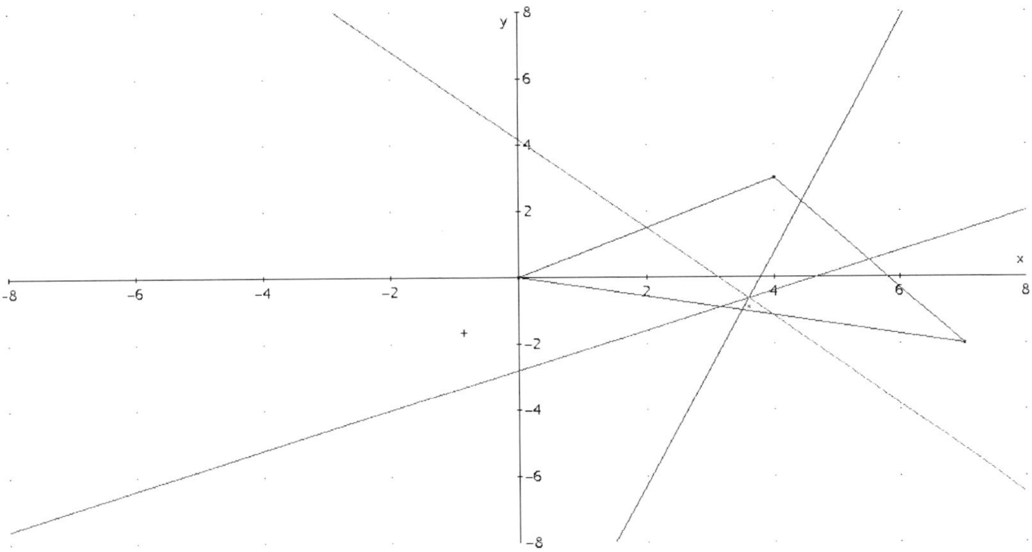

Figure 10. An example of points with the maximum number of positions of weak equilibrium (t_1, t_2) with t_1, t_2 points of the set.

We consider the triangle defined by: $v_1 = (0,0)$, $v_2 = (4,3)$, $v_3 = (7,-2)$, and we also consider the point $v_4 = \left(\frac{18}{5}, -\frac{9}{10}\right)$, in the triangle (see figure 10). It can be checked that $C_{4,1}$ is $\{v_4\}$, and the 4 points of the set are in $C_{4,2}$ as we have noted in the remark of proposition 16, so we have the maximum possible number of points of the set in $C_{4,1}$, $C_{4,2}$. It can be seen (figure 10) that in this example every position (t_1, t_2) with t_1, t_2 belonging to the set is a position of weak equilibrium, so we have the maximum of 10 positions of weak equilibrium with the two parties in positions of the voters (6 with different policies: $t_1 \neq t_2$).

b) Now we see an example of $n = 5$ points in general position that attains the maximum number of points of the set in $C_{5,2}$, $C_{5,3}$, so it is an example with the maximum number of positions of weak equilibrium (t_1, t_2), with t_1, t_2 belonging to the set, for odd n.

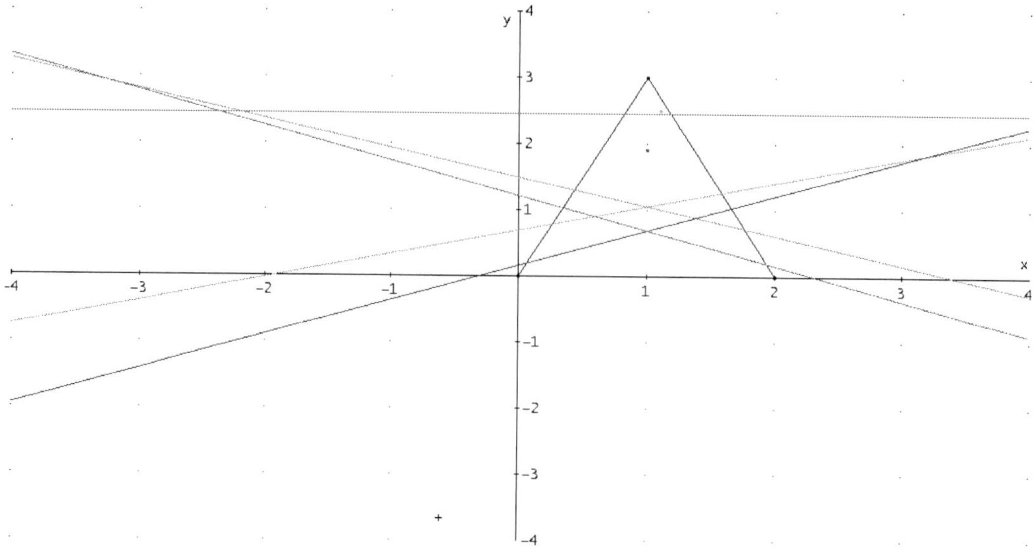

Figure 11. An example of points with the maximum number of positions of weak equilibrium (t_1, t_2) with t_1, t_2 points of the set, when n is an odd number.

We consider the triangle defined by $v_1 = (0,0)$, $v_2 = (2,0)$ and $v_3 = (1,3)$, and the points inside the triangle: $v_4 = \left(\frac{11}{10}, \frac{5}{2}\right)$, $v_5 = \left(1, \frac{19}{10}\right)$ (see figure 11). In this example it is attained the maximum number of two points of the set in $C_{5,2}$ (v_4 and v_5), and the maximum number of two points of the set in $C_{5,3}$.

We can check (figure 11) that $\prod^1(v_4, v_1) = \prod^1(v_4, v_2) = 3$, $\prod^1(v_5, v_1) = \prod^1(v_5, v_2) = \prod^1(v_5, v_3) = 3$, so the conditions on the payoffs are fulfilled, and we have 8 positions of weak equilibrium (t_1, t_2) with t_1, t_2 points of the set.

Remarks:

- The position (v_4, v_3) is not a position of weak equilibrium, since $\prod^1(v_4, v_3) = 4$.
- It can we proven that it is impossible to find an example of a set of 5 points with 9 positions of weak equilibrium (t_1, t_2) with t_1, t_2 points of the set.

c) Now we see an example of $n = 6$ points that attains the maximum number of points of the set in $C_{6,1}$, $C_{6,2}$, $C_{6,3}$, and in all the possible positions of weak equilibrium (t_1, t_2) with t_1, t_2 points of the set, the conditions in the payoffs are satisfied:

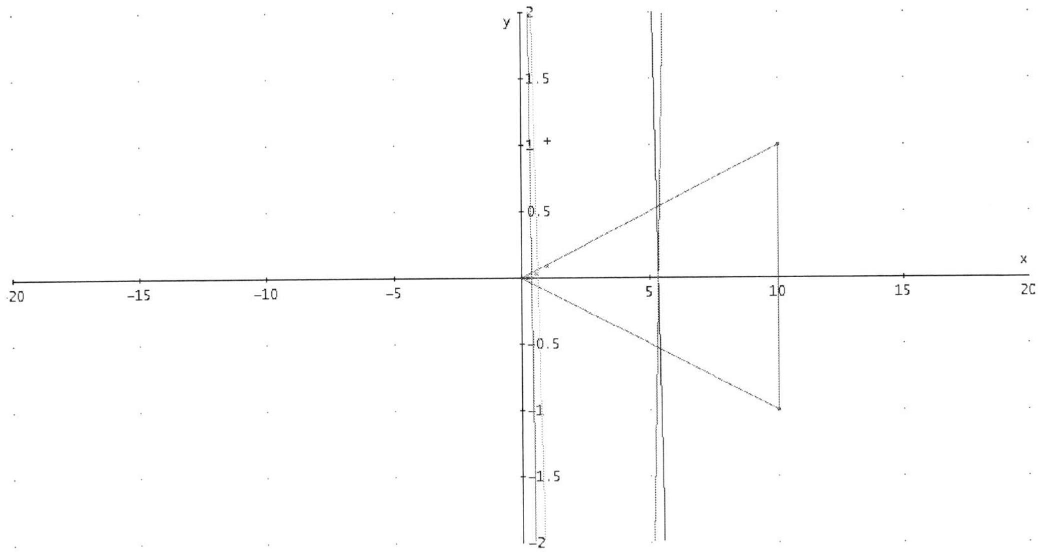

Figure 12. An example of points with the maximum number of positions of weak equilibrium (t_1, t_2) with t_1, t_2 points of the set, when n=6 (bisectors deformed by the scale).

Consider the points $v_1 = (0,0)$, $v_2 = (10,1)$, $v_3 = (10,-1)$, and the points $v_4 = \left(\frac{1}{4}, 0\right)$, $v_5 = \left(1, \frac{3}{35}\right)$, inside the triangle generated by v_1, v_2, v_3. We search for a point v_6 in $C_{5,2}$, that it can be checked to be the triangle enclosed by the segments $v_5 - v_3$, $v_4 - v_2$ and $v_4 - v_5$. Hence, we can choose $v_6 = \left(\frac{14}{25}, \frac{4}{125}\right)$ (see figure 12).

So we have that $C_{6,3}$ contains the 6 points of the set, in $C_{6,2}$ it is attained the maximum of 3 points of the set: v_4, v_5, v_6, (the points of the set that does not belong to the boundary of the convex hull of the 6 points).

Moreover, since we have located v_6 in $C_{5,2}$, the maximum of one point of the set in $C_{6,1}$ (v_6) is attained.

We can check (figure 12) that $\Pi^1(v_4, v_5) = \Pi^2(v_4, v_5) = 3$, $\Pi^1(v_6, v_1) = 4, \Pi^2(v_6, v_1) = 2$, $\Pi^1(v_6, v_2) = 4, \Pi^2(v_6, v_2) = 2$ and $\Pi^1(v_6, v_3) = 4, \Pi^2(v_6, v_3) = 2$.

Hence, in every possible position of weak equilibrium the conditions in the payoffs are satisfied, so we have the maximum of 6 positions of weak equilibrium (t_1, t_2) with t_1, t_2 points of the set, $t_1 \neq t_2$, as in the case $n = 4$ (but there are only three positions of weak equilibrium of the type (t, t), with t a point of the set: the positions (v_i, v_i) with

$i = 4,...,6$, so we have a total of 9 positions of weak equilibrium (t_1, t_2) with t_1, t_2 points of the set, a number lower that in the case $n = 4$).

5. Conclusion

There are a number of localization studies in the framework of the public economy that study various problems using discrete or continuous approaches (Eiselt et al., 1993; Ghosh and Harche, 1993; Hodgson, 1981; Hakimi, 1986; Hakimi, 1990; Mehrez and Stulman, 1982; Church, 1984). In the equilibrium analysis of most of these competitive multidimensional games, it is found that such positions do not exist except for singular cases, so there exist no positions for the players that guarantee that the competitor cannot increase his gain by moving.

This chapter determined the Nash equilibrium of a competitive political game. The scenario presented is a discrete version of the Voronoi game in computational geometry, and also a discrete version of the Downs model in political economics. Furthermore, we have stated the conditions that must be satisfied by possible equilibrium positions. The results obtained have been compared with those established for the bidimensional continuous Downs game.

To associate the model with a real political situation, we have divided the voter population into a finite number of types represented by specific points on a plane. Each party chooses a point on the plane representing their offered policy, and receives the maximum payoff when it minimizes the Euclidean distance to as many voters as possible. This treatment, when coupled with a geometric development for determining the equilibrium position, represents a new insight into the solution of such games. Despite this simplification of the voter population, we have obtained results similar to those presented by works where voter types are represented by a continuum.

The study has been developed in two cases, one in which all the types have the same weight, and other in which they have different weights. Except for the particular case where all the types are aligned along a single line of the plane, an equilibrium, if it exists, is attained only when both parties choose to offer the same policy to their voters. That is to say, the two parties will converge to essentially the same political program in order to maximize the number of voters.

Although in this paper we worked with a simplified (two-party) model, nowadays this treatment is adequate for the majority of countries. In most democracies there are two parties that represent the vast majority of voters, and we can observe that in general their policy offerings become more similar over time (they tend towards the equilibrium position). Our model succeeds in representing this fact.

Despite this, the problem of nonexistence of equilibrium in the majority of situations has appeared. To escape from this situation, a weakened form of equilibrium has been defined, that ensures to each player that the other cannot improve his payoff by more than one unit if he changes his position. This new definition of equilibrium can be useful in cases with no Nash equilibrium.

Indeed, the positions of weak equilibrium are currently regions of the plane, so they yield infinite possibilities for the parties to move. By contrast, the only position for the parties in the Nash equilibrium is to adopt the same policy, which is a meaningless resolution.

To apply the theory, some specific examples have been presented, in which the parties have been located in positions of the voters. The maximum number of positions with this new equilibrium has been found, and sets where this maximum number is attained are given.

This new equilibrium can be an improvement on the treatment of a vaste number of competition studies about different subjects, not only in politics, where a discrete view fits and there does not exist positions of Nash equilibrium. In most of these cases, the variation of one unit in the gain (the only difference between the two equilibria) is not relevant.

The study of the existence and locations of the positions of equilibrium and weak equilibrium has been developed here with a wider scope by applying techniques from computational geometry such as the intersection of convex hulls, because of the discrete nature of the game as presented. So, this work proves the relationship between two scientific areas as Political Economy and Computational Geometry.

References

Abellanas, M., Lillo, I., López, M., Rodrigo, J. (2006) Electoral strategies in a dynamical democratic system: geometric models. *European Journal of Operational Research*, **175**, 870–878.

Ahn, H.-K., Cheng, S.-W., Cheong, O., Golin, M., Van Oostrum, R. (2004). Competitive Facility Location: The Voronoi Game, *Theoretical Computer Science*, **310**, 457-467.

Ansolabehere, S., Snyder, J. (2000). Valence politics and equilibrium in spatial election models. *Public Choice*, **103**, 327–336.

Aurenhammer, R., Klein, R. (2000). Voronoi diagrams. In: Sack, J.-R. and Urrutia, J. (eds.), *Handbook of Computational Geometry*. Elsevier , Amsterdam.

Banks, J., Sundaram, R.K. (1993). Moral hazard and adverse selection in a model of repeated elections. In: Barnett, W.A., Hinich, M.J., and Schoefield, N.J. (eds.), Political Economy: Institutions, Competition and Representation. Cambridge: Cambridge University Press.

Church, R. (1984). The planar maximal covering location problem. *Journal of Regional Science*, **24**(2), 185–201.

de Berg, M., van Kreveld, M., Overmars, M., Schwarzkopf, O. (1997). *Computational Geometry: Algorithms and Applications* (2nd ed.), Springer, New York.

Downs, A. (1957). *An Economic Theory of Democracy*. Harper and Row.

Eisets, H.A., Laporte, G., Thisse, J.H.. (1993). Competitive location models: a framework and bibliography. *Transportation Science*, **27,** 44–54.

Enelow, J., Hinich, M. (1982). Nonspatial candidate characteristics and electoral competition. *Journal of Politics*, **44**, 115–130.

Erdõs, P., Lovász, L., Simmons, A. and Strauss, E. G. (1973). Dissection graphs of planar point sets, in *A Survey of Combinatorial Theory*, J. N. Srivastava, editor, North-Holland, 139-154.

Fekete, S. Meijer, H. (2003). The One-Round Voronoi Game Replayed. Workshop on Algorithms and Data Structures, *Springer Lecture Notes in Computer Science*, **2748**, 150-161.

Ghosh, A., Harche, F. (1993). Location-allocation models in the private sector: progress, problems and prospects. *Location Science*, **1**, 81–106.

Hakimi, S.L. (1986). P-median theorems for competitive location. *Annals of Operations Research*, **5**, 79–88.

Hakimi, S.L. (1990). Locations with spatial interaction: competitive locations and games. In: Mirchandani PB, Francis RL, editors. *Discrete location theory*. New York: Wiley.

Hinich, M.H., Munger, M.G. (1995). Ideology and the Theory of Political Choice. Ann Arbor: University of Michigan Press.

Hodgson, M.J. (1981). A location-allocation model maximizing consumers' welfare. *Regional Studies*, **15**, 493–506.

Kramer, G.H. (1973). On a class of equilibrium conditions for majority rule. *Econometrica*, **42**, 285–297.

Laver, M., Shepsle, K.A. (1996). *Making and Breaking Governments*. Cambridge: Cambridge University Press.

Lodregan, J., Romer, T. (1993). Polarization, incumbency, and the personal vote. In Barnett, W.A., Hinich, M.J., and Schoefield, N.J. (eds.), *Political Economy: Institutions, Competition and Representation*. Cambridge: Cambridge University Press.

McKelvin, R.D. (1976). Intransitivities in multidimensional voting models and implications for agenda control. *Journal of Economic Theory*, **12**, 472–482.

- McKelvey, R.D. (1986). Covering, dominance, and the institution-free properties of social choice. *American Journal of Political Science*, **30**, 283–314.

Mehrez, A., Stulman, A. (1982). The maximal covering location problem with facility placement on the entire plane. *Journal of Regional Science*, **22**(3), 361–365.

Nash, J. (1951). Non-cooperative games. *Annals of Mathematics*, **54**, 286–295.

Okabe, A., Boots, B., Sugihara, K., Chiu, S. (2000). *Spatial Tessellations: Concepts and Applications of Voronoi Diagrams.* John Wiley & Sons, Chichester.

Persson, T., Tabellini, G. (1999). Political Economics and Public Finance. NBER Working Papers Series. *Handbook of Public Economics*, vol III.

Plott, C.R. (1967). A notion of equilibrium and its possibility under majority rule. *American Economic Review*, **57**, 787–806.

Roemer, J. (2001). *Political Competition.* Harvard University Press.

Serra, D., Revelle, C. (1994). Market capture by two competitors: the preemptive location problem. *Journal of Regional Science*, **34**(4), 549–561.

Smid, M. (1997). Closest-point problems in computational geometry, In Sack, J.-R., and Urrutia, J. (eds.) *Handbook of Computational Geometry*. Elsevier, Amsterdam.

Wendell, R.E., McKelvery, R.D. (1981). New perspectives in competitive location theory. *European Journal of Operational Research*, **6**, 174-182.

In: Game Theory: Strategies, Equilibria and Theorems ISBN 978-1-60456-844-8
Editors: Ingrid N. Haugen et al, pp. 183-216

Chapter 6

LEARNING AUTOMATA-BASED SOLUTIONS TO THE GOORE GAME AND ITS APPLICATIONS*

B. John Oommen[1,2†] ***and Ole-Christoffer Granmo***[2‡]
[1]School of Computer Science, Carleton University,
Ottawa, Canada : K1S 5B6
[2]Department of Information and Communication Technology,
University of Agder, Grooseveien 36, N-4876 Grimstad, Norway

Abstract

The Goore Game[1] (GG) introduced in [1] has the fascinating property that it can be resolved in a completely distributed manner with no inter-communication between the players. The game has recently found applications[2] in many domains, including the field of sensor networks and Quality-of-Service (QoS) routing. In actual implementations of the solution, the players are typically replaced by Learning Automata (LA). In this Chapter, we shall first briefly survey the field of LA and report the LA-based solutions to the GG. The problem with the existing reported approaches is that the accuracy of the solution achieved is intricately related to the number of players participating in the game - which, in turn, determines the resolution. In other words, an arbitrary accuracy can be obtained only if the game has an infinite number of players, which renders a practical solution infeasible. Thus, we shall describe some of the recent advances, and show how we can attain an unbounded accuracy for the GG by utilizing no more than three stochastic learning machines, and by recursively pruning the solution space to guarantee that the retained domain contains the solution to the game with a probability as close to unity as desired. The Chapter contains the formal algorithms, the proofs of the respective convergence results, and simulation results demonstrating its power. It also conjectures on how the solution can be applied to some of the application domains.

---

*This work was partially supported by NSERC, the Natural Sciences and Engineering Research Council of Canada.

†E-mail address: oommen@scs.carleton.ca.
‡E-mail address: ole.granmo@uia.no

[1]This game is also referred to as the Gur Game in the related literature.

[2]A brief survey of these applications will be included in the Chapter.

Keywords: Learning Automata, Intelligent Game Playing, Goore Games, Sensor Networks and Quality-of-Service Routing

1. Overview of the Chapter

Learning Automata (LA) have been used to model biological learning systems and to find the optimal action that is offered by a Random Environment. Learning is accomplished by explicitly interacting with the Environment and processing its responses to the actions that are chosen, while gradually converging toward an ultimate goal. They have also been used to play games of various sorts where the parameters of the game are unknown and stochastic. In this Chapter, we intend to study its application in solving the so-called Goore Game[3] (GG) introduced in [44]. The GG has the fascinating property that it can be resolved in a completely distributed manner with no inter-communication between the players. The game has recently found applications[4] in many domains, including the field of sensor networks and Quality-of-Service (QoS) routing. In actual implementations of the solution, the players are typically replaced by LA. The problem with the existing reported approaches is that the accuracy of the solution achieved is intricately related to the *number* of players participating in the game – which, in turn, determines the resolution. In other words, an arbitrary accuracy can be obtained only if the game has an *infinite* number of players, which renders a practical solution infeasible. In this Chapter, we show how we can attain an unbounded accuracy for the GG by utilizing no more than *three* stochastic learning machines, and by recursively pruning the solution space to guarantee that the retained domain contains the solution to the game with a probability as close to unity as desired. The Chapter contains the formal algorithms, the proofs of the respective convergence results, and simulation results demonstrating its power. It also conjectures on how the solution can be applied to some of the application domains.

2. Introduction

One of the most fascinating games studied in the field of artificial games is the Goore Game (GG). We describe it using the following informal formulation given in [24].

> *Imagine a large room containing N cubicles and a raised platform. One person (voter) sits in each cubicle and a Referee stands on the platform. The Referee conducts a series of voting rounds as follows. On each round the voters vote "Yes" or "No" (the issue is unimportant) simultaneously and independently (they do not see each other) and the Referee counts the fraction, λ, of "Yes" votes. The Referee has a uni-modal performance criterion $G(\lambda)$, which is optimized when the fraction of "Yes" votes is exactly λ^*. The current voting round ends with the Referee awarding a dollar with probability $G(\lambda)$ and assessing a dollar with probability $1 - G(\lambda)$ to every voter independently. On*

[3] This game is also referred to as the *Gur Game* in the related literature.

[4] A brief survey of these applications will be included in the Chapter.

the basis of their individual gains and losses, the voters then decide, again independently, how to cast their votes on the next round.

The game has many interesting and fascinating features which render it both non-trivial and intriguing. These are listed below:

1. The game is a non-trivial *non-zero-sum* game.
2. Unlike the games traditionally studied in the AI literature (like Chess, Checkers, Lights-Out etc.) the game is essentially a *distributed* game.
3. The players of the game are ignorant of all of the parameters of the game. All they know is that they have to make a choice, for which they are either rewarded or penalized. They have no clue as to how many other players there are, how they are playing, or even of how/why they are rewarded/penalized.
4. The stochastic function used to reward or penalize the players can be completely arbitrary, as long as it is uni-modal.
5. The most fascinating feature of this game [24] is that if each voter updates its action based on either a Tsetlin automaton with large memory, or an absolutely expedient[5] algorithm, then the entire group will asymptotically optimize the Referee's performance criterion.

2.1. Applications to the Goore Game

The literature concerning the GG is sparse. It was initially studied in the general learning domain, and, as far as we know, was for a long time merely considered as an interesting pathological game. Recently, however, the GG has found important applications within two main areas, namely, QoS (Quality of Service) support in wireless sensor networks [14] and within cooperative mobile robotics as summarized in [13].

The GG has found applications within the field of sensor networks, as explained briefly here. Consider a base station that collects data from a sensor network. The sensors of the network are battery driven and have been dropped from the air, leaving some of them non-functioning. The functioning sensors can either be switched on or off, and since they are battery-driven, it is expedient that they should be turned off whenever possible. The base station, on the other hand, has been set to maintain a certain resolution (i.e., QoS), and therefore requires that Q sensors are switched on. Unfortunately, it does not know the number of functioning sensors, and it is only able to contact them by means of a broadcast, leaving it unable to address them individually. This leaves us with the following challenge: *How can the base station turn on exactly Q sensors, only by means of its limited broadcast capability?*

Iyer *et al.* [17] proposed a scheme where the base station provided broadcasted QoS feedback to the sensors of the network. Using this model, the above problem was solved by modeling it as a GG [45]. From the GG perspective, a sensor is seen as a voter that

[5] A brief survey - which will provide a fundamental knowledge of the field of Learning Automata - is included in this Chapter. More detailed reviews of this material can be found in [19,23,24,26,36,42].

chooses between transmitting data or remaining idle in order to preserve energy[6]. Thus, in essence, each sensor takes the role of a GG player that either votes "On" or "Off", and acts accordingly. The base station, on the other hand, is seen as the GG Referee with a unimodal performance function $G(\cdot)$ whose maximum is found at Q normalized by the total number of sensors available. The "trick" is to let the base station (1) count the number of sensors that have turned on, and (2) use the broadcast mechanism to distribute, among the sensors, the corresponding reward based on the probability obtained from $G(\cdot)$. The application of the GG solution to the field of sensor networks is thus both straightforward and obvious.

Furthermore, Tung and Kleinrock [45] have demonstrated[7] how the GG can be used for coordinating groups of mobile robots (also called "mobots") that have a restricted ability to communicate. The main example application described in [45] consists of a fixed number of mobots[8] that can either (1) collect pieces of ore from a landscape, or (2) sort already collected ore pieces. The individual mobots vary with respect to how fast they collect and how fast they sort these pieces of ore. In this context, the GG is used to make sure that the mobots choose their action so as to maximize the throughput of the overall collection and sorting system.

Other possible cooperative robotics applications include controlling a moving platform and guarding a specified perimeter [13]. In all of these cases, the solution to the problem in question would essentially utilize the solution to the GG in a plug-and-play manner.

2.2. Salient Aspects of the Chapter

2.2.1. Contributions of the Chapter

The contributions of the Chapter are the following:

1. We report the first solution to the GG which needs only a finite number of LA. Indeed, without loss of generality, the number of LA can be as small as 3.

2. We report the first GG solution which is arbitrarily accurate.

3. The solution we propose is recursive. To the best of our knowledge, there has been no other reported recursive solution.

4. The solution that we propose is "fast". Although this is a relative term, it turns out that, usually, each epoch of the recursion converges within a few hundred iterations, and the accuracy of the solution increases exponentially with the number of recursive calls. It is thus, arguably, the first reported realistic solution to this intriguing game.

---

[6]Neither Iyer *et al.* [17], Tung *et al* [45] or we consider the computational power required to run the LA on these sensors. But our belief is that this is marginal because it essentially involves maintaining a single probability and performing at most two multiplications per broadcast.

[7]The results that Tung and Kleinrock have reported are quite impressive. Our endeavour, here, is to argue that any solution that they would advocate will be achieved much more efficiently and accurately using the strategy presented in this Chapter.

[8]This is, indeed, a direct application of the GG. What is elegant about this problem is that we do not need a specific entity to serve the purpose of the Referee - the composition of the final mixture solution automatically dictates the Reward/Penalty signal passed on to the mobots.

The problem we study is akin to the ones studied in [9–11, 31, 33, 34, 37] for the point location problem. The solution we propose is related, in principle, to the tertiary and d-ary recursive search mechanisms earlier proposed [31,33,34]. But unlike the solutions reported in [31, 33, 34], the solution here is far more consequential because the system does not rely on a Teacher or "Oracle" instructing the LA which way it should move. This will be clarified later.

2.2.2. Applications of the Chapter

To motivate the problem, we present some of its straightforward applications. First of all, without belaboring the point, we include those areas mentioned above for which the GG has found *direct* applications [17, 45]. Besides this, the entire field of LA and stochastic learning has had a myriad of applications [19, 23, 24, 36, 42], which (apart from those listed in these books) include solutions for problems in network and communications [22,28,32,35], network call admission, traffic control, quality of service routing [5,6], distributed scheduling [39], training hidden Markov models [18], neural network adaptation [21], intelligent vehicle control [46], and even graph partitioning [30]. In all these cases, our solution provides a potential solution if the problem being solved can be modeled as an optimization problem involving a uni-modal function. In addition, we believe that our solution will also be applicable in a variety of areas where LA, which in their virgin form seek for the best action from a finite set of actions, have not found direct applications. Indeed, this is because the results involving LA, which can learn from an infinite action set, is sparse [37]. Thus, we suggest that our strategy to solve the GG can be invoked to learn the best parameter to be used in any given optimization algorithm - if the efficiency of the solution (measured as a function of the parameter) can be modeled as a uni-modal function. The outstanding feature of such a strategy is that the solution can be rendered in a completely distributed manner, and the information provided to the LA can be kept to the minimum.

2.2.3. Organization of the Chapter

The organization of the Chapter is as follows: After Sections 3., 4. and 5. that briefly survey the field of LA and their relevance to the GG, we proceed in Sections 6. and 7. to present a kernel solution (with the formal proofs) for the scenario when the GG uses d LA. The implementation details of the scheme are described in Section 8., where we also present the results obtained from experimentally testing the scheme on a variety of scenarios. Section 9. concludes the Chapter.

3. Learning Automata and the Goore Game

Before we proceed, it is pertinent to mention that a *sequential* solution of the game can probably[9] be obtained by using regret-minimizing "bandit" algorithms [7, 15, 16]. Using these algorithms we believe that we can attain to a solution in which each voter guarantees a performance approaching that of the *best* fixed action that they could have played in

[9] We add this qualifier here, because since the GG is not generally studied in this context, the literature reports no such "sequential" solution.

hindsight. Consequently, the overall *history* of play will approach a correlated equilibrium from which the optimal λ can be inferred. Indeed, if we are allowed to consider the history of the play, we could possibly devise a scheme which converges to the value of λ^* that maximizes $f(\lambda)$ - which can be perceived to be a solution of the GG.

However, our model is quite different, and so our goal is also distinct. First of all, the model of computation is different from the one used in typical "bandit" problems. In our model, each player communicates ONLY with the Referee and the Referee ONLY responds with a stochastic response (1 with probability f(x), where 'x' is the ratio of the number of "Yes" votes to the total number of players). From this perspective, as far as we know, there is no known updating rule by which the 2-armed bandit(s) will lead to the optimal solution when they do not communicate and do not use their history of play. However, by mapping each bandit algorithm to a LA and using an absolutely expedient LA rule, the distributed vote converges to the optimal solution. But here, utilizing d 2-armed bandits can lead to an accuracy determined by the resolution, namely of the order of $\frac{1}{d}$. Thus, our aim is that we converge to a solution that is not "inferred" by recording the "history of play", but rather, after we converge, we want a certain number (say, k) of the d 2-armed Bandit algorithms to consistently vote "Yes"' at *every* time instant - where $f(\frac{k}{d}) > f(\frac{i}{d})$ for all $i \neq k$.

Thus, there are, in essence, two high-level goals in our investigation: The first involves analyzing the ability of well-motivated learning algorithms (the families of LA) to achieve global optima in the interesting GG setting, and the second is to develop a fully *distributed* algorithm to achieve this specific objective.

With this as a background, we proceed to explain more specifically the tools at our disposal (namely, LA), and the solution that we propose.

4. Some Fundamentals of LA

4.1. Learning Automata and Cybernetics

What is a *Learning Automaton*[10]? What is "Learning" all about? What are the different types of Learning Automata (LA) available? How are LA related to the general field of *Cybernetics*? These are some of the fundamental issues that this section attempts to describe, so that we can understand the potential of the mechanisms, and their capabilities as primary tools which can be used to solve a host of very complex problems.

The Webster's dictionary defines "Cybernetics" as *the science of communication and control theory that is concerned especially with the comparative study of automatic control systems (as the nervous system, the brain and mechanical-electrical communication systems)*. The word "Cybernetics" itself has its etymological origins in the Greek root *kybernan* meaning "to steer" or "to govern". Typically, as explained in the *Encyclopaedia Britannica*, "Cybernetics is associated with models in which a monitor compares what is happening to a system at various sampling times with some standard of what should be happening, and a controller adjusts the system's behaviour accordingly". Of course, the goal of the exercise is to design the "controller" so as to appropriately adjust the system's behavior. Modern cybernetics is an interdisciplinary field, which philosophically encompasses an en-

[10]Singular: Automaton, Plural: Automata.

semble of areas including neuroscience, computer science, cognition, control systems and electrical networks.

The linguistic meaning of *automaton* is a self-operating machine or a mechanism that responds to a sequence of instructions in a certain way, so as to achieve a certain goal. The automaton either responds to a pre-determined set of rules, or adapts to the environmental dynamics in which it operates. The latter types of automata are pertinent to this Chapter, and are termed as *adaptive automata*. The term *learning* in Psychology means the act of acquiring knowledge and modifying one's behavior based on the experience gained. Viewed from this perspective, the learning mechanism which we work with, i.e., the adaptive automaton central to this Chapter, adapts to the responses from the Environment through a series of interactions with it, as will be explained presently. It then attempts to learn the best action from a set of possible actions that are offered to it by the random stationary or non-stationary Environment in which it operates. The Automaton, thus, acts as a decision maker to arrive at the best action.

Well then, what do "Learning Automata" have to do with "Cybernetics"? The answer to this probably lies in the results of the Russian pioneer Tsetlin [43] and [44]. Indeed, when Tsetlin first proposed his theory of learning, his aim was to use the principles of automata theory to model how biological systems could learn. Little did he guess that his seminal results would lead to a completely new paradigm for learning, and a sub-field of cybernetics.

The operations of the LA can be best described through the words of the pioneers Narendra and Thathachar [24]: "... a decision maker operates in the random environment and updates its strategy for choosing actions on the basis of the elicited response. The decision maker, in such a feedback configuration of decision maker (or automaton) and environment, is referred to as the learning automaton. The automaton has a finite set of actions, and corresponding to each action, the response of the environment can be either favorable or unfavorable with a certain probability" ([24], pp. 3).

4.2. History, Literature Sources and Applications of LA

The first studies with LA models date back to the studies by mathematical psychologists like Bush and Mosteller [12], and Atkinson *et al.* [4]. In 1961, the Russian mathematician, Tsetlin [43] [44] studied deterministic LA in detail. Varshavskii and Vorontsova [47] introduced the stochastic variable structure versions of the LA. Tsetlin's deterministic automata [43], [44] and Varshavskii and Vorontsova's stochastic automata [47] were the major initial motivators of further studies in this area. Following them, several theoretical and experimental studies have been conducted by several researchers: K. Narendra, M. A. L. Thathachar, S. Lakshmivarahan, M. Obaidat, K. Najim, A.S. Poznyak, N. Baba, L. G. Mason, G. Papadimitriou, and B. J. Oommen, just to mention few.

A complete study of the theory and applications of LA can be found in excellent books by Lakshmivarahan [19], Narenda and Thathachar [24], Najim *et al.* [23] and Poznyak *et al.* [36]. Besides these, a recent issue of the *IEEE Transactions on Systems, Man and Cybernetics* [26] (also see [27]), has been dedicated entirely to the study of LA, and a more recent book [42] describes the state of the art when it concerns networks and games of LA. Some of the fastest reported LA belong to the the family of estimator algorithms whose

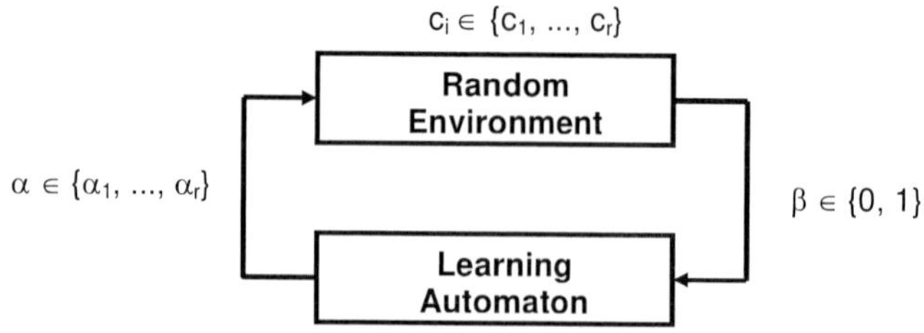

Figure 1. The Automaton – Environment Feedback Loop.

study was initiated by Thathachar and Sastry, and followed by others [1–3, 20, 26, 29].

LA, thus, find applications in optimization problems in which an optimal action needs to be determined from a set of actions. In this context, learning might be of best help only when there are high levels of *uncertainty* in the system in which the automaton operates. In systems with low levels of uncertainty, LA-based learning may not be a suitable tool of choice [24].

4.3. The LA-Environment Model

As mentioned earlier, LA [19,23,24,26,36,42] have been used to model biological learning systems and to find the optimal action that is offered by a Random Environment. Learning is accomplished by explicitly interacting with the Environment and processing its responses to the actions that are chosen, while gradually converging toward an ultimate goal. We explain now how this is done, and here, the origins of LA can be best described by the words of the pioneers Narendra and Thathachar [24]:

> *The concept of learning automaton grew out of a fusion of the work of psychologists in modeling observed behavior, the efforts of statisticians to model the choice of experiments based on past observations, the attempts of operation researchers to implement optimal strategies in the context of the two-armed bandit problem, and the endeavors of system theorists to make rational decisions in random environments.*

In short, the process of learning is based on a learning loop involving the two entities: the Random Environment (RE), and the LA, as described in Figure 1. The RE offers the automaton with a set of possible actions $\{\alpha_1, \alpha_2, \ldots, \alpha_r\}$ to choose from. The automaton chooses one of those actions, say α_i, which serves as an input to the RE. Since the RE is "aware" of the underlying penalty probability distribution of the system, depending on the *penalty probability* c_i corresponding to α_i, it "prompts" the LA with feedback β which is a reward (typically denoted by the value "0"), or a *penalty* (typically denoted by the value "1"). The reward/penalty information (corresponding to the action) provided to the LA helps it to choose the subsequent action. By repeating the above process, through a series of Environment-Automaton interactions, the LA finally attempts to learn the *optimal* action from the Environment.

The LA is called a Fixed Structure Stochastic Automaton (FSSA) if its structure does not change with time, and a Variable Structure Stochastic Automaton (VSSA) if the structure changes with time. In the latter case, the LA choose their current action based on the current value of the so-called action probability vector, and the particular VSSA is fully described in terms of the function which updates *this* vector.

Note that in its most basic version, a LA forms both a theoretically founded and a computationally efficient solution to the well known Two-Armed Bandit Problem which has also been addressed from the perspectives of other fields, such as operation research. However, because of the solid theoretical basis that has been established within the LA field during the last decades, and particularly due to the results concerning games, we have chosen LA as the starting point for our work on the GG. Still, when it comes to solving the GG, it will be clear that our work addresses a fundamental weakness concerning *any* solution to the Two-Armed (or for that matter, Multi-Armed) Bandit Problem[11].

4.4. Learning Criteria

We now provide a few important definitions used in the field. $P(t)$ is referred to as the action probability vector, where, $P(t) = [p_1(t), p_2(t), \ldots, p_r(t)]^T$, in which each element of the vector, $p_i(t)$, satisfies:

$$p_i(t) = Pr[\alpha(t) = \alpha_i],\ i = 1, \ldots, r, \text{ such that } \sum_{i=1}^{r} p_i(t) = 1\ \forall t. \tag{1}$$

Given an action probability vector, $P(t)$ at time t, the *average penalty* is:

$$\begin{aligned} M(t) &= E[\beta(t)|P(t)] = Pr[\beta(t) = 1|P(t)] \\ &= \sum_{i=1}^{r} Pr[\beta(t) = 1|\alpha(t) = \alpha_i]\ Pr[\alpha(t) = \alpha_i] = \sum_{i=1}^{r} c_i p_i(t). \end{aligned} \tag{2}$$

The average penalty for the "pure-chance" automaton is given by:

$$M_0 = \frac{1}{r} \sum_{i=1}^{r} c_i. \tag{3}$$

The average penalty at time 't', $E[M(t)]$, is given by:

$$E[M(t)] = E\{E[\beta(t)|P(t)]\} = E[\beta(t)]. \tag{4}$$

As $t \mapsto \infty$, if the average penalty $E[M(t)] < M_0$, at least asymptotically, the automaton is generally considered to be better than the pure-chance automaton. A LA that performs better than by pure-chance is said to be *expedient*.

Definition 1: A LA is said to be *absolutely expedient* if $E[M(t+1)|P(t)] < M(t)$, implying that $E[M(t+1)] < E[M(t)]$.

Definition 2: A LA is considered *optimal* if $\lim_{t \mapsto \infty} E[M(t)] = c_l$,where $c_l = \min_i\{c_i\}$.

Definition 3: A LA is considered *ϵ-optimal* if:

$$\lim_{n \mapsto \infty} E[M(t)] < c_l + \epsilon, \tag{5}$$

[11]If a solution to either of these bandit problems is used as the kernel to solve the GG, it turns out that we will have to use an arbitrarily large number of players to obtain an arbitrarily small error.

where $\epsilon > 0$, and can be arbitrarily small, by a suitable choice of some parameter of the LA. It should be noted that no optimal LA exist. Marginally sub-optimal performance, also termed above as ϵ-optimal performance, is what LA researchers attempt to attain. Further, every *absolutely expedient* LA is also *ϵ-optimal.*

4.5. The Tsetlin Automaton

When the GG was first investigated, the pioneer Tsetlin utilized his so-called Tsetlin automaton to solve it. Later, as the area of LA developed, it was shown that all *absolutely expedient* LA could also solve the GG efficiently. To put present-day solutions in the right perspective, we briefly survey the Tsetlin automaton and an *absolutely expedient* LA, the continuous Linear Reward-Inaction scheme (L_{RI}), central to the Chapter.

The Tsetlin Automaton was the first learning automaton presented in the literature [44]. It is a deterministic FSSA, denoted by $L_{2N,2}$, with $2N$ states and 2 actions, i.e. N states for each action. Furthermore, its structure can easily be extended to deal with r $(2 < r < \infty)$ actions. As any learning automaton, its goal is to incorporate knowledge from the past behavior of the system in its decision rule for choosing the next sequence of actions. To achieve this, the automaton calculates the number of successes and failures received for each action, and switches to the alternate action only when it receives a sufficient number of failures, depending on its current state. In order to describe the behavior of this type of automaton, its output and state transition functions will be described below. The output function of the automaton is simple: if the automaton is in a state q_i $(1 \leq i \leq N)$, it chooses action α_1, and if it is in a state q_i $(N+1 \leq i \leq 2N)$ it chooses action α_2. Since each action has N states associated with it, N is called the memory associated with each action and the automaton is said to have a total memory of $2N$. The state transitions are illustrated by the two graphs presented in Figure 2, one for a favorable response and the other for an unfavorable response. If the environment replies with a reward (favorable response), the automaton moves deeper into the memory of the corresponding action. If the environment replies with a penalty (unfavorable response), the automaton moves towards the outside boundary of the memory of the corresponding action. The deepest states in memory are referred to as the most internal states, or the end states.

The Tsetlin automaton has been analyzed using the theory of ergodic Markov chains [24] and shown[12] to be ϵ-optimal whenever the minimum penalty probability is less than 0.5. The problem, however, with the Tsetlin (and the other reported fixed-structure LA) is that they are unreasonably slow, as will be explained presently.

4.6. Variable Structure Learning Automata

Unlike the FSSA, *Variable Structure Stochastic Automata* (VSSA) are the ones in which the state transition probabilities are not fixed. In such automata, the state transitions or the action probabilities themselves are updated at every time instant using a suitable scheme. The transition probabilities f_{ij}^{β} and the output function g_{ij} vary with time, and the action probabilities are updated on the basis of the input. These automata are

[12]The explicit expressions for the asymptotic penalty are found in [24] and are omitted as they are not needed here.

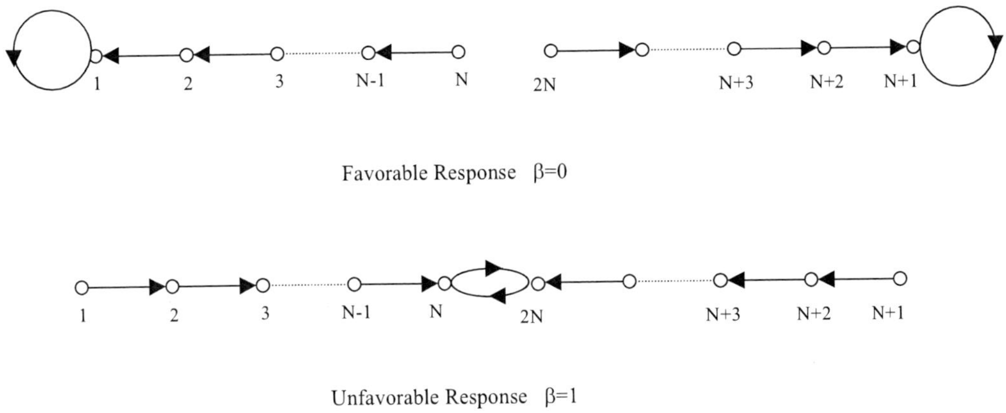

Figure 2. State transition graphs for the Tsetlin automaton $L_{2N,2}$.

discussed here in the context of linear schemes. But the concepts discussed below can be extended to non-linear updating schemes as well. The types of automata that update transition probabilities with time were introduced in 1963 by Varshavskii and Vorontsova [47]. A VSSA depends on a random number generator for its implementation. The action chosen is dependent on the action probability distribution vector, which is, in turn, updated based the reward/penalty input that the automaton receives from the RE.

Definition 2.6 A VSSA is a quintuple $\langle Q, A, B, T \rangle$, where Q represents the different states of the automaton, A is the set of actions, B is the set of responses from the environment to the LA, G is the output function, and T is the action probability updating scheme $T : [0,1]^r \times A \times B \mapsto [0,1]^r$, such that

$$P(t+1) = T(P(t), \alpha(t), \beta(t)), \tag{6}$$

where $P(t)$ is the action probability vector.

Normally, VSSA involve the updating of both the state and action probabilities. For the sake of simplicity, in practice, it is assumed that in such automata, each state corresponds to a distinct action, in which case the action transition mapping G becomes the identity mapping, and the number of states, s, is equal to the number of actions, r ($s = r < \infty$).

VSSA can be analyzed using a discrete-time Markov Process, defined on a suitable set of states. If a probability updating scheme T is time invariant, $\{P(t)\}_{t \geq 0}$ is a discrete-homogenous Markov process, and the probability vector at the current time instant $P(t)$, (along with $\alpha(t)$, and $\beta(t)$) completely determines $P(t+1)$. Hence, each distinct updating scheme, T, identifies a different type of learning algorithm, as follows:

- *Absorbing algorithms* are the ones in which the updating scheme, T is chosen in such a manner that the Markov process has absorbing states.
- *Non-absorbing algorithms* are the ones in which the Markov process has no absorbing states.
- *Linear algorithms* are the ones in which $P(t+1)$ is a linear function of $P(t)$.

- *Nonlinear algorithms* are the ones in which $P(t+1)$ is a non-linear function of $P(t)$.

In a VSSA, if a chosen action α_i is rewarded, the probability for the current action is increased, and the probabilities for all the other actions are decreased. On the other hand, if the chosen action α_i is penalized, the probability of the current action is decreased, whereas the probabilities for the rest of the actions could, typically, be increased. This leads to the following different types of learning schemes for VSSA:

- **Reward-Penalty (RP):** In both the cases, when the automaton is rewarded as well as penalized, the action probabilities are updated.

- **Inaction-Penalty (IP):** When the automaton is penalized, the action probability vector is updated, whereas when the automaton is rewarded, the action probabilities are neither increased nor decreased.

- **Reward-Inaction (RI):** The action probability vector is updated whenever the automaton is rewarded, and is unchanged whenever the automaton is penalized.

A LA is considered to be a *continuous automaton* if the probability updating scheme T is continuous, i.e., the probability of choosing an action can be any real number in the closed interval $[0, 1]$.

In a VSSA, if there are r actions operating in a stationary environment with $\beta = \{0, 1\}$, a general action probability updating scheme for a continuous automaton is described below. We assume that the action α_i is chosen, and thus, $\alpha(t) = \alpha_i$. The updated action probabilities can be specified as:

$$\begin{aligned} \text{For } \beta(t) &= 0, \ \forall j \neq i\,, \quad p_j(t+1) = p_j(t) - g_j(P(t)) \\ \text{For } \beta(t) &= 1, \ \forall j \neq i\,, \quad p_j(t+1) = p_j(t) + h_j(P(t)) \end{aligned} \tag{7}$$

Since $P(t)$ is a probability vector, $\sum_{j=1}^{r} p_j(t) = 1$. Therefore,

$$\begin{aligned} \text{when } \beta(t) &= 0, \quad p_i(t+1) = p_i(t) + \sum_{j=1, j\neq i}^{r} g_j(P(t)), \\ \text{and when } \beta(t) &= 1, \quad p_i(t+1) = p_i(t) - \sum_{j=1, j\neq i}^{r} h_j(P(t)). \end{aligned} \tag{8}$$

The functions h_j and g_j are *nonnegative* and *continuous* in $[0, 1]$, and obey:

$$\begin{aligned} &\forall i = 1, 2, \ldots, r, \quad \forall P \in (0,1)^R, \ 0 < g_j(P) < p_j, \\ &\text{and} \qquad 0 < \sum_{j=1, j\neq i}^{r} [p_j + h_j(P)] < 1 \end{aligned} \tag{9}$$

For *continuous linear* VSSA, the following four learning schemes are extensively studied in the literature:

- The Linear Reward-Inaction Scheme (L_{RI})

Table 1. Properties of the Continuous Learning Schemes.

Learning Scheme	Learning Parameters	Usefulness (Good/Bad)	Optimality	Ergodic/Absorbing (When Useful)
L_{RI}	$a > 0; b = 0$	Good	ϵ-optimal as $a \mapsto 0$	Absorbing (Stationary E)
L_{IP}	$a = 0, b > 0$	Very Bad	Not even Expedient	Ergodic (Non-stationary E)
L_{RP} (Symmetric)	$a = b; a, b > 0$	Bad	Never ϵ-optimal	Ergodic (Non-stationary E)
$L_{R-\epsilon P}$	$a > 0, b << a$	Good	ϵ-optimal as $a \mapsto 0$	Ergodic (Non-stationary E)

- The Linear Inaction-Penalty Scheme (L_{IP})
- The Symmetric Linear Reward-Penalty Scheme (L_{RP})
- The Linear Reward-ϵ-Penalty Scheme ($L_{R-\epsilon P}$).

They are explained for the 2-action case; their extension to the r-action case, where $r > 2$, are straightforward, and can be found in [24]. Their properties are, however, tabulated in Table 1.

For a 2-action LA, let

$$\begin{aligned} g_i(P(t)) &= a\, p_j(t) \\ \text{and } h_j(P(t)) &= b\,(1 - p_j(t)) \end{aligned} \tag{10}$$

In Equation (10), a and b are called the reward and penalty parameters, and they obey the following inequalities: $0 < a < 1, 0 \leq b < 1$. Equation (10) will be used further to develop the action probability updating equations.

The above-mentioned linear schemes are quite popular in LA because of their analytical tractability. They exhibit significantly different characteristics as can be seen in Table 1. Of particular interest to this Chapter is the L_{RI} scheme[13] explained in greater detail in the next sub-section. The so-called symmetry conditions for the functions $g(\cdot)$ and $h(\cdot)$ to lead to absolutely expedient LA are also derived in [19] and [24].

4.7. The Continuous Linear Reward-Inaction Scheme (L_{RI})

The L_{RI} scheme was first introduced by Norman [25], and then by Shapiro and Narendra [40]. It is based on the principle that whenever the automaton receives a favorable response (i.e., reward) from the environment, the action probabilities are updated, whereas if the

[13]The scheme is not suitable for non-stationary environments. On the other hand, for stationary random environments, the L_{RI} scheme is both absolutely expedient, and ϵ-optimal [24]. The L_{IP} and L_{RP} schemes are devised similarly, and are omitted from further discussions. They, and their respective analysis can be found in [24].

automaton receives an unfavorable response (i.e., penalty) from the environment, the action probabilities are unaltered.

The probability updating equations for this scheme are characterized by a parameter θ $(0 \ll \theta < 1)$ and can be simplified to be as below [24]:

$$p_1(t+1) = p_1(t) + (1-\theta) \times (1 - p_1(t)) \quad \text{if} \quad \alpha(t) = \alpha_1, \text{and } \beta(t) = 0 \tag{11}$$

$$p_1(t+1) = \theta \times p_1(t) \quad \text{if} \quad \alpha(t) = \alpha_2, \text{and } \beta(t) = 0 \tag{12}$$

$$p_1(t+1) = p_1(t) \quad \text{if} \quad \alpha(t) = \alpha_1 \text{ or } \alpha_2, \text{ and } \beta(t) = 1 \tag{13}$$

We see that if action α_i is chosen, and a reward is received, the probability $p_i(t)$ is increased, and the other probability $p_j(t)$ (i.e., $j \neq i$) is decreased. If either α_1 or α_2 is chosen, and a penalty is received, $P(t)$ is unaltered.

Equation (11) shows that the L_{RI} scheme has the vectors $[1, 0]^T$ and $[0, 1]^T$ as two absorbing states - one of which it converges to. Therefore, the convergence of the L_{RI} scheme is dependent on the nature of the initial conditions and probabilities.

We first state a fundamental result for the L_{RI} learning scheme which we will repeatedly allude to, in the rest of the Chapter.

Lemma 1. An L_{RI} learning scheme with parameter $0 \ll \theta < 1$ is ϵ-optimal whenever an optimal action exists. In other words, if α_k is the optimal action, $\lim_{\theta \to 1} \lim_{n \to \infty} p_k(n) \to 1$.

The above result is well known [19, 24, 38]. Thus, we are guaranteed that for any L_{RI} scheme with the two actions $\{\alpha_0, \alpha_1\}$, if $\exists k \in \{0, 1\}$ such that $c_k < c_{1-k}$, then the action α_k is optimal, and for this action $p_k(n) \to 1$ as $n \to \infty$ and $\theta \to 1$.

4.8. Remarks on the Accuracy/Rate of Convergence

We now give a brief categorization of LA in terms of their rates of convergence. The results that we cite below can be seen in greater detail in [24], and so this subsection is necessarily brief.

FSSA, typically, converges slowly - requiring *tens of thousands* of iterations to reach values close to the asymptotic value. This is because the LA can linger in an inferior action for a long time, and transfer to a superior action, typically, at the boundary states. The convergence of these LA is studied by means of an eigenvalue analysis of the underlying Markov chain.

VSSA, on the other hand, are typically about an order of magnitude faster than FSSA (requiring only *thousands* of iterations to solve problems identical to the ones mentioned in the previous paragraph). This is because the LA can switch between the actions at every time instant, thus, rendering it with the potential of quickly visiting inferior actions, and determining that is futile to visit them more often. In particular, the L_{RI} scheme has increased accuracy as the parameter θ tends to unity. The closer θ is to unity, the smaller are the changes in the action probability vector, and thus we have a more conservative learning phenomenon. Unfortunately, this brings along the disadvantage of a slower convergence.

5. Known LA Solutions to the Goore Game

5.1. Modeling the LA Solutions to the GG

We assume that we are dealing with a "team" of d LA, $\{A^1, A^2, \ldots A^d\}$. In terms of notation, we assume that the actions offered to each LA, A^j, from the Environment in question are $\{\alpha_0^j, \alpha_1^j\}$, and that the corresponding penalty probabilities are $\{c_0^j, c_1^j\}$ respectively. Similarly, we let $P_k^j(n)$ represent the component of the action probability vector of A^j for action α_k^j, where n represents the discretized time index.

In the interest of simplicity, throughout this Chapter, we shall assume that the individual LA used is the well-known L_{RI} scheme with parameter θ [19, 23, 24, 26, 36, 42]. Alternatively, the Tsetlin automaton or any other absolutely expedient (or probably ϵ-optimal scheme – including those belonging to the estimator families cited above) can be used just as effectively.

Before we describe how LA are used to solve the GG, a description of the modeling required is mandatory, and we undertake that here.

The Environment: The Environment serves a dual purpose. It is used to model the Referee, and to also take into consideration the characteristics of the unknown function, $G(\cdot)$ to be optimized. It receives as its input the decisions of all the automata/players (whom we model presently). It then computes the ratio of the number of "Yes" votes to the total number of players, which is the quantity λ, and evaluates the function at this point, $G(\lambda)$. Since this evaluation leads to a real number in the interval $[0, 1]$, it is now able to either reward or penalize the individual automata based on $G(\lambda)$.

The Automata: Modeling the set of players using LA is fairly trivial. First of all, each automaton quite simply represents a player. Its input is the response from the Environment (described above). Its output is a "Yes"/"No"-decision based on the current action-probability vector if it is a VSSA, or based on the state it is in if it is a FSSA. The updating is done based on the rule that the LA follows. In the case of the FSSA, this is achieved by updating the state index, and in the case of VSSA by updating the action probability vector.

The Optimization Function: The function, $G(\cdot)$, to be optimized plays a crucial role in the operation of the Environment. First of all, the form and parameters of the function are unknown to every single LA. However, the Referee who has access to the function is able to evaluate it, and respond accordingly with a corresponding Reward or Penalty. Notice again that each automaton will get an independent Reward/Penalty based on $G(\cdot)$, because the Referee rewards or penalizes them independently as per λ, the ratio of "Yes"-votes to the total number of players

Although this completes the entire modeling phase, the reader should observe that all the automata, although operating independently and without mutual communication, interact with the *same* Referee, and thus have a common Environment. Observe that this is almost the converse of the multi-teacher model of LA [8] in which a single LA learns from multiple Environments - here we have a single Teacher who is attempting to teach all the LA in a non-communicating distributed manner.

Although this description is quite comprehensive, the reader should observe that if we have d players represented by LA, each of them can converge only as fast as their underlying updating rule permits. However, the fact that d is finite, further limits the accuracy of the solution to the Goore game. Since λ is in the open interval $(0, 1)$, the maximum accuracy that d LA can lead to is bounded by $\frac{1}{d}$. To get a larger accuracy, d has to be increased indefinitely[14], and θ increased towards unity, rendering a slower convergence on both fronts. This is the fundamental problem of the reported LA-based GG solutions that we intend to overcome[15].

5.2. Fundamental Convergence Properties of the Goore Game

Let $G(\cdot)$ be an arbitrary uni-modal function from $[0, 1] \rightarrow [0, 1]$ known to the Referee interacting with a team of d LA, $\{A^1, A^2, \ldots A^d\}$. Each LA, A^j, independently chooses an action $\alpha^j(n)$ which is either α_0^j or α_1^j, for which it receives, from the Referee, a response $\beta^j \in \{0, 1\}$, (with $\beta^j = 0$ being regarded as a *Reward*) as per:

$$\beta^j(n) = 0 \qquad \textbf{w.p.} \qquad G\left(\frac{\sum_{j=1}^{d} \alpha^j(n)}{d}\right). \tag{14}$$

Then the following property, referred to as the Learning Automata - Goore Game (LA-GG) Property, is true.

Theorem 1. LA-GG Property

If each LA, A^i, receives its feedback signals from the Referee as per Equation (14), and uses the L_{RI} learning scheme with parameter $0 \ll \theta < 1$ to update its learning model, then each LA converges so that the collective behavior of the team optimizes the unknown function $G(\cdot)$. Thus, if $k^+ = \lim_{n\rightarrow\infty} \sum_{j=1}^{d} \alpha^j(n)$,

$$G\left(\frac{k^+}{d}\right) > G\left(\frac{k}{d}\right) \quad \forall \quad k \neq k^+. \tag{15}$$

Proof. The proof is cited in [24, 41] and is omitted here as it is a fundamental result. □

Example 1.

Let us suppose that the function $G(\cdot)$ used by the Referee is $G(x) = 0.7 \times e^{-\frac{(0.9123-x)^2}{0.0625}}$ and that 5 LA are participating in the game. Then, the number of LA who vote "Yes" could be in the set $\{0, 1, 2, 3, 4, 5\}$, with their corresponding G values being:

$G(0) = 0.000015$, $G(0.2) = 0.000208$, $G(0.4) = 0.01050$, $G(0.6) = 0.14702$, $G(0.8) = 0.57209$, $G(1.0) = 0.61895$.

Observe that although $G(\cdot)$ has its maximum value at 1.0 within the *discretized* domain, the maximum of the function itself occurs at 0.9123. Theorem 1 claims that if each LA is an L_{RI} scheme with parameter θ being arbitrarily close to unity, *all* of the 5 LA will converge to a "Yes" vote (i.e., to a value of $\alpha = 1$) with a probability as close to unity as desired. □

---

[14] Ironically, increasing the number of players also decreases the rate of convergence of the overall GG solution.

[15] Tung and Kleinrock [45] have also shown that if the Tsetlin LA is used, one can obtain an increased accuracy only by increasing both the number of LA *and* their associated number of states.

5.3. Problems with Reported LA Solutions to the GG

The above solution to the GG is indeed both fascinating and intriguing. Without a knowledge of the function $G(\cdot)$, any understanding of how their partners decide, or even a perception of how the Referee "manufactured" their respective responses, the LA converge to the optimal solution *without* any communication. However, in its current form, the above solution is extremely limited. The main handicap associated with using it in real-life applications concerns the *accuracy* of the solution obtained. Indeed, this accuracy is intricately linked to the number of LA used. The situation is quite ironic. If the number of LA involved in the game is d, the precision of the solution is bounded by $\frac{1}{d}$, and thus the solution can be arbitrarily accurate only as d is increased indefinitely. Unfortunately, when this occurs, the team of LA converge extremely slowly, rendering the solution infeasible.

In this Chapter, we shall show how we can obtain an arbitrarily accurate solution to the GG by just using a finite set of LA. Indeed, we shall see that, without loss of generality, the number of LA can be as small as 3.

6. Continuous Goore Game with Adaptive d-ary Search

The solution presented in this Chapter is based on a strategy, the so-called Continuous Goore Game with Adaptive d-ary Search (CGG–AdS) strategy. Unlike the work of [31,34], we do not require that the LA are instructed by an "Oracle" about the direction in which they have to move (within the solution space) to attain to the final solution. The basic idea behind the CGG–AdS solution is to use d LA to play the GG, and then to use the results of *their* solution to systematically explore a *sub*-interval of the current interval for the solution. This exploration is a series of estimates, each one more accurate than the previous one.

In CGG–AdS, the given search interval is divided into d partitions representing d disjoint sub-intervals. In each interval, the LA utilize their distributed learning capabilities (with no mutual communication) to attain to a consensus as to where the optimal point lies. To achieve this, unlike the works of [31,34], we do *not* explore each of the d partitions of the interval. Rather, the LA-GG property given by Theorem 1 is independently invoked using the L_{RI} ϵ-optimal fast converging two-action LA, where the two actions are those of voting "Yes" or "No" to the *current* hypothesis. Based on the collective response, the Referee then prunes the space *without informing the LA*, and eliminates at least one of the sub-intervals from being searched further. The search is then recursively invoked within the remaining pruned contiguous interval until the search interval is at least as small as the required resolution of estimation. This elimination process essentially utilizes the ϵ-optimality property of the underlying automata and the monotonicity of the intervals to guarantee the convergence, as stated in Theorem 1.

As mentioned earlier, in the interest of simplicity, we shall assume that the individual LA used is the well-known L_{RI} scheme with parameter θ, although, as discussed, other schemes will perform just as efficiently. Also, to simplify matters, we shall derive most of the results for the general case of any d, but for the ease of implementation, *some* of the algorithmic aspects are specified for specific "small" instantiations of d.

6.1. Notations and Definitions

Let $\Delta(t) = [\sigma, \gamma)$ s.t. $\sigma \leq \lambda^\star < \gamma$ be the current search interval at epoch t, containing $\lambda^\star$, whose left and right (smaller and greater) boundaries on the real line are σ and γ respectively. $\Delta(0)$ is initialized to be the unit interval. We partition $\Delta(t)$ into d equi-sized disjoint partitions[16] Δ^j, $j \in \{1, 2, \ldots d\}$, such that, $\Delta^j = [\sigma^j, \gamma^j)$. To formally describe the relative locations of intervals we define an interval relational operator $\prec$ such that, $\Delta^j \prec \Delta^k$ iff $\gamma^j < \sigma^k$. Since points on the real interval are monotonically increasing, we have, $\Delta^1 \prec \Delta^2 \ldots \prec \Delta^d$.

To relate the various intervals to $\lambda^\star$, we introduce the following relational operators.

$\lambda^\star \textcircled{<} \Delta^j$	iff	$\lambda^\star < \sigma^j$.	i.e., $\lambda^\star$ is to the left of the interval Δ^j.
$\lambda^\star \textcircled{>} \Delta^j$	iff	$\lambda^\star > \gamma^j$.	i.e., $\lambda^\star$ is to the right of the interval Δ^j.
$\lambda^\star \textcircled{=} \Delta^j$	iff	$\sigma^j \leq \lambda^\star < \gamma^j$.	i.e., $\lambda^\star$ is contained in the interval Δ^j.
$\lambda^\star \textcircled{\leq} \Delta^j$	iff	$\lambda^\star \textcircled{<} \Delta^j$ or $\lambda^\star \textcircled{=} \Delta^j$	i.e., $\lambda^\star$ is either to the left of or inside the interval Δ^j.
$\lambda^\star \textcircled{\geq} \Delta^j$	iff	$\lambda^\star \textcircled{>} \Delta^j$ or $\lambda^\star \textcircled{=} \Delta^j$	i.e., $\lambda^\star$ is either to the right of or inside the interval Δ^j.

These operators can trivially be shown to satisfy the usual laws of transitivity.

6.2. Construction of the Learning Automata

In the CGG–AdS strategy, we define d 2-action L_{RI} automata $\{\mathcal{A}^j = (\Sigma^j, \Pi^j, \beta^j, \Upsilon^j, \Omega^j)\}$ where Σ^j is the set of actions - representing "Yes" or "No" decisions, Π^j is the set of action probabilities, β^j is the set of feedback inputs from the Environment, Υ^j is the set of action probability updating rules, and Ω^j is the set of possible decision outputs of the automata at the end of each epoch. The Referee has, in its possession, a secret arbitrary uni-modal function $G(\cdot)$ from $[0, 1] \to [0, 1]$. The Environment, E , for each LA, is governed by the response of the Referee, who, unknown to the LA, rewards or penalizes them based on the values of the function $G(\cdot)$ within the current interval of interest. It, in a distributed manner, governs the overall search strategy by providing the responses to the LA, and additionally enhancing CGG–AdS, by using a Pruning Decision Rule (PDR)[17], Λ, to prune the search interval. It achieves this by utilizing the LA-GG property and the decisions, Ω^j, made in the previous epoch by the d LA. Thus $\mathcal{A}^j$, $j \in \{1, \ldots d\}$, together with E and Λ completely define the CGG–AdS strategy. These are formalized below.

1. *The set of actions of the automaton*: (Σ^j)
   The two actions of the automaton are α_k^j, for $k \in \{0, 1\}$, where, $\alpha_0^j = 0$ corresponds to the LA casting a "No" vote, and $\alpha_1^j = 1$ corresponds to the LA casting a "Yes" vote.

2. *The action probabilities*: (Π^j)
   $P_k^j(n)$ represent the probabilities of selecting the action α_k^j, for $k \in \{0, 1\}$, at step n. Initially, $P_k^j(0) = 0.5$, for $k = 0, 1$.

---

[16]The equi-partitioning is really not a restriction. It can easily be generalized.

[17]This rule is also referred to as the Pruning Table.

3. *The feedback inputs from the Environment to each automaton*: (β^j)
Each LA receives, from the Referee, a response $\beta^j \in \{0, 1\}$, (with $\beta^j = 0$ being regarded as a *Reward*) as per Equation (14) given below:

$$\beta^j(n) = 0 \qquad \textbf{w.p.} \qquad G\left(\frac{\sum_{j=1}^{d} \alpha^j(n)}{d}\right).$$

4. *The action probability updating rules*: (Υ^j)
First of all, since we are using the L_{RI} scheme, we ignore all the penalty responses. Upon reward, we obey the following updating rule :
If α_k^j for $k \in \{0, 1\}$was rewarded then,

$$P_{1-k}^j(n+1) \leftarrow \theta \times P_{1-k}^j(n)$$
$$P_k^j(n+1) \leftarrow 1 - \theta \times P_{1-k}^j(n)$$

where $0 \ll \theta < 1$ is the L_{RI} reward parameter.

5. *The decision outputs at each epoch*: (Ω^j)
From the action probabilities we infer the decision Ω^j of the L_{RI} automaton, $\mathcal{A}^j$, after a fixed number N_∞, of iterations. Typically, N_∞ is chosen so as to ensure (with a very high probability) that the team of automata will have converged[18]. Ω^j indicates that the particular LA, $\mathcal{A}^j$, has decided that it should vote either "Yes" or "No" with an arbitrary high accuracy. The set of values that Ω^j can take and the preconditions are given by:

$$\Omega^j = \begin{cases} No & \text{If} \quad P_0^j(N_\infty) \geq 1 - \epsilon. \\ Yes & \text{If} \quad P_1^j(N_\infty) \geq 1 - \epsilon. \end{cases}$$

6. *The Pruning Decision Rule (PDR) for pruning the search space*: (Λ)
Since the actions chosen by each LA can lead to one of the two possible decisions, namely Yes or No, the number of "Yes" votes can be any integer in the set $\{0, 1, \ldots, d\}$. Once the team of automata have made a decision regarding where they reckon $\lambda^\star$ to be (by virtue of their votes), the CGG–AdS reduces the size of the search interval by a factor of at least $\frac{2}{d}$. If k^+ is the number of "Yes" votes received, the new pruned search interval, Δ^{new}, for the subsequent learning phase (epoch) is generated according to the PDR, Λ, for the specific value of d, defined as follows:

$$\Delta(t+1) = \begin{cases} \Delta^1 & \text{If} \quad k^+ = 0. \\ \Delta^m \cup \Delta^{m+1} & \text{If} \quad k^+ = m; m \neq 0, d. \\ \Delta^d & \text{If} \quad k^+ = d. \end{cases} \tag{16}$$

The PDR (generally, synonymously and equivalently given as a table), Λ, is shown in Tables 2 and 3 for the cases of $d = 3$ and $d = 4$ respectively. Clearly, the table "prunes" the size of the interval, because $\Delta(t+1)$ at the next epoch is, at most, of length $\frac{2}{d}$.

[18]This is always guaranteed if we use an absolutely expedient scheme in which the unit vectors are absorbing barriers [19,23,24,26,36,42].

Table 2. The decision table, (Λ), to prune the search space of CGG–AdS for $d = 3$ based on the LA outputs Ω^j. k^+ is the number of LA who vote "Yes".

k^+	**New Sub-interval :** Δ^{new}
0	Δ^1
1	$\Delta^1 \cup \Delta^2$
2	$\Delta^2 \cup \Delta^3$
3	Δ^3

Table 3. The decision table, (Λ), to prune the search space of CGG–AdS for $d = 4$ based on the LA outputs Ω^j. k^+ is the number of LA who vote "Yes".

k^+	**New Sub-interval :** Δ^{new}
0	Δ^1
1	$\Delta^1 \cup \Delta^2$
2	$\Delta^2 \cup \Delta^3$
3	$\Delta^3 \cup \Delta^4$
4	Δ^4

We shall now derive the fundamental properties of CGG–AdS.

7. Convergence Properties of CGG–AdS

We consider here the convergence results concerning CGG–AdS for the general GG. Lemmas 2 and 3 essentially use the ϵ-optimality property of L_{RI} automata to prove that they produce the correct decision output for each partition. These are then used in Theorem 2 to prove that the formula used to create the PDR is correct. This, thus, establishes that after elimination of one or more partitions, the remaining interval still contains $\lambda^\star$with an arbitrarily high probability, thereby assuring convergence. It should be borne in mind that these are still probabilistic results, although the probability is shown to be potentially as close to unity as we want, provided that we choose the parameters for the L_{RI} automata appropriately.

Lemma 2. Consider an arbitrary GG with a Referee providing responses as per Equation (14), and the LA working with an L_{RI} scheme with a parameter θ which is arbitrarily close to unity. Then, for $1 \leq m \leq d$, the following is true:

$$\text{If } (\lambda^* \ominus \Delta^m), \text{Then } Pr\left[(k^+ = m - 1) \; or \; (k^+ = m)\right] \to 1.$$

Proof. Consider the case when $\lambda^* \ominus \Delta^m$. The uni-modality of $G(\cdot)$ implies that G attains its maximum for a value of λ that lies between $\frac{m-1}{d}$ and $\frac{m}{d}$. We now have two scenarios. If $G(\lambda^*)$ is closer to $G\left(\frac{m-1}{d}\right)$, by the LA-GG property, $(m - 1)$ of the d LA will converge to a "Yes" vote. Conversely, if $G(\lambda^*)$ is closer to $G\left(\frac{m}{d}\right)$, m of the d LA will converge to a "Yes" vote. In either of these scenarios, since the LA are of an absorbing sort, all of them will converge to an absorbing unit vector, and hence, the number of LA converging to the "Yes" vote will be either m or $m+1$. Hence the result. □

Lemma 3. Consider an arbitrary GG with a Referee providing responses as per Equation (14), and the LA working with an L_{RI} scheme with a parameter θ which is arbitrarily close to unity. Then the following is true:

$$\begin{array}{ll} \text{If } (k^+ = 0), & \text{Then } Pr\left[(\lambda^* \ominus \Delta^1)\right] \to 1. \\ \text{If } (k^+ = m, m \neq 0, d), & \text{Then } Pr\left[(\lambda^* \ominus \Delta^m) \text{ } or \text{ } (\lambda^* \ominus \Delta^{m+1})\right] \to 1. \\ \text{If } (k^+ = d), & \text{Then } Pr\left[(\lambda^* \ominus \Delta^d)\right] \to 1. \end{array}$$

Proof. Consider the case when none of the LA decide on a "Yes" vote, yielding $k^+ = 0$. The uni-modality of $G(\cdot)$ and the LA-GG property imply that G attains its maximum for a value of λ that lies between 0 and $\frac{1}{d}$. Furthermore, since $k^+ = 0$,

$$|G(\lambda^*) - G(0)| < |G(\lambda^*) - G\left(\tfrac{1}{d}\right)|,$$

which proves the first claim.

The case when all of the LA decide on a "Yes" vote, leads to the scenario $k^+ = d$. Again, the uni-modality of $G(\cdot)$ and the LA-GG property imply that G attains its maximum for a value of λ that lies between $\frac{d-1}{d}$ and 1. Again, since $k^+ = d$,

$$|G(\lambda^*) - G(1)| < |G(\lambda^*) - G\left(\tfrac{d-1}{d}\right)|.$$

This proves the last claim.

For Claim 2, consider the case when $k^+ = m$. By virtue of the LA-GG property and the uni-modality of $G(\cdot)$, this means that the maximum value of G lies between to $\frac{m-1}{d}$ and $\frac{m+1}{d}$. Now, we observe that, since $k^+ = m$,

$$|G(\lambda^*) - G\left(\tfrac{m}{d}\right)| < |G(\lambda^*) - G\left(\tfrac{m-1}{d}\right)|, \text{ and}$$

$$|G(\lambda^*) - G\left(\tfrac{m}{d}\right)| < |G(\lambda^*) - G\left(\tfrac{m+1}{d}\right)|.$$

We now have two scenarios where λ^* is either to the left of $\frac{m}{d}$ or to its right. Clearly, in the former case, $(\lambda^* \ominus \Delta^m)$, and in the latter, $(\lambda^* \ominus \Delta^{m+1})$. The result follows. □

Theorem 2. Consider an arbitrary GG with a Referee providing responses as per Equation (14), and the LA working with an L_{RI} scheme with a parameter θ which is arbitrarily close to unity. Then:

1. The rules specified in Section 6.2. defining the construction of the PDR are valid.
2. The search domain for the solution of the GG reduces at each step of the recursion by a factor of at least $\frac{2}{d}$.
3. The unknown λ^* is always contained (w. p. 1) in the new search-interval Δ^{new} resulting from the application of the PDR specified in Section 6.2..

Proof. The proof can be argued to be a consequence of Lemmas 2 and 3. By considering the various cases in which the team of LA cast k^+ "Yes" votes, and by examining where k^+ is with regard to $\lambda^\star$, the reader can easily observe that every single case is considered in the construction of the PDR. Thus, Claims 1 and 2 are proven.

Claim 3 can also be proven in a similar manner, but to strengthen the argument, we allude to the transitivity of the regions, Δ^i, (i.e., that $\Delta^1 \prec \Delta^2 \ldots \prec \Delta^d$). The reason for invoking such a transitivity argument is the following: Consider the case when $k^+ = i$, (where $i \neq 0, d$). In this case, we can conclude that the probability of the maximum being in Δ^i or Δ^{i+1} approaches unity. This, in turn, implies that the probability of the maximum being in Δ^{i+2} approaches zero (if $i + 2 < d$). But more importantly, while the probability of the maximum being in Δ^{i+3} (if $i + 3 < d$) approaches zero, due to the uni-modality of $G(\lambda)$, this probability approaches zero even more rapidly than the probability of the maximum being in Δ^{i+2} approaches zero. This argument implies that the probability mass of the maximum being in the region of interest is arbitrarily close to unity, and indeed, closer to unity than Lemmas 2 and 3 seem to suggest.

We formalize this as follows. Consider the case encountered for the PDR when $\Omega^1 = No, \Omega^2 = No \ldots \Omega^d = No$, implying that $k^+ = 0$. Appealing to Lemma 3 for each of the automata outputs, we get

$$Pr\left[\lambda^\star \ominus \Delta^1\right] \to 1,$$

$$Pr\left[\lambda^\star \ominus \Delta^2\right] \to 0,$$

$$Pr\left[\lambda^\star \ominus \Delta^3\right] \to 0 \text{ with } Pr[\lambda^\star \ominus \Delta^3] < Pr[\lambda^\star \ominus \Delta^2],$$

$$Pr\left[\lambda^\star \ominus \Delta^j\right] \to 0 \text{ with}$$
$$Pr[\lambda^\star \ominus \Delta^j] < Pr[\lambda^\star \ominus \Delta^{j-1}] \ldots < Pr[\lambda^\star \ominus \Delta^3] < Pr[\lambda^\star \ominus \Delta^2] \text{ for } 3 < j \leq d.$$

A consequence of the above d relationships is the fact that the probability mass is concentrated in an arbitrarily small interval around Δ^1, leading to the predicate, $Pr\left[\lambda^\star \ominus \Delta^1\right] \to 1$. Thus, the partition Δ^1 that remains after pruning , will continue to contain $\lambda^\star$ with as high a probability as we want.

Similar arguments can be repeated for each of the other possible decisions made by the LA, and are omitted in the interest of brevity. □

Remark 1: It is easy to see that the decision tables, Tables 2 and 3, used to prune the search space of CGG–AdS for the case when $d = 3$ and $d = 4$ are correct.

Remark 2: It should be mentioned that the smallest number of LA that can be used effectively to achieve this recursive pruning is actually 3. This is because the decision table obtained for the special case of utilizing $d = 2$ LA could sometimes lead to scenarios when the size of the space being examined is never reduced – thus leading to an "infinite" loop.

Remark 3: In terms of computational complexity, it is clear that in the worst case, the length of the interval being examined after J recursive invocations for $d = 3$ LA is $(\frac{2}{3})^J$. Thus the number of recursive steps required to achieve an accuracy of δ is bounded by J_{max}, where,

$$J_{max} = \frac{\log\ \delta}{\log\ (\frac{2}{3})}$$

Notice that for each step of the recursion, the LA would require N_∞ calls to evaluate the function to be optimized.

With these results in place, we are now ready to construct a mechanism that can learn the optimal solution $\lambda^\star$ for the GG.

8. Implementation and Evaluation of CGG–AdS Scheme

The CGG–AdS strategy is fairly simple to implement, because it uses a straightforward partitioning of the search interval, a simple decision table for elimination, and the well known L_{RI} learning algorithm for playing the Goore Game. In this Section we present the pseudo-code for the overall learning strategy as well as that of the L_{RI} automata playing it. We also present a sample trace (for $d = 3$) to demonstrate its correct convergence. Finally, we present numerical results to augment our analysis presented in the previous sections.

8.1. Implementation of the CGG–AdS Strategy

The CGG–AdS strategy has been implemented and tested with a wide range of inputs. The pseudo-code for the algorithms, and sample traces are presented in Figures 3, 9 and 10 respectively, to illustrate the workings of the CGG–AdS strategy, where the latter is given for the case when $d = 3$.

The pseudo-code in Figure 3 shows the recursive organization of the search, including the systematic pruning of the search interval. Each pruning decision is obtained by consulting Table 2, after observing the outcome of an L_{RI} GG that has been projected into the *current* search interval. The algorithm is then recursively invoked. The recursion is terminated when the width of the interval is less than twice the desired accuracy. As exemplified in Figures 9 and 10, it is the projection of the L_{RI} solution to the GG into increasingly smaller search intervals that allows unbounded solution precision.

Although we have done numerous experiments, we present here four specific examples, to highlight two crucial issues. In the experiments which we report, we first used a Gaussian performance criterion function $G(\lambda) = ae^{-(\frac{\lambda^*-\lambda}{b})^2}$, allowing the magnitude and peakedness of the criterion function to be controlled by the parameters a and b respectively. Subsequently, we utilized a doubly-exponential distribution to model non-differentiable and asymmetric functions:

$$G(\lambda) = \begin{cases} ae^{-b_1(\lambda^*-\lambda)} & \lambda \leq \lambda^* \\ ae^{-b_2(\lambda-\lambda^*)} & \lambda > \lambda^* \end{cases}$$

This permitted us to simulate a wide variety of environments in which the derivative on either side of the mean was unequal, and when the latter was undefined at the peak.

In the first experiment which we report, we considered the case when $G(\lambda)$ attains it maximum at 0.9123 - which was exactly the solution for the example given in [31]. This was done to highlight the difference between our recursive GG solution, and the solution presented earlier for the stochastic point location problem. Although the solutions reported in [31, 33, 34] were novel (and in the case of [31], it still remains the only known solution) the LA solution to the GG presented here do not have the luxury of a Teacher/"Oracle" to assist them. Secondly, each LA in the case of the results of [31, 34] have 3 possible decisions, and thus the size of the possible set of decisions is 3^d, which represents a lot of information. In our present case, this is significantly reduced by the pruning to $O(d)$. Here, the number of possible solutions is significantly less – merely $O(d)$, and the reduction that the pruning can achieve is even less significant. Finally, and most importantly, each LA in [31,34] has the advantage of knowing that if the solution is likely to be to the "Left" of a certain region Δ^i, it is even more likely to be to the left of a region Δ^j, where $j > i$. Our current solution has to infer all this - and that in a distributed manner - without knowing how their partners performed or how and why they got a penalty/reward.

Figure 9 shows the trace of execution of the CGG–AdS algorithm for the case when $d = 3$. In this example run, the initial search interval is $[0, 1)$ and $\lambda^\star$ is 0.9123, and the parameters a and b were set to 0.7 and 0.035 respectively - which means that the optimal value of $G(\lambda)$ is 0.7. The search terminated when the width (i.e., the resolution) of the interval was ≤ 0.0002. The reward factor θ of the automata was 0.9999 and $\epsilon = 0.05$. In every invocation of CGG–AdS, the results of the automata are given as the optimal number of "Yes" votes, k^+. Note that at Step 18 in Figure 9, the width of the interval $[0.9121, 0.9123]$ is 0.0002, at which point the estimated value for $\lambda^\star$ is the mid-point of the interval $[0.9121, 0.9123]$, which is 0.9122. We note that at this resolution, our scheme is very close to optimizing the performance criterion function because $G(0.9122) \approx 0.69999$. The corresponding problem in the case of the solution in [31] converged after 10 recursive steps.

It should be mentioned that the traditional LA solution to the GG would require $10,000$ LA to attain this level of precision. Further, as stated earlier, when we are dealing with such a large number of LA, to ensure that each of them converges, we have to assign the value of θ for all of them to be very close to unity - which will require that they individually will converge much slower [24]. Hence, the power of our strategy!

Figure 10 shows the trace of execution of the CGG–AdS algorithm for another case for $d = 3$, and when $\lambda^\star$ is 0.3139, and the parameters a and b were set to 0.7 and 0.03 respectively - which means that the optimal value of $G(\lambda)$ is 0.7. Here, the search terminated when the width (i.e., the resolution) of the interval was ≤ 0.0002 when the reward factor θ of the automata was 0.99975 and $\epsilon = 0.05$. Note that at Step 19 in Figure 10, the algorithm terminates when the width of the interval $[0.3139, 0.3140]$ is less than the specified resolution (0.0002). The estimated value for $\lambda^\star$ is the mid-point of the interval $[0.3139, 0.3140]$, which is 0.31395. Observe that at this resolution, our scheme is very close to optimizing the performance criterion function because $G(0.31395) \approx 0.70000$. In this case too, it should be mentioned that if we wanted to attain an accuracy of 10^{-4}, the traditional LA solution to the GG would require $10,000$ LA to attain this level of precision.

The CGG–AdS was experimentally evaluated to verify the validity of our analytic results and to examine its rate of convergence for various numbers of automata d too. For the first example of Figure 9, we report for brevity's sake, only the results for resolution 0.0002, when $\lambda^\star = 0.9123$, $a = 0.7$, $b = 0.035$, and $d = 3, 5, 9$. For these values, an ensemble of several independent replications with different random number streams were performed to minimize the variance of the reported results. Surprisingly, the CGG–AdS scheme seemed to follow essentially the same pruning strategy for each random number stream, resulting in a minimal variance within the ensemble. Therefore, we only report the results of a single replication here, thereby accurately representing the complete ensemble. The results are

Procedure Search(Δ)

Input : Δ: Search interval $[\sigma, \gamma)$ containing $\lambda^\star$. *Resolution:* The size of the smallest significant interval containing λ^*. Its magnitude determines the accuracy of the final estimate and is used to terminate the recursion.The function *MidPointOfInterval* returns the mid-point of the specified interval and the function *PartitionInterval* partitions the given interval into d sub-intervals. These are trivial, and is thus not described here.

Output : The estimate of $\lambda^\star$.

```
Method :
Begin
    If (WidthOfInterval(Δ) ≤ Resolution) Then
        Return (MidPointOfInterval(Δ))                                /* Terminate Recursion */
    Else
        {Δ^0, ..., Δ^d} := PartitionInterval(Δ)
        k^+ := ExecuteGooreGame({Δ^0, ..., Δ^d})
        Δ^new := ChooseNewSearchInterval({Δ^0, ..., Δ^d} , k^+, Decision-Table)
        Search(Δ^new)                                                  /* Tail Recursion */
    EndIf
END Procedure Search
```

Figure 3. Algorithm CGG–AdS: Overall search strategy.

summarized in Figure 5.

The analogous results for the example of Figure 9 are given in Figure 6, and the results for the doubly-exponential function are found in Figures 7 and 8 respectively. Figure 7 covers the symmetric case with $a = 0.7$ and $b_1 = b_2 = 25$, while Figure 8 deals with the asymmetric case in which $a = 0.7$, $b_1 = 25$, and $b_2 = 100$. As seen, asymmetry and non-differentiability do not significantly affect the performance of our scheme. However, we have observed that although the number of recursive calls in both the doubly-exponential case and the Gaussian case is approximately the same, the individual L_{RI} scheme seems to converge much faster *in each epoch* for the doubly-exponential function.

We first note that as the solution resolution increases at each recursion step, the accuracy of the $\lambda^\star$estimates does not increase monotonically, as, perhaps, could have been expected. Instead, the estimates fluctuate noticeably, however, with decreasing variance. This fluctuation is not a result of the random properties of our algorithm. Rather, the fluctuations form a fixed pattern governed by the (artificially) fixed sub-partitions Δ^i of each search interval, Δ. As a result, the patterns repeat themselves from one simulation run to another. However, for larger number of automata, the positioning of the sub-partitions seems to become less significant, as seen in Figure 5 for $d = 9$.

The reader should also note that at any given recursion step, the speed of convergence seems to be determined by the magnitude that the best available estimate λ^+ differs from the inferior estimates. Thus, for instance, a function G with a $G(\lambda)$ that is flat around the optimal value λ^* may slow down convergence when the search interval falls within the flat area. However, such a flat search interval means that all of the candidate estimates are close to the optimal value of the performance criterion function, and accordingly, the search could be terminated without a significant loss of accuracy.

Procedure ExecuteGooreGame($\{\Delta^0, \ldots, \Delta^d\}$)

Input: The partitioned search interval $\Delta = [\sigma, \gamma)$; the parameters θ and ϵ of the L_{RI} scheme; the performance criterion function $G(\lambda)$ of the Environment. The functions *ChooseAction* and *GetFeedBack* are trivial from an LA perspective, and are hence not explained in detail. If the user opts to use any other ϵ-optimal scheme, for example, from the family of estimator algorithms, he should replace the updating equations in this module. Finally, the reader will observe that there is some duplication of statements in the "If-Then-Else" blocks. This is done just to improve the readability.

Output: A decision k^+ from the set $D = \{0, \ldots, d\}$. The decision represents the optimal number of "Yes" votes among D with respect to optimizing the performance criterion function $G(\lambda)$.

```
Method :
Begin
    For i := 1 To d Do
        P_0^i := P_1^i := 0.5                                        /* Initialization */
    While ε < P_0^i < 1 − ε For Any i ∈ {1,...,d} Do   /* Repeat until all of the d automata have converged */
        k := 0
        For i := 1 To d Do                 /* Collect voting results and count the number of "Yes" votes */
            y_i := ChooseAction(Δ^i)
            k := k + y_i
        EndFor
        For i := 1 To d Do
            If (y_i = 0) Then                                  /* α_0^i is the chosen action (voting no) */
             β := GetFeedBack(σ^k) /* Left boundary σ^k of interval Δ^k projects k into the current search interval */
             If (β = 0) Then                                       /* Oracle has rewarded the choice */
                P_1^i := θ.P_1^i; P_0^i := 1 − P_1^i
             EndIf
            Else                                                /* α_1^i is the chosen action (voting "Yes") */
             β := GetFeedBack(σ^k) /* Left boundary σ^k of interval Δ^k projects k into the current search interval */
             If (β = 0) Then                                       /* Oracle has rewarded the choice */
                P_0^i := θ.P_0^i; P_1^i := 1 − P_0^i
             EndIf
            EndIf
        EndFor
    EndWhile
    k+ := 0      /* Count and return the number of automata that have converged to voting mainly "Yes" (k+) */
    For i := 1 To d Do
        If (P_1^i ≥ 1 − ε) Then
            k+ := k+ + 1
        EndIf
    EndFor
    Return (k+)
End Procedure ExecuteGooreGame
```

Figure 4. Procedures for L_{RI} Automaton's Execution and Pruning of the GG Search Space.

Thus, it is not unfair to state that the results of our experiments are truly conclusive and confirm the power of the CGG–AdS scheme.

Surprisingly, particularly peaked functions, $G(\lambda)$, may also in some cases slow the convergence down. This is the case when the peak falls within the middle area of a sub-partition, leaving the boundary points of all the sub-partitions to be on the extreme ends of the peak tails[19].

[19]We believe that the slow convergence in this case can be reduced by sampling a value randomly from the candidate intervals Δ^i instead of using the left boundary σ_i to generate the environmental feedback. However, a closer analysis of such a strategy is open, and is currently being investigated.

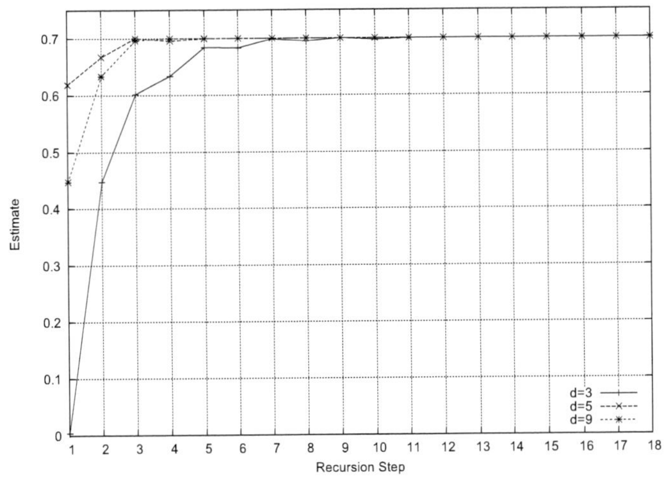

Figure 5. Convergence of estimates for $d = 3, 5, 9$ when the criterion function is Gaussian. The unknown parameter $\lambda^{\star} = 0.9123$ and the optimal criterion functions value is 0.7.

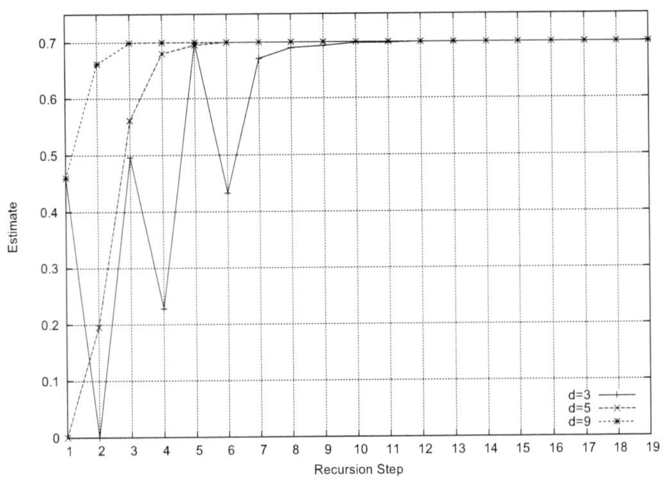

Figure 6. Convergence of estimates for $d = 3, 5, 9$ when the criterion function is Gaussian. The unknown parameter $\lambda^{\star} = 0.9123$ and the optimal criterion functions value is 0.7.

9. Conclusions

In this Chapter, we have considered a fascinating game, called the Goore Game (GG) introduced in [44], and which has recently found applications in many domains, including the field of sensor networks and Quality-of-Service (QoS) routing. The GG has the fascinating property that it can be resolved in a completely distributed manner with no intercommunication between the players. The literature reports that in actual implementations of the solution, the players are typically replaced by Learning Automata (LA). The existing reported approaches have a fundamental “ailment” : The accuracy of the solution achieved

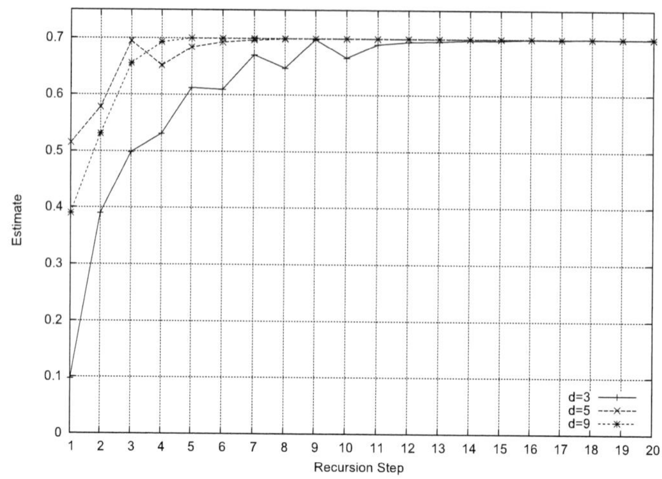

Figure 7. Convergence of estimates for $d = 3, 5, 9$ when the criterion function is doubly-exponential and symmetric. The unknown parameter $\lambda^\star = 0.9123$ and the optimal criterion functions value is 0.7.

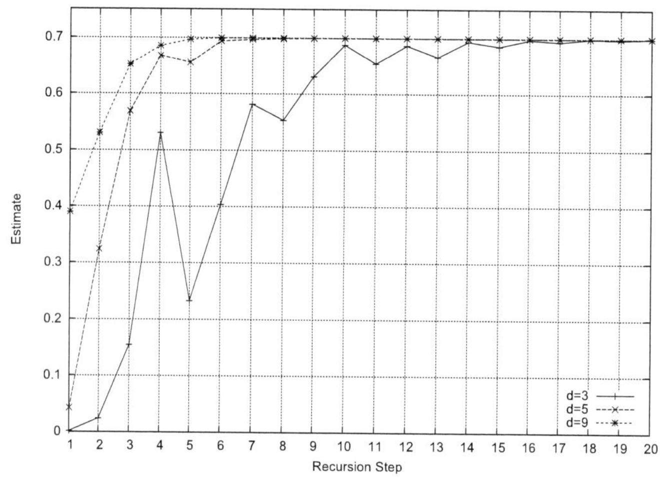

Figure 8. Convergence of estimates for $d = 3, 5, 9$ when the criterion function is doubly-exponential and asymmetric. The unknown parameter $\lambda^\star = 0.9123$ and the optimal criterion functions value is 0.7.

is intricately related to the *number* of players participating in the game – which, in turn, determines the resolution. In other words, an arbitrary accuracy can be obtained only if the game has an *infinite* number of players, and thus a practical solution is infeasible. In this Chapter, we showed how we can attain unbounded accuracy for the GG by utilizing at most d LA, and by recursively pruning the solution space to guarantee that the retained domain contains the solution to the game with a probability as close to unity as desired. Indeed, d can be made as small as *three*. The Chapter contains the formal algorithms, the proofs of the respective convergence results, and simulation results demonstrating its power. Indeed, we believe that we have presented here the first practical implementation of the GG.

Step 1: $\Delta = [0.0, 1.0)$
Partitions: $\Delta^1 = [0.0, 0.3333)$ $\Delta^2 = [0.3333, 0.6667)$ $\Delta^2 = [0.6667, 1.0)$
Results: $k^+ = 3$ Conclusion: New Search Interval is $\Delta^3 = [0.6667, 1.0)$
Step 2: $\Delta = [0.6667, 1.0)$
Partitions: $\Delta^1 = [0.6667, 0.7778)$ $\Delta^2 = [0.7778, 0.8889)$ $\Delta^3 = [0.8889, 1.0)$
Results: $k^+ = 2$ Conclusion: New Search Interval is $\Delta^2 \cup \Delta^3 = [0.7778, 1.0)$
Step 3: $\Delta = [0.7778, 1.0)$
Partitions: $\Delta^1 = [0.7778, 0.8519)$ $\Delta^2 = [0.8519, 0.9259)$ $\Delta^3 = [0.9259, 1.0)$
Results: $k^+ = 2$ Conclusion: New Search Interval is $\Delta^2 \cup \Delta^3 = [0.8519, 1.0)$
Step 4: $\Delta = [0.8519, 1.0)$
Partitions: $\Delta^1 = [0.8519, 0.9012)$ $\Delta^2 = [0.9012, 0.9506]$ $\Delta^3 = [0.9506, 1.0)$
Results: $k^+ = 1$ Conclusion: New Search Interval is $\Delta^1 \cup \Delta^2 = [0.8519, 0.9506)$
Step 5: $\Delta = [0.8519, 0.9506)$
Partitions: $\Delta^1 = [0.8519, 0.8848)$ $\Delta^2 = [0.8848, 0.9117)$ $\Delta^3 = [0.9117, 0.9506)$
Results: $k^+ = 2$ Conclusion: New Search Interval is $\Delta^2 \cup \Delta^3 = [0.8848, 0.9506)$
Step 6: $\Delta = [0.8848, 0.9506)$
Partitions: $\Delta^1 = [0.8848, 0.9067)$ $\Delta^2 = [0.9067, 0.9287)$ $\Delta^3 = [0.9287, 0.9506)$
Results: $k^+ = 1$ Conclusion: New Search Interval is $\Delta^1 \cup \Delta^2 = [0.8848, 0.9287)$
Step 7: $\Delta = [0.8848, 0.9287)$
Partitions: $\Delta^1 = [0.8848, 0.8994)$ $\Delta^2 = [0.8994, 0.9140)$ $\Delta^3 = [0.9140, 0.9287)$
Results: $k^+ = 2$ Conclusion: New Search Interval is $\Delta^2 \cup \Delta^3 = [0.8994, 0.9287)$
Step 8: $\Delta = [0.8994, 0.9287)$
Partitions: $\Delta^1 = [0.8994, 0.9092)$ $\Delta^2 = [0.9092, 0.9189)$ $\Delta^3 = [0.9189, 0.9287)$
Results: $k^+ = 1$ Conclusion: New Search Interval is $\Delta^1 \cup \Delta^2 = [0.8994, 0.9189)$
Step 9: $\Delta = [0.8994, 0.9189)$
Partitions: $\Delta^1 = [0.8994, 0.9059)$ $\Delta^2 = [0.9059, 0.9124)$ $\Delta^3 = [0.9124, 0.9189)$
Results: $k^+ = 2$ Conclusion: New Search Interval is $\Delta^2 \cup \Delta^3 = [0.9059, 0.9189)$
Step 10: $\Delta = [0.9059, 0.9189)$
Partitions: $\Delta^1 = [0.9059, 0.9102)$ $\Delta^2 = [0.9102, 0.9146)$ $\Delta^3 = [0.9146, 0.9189)$
Results: $k^+ = 1$ Conclusion: New Search Interval is $\Delta^1 \cup \Delta^2 = [0.9059, 0.9146)$
Step 11: $\Delta = [0.9059, 0.9146)$
Partitions: $\Delta^1 = [0.9059, 0.9088)$ $\Delta^2 = [0.9088, 0.9117)$ $\Delta^3 = [0.9117, 0.9146)$
Results: $k^+ = 2$ Conclusion: New Search Interval is $\Delta^2 \cup \Delta^3 = [0.9088, 0.9146)$
Step 12: $\Delta = [0.9088, 0.9146)$
Partitions: $\Delta^1 = [0.9088, 0.9107)$ $\Delta^2 = [0.9107, 0.9127)$ $\Delta^3 = [0.9127, 0.9146)$
Results: $k^+ = 2$ Conclusion: New Search Interval is $\Delta^2 \cup \Delta^3 = [0.9107, 0.9146)$
Step 13: $\Delta = [0.9107, 0.9146)$
Partitions: $\Delta^1 = [0.9107, 0.9120)$ $\Delta^2 = [0.9120, 0.9133)$ $\Delta^3 = [0.9133, 0.9146)$
Results: $k^+ = 1$ Conclusion: New Search Interval is $\Delta^1 \cup \Delta^2 = [0.9107, 0.9133)$
Step 14: $\Delta = [0.9107, 0.9133)$
Partitions: $\Delta^1 = [0.9107, 0.9116)$ $\Delta^2 = [0.9116, 0.9124)$ $\Delta^3 = [0.9124, 0.9133)$
Results: $k^+ = 2$ Conclusion: New Search Interval is $\Delta^2 \cup \Delta^3 = [0.9116, 0.9133)$
Step 15: $\Delta = [0.9116, 0.9133)$
Partitions: $\Delta^1 = [0.9116, 0.9122)$ $\Delta^2 = [0.9122, 0.9127)$ $\Delta^3 = [0.9127, 0.9133)$
Results: $k^+ = 1$ Conclusion: New Search Interval is $\Delta^1 \cup \Delta^2 = [0.9116, 0.9127)$
Step 16: $\Delta = [0.9116, 0.9127)$
Partitions: $\Delta^1 = [0.9116, 0.9120)$ $\Delta^2 = [0.9120, 0.9123)$ $\Delta^3 = [0.9123, 0.9127)$
Results: $k^+ = 1$ Conclusion: New Search Interval is $\Delta^1 \cup \Delta^2 = [0.9116, 0.9123)$
Step 17: $\Delta = [0.9116, 0.9123)$
Partitions: $\Delta^1 = [0.9116, 0.9118)$ $\Delta^2 = [0.9118, 0.9121)$ $\Delta^3 = [0.9121, 0.9123)$
Results: $k^+ = 3$ Conclusion: New Search Interval is $\Delta^3 = [0.9121, 0.9123)$
Step 18: $\Delta = [0.9121, 0.9123)$
Partitions: $\Delta^1 = [0.9121, 0.9122)$ $\Delta^2 = [0.9122, 0.9123)$ $\Delta^3 = [0.9123, 0.9123)$
Results: $k^+ = 1$ Conclusion: New Search Interval is $\Delta^1 \cup \Delta^2 = [0.9121, 0.9123)$

Figure 9. Trace of the execution of an example run of CGG–AdS algorithm for the case when $d = 3$ and the function to be optimized is $G(\lambda) = ae^{-(\frac{\lambda^* - \lambda}{b})^2}$, with a=0.7 and b=0.035. The game converges after 19 recursive epochs.

Step 1: $\Delta = [0.0, 1.0)$
Partitions: $\Delta^1 = [0.0, 0.333)$ $\Delta^2 = [0.3333, 0.6667)$ $\Delta^3 = [0.6667, 1.0)$
Results: $k^+ = 1$ Conclusion: New Search Interval is $\Delta^1 \cup \Delta^2 = [0.0, 0.6667)$
Step 2: $\Delta = [0.0, 0.6667)$
Partitions: $\Delta^1 = [0.0, 0.2222)$ $\Delta^2 = [0.2222, 0.4444)$ $\Delta^3 = [0.4444, 0.6667)$
Results: $k^+ = 1$ Conclusion: New Search Interval is $\Delta^1 \cup \Delta^2 = [0.0, 0.4444)$
Step 3: $\Delta = [0.0, 0.4444)$
Partitions: $\Delta^1 = [0.0, 0.1481)$ $\Delta^2 = [0.1481, 0.2963)$ $\Delta^3 = [0.2963, 0.4444)$
Results: $k^+ = 2$ Conclusion: New Search Interval is $\Delta^2 \cup \Delta^3 = [0.1481, 0.4444)$
Step 4: $\Delta = [0.1481, 0.4444)$
Partitions: $\Delta^1 = [0.1481, 0.2469)$ $\Delta^2 = [0.2469, 0.3457)$ $\Delta^3 = [0.3457, 0.4444)$
Results: $k^+ = 2$ Conclusion: New Search Interval is $\Delta^2 \cup \Delta^3 = [0.2469, 0.4444)$
Step 5: $\Delta = [0.2469, 0.4444)$
Partitions: $\Delta^1 = [0.2469, 0.3128)$ $\Delta^2 = [0.3128, 0.3786)$ $\Delta^3 = [0.3786, 0.4444)$
Results: $k^+ = 1$
Conclusion: New Search Interval is $\Delta^1 \cup \Delta^2 = [0.2469, 0.3786)$
Step 6: $\Delta = [0.2469, 0.3786)$
Partitions: $\Delta^1 = [0.2469, 0.2908)$ $\Delta^2 = [0.2908, 0.3347)$ $\Delta^3 = [0.3347, 0.3786)$
Results: $k^+ = 2$ Conclusion: New Search Interval is $\Delta^2 \cup \Delta^3 = [0.2908, 0.3786)$
Step 7: $\Delta = [0.2908, 0.3786)$
Partitions: $\Delta^1 = [0.2908, 0.3201)$ $\Delta^2 = [0.3201, 0.3493)$ $\Delta^3 = [0.3493, 0.3786)$
Results: $k^+ = 1$ Conclusion: New Search Interval is $\Delta^1 \cup \Delta^2 = [0.2908, 0.3493)$
Step 8: $\Delta = [0.2908, 0.3493)$
Partitions: $\Delta^1 = [0.2908, 0.3103)$ $\Delta^2 = [0.3103, 0.3298)$ $\Delta^3 = [0.3298, 0.3493)$
Results: $k^+ = 1$ Conclusion: New Search Interval is $\Delta^1 \cup \Delta^2 = [0.2908, 0.3298)$
Step 9: $\Delta = [0.2908, 0.3298)$
Partitions: $\Delta^1 = [0.2908, 0.3038)$ $\Delta^2 = [0.3038, 0.3168)$ $\Delta^3 = [0.3168, 0.3298)$
Results: $k^+ = 2$ Conclusion: New Search Interval is $\Delta^2 \cup \Delta^3 = [0.3038, 0.3298)$
Step 10: $\Delta = [0.3038, 0.3298)$
Partitions: $\Delta^1 = [0.3038, 0.3125)$ $\Delta^2 = [0.3125, 0.3212)$ $\Delta^3 = [0.3212, 0.3298)$
Results: $k^+ = 1$ Conclusion: New Search Interval is $\Delta^1 \cup \Delta^2 = [0.3038, 0.3212)$
Step 11: $\Delta = [0.3038, 0.3212)$
Partitions: $\Delta^1 = [0.3038, 0.3096)$ $\Delta^2 = [0.3096, 0.3154)$ $\Delta^3 = [0.3154, 0.3212)$
Results: $k^+ = 2$ Conclusion: New Search Interval is $\Delta^2 \cup \Delta^3 = [0.3096, 0.3212)$
Step 12: $\Delta = [0.3096, 0.3212)$
Partitions: $\Delta^1 = [0.3096, 0.3134)$ $\Delta^2 = [0.3134, 0.3173)$ $\Delta^3 = [0.3173, 0.3212)$
Results: $k^+ = 1$ Conclusion: New Search Interval is $\Delta^1 \cup \Delta^2 = [0.3096, 0.3173)$
Step 13: $\Delta = [0.3096, 0.3173)$
Partitions: $\Delta^1 = [0.3096, 0.3122)$ $\Delta^2 = [0.3122, 0.3147$ $\Delta^3 = [0.3147, 0.3173)$
Results: $k^+ = 2$ Conclusion: New Search Interval is $\Delta^2 \cup \Delta^3 = [0.3122, 0.3173)$
Step 14: $\Delta = [0.3122, 0.3173)$
Partitions: $\Delta^1 = [0.3122, 0.3139)$ $\Delta^2 = [0.3139, 0.3156)$ $\Delta^3 = [0.3156, 0.3173)$
Results: $k^+ = 1$ Conclusion: New Search Interval is $\Delta^1 \cup \Delta^2 = [0.3122, 0.3156)$
Step 15: $\Delta = [0.3122, 0.3156)$
Partitions: $\Delta^1 = [0.3122, 0.3133)$ $\Delta^2 = [0.3133, 0.3144)$ $\Delta^3 = [0.3144, 0.3156)$
Results: $k^+ = 1$ Conclusion: New Search Interval is $\Delta^1 \cup \Delta^2 = [0.3122, 0.3144)$
Step 16: $\Delta = [0.3122, 0.3144)$
Partitions: $\Delta^1 = [0.3122, 0.3129)$ $\Delta^2 = [0.3129, 0.3137)$ $\Delta^3 = [0.3137, 0.3144)$
Results: $k^+ = 3$ Conclusion: New Search Interval is $\Delta^3 = [0.3137, 0.3144)$
Step 17: $\Delta = [0.3137, 0.3144)$
Partitions: $\Delta^1 = [0.3137, 0.3139]$ $\Delta^2 = [0.3139, 0.3142)$ $\Delta^3 = [0.3142, 0.3144)$
Results: $k^+ = 1$
Conclusion: New Search Interval is $\Delta^1 \cup \Delta^2 = [0.3137, 0.3142)$
Step 18: $\Delta = [0.3137, 0.3142)$
Partitions: $\Delta^1 = [0.3137, 0.3139)$ $\Delta^2 = [0.3139, 0.3140)$ $\Delta^3 = [0.3140, 0.3142)$
Results: $k^+ = 2$ Conclusion: New Search Interval is $\Delta^2 \cup \Delta^3 = [0.3139, 0.3142)$
Step 19: $\Delta = [0.3139, 0.3142)$
Partitions: $\Delta^1 = [0.3139, 3140)$ $\Delta^2 = [0.3140, 0.3141)$ $\Delta^3 = [0.3141, 0.3142)$
Results: $k^+ = 0$ Conclusion: New Search Interval is $\Delta^1 = [0.3139, 0.3140)$

Figure 10. Trace of the execution of an example run of CGG–AdS algorithm for the case when $d = 3$ and the function to be optimized is $G(\lambda) = ae^{-(\frac{\lambda^* - \lambda}{b})^2}$, with a=0.7 and b=0.03. The game converges after 19 recursive epochs.

We are currently investigating the application of these results to a variety of potential applications involving neural networks and optimization, and in the application domains related to sensor networks and QoS routing.

The question of how the GG can be solved if the function G is time-varying remains open. We believe, though, that our current solution would still be valid if the function is time-varying but the *position* of the optimum is unchanged. However, if the location of the optimum point changes with time, the question of the convergence of traditional LA to the time-varying optimum is itself unknown.

References

[1] M. Agache. *Estimator Based Learning Algorithms*. M.C.S. Thesis, School of Computer Science, Carleton University, Ottawa, Canada, 2000.

[2] M. Agache and B. J. Oommen. Generalized pursuit learning schemes: New families of continuous and discretized learning automata. *IEEE Transactions on Systems, Man and Cybernetics*, SMC-32(B):738– 749, 2002.

[3] M. Agache and B. J. Oommen. Generalized TSE: A new generalized estimator-based learning automaton. In *Proceedings of CIS, the 2004 IEEE Conference on Cybernetics and Intelligent Systems*, pages 245– 251, Singapore, December 2004.

[4] C. R. Atkinson, G. H. Bower, and E. J. Crowthers. *An Introduction to Mathematical Learning Theory*. John Wiley snd Sons, New York, 1958.

[5] A. F. Atlassis, N. H. Loukas, and A. V. Vasilakos. The use of learning algorithms in atm networks call admission control problem: A methodology. *Computer Networks*, 34:341–353, 2000.

[6] A. F. Atlassis and A. V. Vasilakos. The use of reinforcement learning algorithms in traffic control of high speed networks. *Advances in Computational Intelligence and Learning*, pages 353–369, 2002.

[7] P. Auer, N. Cesa-Bianchi, Y. Freund, and R. E. Schapire. The nonstochastic multi-armed bandit problem. *SIAM Journal on Computing*, 32:48–77, 2003.

[8] N. Baba and Y. Mogami. Learning behaviors of the hierarchical structure stochastic automata operating in the nonstationary multiteacher environment. In *Knowledge-Based Intelligent Information and Engineering Systems: 9th International Conference, KES 2005*, pages 624–634, Melbourne, Australia, September 2005.

[9] R. A. Baeza-Yates, J. C. Culberson, and G. J. E. Rawlins. Searching with uncertainty. In *Proceedings of the Scandinavian Workshop on Algorithms and Theory, SWAT 88*, pages 176–189, 1988.

[10] R. A. Baeza-Yates and R. Schott. Parallel searching in the plane. In *1992 International Conference of the Chilean Computer Society, IC-SCCC 92*, pages 269–279. Chilean Computer Society, 1992.

[11] J. L. Bentley and A. C-C. Yao. An almost optimal algorithm for unbounded searching. *Information Processing Letters*, pages 82–87, August 1976.

[12] R. R. Bush and F. Mosteller. *Stochastic Models for Learning*. John Wiley snd Sons, New York, 1965.

[13] Y. U. Cao, A. S. Fukunaga, and A. Kahng. Cooperative Mobile Robotics: Antecedents and Directions. *Autonomous Robots*, 4(1):7–27, 1997.

[14] D. Chen and P. K. Varshney. QoS Support in Wireless Sensor Networks: A Survey. In *The 2004 International Conference on Wireless Networks (ICWN 2004)*, 2004.

[15] D. P. Foster and R. Vohra. Regret in the on-line decision problem. *Games and Economic Behavior*, 29:7–35, October 1999.

[16] S. Hart and A. Mas-Colell. A general class of adaptive strategies. *Journal of Economic Theory*, 98:26–54, 2001.

[17] R. Iyer and L. Kleinrock. Qos control for sensor networks. In *IEEE International Conference on Communications*, volume 1, pages 517–521, 2003.

[18] J. Kabudian, M. R. Meybodi, and M. M. Homayounpour. Applying continuous action reinforcement learning automata (CARLA) to global training of hidden markov models. In *Proceedings of the International Conference on Information Technology: Coding and Computing , ITCC'04*, pages 638–642, Las Vegas, Nevada, 2004.

[19] S. Lakshmivarahan. *Learning Algorithms Theory and Applications*. Springer-Verlag, 1981.

[20] J.K. Lancôt and B.J. Oommen. Discretized estimator learning automata. *IEEE Transactions on Systems Man and Cybernetics*, SMC-22:1473– 1483, 1992.

[21] M. R. Meybodi and H. Beigy. New learning automata based algorithms for adaptation of backpropagation algorithm pararmeters. *International Journal of Neural Systems*, 12:45–67, 2002.

[22] S. Misra and B. J. Oommen. GPSPA : A new adaptive algorithm for maintaining shortest path routing trees in stochastic networks. *International Journal of Communication Systems*, 17:963– 984, 2004.

[23] K. Najim and A. S. Poznyak. *Learning Automata: Theory and Applications*. Pergamon Press, Oxford, 1994.

[24] K.S. Narendra and M.A.L. Thathachar. *Learning Automata*. Prentice-Hall, 1989.

[25] M. F. Norman. On Linear Models with Two Absorbing Barriers. *Journal of Mathematical Psychology*, 5:225–241, 1968.

[26] M. S. Obaidat, G. I. Papadimitriou, and A. S. Pomportsis. Learning automata: Theory, paradigms and applications. *IEEE Transactions on Systems Man and Cybernetics*, SMC-32:706–709, 2002.

[27] M. S. Obaidat, G. I. Papadimitriou, and A. S. Pomportsis. Efficient fast learning automata. *Information Sciences*, 157:121–133, 2003.

[28] M. S. Obaidat, G. I. Papadimitriou, A. S. Pomportsis, and H. S. Laskaridis. Learning automata-based bus arbitration for shared-medium ATM switches. *IEEE Transactions on Systems, Man, and Cybernetics : Part B*, 32:815–820, 2002.

[29] B. J. Oommen and M. Agache. Continuous and discretized pursuit learning schemes: Various algorithms and their comparison. *IEEE Transactions on Systems, Man and Cybernetics*, SMC-31(B):277– 287, 2001.

[30] B. J. Oommen and E.V. de St. Croix. Graph partitioning using learning automata. *IEEE Transactions on Computers*, C-45:195–208, 1995.

[31] B. J. Oommen, G. Raghunath, and B. Kuipers. Parameter learning from stochastic teachers and stochastic compulsive liars. *IEEE Transactions on Systems Man and Cybernetics*, page To Appear, 2006.

[32] B. J. Oommen and T. D. Roberts. Continuous learning automata solutions to the capacity assignment problem. *IEEE Transactions on Computers*, C-49:608–620, 2000.

[33] B.J. Oommen. Stochastic searching on the line and its applications to parameter learning in nonlinear optimization. *IEEE Transactions on Systems, Man and Cybernetics*, SMC-27:733– 739, 1997.

[34] B.J. Oommen and G. Raghunath. Automata learning and intelligent tertiary searching for stochastic point location. *IEEE Transactions on Systems, Man and Cybernetics*, SMC-28B:947– 954, 1998.

[35] G. I. Papadimitriou and A. S. Pomportsis. Learning-automata-based TDMA protocols for broadcast communication systems with bursty traffic. *IEEE Communication Letters*, pages 107–109, 2000.

[36] A. S. Poznyak and K. Najim. *Learning Automata and Stochastic Optimization*. Springer-Verlag, Berlin, 1997.

[37] G. Santharam, P.S. Sastry, and M.A.L. Thathachar. Continuous action set learning automata for stochastic optimization. *Journal of the Franklin Institute*, 331B5:607–628, 1994.

[38] Y. Sawaragi and N. Baba. A note on the learning behavior of variable-structure stochastic automata. *IEEE Transactions on Systems, Man and Cybernetics*, 3:644–647, 1973.

[39] F. Seredynski. Distributed scheduling using simple learning machines. *European Journal of Operational Research*, 107:401–413, 1998.

[40] I. J. Shapiro and K. S. Narendra. Use of Stochastic Automata for Parameter Self-Optimization with Multi-modal Performance Criteria. *IEEE Transactions on Systems Science, and Cybernetics*, SSC-5:352–360, 1969.

[41] M. A. L. Thathachar and M. T. Arvind. Solution of Goore game using models of stochastic learning automata. *J. Indian Institute of Science*, (76):47–61, January-February 1997.

[42] M. A. L. T. Thathachar and P. S. Sastry. *Networks of Learning Automata : Techniques for Online Stochastic Optimization*. Kluwer Academic, Boston, 2003.

[43] M. L. Tsetlin. On the behaviour of finite automata in random media. *Automation and Remote Control*, 22:1210–1219, 1962.

[44] M.L. Tsetlin. *Automaton Theory and the Modelling of Biological Systems*. Academic Press, New York and London, 1973.

[45] Brian Tung and Leonard Kleinrock. Using Finite State Automata to Produce Self-Optimization and Self-Control. *IEEE Transactions on Parallel and Distributed Systems*, 7(4):47–61, 1996.

[46] C. Unsal, P. Kachroo, and Bay J. S. Simulation study of multiple intelligent vehicle control using stochastic learning automata. *Transactions of the Society for Computer Simulation International*, 14:193–210, 1997.

[47] V. I. Varshavskii and I. P. Vorontsova. On the behavior of stochastic automata with a variable structure. *Automation and Remote Control*, 24:327–333, 1963.

In: Game Theory: Strategies, Equilibria and Theorems ISBN 978-1-60456-844-8
Editors: Ingrid N. Haugen et al, pp. 217-229

Chapter 7

Product Line Competition and ICT Investments

Nobuo Matsubayashi*
Department of Administration Engineering,
Faculty of Science and Technology, Keio University
Hiyoshi 3-14-1 Kohoku-ku, Yokohama, 223-8522, Japan

Abstract

In this paper, we study strategic product line designs under duopoly. While the availability of a rich variety of products may attract consumers, it may also lead to confusion in the purchase decision-making process resulting in consumer dissatisfaction. Firms are thus motivated to offer their products with ICT tools in order to supply consumers with sufficient communication about them. We present a simple game theoretic model to analyze the impact on product line decisions of the cost of developing these tools, which may even be affected even by the rival's product line. We show that the ICT progress in solving consumers' confusion due to a firm's own products enlarges the profit gap between asymmetric firms. In contrast, it is shown that the ICT progress in solving consumers' confusion arising from a comparison with the rival's products reduces the profit gap between them. The cost asymmetry between two firms makes both of these opposite effects more robust.

1. Introduction

The recent progress of information and communication technology (ICT) has dramatically increased the ability of firms to offer a wide variety of products. As a result, not only the limited number of previously popular products but also a large number of low-volume products can be offered online in competition with each other. However, the aggregation of demands for these "niche" products has an unnegligible impact on the market, which is known as "long tail" effect.

However, we should note that firms in the market are not able to benefit from these niche products unless consumers actually find them. Obviously, when a multitude of products are available, it is not profitable for a firm to offer a product that cannot be readily

*E-mail address: nobuo-m@pa2.so-net.ne.jp

found. Also, even if every product is displayed to consumers, when there is insufficient information about product attributes, a rich variety of products may cause consumers' confusion resulting in their dissatisfaction. In fact, studies by Villas-Boas (2004) and Kuksov and Villas-Boas (2005) refer to such a confusion problem and state that the excessive varieties of products have a rather negative effect on consumers' motivation to purchase, unless firms provide sufficient communication with consumers about them. Iyengar and Lepper (2000) also provide empirical evidence for the existence of this negative impact. Van Zandt (2004) refers to such a negative externality in the context of information overload. These observations suggest that it is important for firms to control the length of their product line. In this paper, we explore strategic product line decisions under a duopoly market.

When firms offer a wide variety of products in competition with each other, utilizing ICT tools may help consumers recognize the characteristics of each product. Firms are thus motivated to offer their products with ICT tools to provide sufficient communication about them. For example, the introduction of a search engine on a firm's web site enables a consumer to search for his/her desired products quickly. Another example is an investment in Google's advertising service (e.g., Adsense), which can effectively provide information about a large number of products to many consumers. However, in any case, as the number of products offered to the market increases, solving consumers' confusion becomes a more difficult problem. Therefore, firms must incur the cost of developing these tools in order to offer their product line successfully and this increases with the number of products offered to the market. Based on this motivation, we present a simple game theoretic model to analyze the product line decisions between two firms in consideration of the cost impact of developing the ICT tools. The main purpose of this paper is to investigate how progress in ICT development i.e., a decline in ICT cost affects the equilibrium outcomes.

Our analysis is developed in a microeconomic framework. There are many economic literatures on product line design, for example Moorthy (1984), Kim and Chhajed (2000), Krishnan and Zhu (2006), Netessine and Taylor (2007), and so on. While most of these papers focus on the cannibalization problem faced by the monopoly's product line, several papers such as Desai (2001), Brander and Eaton (1984) and Wernerfelt (1986) study the product line competition between firms. Furthermore, the recent paper of Matsubayashi et al. (2007a) discusses the leader-follower game where two firms sequentially determine their product line in consideration of the negative impact of product variety. However, the common interest to all of the above works is not only the decision of product-line length but also price and quality design of each product as a differentiation tool. Thus for the analytical tractability, their models assume that the maximum length of product line (the maximum number of products offered by each firm) is *exogenously* given at a small level (two for each firm in most papers). In contrast, our model supposes that firms can *endogenously* determine their product-line length under a situation where the price and quality competition between two firms is not at work. In fact, in some markets such as those of books, music CDs and supermarkets, a variety of alternatives is more important for consumers and therefore strategic price and quality decisions of each product is out of firms' scope, see Dobson and Waterson (2005) and Peng and Tabuchi (2007).

Our study is closely related to Draganska and Jain (2005) and Matsubayashi (2007b), where both papers investigate the impact of product-line length as a competitive tool. Draganska and Jain (2005) show that the length of a product line affects consumers' motivation to purchase and thus competitive firms must strategically control their product-line length. On the other hand, Matsubayashi (2007b) explores the strategic product line design under various competition structures and discusses the possibility of attaining the socially optimal number of products in the market. However, in their studies, it is assumed that the cost for recognizing a wide variety of products is incurred by consumers themselves. In contrast, our setting supposes that firms incur this cost through the investments for ICT tools. Specifically, we model the cost structure reflecting the ICT costs that firms must incur for obtaining the consumers' recognition on the whole of its product line. The key point is that the firm's ICT cost is affected by not only the number of its own products (say hereafter, *intra-brand effect*) but also that of the rival's products (say hereafter, *inter-brand effect*), since consumers' confusion would increase according to the total number of products in the market, as mentioned above. This consideration of cost impact for solving the inter-brand effect is the crucial difference from settings in the previous literature.

In addition, we assume in this paper that the cost impacts for solving the intra-brand effect and inter-brand effect are independent, since ICT technologies for solving each confusion are in general different from each other. Specifically, progress in technology for the intra-brand effect such as the introduction of a search engine and progress in the development of web design tools are relatively due to an internal R&D. On the other hand, progress in technology for the inter-brand effect such as a collaboration in developing promotion tools and a support by a third party (e.g. request for customer review through Web 2.0 tools such as SNS, investment to Google's advertising service, and so on) are due to more external factors. Since costs for external factors are common to both firms, the cost asymmetry here is assumed to exist only with respect to the cost for solving the intra-brand effect. We analyze the impacts of the declines in cost for solving each effect separately.

We first analyze the impact of ICT progress for solving the intra-brand effect. The result on the equilibrium product-line length is asymmetric between two firms: the equilibrium number of products offered by the firm with the cost disadvantage (say hereafter, the weaker firm) does not change monotonically with the technological progress for solving the intra-brand effect, while that offered by the firm with the cost advantage (say hereafter, the stronger firm) is always increasing with ICT progress. The asymmetry also holds for the equilibrium profits. Loosely speaking, the decline in ICT cost for the intra-brand effect enlarges the gap between two firms. However, surprisingly, we have the opposite result for the inter-brand case: the decline in ICT cost for the inter-brand effect reduces the gap between two firms. In summary, solving the intra-brand effect and the inter-brand effect have the opposite impact on asymmetric firms. We further show that the asymmetry between two firms makes these impacts more robust.

The rest of this paper is organized as follows: Section 2 introduces our game-theoretic model and derives the Nash equilibrium. In Section 3, we analyze the impact of ICT progress on the equilibrium outcomes through a comparative static approach. Section 4 summarizes our findings. All proofs of the results are given in the Appendix.

2. The Model

In this section, we introduce our game-theoretic model for product line decisions. For a single commodity, say A, we assume a duopoly market consisting of two firms, 1 and

2. Each firm potentially has an ability to develop numerous horizontally differentiated products/items with respect to A by using various tools for differentiation and thus can endogenously choose the length of its product line. Let n_1 and n_2 be the number of products offered by firms 1 and 2, respectively. For the analytical tractability, n_i is allowed to be a non-integer number and we assume that $n_i = 0$ does not immediately imply no purchase from firm i (For example, this would be true if n_i corresponds to the number of mobile phone contents available for download). Consumers have different brand/taste preferences and each consumer is indexed by x. We assume that x is uniformly distributed on the interval $[0, 1]$ with density 1 and firms are located at the opposite ends of this unit line: firm 1 is located at 0 and firm 2 is located at 1. Thus consumer 0 is the most loyal consumer to firm 1 and in contrast consumer 1 is most loyal to firm 2. According to Draganska and Jain (2005), we assume the situation where consumers optimize their utility by selecting the firm that *is likely to offer* the products they want. Based on this, therefore, consumers' utility is assumed to be more affected by *how many* products are offered but not which products are offered. The utility level of consumer x for each firm is specifically given as follows:

$$\begin{aligned} u_1(x) &= V + vn_1 - tx \\ u_2(x) &= V + vn_2 - t(1 - x), \end{aligned}$$

where V is his/her utility level of commodity A at baseline and $v(v > 0)$ and $t(t > 0)$ are constant parameters with respect to product-line length and his/her loyalty, respectively. We suppose that V is sufficiently high, so that every consumer necessarily purchases products from either firm. As indicated in the Introduction, we assume here that the price competition between two firms is not at work and thus utility level is modeled independently of prices of products.

By solving the equation $u_1(x) = u_2(x)$ for x, we can see that the consumer who is indifferent about purchasing from firms 1 and 2 is located at

$$\bar{x} = \frac{t + vn_1 - vn_2}{2t}.$$

Therefore, the demand quantities of firms 1 and 2, denoted by D_1 and D_2, can be described as follows:

$$\begin{aligned} D_1(n_1, n_2) &= \bar{x} = \frac{t + vn_1 - vn_2}{2t}, \\ D_2(n_1, n_2) &= 1 - \bar{x} = \frac{t - vn_1 + vn_2}{2t}. \end{aligned}$$

We next formulate the cost structure in our context. To concentrate our attention on the relation between product line length and ICT investment, we simply employ the following costs for developing ICT tools:

$$\begin{aligned} \psi_1(n_1, n_2) &\equiv \phi n_1^2 + kn_1n_2, \\ \psi_2(n_1, n_2) &\equiv a\phi n_2^2 + kn_1n_2, \end{aligned}$$

where $\phi > 0$, $k \geq 0$ and $0 < a \leq 1$. Roughly speaking, these cost functions reflect the costs that firm $i(i = 1, 2)$ must incur on the whole of its product line in order to obtain consumer recognition. The first term of ψ_i reflects the *intra-brand effect*, that is the ICT cost for solving consumers' confusion arising from a comparison between firm i's own products. On the other hand, the second term reflects the *inter-brand effect*, that is the ICT cost for solving consumers' confusion arising from a comparison with products offered by the rival. We assume the quadratic form of the product line lengths of firms. Alternatively, another function form might be more appropriate for modeling the ICT cost. However, the following two can justify the use of the quadratic cost structure: (i) this is consistent with the cost structure for quality improvement in the existing manufacturing, marketing and product development literature (e.g., Banker et al. 1998, Kim and Chhajed 2002, Villas-Boas 2004, Krishnan and Zhu 2006, Matsubayashi 2008), and (ii) we note that without any guidance it takes O($n_i n_j$) time to compare n_i items with n_j items and thus the effort for the reduction in this time spent would be proportional to $n_i n_j$.

Coefficient parameters ϕ and k represent the difficulty of solving consumer's confusion and both are affected by ICT technology progress. However, we assume that they are independent, since ICT technologies for solving each confusion are in general different from each other. Technology progresses for the intra-brand effect are relatively due to internal factors for example, an improvement of Web design and an introduction of a search engine are within the scope of a firm's own R&D. On the other hand, technological progresses for the inter-brand effect are due to factors that are more external. These might include for example, the direct cooperation of both firms (e.g., collaboration in developing promotion tools) and support by a third party (e.g., a request for customer review through Web 2.0 tools such as SNS, investments to the Google's service for online advertising, and so on), which are out of the scope of a firm's own R&D. Finally, parameter a models the asymmetry between the two firms. However, since the cost for solving the inter-brand effect is usually common to both firms, the cost advantage/disadvantage is assumed to exist only with respect to the cost for solving the intra-brand effect. Without loss of generality, firm 2 has the advantage, which implies $a \leq 1$.

As mentioned above, all products are assumed to be horizontally differentiated with each other and consumers are assumed to be price-inelastic. We thus employ a simple assumption that all prices of products are identical and normalized as 1. The profit functions π_1 and π_2 are now given as follows:

$$\pi_1(n_1, n_2) = \frac{t + vn_1 - vn_2}{2t} - \phi n_1^2 - kn_1 n_2,$$
$$\pi_2(n_1, n_2) = \frac{t - vn_1 + vn_2}{2t} - a\phi n_2^2 - kn_1 n_2.$$

We define a strategic-form game with the strategies n_i and the payoffs $\pi_i(i = 1, 2)$ as G. The first step of our study is to analyze the Nash equilibrium of G. It is obvious that $\pi_i(i = 1, 2)$ are always strictly concave with respect to n_i, which implies that the Nash equilibrium always exists.

3. The Nash Equilibrium and Its Analysis

Theorem 3.1. *The unique Nash equilibrium in pure strategy $n_i^*(i = 1, 2)$ is specifically given as follows:*

1. *When* $\frac{k}{2\phi} < a \leq 1$: $n_1^* = \frac{(2a\phi - k)v}{2t(4a\phi^2 - k^2)}$, $n_2^* = \frac{(2\phi - k)v}{2t(4a\phi^2 - k^2)}$,
2. *When* $a \leq \frac{k}{2\phi}$: $n_1^* = 0$, $n_2^* = \frac{v}{4a\phi t}$.

To focus on the analysis of the case where product-line lengths of both firms are positive, we hereafter assume that $\frac{k}{2\phi} < a \leq 1$. In this case (Case 1), it is straightforward that the equilibrium profits π_i^* are described as follows:

$$\pi_1^* = \frac{1}{2} + \frac{-k^3 + 3\phi k^2 - 4(2-a)a\phi^3}{4t^2(4a\phi^2 - k^2)^2} v^2,$$

$$\pi_2^* = \frac{1}{2} + \frac{-k^3 + 3a\phi k^2 + 4a(1-2a)\phi^3}{4t^2(4a\phi^2 - k^2)^2} v^2.$$

Our main analysis is to investigate how a decline in ICT cost affects the equilibrium profit of firms. However, as mentioned in the previous section, we consider the situation where the technological progresses for solving the intra-brand effect and the inter-brand effect are independent. Therefore, we analyze the impacts of the declines in ϕ and k separately. We perform the analysis through comparative static.

3.1. The Impact of ICT Progress for Solving the Intra-brand Effect

In this subsection, we focus on the impact of ICT progress for solving the intra-brand effect on the equilibrium outcome. We first present the following basic result with regard to the relation between the equilibrium product-line length n_i^* and the parameter ϕ capturing the ICT cost relating to the intra-brand effect.

Theorem 3.2. *For any a such that* $\frac{k}{2\phi} < a \leq 1$,

1. n_1^* *is an inverted U-shaped function of* ϕ *which has its uniquely maximum point at* $\bar{\phi}_1 = \frac{k(1+\sqrt{1-a})}{2a}$.
2. $\frac{\partial n_2^*}{\partial \phi} < 0$ *always holds.*

It can easily be seen that the result is asymmetric between the two firms. Specifically, the equilibrium number of products of the weaker firm does not change monotonically with the technological progress for solving the intra-brand effect, while that of the stronger firm is always increasing with ICT progress. It should be noted that in our context, firms increase their revenue only by increasing the number of products offered. Thus the decline in the ICT cost promotes their intense product line competition, so that the level of product-line length tends to increase. However, as the ICT progress is significant, the asymmetry between two firms plays an important role in their competition. The increase of product offering against the rival is no longer beneficial for the weaker firm, which compels the firm to control the length of its product line. We thus have the "boundary" in the progress level of ICT that determines whether to increase its product offering aggressively. This boundary depends on the level of asymmetry (the value of parameter a). As asymmetry becomes severe, the boundary shifts to higher ϕ ($\frac{\partial \bar{\phi}_1}{\partial a} < 0$), which implies that the weaker firm becomes less aggressive in providing a rich variety of products.

We next show our results with regard to the relation between the equilibrium profits π_i^* and the cost parameter ϕ.

Theorem 3.3. *For any a such that $\frac{k}{2\phi} < a \leq 1$,*

1. $\frac{\partial \pi_1^*}{\partial \phi} > 0$ *always holds.*
2. *If $a \leq \frac{1}{2}$, then $\frac{\partial \pi_2^*}{\partial \phi} < 0$ always holds. Otherwise, π_2^* is a U-shaped function of ϕ which has its uniquely minimum point at $\bar{\phi}_2$, where $\bar{\phi}_2$ is the unique solution in $\frac{k}{2a} < \phi$ for the equation $16(1-2a)a\phi^4 + 12(1+a)k^2\phi^2 - 16k^2\phi + 3k^4 = 0$.*

The result is also asymmetric between the two firms. However, in contrast to the result with regard to the equilibrium product-line length, the equilibrium profit of the firm with the cost advantage does not change monotonically with ICT progress. As seen in Theorem 3.2, the decline in cost due to the technological progress increases the intensity of their product line competition. As seen in many typical competitions (e.g., the usual Bertrand price competition), this intensity is apt to make the equilibrium profit of both firms lower. However, when the cost asymmetry between two firms exists, the significant ICT progress enlarges the difference markedly between them in cost impact for increasing product offering. This makes the weaker firm give up offering a richer product line in competition with the rival, as seen in Theorem 3.2, while the stronger firm easily increases the length of its product line. In other words, the stronger firm succeeds in being monopolistic. Therefore, its equilibrium profit is increasing when there is technological progress. We note that the boundary in the progress level of ICT ($\bar{\phi}_2$) provides the worst circumstance for the stronger firm, which depends on the level of asymmetry. When two firms are completely symmetric ($a = 1$), the boundary does not exist and the equilibrium profit of each firm always decreases with technological progress.

3.2. The Impact of ICT Progress for Solving the Inter-brand Effect

We next investigate the impact of the ICT progresses for solving the inter-brand effect on the equilibrium outcome. As in the previous subsection, we first introduce the basic result with regard to the relation between the equilibrium product line length n_i^* and the parameter k capturing the ICT cost relating to the inter-brand effect.

Theorem 3.4. *For any a such that $\frac{k}{2\phi} < a \leq 1$,*

1. $\frac{\partial n_1^*}{\partial k} < 0$ *always holds.*
2. *n_2^* is a U-shaped function of k which has its uniquely minimum point at $\bar{k}_1 = 2\phi(1 - \sqrt{1-a})$.*

It should indeed be noted that the impact of solving the inter-brand effect is opposite to that of solving the intra-brand effect: the equilibrium product-line length of the weaker firm is always increasing with the progress of ICT in regard to the inter-brand effect, while that of the stronger firm does not change monotonically. The intuition behind this is as follows: the decline in the ICT cost encourages both firms to increase their product variety. However, it can be easily seen from the form of our cost function that as both firms have larger and similar product-line lengths, the inter-brand effect becomes more critical. This has the result of discouraging the stronger firm to offer a richer product line, while the weaker firm is relatively aggressive in extending its product line. This strategic asymmetry between two firms is significant in the situation where the inter-brand effect is relatively high. In contrast, when the inter-brand effect is relatively low, both firms are no longer concerned about the difficulty of solving the problem of consumer confusion arising from the wide variety of their products. Thus they compete with each other in increasing their product-line lengths.

In the following, we show our results with regard to the relation between the equilibrium profits π_i^* and the cost parameter k.

Theorem 3.5. *For any a such that $\frac{k}{2\phi} < a \leq 1$,*

1. *If $a \leq \frac{1}{2}$, then $\frac{\partial \pi_1^*}{\partial k} \leq 0$ always holds. Otherwise, π_1^* is an inverted U-shaped function of k which has its uniquely maximum point at $\bar{k}_2$, where $\bar{k}_2$ is the unique solution in $k < 2a\phi$ for the equation $k^3 - 6\phi k^2 + 12a\phi^2 k + 8a(1-2a)\phi^3 = 0$.*

2. *$\frac{\partial \pi_2^*}{\partial k} \geq 0$ always holds.*

The result stands in sharp contrast to that in Theorem 3.3. Comparing Theorem 3.5 (1) with Theorem 3.3 (2), we can see that in the presence of a relatively high degree of asymmetry in a firms' cost structure, the progress in ICT that solves the inter-brand effect increases the equilibrium profit of the weaker firm, while that for solving the intra-brand effect increases the equilibrium profit of the stronger firm. It should be noted that as the two firms become asymmetric, the boundary $\bar{k}_2$ becomes lower, as well as $\bar{\phi}_2$ becomes higher. On the other hand, the comparison between Theorem 3.5 (2) with Theorem 3.3 (1) shows that the ICT progress for solving the inter-brand effect decreases the equilibrium profit of the stronger firm, while that for solving the intra-brand effect decreases the equilibrium profit of the weaker firm. That is, loosely speaking, the ICT progress for solving the inter-brand effect benefits the weaker firm, while that for solving the intra-brand effect benefits the stronger firm. As seen in Theorem 3.4, the technology progress with regard to the inter-brand effect does not encourage the stronger firm to increase its product line, so that the impact of asymmetry between firms on their competition is reduced. As a result, the difference in the equilibrium profits between them becomes smaller. We should note that a potential asymmetry between two firms here plays an important role. If two firms are relatively symmetric, the technological progress simply promotes the intense competition between them and thus the equilibrium profits of both firms decrease, which is a common observation to that appearing in the intra-brand case.

All results in the theorems 3.2-3.5 are roughly illustrated in Figures 3.1-3.4, which describe the relation of cost parameters and the equilibrium outcomes. In every figure, the shift from solid line to dotted line implies that two firms become more asymmetric (decline in parameter a). The opposite effects can be easily seen from Figures 3.1 and 3.2: the decline in ϕ enlarges the gap of the equilibrium product-line lengths between two firms while that in k reduces it. Furthermore, from Figures 3.3 and 3.4 we can see that this observation is also appropriate with regard to the equilibrium profits. In addition, it can be easily verified that the asymmetry between two firms makes these effects more robust.

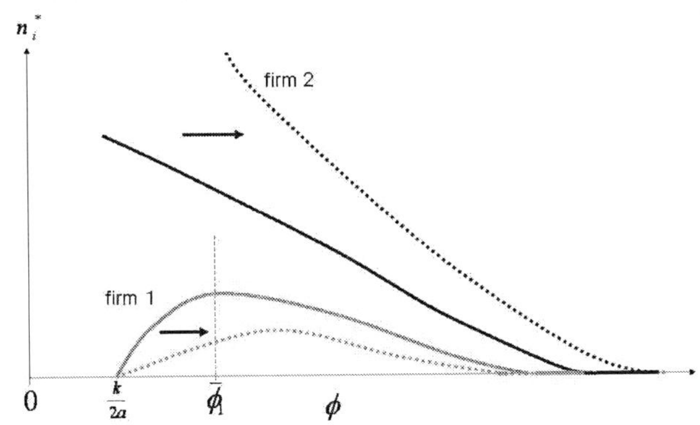

Figure 3.1. The relation of ICT cost for solving the intra-brand effect and the equilibrium product-line lengths.

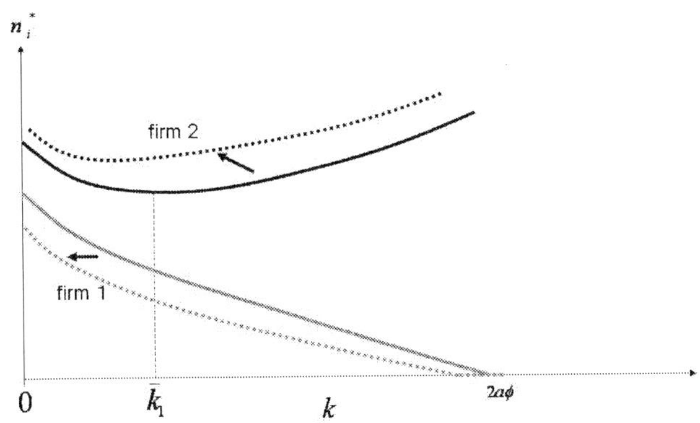

Figure 3.2. The relation of ICT cost for solving the inter-brand effect and the equilibrium product-line lengths.

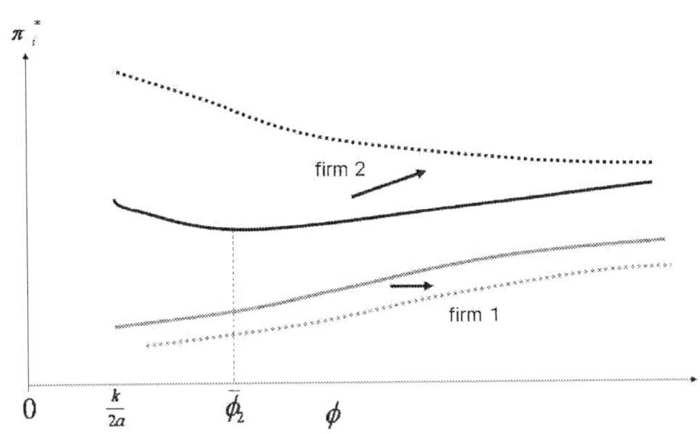

Figure 3.3. The relation of ICT cost for solving the intra-brand effect and the equilibrium profits.

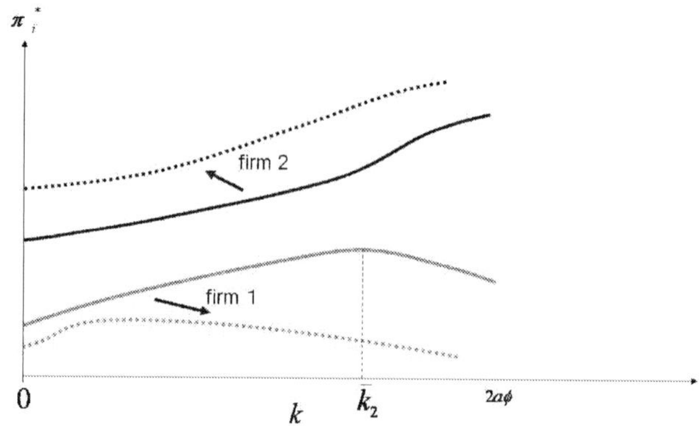

Figure 3.4. The relation of ICT cost for solving the inter-brand effect and the equilibrium profits.

4. Conclusion

In this study, we explored the problem of product line design under a competitive environment. When firms offer a rich variety of products in competition with each other, utilizing ICT tools may help consumers recognize the characteristics of each product. Firms are thus motivated to offer their products with ICT tools in order to provide sufficient communication about them. Utilizing a simple game theoretic model in consideration of the cost impact of developing the ICT tools, we analyzed how ICT progresses affect the equilibrium product-line lengths and profits. It has been shown that the ICT progress for the intra-brand effect enlarges the gap between two asymmetric firms, while that for the inter-brand effect reduces the gap between them. The point for understanding this is that under significant ICT progress for the intra-brand, cost asymmetry is critical for the weaker firm to increase its product offering against the rival, which compels the firm to control the length of its product line. In contrast, under significant ICT progress for the inter-brand, the stronger firm is discouraged from offering a richer product line having become aware that the weaker firm is relatively aggressive for extending its product line. We showed that both of these opposite effects are more robust when two firms become more asymmetric. We can obtain some managerial implications from these results: the stronger firm could benefit if it would somehow succeed in developing ICT for the intra-brand effect, such as an improvement of Web design or a search engine. On the other hand, it might be important for the weaker firm to make efforts to develop ICT for the inter-brand effect, such as Blog, SNS, the Google's service for online advertising, and so on.

Appendix

a. **Proof of Theorem 3.1.** We first derive the best response functions of both firms as $n_1(n_2) = \max(-\frac{k}{2\phi}n_2 + \frac{v}{4\phi t}, 0)$ and $n_2(n_1) = \max(-\frac{k}{2a\phi}n_1 + \frac{v}{4a\phi t}, 0)$. It can be easily verified that they uniquely intersect at the interior point $(n_1^*, n_2^*) = (\frac{(2a\phi-k)v}{2t(4a\phi^2-k^2)}, \frac{(2\phi-k)v}{2t(4a\phi^2-k^2)})$ if and only if $\frac{k}{2\phi} < a(\leq 1)$. Otherwise, they uniquely intersect at $(n_1^*, n_2^*) = (0, \frac{v}{4a\phi t})$.

b. **Proof of Theorem 3.2.** By a direct calculation, we have $\frac{\partial n_1^*}{\partial \phi} = -\frac{av(4a\phi^2-4k\phi+k^2)}{t(4a\phi^2-k^2)^2}$. Since $\phi > \frac{k}{2a}$, $\bar{\phi}_1 = \frac{k(1+\sqrt{1-a})}{2a}$ is the unique feasible solution for $4a\phi^2-4k\phi+k^2 = 0$. We also have that $\frac{\partial n_1^*}{\partial \phi} > 0$ for $(\frac{k}{2a} <)\phi < \bar{\phi}_1$ and $\frac{\partial n_1^*}{\partial \phi} < 0$ for $\phi > \bar{\phi}_1$. On the other hand, we directly derive $\frac{\partial n_2^*}{\partial \phi} = -\frac{v(4a\phi^2-4ak\phi+k^2)}{t(4a\phi^2-k^2)^2}$. However, $4a\phi^2 - 4ak\phi + k^2 = 0$ has no feasible solution for ϕ, which implies that $4a\phi^2 - 4ak\phi + k^2 > 0$ at all $\phi(> \frac{k}{2a})$. The result thus follows.

c. **Proof of Theorem 3.3.** It can be directly derived that $\frac{\partial \pi_1^*}{\partial \phi} = -\frac{-16(2-a)a^2\phi^4+12a(1+a)k^2\phi^2-16ak^3\phi+3k^4}{4t^2(4a\phi^2-k^2)^3}v^2$. Let $f(\phi) \equiv -16(2-a)a^2\phi^4 + 12a(1+a)k^2\phi^2 - 16ak^3\phi + 3k^4$. We can see that $\frac{\partial f}{\partial \phi} = -64(2-a)a^2\phi^3 + 24a(1+a)k^2\phi - 16ak^3$ is decreasing in ϕ for all $\phi > \frac{k}{2a}$, since it can be verified that both solutions for $\frac{\partial^2 f}{\partial \phi^2} = 0$ are less than $\frac{k}{2a}$. In addition, we have $\frac{\partial f(\frac{k}{2a})}{\partial \phi} = \frac{-4(a-4)(a-1)}{a} \leq 0$, which implies $\frac{\partial f}{\partial \phi} \leq 0$ at every $\phi > \frac{k}{2a}$. Furthermore, by $f(\frac{k}{2a}) = \frac{-2(a-1)^2}{a^2}k^4 \leq 0$, we can obtain $f(\phi) < 0$ for any $\phi > \frac{k}{2a}$. Hence, we have $\frac{\partial \pi_1^*}{\partial \phi} > 0$ for any $\phi > \frac{k}{2a}$.

On the other hand, we have $\frac{\partial \pi_2^*}{\partial \phi} = -\frac{16(1-2a)a^2\phi^4+12a(1+a)k^2\phi^2-16ak^3\phi+3ak^4}{4t^2(4a\phi^2-k^2)^3}v^2$. Let $g(\phi) \equiv 16(1-2a)a\phi^4 + 12(1+a)k^2\phi^2 - 16k^3\phi + 3k^4$. We first suppose $a \leq \frac{1}{2}$. Then, since $\frac{\partial^2 g}{\partial \phi^2} = 192a(1-2a)\phi^2 + 24(1+a)k^2 > 0$ and $\frac{\partial g(\frac{k}{2a})}{\partial \phi} = \frac{-a^2-a+2}{2a^2} > 0$, $\frac{\partial g}{\partial \phi} > 0$ holds for any $\phi > \frac{k}{2a}$. Furthermore, $g(\frac{k}{2a}) = \frac{3a^3-5a^2+a+1}{a^3}k^4 \geq 0$ holds for any $0 < a \leq 1$, which implies that $g(\phi)$ is always positive for $\phi > \frac{k}{2a}$. It thus follows that $\frac{\partial \pi_2^*}{\partial \phi} < 0$. We next suppose $\frac{1}{2} < a \leq 1$. In this case, $\frac{\partial^2 g}{\partial \phi^2} = 0$ has two solutions $\phi = -\sqrt{\frac{1+a}{8a(2a-1)}}k$, $\sqrt{\frac{1+a}{8a(2a-1)}}k$. However, we directly have that $\frac{k}{2a} < \sqrt{\frac{1+a}{8a(2a-1)}}k$, $\frac{\partial g(\frac{k}{2a})}{\partial \phi} \geq 0$ and $g(\frac{k}{2a}) \geq 0$, which implies that there exists $\bar{\phi}_2$ such that $g(\phi) > 0$ for $(\frac{k}{2a} <)\phi < \bar{\phi}_2$, $g(\bar{\phi}_2) = 0$ and $g(\phi) < 0$ for $\phi > \bar{\phi}_2$. The result thus follows.

d. **Proof of Theorem 3.4.** By a direct calculation, we have $\frac{\partial n_1^*}{\partial k} = \frac{-k^2+4a\phi k-4a\phi^2}{(4a\phi^2-k^2)^2}$. $f(k) \equiv -k^2 + 4a\phi k - 4a\phi^2$ is maximized at $k = 2a\phi$. However, $f(2a\phi) = 4a\phi^2(a-1) \leq 0$ holds, which implies that $\frac{\partial n_1^*}{\partial k}$ is always negative for $k < 2a\phi$. On the other hand, we have $\frac{\partial n_2^*}{\partial k} = \frac{-k^2+4\phi k-4a\phi^2}{(4a\phi^2-k^2)^2}$. Since $-k^2 + 4\phi k - 4a\phi^2$ is an inverted U-shaped function of k and $k = 2\phi(1-\sqrt{1-a})(\equiv \bar{k}_1)$ is the unique feasible solution for $-k^2 + 4\phi k - 4a\phi^2 = 0$, we derive that $\frac{\partial n_2^*}{\partial k} < 0$ for $k < \bar{k}_1$ and $\frac{\partial n_2^*}{\partial k} > 0$ for $k > \bar{k}_1$.

e. **Proof of Theorem 3.5.** We can directly derive that $\frac{\partial \pi_1^*}{\partial k} = -\frac{k^4-6\phi k^3+12a\phi^2k^2+8a(1-2a)\phi^3 k}{4t^2(4a\phi^2-k^2)^3}v^2$. Let $f(k) \equiv k^3 - 6\phi k^2 + 12a\phi^2 k + 8a(1-2a)\phi^3$. It can be easily verified that $\frac{\partial f}{\partial k} = 0$ has two solutions $k = 2\phi(1-\sqrt{1-a})(< 2a\phi)$, $2\phi(1+\sqrt{1-a})(> 2a\phi)$. We first suppose $a \leq \frac{1}{2}$. Then, since

$f(0) = 8a(1-2a)\phi^3 \geq 0$ and $f(2a\phi) = 8a\phi^3(1-a)^2 > 0$, $f(k)$ is always positive for $k < 2a\phi$. Therefore, $\frac{\partial \pi_1^*}{\partial k} \leq 0$ always holds. We next suppose $\frac{1}{2} < a \leq 1$. Then, since $f(0) < 0$ and $f(2a\phi) > 0$, there exists $\bar{k}_2$ such that $f(k) < 0$ for $k < \bar{k}_2$, $f(\bar{k}_2) = 0$ and $f(k) > 0$ for $(2a\phi >)k > \bar{k}_2$. The result thus follows.

On the other hand, we have $\frac{\partial \pi_2^*}{\partial k} = -\frac{k^4 - 6a\phi k^3 + 12a\phi^2 k^2 - 8a(2-a)\phi^3 k}{4t^2(4a\phi^2 - k^2)^3} v^2$. Let $g(k) \equiv k^3 - 6a\phi k^2 + 12a\phi^2 k - 8a(2-a)\phi^3$. We can easily see that $\frac{\partial g}{\partial k} = 0$ has no feasible solution, which implies that $\frac{\partial g}{\partial k}$ is always positive. However, it follows that $g(0) = -8a(2-a)\phi^3 < 0$ and $g(2a\phi) = -16a(1-a)^2\phi^3 \leq 0$. Therefore, $g(k) \leq 0$ always holds and thus we have $\frac{\partial \pi_2^*}{\partial k} \geq 0$.

Acknowledgments

The author is supported by the Grant-in-Aid for Young Scientists (B) 18710138 of the Ministry of Education, Culture, Sports, Science and Technology of Japan.

References

[1] Banker, R.D., Khosla, I. and Sinha, K.K., 1998. Quality and competition. *Management Science*, **44**(9), 1179-1192.

[2] Brander, J.A. and Eaton, J., 1984. Product line rivalry. *The American Economic Review*, **74**(3), 323-334.

[3] Desai, P.S., 2001. Quality segmentation in spatial markets: when does cannibalization affect product line design? *Marketing Science*, **20**(3), 265-283.

[4] Dobson, P.W. and Waterson, M., 2005. Chain-store pricing across local markets. *Journal of Economics and Management Strategy*, **14**, 93-119.

[5] Draganska, M. and Jain, D.C., 2005. Product-line length as a competitive tool. *Journal of Economics and Management Strategy*, **14**, 1-28.

[6] Kim, K. and Chhajed, D., 2000. Commonality in product design: cost saving, valuation change and cannibalization. *European Journal of Operational Research*, **125**(3), 602-621.

[7] Krishnan, V. and Zhu, W., 2006. Designing a family of development-intensive products. *Management Science*, **52**(6), 813-825.

[8] Kuksov, D. and Villas-Boas, J.M., 2005. When more alternatives lead to less choice. Working Paper.

[9] Iyengar, S.S. and Lepper, M.R., 2000. When choice is demotivating: can one desire too much of a good thing? *Journal of Personality and Social Psychology*, **79**, 995-1006.

[10] Matsubayashi, N., Ishii, Y., Watanabe, K. and Yamada, Y., 2007a. Full-line or specialization strategy? The negative effect of product variety on product line strategy. Working Paper.

[11] Matsubayashi, N., 2007b. Product design rivalry: multiple-attributes approach to differentiation. *Managerial and Decision Economics*, forthcoming.

[12] Matsubayashi, N. and Yamada, Y., 2008. A note on price and quality competition between asymmetric firms. *European Journal of Operational Research*, **187**(2), 571-581.

[13] Moorthy, K.S., 1984. Market segmentation, self-selection, and product line design. *Marketing Science*, **3**(4), 288-307.

[14] Netessine, S. and Taylor, T.A., 2007. Product line design and production technology. *Marketing Science*, **26**(1), 101-117.

[15] Peng, S. and Tabuchi, T., 2007. Spatial competition in variety and number of stores. *Journal of Economics and Management Strategy*, **16**, 227-250.

[16] Van Zandt, T., 2004. Information overload in a network of targeted communication. *Rand Journal of Economics*, **35**(3), 542-560.

[17] Villas-Boas, J.M., 2004. Communication strategies and product line design. *Marketing Science*, **23**(3), 304-316.

[18] Wernerfelt, B., 1986. Product line rivalry: Note. *The American Economic Review*, **76**(4), 842-844.

In: Game Theory: Strategies, Equilibria and Theorems
Editors: Ingrid N. Haugen et al, pp. 231-240
ISBN 978-1-60456-844-8

Chapter 8

TRANSFERABLE UTILITY GAMES WITH INDIVIDUAL GOALS

***M.A. Hinojosa*[1]**and A.M. Mármol*[2]†**
[1]Universidad Pablo de Olavide. Dept. of Economics,
Quantitative Methods and Economic History. Pablo de Olavide University
[2]Dept. Economía Aplicada III. Seville University. Spain

Abstract

The objective of this paper is to propose and analyze a solution concept for the class of transferable utility cooperative games which generalizes the notion of core and is stable in relation to a set of individual goals, in the sense that the players can guarantee the achievement of certain goals in any coalition they join. In other words, individual players impose their own goals in any cooperation group.

We also investigate which are the best individual goals that can be attained in a game and provide a characterization of the allocations associated to those goals as the solutions of a multiobjective linear problem. Finally, we explore the potential of goal programming approaches to obtain compromise allocations for the transferable utility game. These allocations are such that the excesses the players obtain can not be improved simultaneously. In particular, we apply a minimax reference point approach, which is specially appropriate for finding equitable solutions when a group of decision makers is involved in deciding the most preferred solutions as is the case in n-person cooperative games.

1. Introduction

Cooperative games with transferable utility have been extensively explored in the literature, both from the theoretical point of view and from their potential towards applications One of the most interesting solution concepts proposed for this class of games is the concept of core, first formalized in [2]. Allocations in the core are stable in the sense that there is no coalition with both the desire and the power to change the outcome of the game.

*E-mail address: mahinram@upo.es.
†E-mail address: amarmol@us.es.

These outcomes in the core are also fair in the sense that no subgroup of players subsidizes any other subgroup. Therefore, solutions in the core provide incentives for cooperation. Nevertheless, the core has some disadvantages as "the solution" of the game: some games have an empty core and others have too many points in the core.

Several other solution concepts based on different ideas of stability, not generally as strong as that represented by the core, have also been proposed in the literature: stable sets [12], the bargaining set, and the related concepts of kernel and nucleolus (see [5] for a detailed discussion). However, even though the rationale of these concepts is sufficiently justified from the theoretical point of view, in general, they share the difficulty of effective calculation.

The objective of this paper is to propose and analyze a new solution concept for the class of transferable utility cooperative games which is stable in relation to a set of individual goals, in the sense that the players can guarantee the achievement of certain individual goals in any coalition they join. In other words, individual players impose their own goals in any cooperation group. This idea can be seen as a generalization of the notion of core in that it requires the excesses of the coalitions to be above certain bounds, while in the core non-negative excesses are required.

We also investigate which are the best individual goals that can be attained in the game and provide a characterization of the allocations associated to those goals.

In the case of a game a group of individuals is involved in the decision, and the final solution is dependent upon the preferences of the various agents. In these situations a particularly useful way of communicating preferences is by assigning targeted values(goals) to the players' conflicting objectives as well as relative weights and priority levels to attain these goals. Goal programming is one of the most popular multiobjective optimisation techniques, which provides a pragmatic and flexible way to deal with decision makers' preferences (see [1] and [3]).

In this paper we explore the potential of goal programming approaches to obtain compromise allocations for the transferable utility game such that the excesses the players obtain cannot be improved simultaneously. In particular, we apply a minimax reference point approach developed in [9]. This method is especially appropriate for finding equitable outcomes when a group of decision makers is involved in deciding the most preferred solutions as is the case in n-person cooperative games. In addition, the calculations involved can be performed using the existing standard linear programming software packages.

The rest of the paper is organized as follows. In Section 2, we present the model of a game with individual goals, the concept of p-core is introduced and non-dominated vector of goals are characterized. Section 3 presents the goal programming approach to the problem of finding allocations that take into account the individual goals. Finally, Section 4 is devoted to the conclusions.

2. Cooperative Transferable Utility Games with Individual Goals

A cooperative transferable utility (TU) game is represented as a pair (N, v), where $N = \{1, 2, \ldots, n\}$ is the set or players. A nonempty subset $S \subseteq N$ of the player set is called a coalition. The map v is the characteristic function of the game, which associates each $S \subseteq N$ with a positive real number $v(S)$, assuming $v(\emptyset) = 0$.

The characteristic function represents the value of each coalition, that is to say, the amount of utility that a coalition could obtain by itself without cooperating with the players outside the coalition. Therefore, $v(N)$ is the total amount of utility obtained if all the players cooperate. It is assumed that the characteristic function is expressed in units of an infinitely divisible commodity. To assume transferable utility is to require that the payoffs attainable by any particular coalition consist of all individual payoffs that sum to no more than a particular number and that they can be transferred without loss between players.

An interesting question which arises is how $v(N)$ should be allocated among the various players taking into account the values of the coalitions. An allocation is a payoff vector, $x \in \Re^n$, whose components represent the payoff for each player, x_i being the payoff for the i-th player. Therefore the efficiency condition $\sum_{i=1}^n x_i = v(N)$ must hold. We denote the set of allocations of the game by $I^*(N, v)$, that is, $I^*(N, v) = \{x \in \Re^n, \sum_{i=1}^n x_i = v(N)\}$, and by $x(S)$ the overall payoff obtained by coalition S under the allocation x, $x(S) = \sum_{i \in S} x_i$. Consequently $x(N) = v(N)$.

A solution concept for cooperative games associates a set (possibly empty) of allocations with each game. We will be concerned with valuing the differences between what the coalitions obtain with a certain allocation and their values in the game. For each allocation and each coalition, the excess function $e(x, S) = x(S) - v(S)$, measures the excesses of coalition S under allocation x. The excess function plays a central role in the definition of some solution concepts for cooperative games, especially in the concept of core. The core of a game (N, v), denoted by $C(N, v)$, consists of those allocations from which no coalition has any incentive to deviate and abandon the grand coalition. It can be defined as $C(N, v) = \{x \in I^*(N, v) | \, e(x, S) \geq 0, \, \forall \, S \subseteq N, \, S \neq \emptyset\}$.

However, the core can be too wide a set of solutions, or a too restrictive set, or it can even be empty. In practice, it may be interesting not only to determine if there exist allocations with non negative excesses, but also if it is possible to establish certain bounds on the excess function in order to obtain the allocations that achieve a better performance. When the core is empty, we are interested in determining the allocations closest to the core, and if the core is not empty the interest is focused on those subsets of allocations with higher excesses.

This is the underlying idea in the solution concept presented in this paper. The idea of measuring how " close" the game is to having a nonempty core, or how "big" the core of the game is, has already been explored in the literature. In fact, the ε-core, which consists of allocations that are within ε of being in the core or having excesses above ε, that is, $C_\varepsilon(N, v) = \{x \in I^*(N, v) | \, e(x, S) \geq \varepsilon, \, \forall \, S \subseteq N, \, S \neq \emptyset\}$, was introduced in [8], and the least core, which is the intersection of all non-empty ε-cores, analyzed in [4]. A scalar common measure of the excesses for any coalition regardless of the number of individuals in it, is considered in the ε-core. In contrast, a multidimensional measure that takes into account the excesses that the different players can obtain, is used in the solution concept presented herein.

Suppose each player establishes a goal on the excesses, $p_i, i \in N$, and is only interested in allocations that assure he will achieve this goal regardless of which coalition he joins. Formally, let $p \in \Re^n$, $p = (p_1, p_2, \ldots, p_n)$, be a vector of individual goals, and let $p(S) = \sum_{i \in S} p_i$. The amount $p(S)$ represents the overall goal to be attained by a coalition when the players establish the vector of goals p. We want to investigate solution concepts for the game (N, v) that take into account these goals.

Definition 2.1. The p-core for the game (N, v), denoted $C(N, v, p)$, is the set of allocations $C(N, v, p) = \{x \in I^*(N, v) |\, e(x, S) \geq p(S),\ \forall\, S \subset N,\ S \neq \emptyset\}$.

If an allocation is in the p-core, it will not only give every player at least an extra p_i in relation to its characteristic value, $v(i)$, but it will also be stable, in the sense that all the coalitions will obtain an excess that is sufficient to allocate an extra p_i to each of its players. In other words, in these allocations, individual players impose their goals in any cooperation group.

It is easy to see that for $p = 0$, the p-core coincides with the core. However, when $p \leq 0$[1], the concept of p-core is a relaxation of the notion of core, in that the excess function only needs to be maintained above negative levels. When $p \geq 0$, the concept of p-core becomes more restrictive than the core, by requiring that some of the excesses attained are strictly above zero. It is also straightforward to see that if $p \geqq q$, then $C(N, v, p) \subseteq C(N, v, q)$. Consequently, if $p \geqq 0$ then $C(N, v, p) \subseteq C(N, v)$.

This solution concept can be seen as a generalization of the ε-core [8]. However, while in the ε-core each individual attains an extra fixed excess (or goal) in any coalition, the p-core permits different players to attain different goals. The requirement for an allocation to be in a p-core could be interpreted as follows: If, for each individual, joining a coalition entails a cost, p_i, the p-core provides a way of taking into account the costs involved in coalition-forming. On the other hand, a negative p_i may represent a bonus obtained when joining a coalition. Alternatively, we might regard the organization costs as negligible, or as already included in $v(S)$, but view p_i as a threshold below which the requirement of fulfilling group rationality, that is $\sum_{i \in S} x_i \geq v(S)$, is not considered worth the trouble.

We want to investigate which vectors of goals could be attained in the game. We say that a vector is a feasible vector of goals for if at least one allocation exists for which these individual goals are attained.

Definition 2.2. $p \in \mathbb{R}^n$ is a feasible vector of goals for the game (N, v) if there exists $x \in I^*(N, v)$, such that $x \in C(N, v, p)$.

We will focus on those feasible vector of goals that cannot be improved componentwise.

Definition 2.3. $p \in \mathbb{R}^n$ is a non-dominated vector of goals in the game (N, v), if $C(N, v, p) \neq \emptyset$ and $C(N, v, q) = \emptyset$, for all $q \geq p$. We will denote the set of non-dominated vectors of goals by $\mathcal{P}(N, v)$.

It is worth noting that $\mathcal{P}(N, v)$ is never an empty set. Even when the core is empty, there must exist feasible vectors of individual goals whose components cannot be improved simultaneously.

The next question is how to determine the set of non-dominated vector of goals. For a game (N, v), the following linear problem is considered:

[1] For vectors $x, y \in \Re^n$, we write $x \geqq y$ if $x_i \geq y_i$, $\forall i = 1, \ldots, n$, $x \geq y$ if $x \geqq y$ and $x \neq y$, and $x > y$ if $x_i > y_i$, $\forall i = 1, \ldots, n$.

$$\begin{aligned} & min \quad \sum_{i=1}^{n} x_i \\ & s.t. \quad x(S) \geq v(S) \quad \forall S \subset N \end{aligned} \tag{2.1}$$

Denote by $m^*(N, v)$ the minimum value for problem (2.1) and by $M(N, v)$ the set of its optimal solutions. Let $E = v(N) - m^*(N, v)$.

The next result characterizes the non-dominated vectors of goals as those vectors whose components sum up the quantity E.

Proposition 2.4. *$p \in \mathbb{R}^n$ is a non-dominated vector of goals in the game (N, v) if and only if $\sum_{i=1}^n p_i = E$.*

Proof. If p is non-dominated, then p is feasible, that is, $\exists x \in I^*(N, v)$ such that $x(S) \geq v(S) + p(S)$, $\forall S \subset N$. Consider $\bar{x} = x - p$. It is easy to see that $\bar{x}(S) \geq v(S)$, and therefore, $\sum_{i=1}^n \bar{x}_i \geq m^*(N, v)$. As $\sum_{i=1}^n \bar{x}_i = \sum_{i=1}^n x_i - \sum_{i=1}^n p_i$, it follows that $\sum_{i=1}^n p_i \leq E$.

Suppose now that $\sum_{i=1}^n p_i < E$, and consider $y \in M(N, v)$. It follows that $\sum_{i=1}^n y_i = m^*(N, v)$ and $y(S) \geq v(S), \forall S \subset N$. Let $q_i = p_i + \frac{E - \sum_{i=1}^n p_i}{n}$, and $\bar{y}_i = y_i + q_i$, $i = 1, \ldots, n$. Then, $\bar{y}(S) = y(S) + q(S) \geq v(S) + q(S)$ and $\bar{y}(N) = y(N) + p(N) + v(N) - m^*(N, v) - p(N) = v(N)$. Thus $\bar{y} \in C(N, v, q)$ with $q > p$, what is a contradiction with p being a non-dominated vector of goals.

Conversely, if $\sum_{i=1}^n p_i = E$, consider $x \in M(N, v)$. In this case, $x(S) \geq v(S)$ and $x(N) = m^*(N, v)$. Consider $y = x + p$, it follows that $y(S) \geq v(S) + p(S)$ and $y(N) = v(N)$. Therefore, $y \in C(N, v, p)$. To prove that p is a non-dominated vector of goals, suppose that there exists $q \geq p$ with $C(N, v, q) \neq \{\emptyset\}$. Let $z \in C(N, v, q) \subseteq C(N, v, p)$, and consider $x^* = z - q$ which is a feasible solution for problem (2.1). But $\sum_{i=1}^n x_i^* = \sum_{i=1}^n z_i - \sum_{i=1}^n q_i = v(N) - \sum_{i=1}^n q_i < v(N) - \sum_{i=1}^n p_i = v(N) - E = m^*(N, v)$, which contradicts $m^*(N, v)$ being the minimum for problem (2.1). □

As a consequence of this result, it follows that the p-cores associated to non-dominated vectors of goals can be generated by adding vectors whose components add up to E to the optimal solutions of problem (2.1), as stated in the following result.

Corollary 2.5. *$p \in \mathbb{R}^n$ is a non-dominated vector of goals in the game (N, v) if and only if*

$$C(N, v, p) = \{\, y = x + p \,|\, x \in M(N, v) \,\}.$$

Proof. : First, consider $x \in M(N, v)$, and $y = x + p$. For all $S \subset N$, $y(S) = x(S) + p(S) \geq v(S) + p(S)$ holds. Since $y(N) = x(N) + p(N) = m^*(v, N) + E = v(N)$, then $y = x + p \in C(N, v, p)$.

Now, consider $y \in C(N, v, p)$ and $x = y - p$, it follows that $x(S) = y(S) - p(S) \geq v(S), \forall S \subset N$ and $x(N) = y(N) - p(N) = v(N) - E = m^*(N, v)$. Therefore, $x = y - p \in C(N, v)$. Conversely, if $C(N, v, p) = \{\, y = x + p \,|\, x \in M(N, v) \,\}$, then $p(N) = E$ and p is a non-dominated vector of goals. □

Table 1. Characteristic function and excesses

S	{1}	{2}	{3}	{1,2}	{1,3}	{2,3}	N
v(S)	1	1	1	3	3	3	5
p(S)	-0.5	0.5	0.5	0	0	1	

Example 2.6. Consider a cooperative game, (N, v), with three players and a vector of individual goals, $p = (-0.5, 0.5, 0.5)$. The characteristic function and the excesses required for each coalition accordingly with vector p are represented in Table 1.
For this game $m^*(N, v) = 4.5$ and it follows from Proposition 2.4 that vector $p = (-0.5, 0.5, 0.5)$ is a non-dominated vector of goals.

Note that the core is empty if and only if $E < 0$. If $E \geq 0$, then some allocations in $C(N, v)$ can be generated by adding to an element of $M(N, v)$ a non-negative vector, p, whose components add up to E. Moreover, this vector, p, is a non-dominated vector of goals. Nevertheless, the whole core, $C(N, v)$, cannot be generated by adding only nonnegative vectors to the optimal solutions of problem (2.1), as shown in the next example.

Example 2.7. For the game of Example 2.6, $M(N, v) = \{(1.5, 1.5, 1.5)\}$, and therefore, core allocations can be generated by adding a nonnegative vector whose components sum up 0.5 to $(1.5, 1.5, 1.5)$. For instance, for $p = (0.25, 0, 0.25)$, $(1.5, 1.5, 1.5) + p = (1.75, 1.5, 1.75) \in C(N, v)$. But there are others elements in the core that can only be obtained by adding vectors with negative components. For instance, $(1.5, 1.5, 1.5) + (-0.5, 0.5, 0.5) = (1, 2, 2) \in C(N, v)$.

It is also worth pointing out that, once an allocation in a p-core with a feasible p is found, there is not guarantee that this allocation is in a q-core where q is non-dominated and dominates p.

Example 2.8. For the game of Example 2.6, $p = (-0.3, -0.3, -0.3)$ is a feasible vector of goals and $(1, 2, 2)$ is in the corresponding p-core. Nevertheless, the only way to represent $(1, 2, 2)$ in terms of a non-dominated vector of goals is $(1, 2, 2) = (1.5, 1.5, 1.5) + (-0.5, 0.5, 0.5)$, and $(-0.5, 0.5, 0.5)$ does not dominate $p = (-0.3, -0.3, -0.3)$.

3. Finding Allocations by a Goal Programming Approach

The result established in Proposition 2.4 is not effective in the selection of a final allocation in the game. In practice it could prove more useful to fix a reference vector of goals to direct the search towards certain allocations in the appropriate p-core.

In this section we will focus on finding allocations that attain non-dominated goals close to the reference goal in the sense defined by the appropriate norm. The reference vector of goals can be provided by the players without prior knowledge of its feasibility, nor its dominated/non-dominated status. However, the approach that we present herein will lead to a allocation associated to a non-dominated vector of goals whose distance to the reference goal is minimum.

Given a vector of individual reference goals, $\hat{p} \in \mathbb{R}^n$, we assume that the best compromise allocation of the game is one in a p-core with a non-dominated p dominating $\hat{p}$ and, if the reference goal is not attainable, then one close as possible. In order to deal with the deviations in relation to the goals, we introduce the nonnegative deviation variables d_i^+ and d_i^-, $i \in N$, which represent the degree to which the excess of agent i in every coalition overachieves/underachieves its goal. Thus we can write

$$p_i = \hat{p}_i + d_i^+ - d_i^-$$

We also denote $d^+(S) = \sum_{i \in S} d_i^+$ and $d^-(S) = \sum_{i \in S} d_i^-$, which represent the positive and negative overall deviations for coalition S.

A first requirement for an allocation to attain goals close to the reference goal, is that the values of the undesirable deviation variables, d_i^-, cannot be improved simultaneously. Formally, let $\hat{x}$ be an allocation for (N, v), such that $\hat{x} \in C(N, v, p)$, with $\hat{p}$ a vector of reference goals, and $\hat{d}^-$ be the corresponding vector of undesirable deviational variables.

Definition 3.1. The allocation $\hat{x} \in I^*(N, v)$ is goal-efficient with respect to $\hat{p}$ if there does not exist another $x \in I^*(N, v)$ whose d^- satisfies $d^- \leq \hat{d}^-$.

The characterization of goal-efficient allocations in terms of a multiobjective optimization problem is straightforward.

Proposition 3.2. *The allocation $\hat{x} \in I^*(N, v)$ is goal-efficient with respect to $\hat{p}$ if and only if $\hat{d}^-$ is an efficient solution of problem:*

$$\begin{array}{lll} min & d_1^-, \ldots, d_n^- & \\ s.t.: & e(x, S) \geq \hat{p}(S) + d^+(S) - d^-(S) & \forall S \subset N \\ & x(N) = v(N) & \\ & d_i^+, d_i^- \geq 0 & \forall i \in N \end{array} \tag{3.2}$$

Goal-efficient allocations are not necessarily associated to non-dominated vector of goals. If some of the goals are attained, that is, if $\hat{d}_i^- = 0$, for some i, the uniqueness of the allocations associated to $\hat{d}$ is not guaranteed, and it is necessary to move along higher values of the desirable deviational variables $\hat{d}_i^+$, to reach non-dominated vectors of goals and the corresponding allocations.

The choice of the reference goal implies a previous decision of the players. A possibility is to consider the reference point as the vector of characteristic values of the players, $\hat{p}_i = 0$, $i = 1, \ldots, n$. In this case, the problem becomes the search for those allocations close to the core (or inside the core) that are associated to non-dominated levels of goals.

3.1. Min-Max Goal Programming Approach

A weighted min-max goal programming approach to obtain allocations associated to non-dominated goals is now proposed. This method is especially suitable for these types of problems in which several individuals are involved, as it permits allocations with the best deviations from the goals to be found by taking different weights for different players into account.

Consider the following min-max problem

$$\begin{array}{lll} \min & \max_i\{\omega_i d_i^-\} & \\ s.t.: & e(x,S) \geq \hat{p}(S) + d^+(S) - d^-(S) & \forall S \subset N \\ & x(N) = v(N) & \\ & d_i^+, d_i^- \geq 0 & \forall i \in N \end{array} \tag{3.3}$$

where ω_i are the weights assigned to the deviations of the different players with respect to their goals.

An equivalent linear formulation is

$$\begin{array}{lll} \min & \varepsilon & \\ s.t.: & \omega_i d_i^- \leq \varepsilon & \forall i = 1,2,\ldots,n \\ & e(x,S) \geq \hat{p}(S) + d^+(S) - d^-(S) & \forall S \subset N \\ & x(N) = v(N) & \\ & d_i^+, d_i^- \geq 0 & \forall i \in N \end{array} \tag{3.4}$$

A choice of weights consisting of $\omega_i = 1$, $\forall i \in N$ leads to equitable solutions, in the sense that the importance of the undesirable deviations with respect to the goals is the same for every player. Another interesting choice could be $\omega_i = \frac{v(\{i\})}{\sum_{i=1}^n v(\{i\})}$, $i \in N$. This selection of weights implies that the bigger the characteristic function for the individual player, the more important the corresponding deviation from its goal.

A main advantage in using this approach is that problem (3.4) is a linear problem easily solved by conventional linear programming software. Its resolution is the first step to finding allocations associated to equitable non-dominated goals.

If in the solution of problem (3.4), there is no multiplicity in the values of the deviation variables, d_i^+, $i = 1,2,\ldots,n$, then the corresponding vector p, $p_i = \hat{p}_i + d_i^+ - d_i^-$, is a non-dominated vector of goals for the players in the game (N,v). Moreover, p is the non-dominated vector of goals closest to the reference point $\hat{p}$ (based on a ∞-norm). On the other hand, if the deviation variables are not uniquely obtained from problem (3.4), then in order to obtain the equitable non-dominated vector of goals and the corresponding allocations the following problem is considered:

$$\begin{array}{lll} max & \sum_{i=1}^n \alpha_i y_i & \\ s.t.: & p_i - p_i^* \geq y_i & \forall i \in N \\ & e(x,S) \geq p(S) & \forall S \subset N \\ & x(N) = v(N) & \end{array} \tag{3.5}$$

where $p^* = (p_1^*, \ldots, p_n^*)$ is obtained from an optimal solution of problem (3.4), $p_i^* = \hat{p}_i + d_i^+ - d_i^-$, and y_i are auxiliary variables to be maximized. Solving the above problem with any $\alpha_i > 0$, $i = 1,2,\ldots,n$, if all the optimal auxiliary variables are zero, then p^* is a non-dominated vector of goals. Otherwise, the resultant optimal solution, $\bar{p}$, of problem (3.5) is a non-dominated vector of goals dominating p^* (see [10], and [9]).

Example 3.3. Consider the game represented in Table 2.

Table 3 shows the allocations obtained and the non-dominated vector of goals attained when the players establish different reference goals.

Table 2. Characteristic function

S	{1}	{2}	{3}	{1,2}	{1,3}	{2,3}	N
$v(S)$	1	2	1	3	3	4	6

Table 3. Reference goals, goals attained and allocations

	$\hat{p}$	p	x
(1)	$(0.5, 0.5, 0.5)$	$(1/3, 1/3, 1/3)$	$(4/3, 7/3, 7/3)$
(2)	$(2, 1, 1)$	$(1, 0, 0)$	$(2, 2, 2)$
(3)	$(0, 0.5, 0.25)$	$ch\left\{\begin{array}{c}(0.25, 0.5, 0.25)\\(0, 0.75, 0.25)\\(0, 0.5, 0.5)\end{array}\right\}$	$ch\left\{\begin{array}{c}(1.25, 2.5, 2.25)\\(1, 2.75, 2.25)\\(1, 2.5, 2.5)\end{array}\right\}$

In cases (1) and (2) the reference vector of goals is non-feasible. In both cases, the corresponding non-dominated vector of goals is unique and is obtained by subtracting the same quantity from the reference vector to achieve the level of the non-dominated vectors. In case (3) the reference goal is feasible. When solving the min-max problem (3.4) for $\hat{p} = (0, 0.5, 0.25)$, an optimal solution with $d^- = (0, 0, 0)$ and $d^+ = (0, 0, 0)$ is obtained, and therefore, problem (3.5) has to be solved in order to obtain the non-dominated vector of goals closest to $\hat{p}$, and the corresponding allocations. In this case, the whole set of non-dominated vectors of goals can be obtained as the convex hull of the three extreme solutions of problem (3.4).

4. Conclusion

The conflicting interests of the various players in a game can be interpreted as the multiple objectives of a decision problem. From this point of view, game theory is naturally linked to multiobjective optimization, and multiobjective approaches can be used to analyse and solve a variety of models of game theory.

In particular, the formalization of the idea of simultaneously maximising the goals that the individuals can achieve regardless of the coalition the player joins, leads us to a new type of quasicore for TU cooperative games, the p-core. In contrast to the core, which can be empty, this concept provides a set of solutions for every game.

Goal programming techniques become an interesting tool to deal with the choice of the appropriate allocation in the game in when the agents involved in the decision process can communicate their preferences by assigning goals to the objectives, as well as relative weights for attaining these goals.

Acknowledgements: This research has been partially financed by the Spanish Ministry of Science and Technology project SEJ2007-62711 and by the Consejería de Innovación de la Junta de Andalucía, project P06-SEJ-01801.

References

[1] Charnes A., Cooper W.W. (1977) Goal Programming and Multiple Objective Optimisation - Part I, *European Journal of Operational Research.* **1**: 39-54.

[2] Gillies D.B. (1953) *Some Theorems on n-Person Games.* Ph.D. dissertation, Department of Mathematics, Princeton University.

[3] Hillier F.S., Lieberman G.J. (1990) *Introduction to Operations Research.* Fifth Edition. McGraw-Hill, New York.

[4] Maschler M., Peleg B., Shapley L.S. (1979) Geometric Properties of the Kernel, Nucleolus and Related Solution Concepts. *Mathematics Operations Research.* **4**: 303-338.

[5] Maschler M. (1992) The Bargaining Set, Kernel and Nucleolus. In: *Handbook of Game Theory with Economics Applications Aumann and Hart* (Eds). Vol I. Amsterdam. Elservier Science Publishers.

[6] Owen G. (1975) The Core of Linear Production Games. *Mathematical Programming.* **9**: 357-270.

[7] Schmeidler D. (1969) The Nucleolus of a Characteristic Function Game. *SIAM Journal of Applied Mathematics.* **16**: 1163-1170.

[8] Shapley L., Shubik M. (1966) Quasi-cores in a Monetary Economy with Nonconvex Preferences. *Econometrica.* **34**: 805-827.

[9] Yang J.B. (1999) Gradient Projection and Local Region Search for Multiobjective Optimisation. *European Journal of Operational Research.* **112**: 432-459.

[10] Yang J.B., Chen C., Zhang Z.J. (1990) The Interactive Step Trade-Off Method (ISTM) for Multiobjective Optimization. *IEEE Transactions on Systems, Man., and Cybernetics.* **20**(3): 688-695.

[11] Young H.P. (1985) Cost Allocation. In: Young(eds) Fair allocation. *Proceedings of Symposia in applied Math.* **33**: 69-94. American Mathematical Society, Providence, Rhode Island.

[12] von Neumann J., Morgenstern O. (1944) *Theory of Games and Economic Behaviour.* Princeton, New Jersey: Princeton University Press.

In: Game Theory: Strategies, Equilibria and Theorems
Editors: Ingrid N. Haugen et al, pp. 241-266
ISBN: 978-1-60456-844-8

Chapter 9

Excessive Expenditure in Two-stage Contests: Theory and Experimental Evidence

Wilfred Amaldoss *
Duke University
Amnon Rapoport
University of Arizona

Abstract

Budget-constrained and financially motivated members of independent groups participated in a series of two-stage contests to win a single, commonly valued, and exogenously determined prize. We present and test an equilibrium model that, in addition to the utility of receiving the prize, incorporates 1) a non-pecuniary utility of winning each stage of the contest, and 2) allows for misperception of the probability of winning, which is determined by Tullock's contest success function. The equilibrium solution accounts for the major finding of excessive aggregate expenditures in stage 1 of the contest. We then test a Cognitive Hierarchy model that attributes individual differences in stage 1 expenditures to different levels of depth of reasoning. Although the explanatory power of this model is limited, it emphasizes the importance of the non-pecuniary utility of winning in accounting for the excessive stage 1 expenditures.

Keywords: Two-stage contests, budget constraints, equilibrium analysis, experimental study

JEL Classification: C72, C78, D81

1. Introduction

Contests are economic or political interactive decision making situations in which agents compete with one another over monopoly rights, monetary prizes, power, or influence by

* E-mail address: Wilfred.amaldoss@duke.edu. Corresponding Author: Wilfred Amaldoss, Duke University, Fuqua School of Business, Dept. of Marketing, Box 90120, Durham, NC 27708

expending resources like money or effort. They vary from one another on multiple dimensions including group size, number of groups, number of prizes, number of inter-related stages, symmetric vs. asymmetric agents, simultaneous vs. sequential decisions, information structure, and other rules that govern the interaction. A variety of models have been proposed for different classes of contests, many of them extending Tullock's (1967) seminal model in which contestants vie for a single prize through the expenditure of resources and their probability of winning the prize increases monotonically in their level of expenditure (see, e.g., Nitzan, 1994, for an early review). As rent-seeking behavior in the field (e.g., sport competitions, political competitions, R&D contests) is difficult to observe and document, several researchers have turned to experimental testing of the implications of these various contest models (Anderson & Stafford, 2003; Davis & Reilly, 1998; Millner & Pratt, 1989, 1991; Önçüler & Croson, 2004; Parco, Rapoport & Amaldoss, 2005; Potters, de Vries, & van Winden, 1998; Schmitt, Shupp, Swope, & Cardigan (in press); Schmitt, Shupp, & Walker, 2003; Shogren & Baik, 1991; Vogt, Weinmann, & Yang, 2002; Weimann, Yang, & Vogt, 2000).

Previous Experimental Research. A major finding of these experiments, almost all focusing on single-stage contests, is that aggregate rent-seeking behavior of risk-neutral contestants significantly exceeds the equilibrium predictions. Millner and Pratt (1991) conducted an experiment designed to test predictions derived from a model by Hillman and Katz (1984) that more risk-averse agents dissipate a larger share of the rent. In contrast to the model's predictions, they concluded that more risk-averse subjects dissipate less of the rent, although there is excessive rent-seeking. Millner and Pratt (1989) reported similar results. Davis and Reilly (1998) conducted an experiment in which they compared behavior in a variety of repeated contests and all-pay auctions. They concluded that the equilibrium solution was flawed as a guide for predictions: "Collectively, the agents tend to dissipate more rents than Nash equilibrium predictions in all auctions—an outcome that diminishes, but does not disappear with experience (1998, pp. 110-111)." Anderson and Stafford (2003) tested a model proposed by Gradstein (1995) by varying the cost heterogeneity of the subjects and entry fee. They, too, reported that rent-seeking expenditures significantly exceeded the equilibrium predictions. When the agent's probability of winning the prize was proportional to her expenditure, Potters et al. (1998) also reported over-expenditure relative to the equilibrium prediction. Schmitt et al. (2004) and Önçüler & Croson (2004) reported similar findings, the former in a two-stage game with carryover in which rent-seeking expenditures in period t increase the efficacy of rent-seeking expenditures in period $t+1$, and the latter in a two-stage contest under risk. None of these studies has proposed a general explanation for the excessive stage 1 expenditures.

Two studies by Shogren and Baik (1991) and Vogt et al. (2002) have failed to report excessive expenditure. Both of these studies have unique features that differentiate them from the other studies mentioned above. The former paper only reports the results of the final ten periods. It is possible (see, e.g., Davis & Reilly, 1998, Parco et al., 2005) that excessive expenditures did occur in the early periods and behavior gradually converged to equilibrium play. The latter study by Vogt et al. used a contest success function that was highly discriminative ($r=8$), closer to all-pay auction, and required within each period sequential rather than simultaneous decisions as in all previous studies.

The present study builds on a previous study by Parco et al. (PRA, 2005) that investigated expenditures in *two-stage contests* with *budget-constrained* agents competing to win an exogenously determined fixed prize. The combination of two stages of the contest with a budget constraint would expect to reduce stage 1 expenditures as the contestant must maintain a fraction of the budget for expending on stage 2, conditional on winning stage 1. Varying the prize value in a within-subject design, PRA had their subjects first compete in stage 1 within their own groups by expending a portion of their budget. Winners from each group were chosen probabilistically by Tullock's proportional contest success function. In stage 2, the winners—one from each group--competed with each other for the prize by expending additional resources from the portion of the budget remaining to them after stage 1. The winner of stage 2 was chosen in the same manner. As in most of the previous experiments cited above, PRA observed significant over-expenditure in stage 1 compared to the subgame perfect equilibrium predictions. Similarly to Davis and Reilly, they also found that mean stage 1 expenditures decreased steadily with experience in the direction of equilibrium play.

The present study has two main purposes. The first goal is to study two-stage budget-constrained contests with a larger number of groups and larger group size. Parco et al. limited themselves to the special case of two dyads. Therefore, in their game, at each stage of the contest a contestant had to face only a single competitor. Parco et al. motivated their investigation with the example of political races (congress members, senators, state governors), where budget-constrained candidates first expend resources to secure their party nomination and then the winners expend additional resources in a between-party competition. However, typical of these races is that each group of candidates in stage 1 includes more than two candidates (e.g., several Republicans competing for their party nomination), and even in stage 2 the competition often includes more than two winners (e.g., Democrat, Republican, Liberal, or Independent competing for the position of a state governor). This is also the case in most two-stage sport competitions. The present study reports the results of two new experimental conditions, one with three groups of eight members each, and the other with eight groups of three members each, thereby significantly extending the experimental analyses of two-stage contests with budget constraints.

The second goal is to test a model of expenditures in two-stage contests (PRA, 2005), which assumes that, in addition to the pecuniary utility associated with receiving the prize, agents derive a non-pecuniary utility from winning each stage of the contest. In addition, and in line with results from studies of individual decision making under risk, the model allows for misperception of the probability of winning either stage of the contest by postulating a non-linear weighting function (e.g., Prelec, 1998; Tversky & Kahneman, 1992; Wu & Gonzalez, 1996).

Section 2 describes a model of two-stage contests with symmetric and budget-constrained agents. It then derives point predictions for the game parameters investigated in the present study. A major feature of these predictions is that they are parameter-free. Section 3 describes the experimental method and design. The equilibrium solutions of Stein and Rapoport (SR) and of PRA, that are nested in the more general model, are separately tested in Section 4. The PRA model outperforms the SR model and accounts for the aggregate expenditures. The results suggest that the non-pecuniary utility of winning, rather than misperception of the probabilities of winning, is critical for the good performance of the PRA equilibrium solution. Whereas equilibrium solutions are about individual, not aggregate, behavior, previous experimental studies of contests have largely ignored individual differences. In Section 5 we attempt to account for the individual differences, admittedly with qualified success, by testing the Cognitive Hierarchy model of Camerer et al. (2004), which postulates a hierarchy of subjects in terms of their depth of reasoning. Tests of this model also highlight the critical role played by the non-pecuniary utility of winning in the subjects' expenditure decisions. Section 6 concludes.

2. A Class of Two-Stage Contest with Budget Constraints

The Model

N symmetric agents are assumed to compete with one another in a two-stage contest for an exogenously determined and commonly known prize. The N agents are assumed to be risk-neutral and they assign the same valuation r to the prize. Initially, the N players are divided into k equal-size groups of m members each (thus, $mk=N$). Agents begin stage 1 of the contest with a fixed, positive, and commonly known budget denoted by e_0. Without loss of generality assume that $e_0=1$. In stage 1, the m members of each group compete with one another to choose a winner from their group by expending resources subject to the budget constraint e_0. Each group chooses and then sends a single winner (finalist) to stage 2 of the contest. The k finalists—one from each group—then compete with one another in the second and final stage for the prize r. They do so under the constraint that their expenditures in stage 2 cannot exceed what remains from the initial budget e_0 after subtracting their individual expenditures in stage 1. The individual expenditures in stages 1 or 2 are not recoverable. The major focus of this model is on the allocation of resources between the two stages of the contest when the budget constraint is either binding or not. Gubernatorial contests in the US, where budget-constrained candidates first individually contest for the party nomination, and then the winners of stage 1—one from each party--move to the second and final stage to compete for the position exemplifies this kind of contest.

Consider a designated agent h ($h=1, 2,\ldots, m$) of group j ($j=1,2,\ldots,k$) who expends a_h on stage 1 ($0\leq a_h\leq e_0$). Assume that the probability that player h wins the stage 1 competition in her group depends on her expenditure *relative* to the total expenditures of the m members of her group. Following Tullock (1967, 1980) and the vast literature on contests, this probability is computed from the ratio $a_h^{\alpha} / \sum_{i=1}^{m} a_i^{\alpha}$.

Consider next the k finalists, one from each group, and denote the expenditure of finalist j in stage 2 by b_j ($0\leq b_j\leq(e_0-a_h)$). Invoking again Tullock's logit-form contest rule, the probability that finalist j wins the competition on stage 2 (and receives the prize r) is *relative* to her expenditure in stage 2: $b_j^{\beta} / \sum_{i=1}^{k} b_i^{\beta}$. Tullock (1980) and subsequently Nitzan (1994) interpreted the parameters α ($\alpha>0$) and β ($\beta>0$) as the "marginal returns to lobbying outlays," whereas Hirshleifer (1995) interpreted them as "decisiveness parameters." They could best be viewed as institutional parameters set by the designer of the contest for its preliminary and final stages. When α and β tend to zero, the probabilities that player h wins the contest on stage 1 and player j wins the contest on stage 2 tend to $1/m$ and $1/k$, respectively. When α and β tend to infinity, the contests on both stages become fully discriminatory in the sense that the contestant expending the most on either stage wins the contest on this stage with certainty. And when they are equal to 1, as they are in the present study, the probability of winning is *proportional* to the expenditure level. Introducing two parameters (α and β), one for each stage, rather than the same parameter for both stages, increases the generality of the contest model by allowing the institutional parameters to vary from stage to stage, as they often are in real multi-stage contests.

Utility of Winning. In our model, the choice of the winner at stage 1 is determined probabilistically without any guarantee that a player who expends more than any of the other m-1 members of her group will be the winner. The same is true about choosing the ultimate winner in stage 2. Given this uncertainty, which is characteristic of many multi-stage tournaments (e.g., Poker, Backgammon), PRA have been the first to suggest that agents derive additional *intrinsic* utility from winning the competition at each stage of the contest (all N agents in stage 1 and only the k finalists in stage 2). Both experienced and inexperienced Poker players report a high degree of satisfaction from winning the game even when it is played for very low stakes. This may particularly be the case when the agents are inexperienced, as subjects are in our experiment, and therefore consider winning as a reward by itself. Parco et al. further conjectured that the non-pecuniary utility of winning stage 1, denoted by ω_1, increases in the number of contestants in stage 1 (m) and in the size of the reward (r). They also assumed that the number of competing groups (k) dampens the excitement of winning stage 1. To capture these three effects, PRA assumed the functional form $\omega_1 = \frac{m}{k}\sqrt{r}$. In proposing this functional form they were primarily interested in not adding any new parameters, thereby allowing their model to include the equilibrium solution of Stein and Rapoport (in which winning carries no additional utility) as a special case (by setting ω_1=0). In stage 2 of the contest, the finalist is assumed to derive a non-pecuniary utility, ω_2, in addition to the utility of winning the prize. This utility is conjectured to increase in the number of contestants in stage 2 (k) and size of the reward (r), but decrease in the number of competitors in stage 1 (m). Specifically, PRA assume a similar functional form $\omega_2 = \frac{k}{m}\sqrt{r}$ that only reverses the roles of k and m.

Misperception of Probabilities. Studies of individual decision behavior provide ample evidence that inexperienced subjects misperceive probabilities in a systematic way. Alternative probability weighting functions have been proposed to account for these systematic deviations (e.g., Prelec, 1998, Wu & Gonzales 1996). Following Tversky and Kahneman (1992), PRA chose a one-parameter probability weighting function, namely, $w(P) = \frac{P^\gamma}{\left(P^\gamma + (1-P)^\gamma\right)^{\frac{1}{\gamma}}}$ where 0<γ<1. This function over-weights low probabilities and under-weights high probabilities. Specifically, it is regressive and *S*-shaped. Across several studies, the fixed point at which $w(P)=P$ has been found to be approximately $^1/_e$ (see Prelec, 1998, for a brief review). The fixed point for Tversky and Kahneman's (1992) probability weighting function is 0.34 for gains (see Prelec, 1998). This implies that $\gamma \approx 0.61$.

Recall that the utility of winning is endogenous to the model parameters. PRA suggested that by setting $\gamma = 0.61$ their model could account for the misperception of probabilities without the inclusion of additional parameters. In summary, they incorporate two different and independent psychological factors in their two-stage contest model *without adding any free parameters.*

Equilibrium

The expected utility of a risk-neutral player *h* for the two stages of the contest is given by

$$E_l(a_h, b_j) = (e_0 - a_h) + w(P_{1h})[(r + \omega_2)\times w(P_{2j}) - b_j + \omega_1], \tag{1}$$

where we define $P_{1h}= a_h^{\alpha} / \sum_{i=1}^{m} a_i^{\alpha}$ and $P_{2j}= b_j^{\beta} / \sum_{i=1}^{k} b_i^{\beta}$. Because the *N* players are symmetric, we only need to solve for the equilibrium strategy for any one player.

The budget constraint is not binding if $0<a_h+b_j<1$. It is binding if $a_h+b_j=1$. Stein and Rapoport have shown that there is a critical prize value, denoted here by $r(\omega_1,\omega_2)$, that separates between these two cases. We treat these two cases separately.

<u>Case 1:</u> $0 < r < r(\omega_1, \omega_2)$ (budget constraint is not binding).

PRA constructed the following equilibrium expenditures for stages 1 and 2, if the budget constraint is not binding:

$$\begin{aligned} a_h = \frac{1}{km}(& \left(\left(\frac{1}{k}\right)^{\gamma} + \left(\frac{k-1}{k}\right)^{\gamma}\right)^{-\frac{1+\gamma}{\gamma}} \left(\left(\frac{1}{m}\right)^{\gamma} + \left(\frac{m-1}{m}\right)^{\gamma}\right)^{-\frac{1+\gamma}{\gamma}} \times \\ & \left(\left(\frac{m-1}{m^2}\right)^{\gamma} + \left(\frac{1}{m}\right)^{2\gamma} - \left(\frac{1}{m}\right)^{2\gamma-1} + \left(\left(\frac{m-1}{m^2}\right)^{\gamma} + \left(\frac{1}{m}\right)^{2\gamma}\right)(m-1)\gamma\right) \times \\ & (\left(\left(\frac{1}{k}\right)^{\gamma} + \left(\frac{k-1}{k}\right)^{\gamma}\right)^{\frac{1+\gamma}{\gamma}} k\omega_1 - \\ & \left(\left(\frac{k-1}{k^2}\right)^{\gamma} + \left(\frac{1}{k}\right)^{2\gamma} - 2\left(\frac{1}{k}\right)^{2\gamma-1} + \left(\frac{k-1}{k^2}\right)^{\gamma} k(\gamma-1) - \left(\left(\frac{k-1}{k^2}\right)^{\gamma} + \left(\frac{1}{k}\right)^{2\gamma} - \left(\frac{1}{k}\right)^{2\gamma-1}\right)\gamma\right)\omega_2 - \\ & r\left(\left(\frac{k-1}{k^2}\right)^{\gamma} + \left(\frac{1}{k}\right)^{2\gamma} - 2\left(\frac{1}{k}\right)^{2\gamma-1} - \left(\frac{k-1}{k^2}\right)^{\gamma} k(1-\gamma) - \gamma\left(\left(\frac{k-1}{k^2}\right) + \left(\frac{1}{k}\right)^{2\gamma} - \left(\frac{1}{k}\right)^{2\gamma-1}\right)\right))) \end{aligned} \tag{2}$$

and

$$b_j = \frac{\left(\left(\frac{k-1}{k^2}\right)^{\gamma} + \left(\frac{1}{k}\right)^{2\gamma} - \left(\frac{1}{k}\right)^{2\gamma-1} + \left(\left(\frac{k-1}{k^2}\right)^{\gamma} + \left(\frac{1}{k}\right)^{2\gamma}\right)(k-1)\gamma\right)(r+\omega_2)\left(\left(\frac{1}{k}\right)^{\gamma} + \left(\frac{k-1}{k}\right)^{\gamma}\right)^{-\frac{1+\gamma}{\gamma}}}{k}. \quad (3)$$

If the probability weighting function is the identity function ($\gamma = 1$), then Equations (2) and (3) reduce to

$$a_h = \frac{(m-1)[r + \omega_1 k^2 + \omega_2]}{k^2 m^2} \quad (4)$$

and

$$b_j = \frac{(k-1)(r+\omega_2)}{k^2}. \quad (5)$$

Further, if no additional utility is gained from winning either stage of the contest ($\omega_1 = \omega_2 = 0$) and the probabilities in the contest success function are perceived correctly ($\gamma = 1$), then Equations (4) and (5) are further reduced to those reported originally by Stein and Rapoport by setting $\alpha = \beta = 1$.

The equilibrium solution (Equations 4 and 5) only holds if $0 < a_h + b_j < 1$, a condition that occurs for only certain values of r. Expenditures in each stage of the contest are seen to increase in the prize value r. To determine the range of the feasible values of r, we must add a_h and b_j and require the sum to be smaller than 1. We then obtain $0 < r < r_c(\omega_1, \omega_2)$. If $\gamma = 1$, then $r(\omega_1, \omega_2)$ is given by the following implicit function:

$$\frac{k^2(m^2 - \omega_1(m-1)) - \omega_2(m^2(k-1) + m - 1)}{(m-1) + m^2(k-1)} - r = 0, \quad (6)$$

and the expected utility in equilibrium is:

$$E_1(a_h, b_j) = 1 + \frac{r + \omega_2}{m^2 k^2} + \frac{\omega_1}{m^2}. \quad (7)$$

It can be verified that the equilibrium is subgame perfect. The proof is similar to that in Stein and Rapoport and is, therefore, omitted.

Case 2: $r \geq r_c(\omega_1, \omega_2)$ (budget constraint is binding).

If $a_h + b_j = 1$, then the solution occurs on the boundary. This requires that $r \geq r_c(\omega_1, \omega_2)$. We use the equality $a_h + b_j = 1$ to eliminate a_h, and then solve for b_j. If $\gamma = 1$, then after some algebra we obtain a quadratic equation in b_j:

$$a(b_j)^2+b(b_j)+c=0,$$

where the coefficients of the quadratic equation reduce to

$$a = (m-1)^2k^2$$
$$b = kr(m-1) + mr(k-1) - k^2m(m-1) + k^2(m-1)\omega_1 + [k(m-1) + m(k-1)]\omega_2$$
$$c = mr(k-1) + m(k-1)\omega_2.$$

The solution of this quadratic equation is given by

$$b_j = \begin{cases} \dfrac{-b+\sqrt{b^2-4ac}}{2a}, & if\,\alpha \neq m \\ -c/b, & if\,\alpha = m \end{cases}. \quad (8)$$

Once the value of b_j is computed from Equation (8), then the equilibrium expenditure for stage 1 is determined from $a_h+b_j=1$. The expected utility of the game in equilibrium is:

$$E_1(a_h, b_j) = 1 + (r + \omega_2)/(m^2k^2) + \omega_1/m^2.$$

Following Stein and Rapoport, it can again be shown that the equilibrium solution for case 2 is subgame perfect.

3. Experimental Method and Design

The present study was designed to test the equilibrium solution in the case where the probability of winning each stage of the contest is directly proportional to the expenditure: $\alpha=\beta=1$. The parameter values in our experiment are $r\in\{6, 45\}$ and $m\in\{3, 8\}$. To keep the total number of subjects participating in each experimental session fixed at $N=24$, we set $k=3$ when $m=8$ and $k=8$ when $m=3$.

Experimental Design. The experiment employed a 2×2 factorial design. The stage 1 group size $m\in\{3, 8\}$ was a between-subject factor, whereas the prize value $r\in\{6, 45\}$ was a within-subject factor. Thus, the four conditions (treatments) studied in the present study are *m3r6*, *m3r45*, *m8r6*, and *m8r45* where $k=N/m$. Data were collected from two groups for each level of m.

Subjects. Ninety-six undergraduate students of business administration participated in four separate experimental sessions each including 24 subjects. The subjects were recruited by advertisements posted on bulletin boards and class announcements promising monetary reward contingent on performance in a group decision-making experiment. Both male and female students responded in nearly equal proportions. Each session lasted about 2 hours. The mean payoff per subject was $22.67. In addition, all the subjects received a $5.00 show-up bonus for their participation.

Procedure. At the beginning of each session, the subjects drew poker chips from a bag containing chips numbered 1 through 24 to randomly determine their seat assignment in the laboratory. Subjects were then seated in their designated cubicles and received written instructions (Appendix). They proceeded to read the instructions at their own pace. When all the subjects completed reading the instructions, the supervisor entertained questions from individual subjects. Very few questions were actually asked.

The subjects in each session participated in sixty trials. The two reward values were counter-balanced, with r=6 in trials 1-30 and r=45 in trials 31-60 in the first group, and with r=45 in trials 1-30 and r=6 in trials 31-60 in the second group. At the beginning of each trial, subjects were randomly assigned to one of the k groups, each including m players each. Random assignment on each trial was intended to prevent reputation effects. At the beginning of each trial, the subjects were only informed of the trial number (1-60), the initial budget for the trial (same for all players), and the prize value r. The initial budget was set at e_0=\$1.00 (experimental dollar), and the prize values were accordingly set at r=6 and r=45.

The contest was framed as a two-stage tournament. On the first stage (called "semi-finals" in the instructions), each of the m cohort members was asked to specify privately his or her expenditure for this stage. The winner of each group was chosen randomly by the computer that implemented Tullock's contest rule. The rule was explained and exemplified in detail (see Appendix). Once the k winners of stage 1—one from each group—were thus chosen, a computer screen informed the winners of this fact and privately displayed their remaining budget for stage 2 (called the "finals" stage). The m-1 players who did not advance to stage 2 received no information until the end of the trial. The k finalists were asked to type in their expenditures for stage 2 (without exceeding their remaining budget), and then the proportional contest rule was implemented a second time to determine the ultimate winner.

At the end of stage 2, the computer displayed the decisions of all the N members at each stage of the contest and the outcomes of each stage. This information was displayed as a game tree (see Outcome Screen in the Appendix). Thus, whether or not they proceeded to stage 2, all the N subjects in a session received the same outcome information at the end of each trial. Once all the subjects completed reviewing the Outcome Screen, they pressed a "continue" button. When all the 24 players in the session pressed the "continue" button, the experiment proceeded to the next trial.

At any time during the trial the subjects could review their own results from previous trials by clicking on a button labeled "Review Previous Trials." At the end of the experiment, 10 of the 60 trials were chosen randomly for payment. The subjects were paid their cumulative earnings for these payoff trials and dismissed.

4. Test of the Equilibrium Solutions

In this section we summarize the observed expenditures of the subjects, highlight some empirical regularities, and then report how well two different equilibrium solutions account for the behavior of our subjects.

Table 1. Observed Stage 1 and Stage 2 Expenditures

Condition	Prize	Stage 1 Expenditure (a_h)			Stage 2 Expenditure (b_j)		
		Group 1	Group 2	Mean over Groups	Group 1	Group 2	Mean over Groups
m=3, k=8	r=6	0.393	0.342	0.368	0.521	0.561	0.541
	r=45	0.419	0.417	0.418	0.538	0.497	0.517
m=8, k=3	r=6	0.705	0.514	0.610	0.256	0.388	0.322
	r=45	0.727	0.601	0.664	0.213	0.341	0.277

Table 1 presents the aggregate mean expenditures in stages 1 and 2 of the contests for each group (session) separately. Column 2 displays the prize value r and columns 3-5 the mean stage 1 expenditures of Group 1, Group 2, and the overall mean. Similarly, Columns 6-8 show the mean stage 2 expenditures observed in Group 1, Group 2, and the overall mean. We conducted an ANOVA with m as a between-subject factor and r as a within-subject factor. The null hypothesis that the group size m has no effect on stage 1 mean expenditures was soundly rejected ($F_{(1, 92)}=150.22$, $p<0.001$). For $m=3$, the observed stage 1 expenditures were 0.368 and 0.418 for $r=6$ and $r=45$, respectively. The actual expenditures are higher when $m=8$, with the mean expenditures equal to 0.610 and 0.664 in conditions *m8r6* and *m8r45*, respectively. The second null hypothesis that the prize value has no effect on stage 1 mean expenditures was also soundly rejected ($F_{(1,92)}=19.6$, $p<0.001$). Subjects expended more on stage 1 when $r=45$ than when $r=6$.

Table 2. Stage 1 and Stage 2 Expenditures by Winners and Losers

Condition	Prize	Stage 1 Expenditure (a_h)				Stage 2 Expenditure (b_j)			
		Group 1		Group 2		Group 1		Group 2	
		Winner	Loser	Winner	Loser	Winner	Loser	Winner	Loser
m=3, k=8	r=6	0.451	0.366	0.402	0.312	0.565	0.515	0.587	0.557
	r=45	0.438	0.409	0.483	0.384	0.531	0.539	0.507	0.495
m=8, k=3	r=6	0.744	0.699	0.589	0.504	0.312	0.229	0.443	0.361
	r=45	0.773	0.721	0.652	0.594	0.232	0.204	0.406	0.308

In equilibrium, all players in stage 1 should expend the same amount of capital. Therefore, the stage 1 expenditures of both winners and losers in stage 1 should be the same. Table 2 reports the observed mean expenditures of both winners and losers by stage, condition, and prize value. Consistent with the results presented in Table 1, the mean expenditures of winners and losers in stage 1 increase with the prize value. On comparing the expenditures for $m=3$ with those for $m=8$, we find that the mean stage 1 expenditures of winners and losers increase with m. However, the mean expenditures of winners and losers are not the same. The winners of stage 1 consistently expended more than the losers ($p<0.01$).

Winners in conditions *m3r6*, *m3r45*, *m8r6*, and *m8r45* expended, on average, 26%, 16%, 11% and 8% over the losers, respectively. Even in stage 2, where equilibrium play also calls for equal expenditures, winners in conditions *m3r6*, *m8r6*, *m8r45* expended more than the losers ($p<0.01$). The difference in mean expenditures between winners and losers, however, is not significant in condition *m3r45* ($p>0.2$).

Next, we compare the mean expenditures to the equilibrium predictions of two different models. We begin by testing the SR equilibrium, where players are assumed to be risk-neutral, non-pecuniary utility of winning is set at zero, and the probabilities of winning are perceived correctly. Then, we test the equilibrium predictions of the PRA model where players are allowed to misperceive probabilities and derive additional utility from winning.

Stein and Rapoport Model. The equilibrium stage 1 and stage 2 expenditures under the SR model are presented in Columns 4 and 9 of Table 3. Although the observed behavior is in qualitative agreement with the equilibrium predictions, the actual expenditures are considerably higher than the point predictions of the SR solution. When $m=3$, the equilibrium predictions are 0.021 and 0.347 for $r=6$ and $r=45$, respectively. The actual mean expenditures in conditions *m3r6* and *m3r45* are significantly higher ($r=6$: observed mean = 0.368, $t=10.67$, $p<0.001$; $r=45$: observed mean = 0.418, $t=2.28$, $p<0.03$). Similarly, the corresponding equilibrium stage 1 predictions when $m=8$ are 0.115 and 0.482 for $r=6$ and $r=45$, but the actual expenditures in conditions *m8r6* and *m8r45* are significantly higher ($r=6$: observed mean = 0.610, $t=12.04$, $p<0.001$; $r=45$: observed mean = 0.664, $t=4.99$, $p<0.001$). These results are consistent with the ones we cited earlier. To visually appreciate the discrepancy between observed and predicted stage 1 expenditures, Fig. 1 plots the deviations. The figure shows that the deviations increase with m and decrease with r.

Given the significant and systematic deviations from equilibrium play in stage 1, a comparison between observed and predicted expenditures in stage 2 is clearly meaningless. The only observation worth making is that even in condition *m3r6*, where in equilibrium players should expend only about 68% (2.1% plus 65.6%) of their budget across both stages, in actuality the budget constraint is practically binding. This, of course, is due to the considerable over-expending of resources by subjects in stage 1.

PRA model. Table 3 also presents the predictions of the PRA model ("Full model"). The actual stage 1 mean expenditures are quite closely aligned with the model prediction except in condition *m3r6*. (As is shown below, the discrepancy in this case is due to a single session.) Subjects in condition *m3r6* should expend 0.301 in stage 1 of the contest, but the actual expenditure is significantly higher (observed mean = 0.368, $t=3.58$, $p<0.01$). On average, subjects in condition *m3r45* were predicted to spend 0.437 in stage 1, and we cannot reject the null hypothesis that the actual and predicted expenditures are the same (observed mean = 0.418, $t=0.78$, $p>0.2$). In equilibrium, subjects in conditions *m8r6* and *m8r45* should expend 0.612 and 0.680, respectively, in stage 1 of the game. The actual expenditures in these two conditions are once again not statistically different from the model predictions ($r=6$: observed mean = 0.610, $t=0.07$, $p>0.2$; $r=45$: observed mean = 0.664, $t=0.55$, $p>0.2$).

Table 3. Model Comparison

Condition	Prize	Stage 1 Expenditure (a_h)					Stage 2 Expenditure (b_j)				
		Mean over groups	SR Model	PRA Model			Mean over groups	SR Model	PRA Model		
				Full Model	Restricted Models				Full Model	Restricted Models	
					No Utility of Winning	No Misperception of Probability				No Utility of Winning	No Misperception of Probability
m=3, k=8	r=6	0.368	0.021	0.275	0.083	0.301	0.541	0.656	0.725	0.530	0.699
	r=45	0.418	0.347	0.437	0.365	0.451	0.517	0.653	0.563	0.347	0.635
m=8, k=3	r=6	0.610	0.115	0.612	0.117	0.651	0.322	0.885	0.388	0.694	0.349
	r=45	0.664	0.482	0.680	0.461	0.698	0.277	0.518	0.320	0.539	0.302

These results prompted us to delve further into what is driving the PRA model to outperform the SR model by considering two nested models. Recall that the PRA model allows for both misperception of probability and non-pecuniary utility of winning. First, consider the case where players are not assumed to derive additional utility from winning ($\omega_1 = \omega_2 = 0$). The corresponding equilibrium predictions for stage 1 expenditures are presented in Column 6 of Table 3. We note that the equilibrium predictions of this nested model are substantially lower than the actual expenditures ($p<0.01$). Misperception of probability by itself cannot help the PRA model to account for the excessive expenditures in stage 1. Next, consider the other case where players are restricted to accurately perceive probabilities. The equilibrium predictions corresponding to this nested model are presented in Column 7. We cannot reject the null hypothesis that actual and predicted expenditures are the same in conditions *m3r45*, *m8r6* and *m8r45* ($p>0.18$). This finding implies that the non-pecuniary utility from winning is the key force in the PRA model.

Parco et al. claimed that both factors are necessary to account for the mean expenditures. In their experiment, the utility of winning remained constant in both stages of the game as $m=k$. In contrast, in our experiment, where $m \neq k$, the incremental utility from winning systematically changes in each stage of the contest and helps to better account for the data even when we set $\gamma = 1$. Subjects in our experiments played the same game for the first thirty trials and then played another game in the next thirty trials. However, in the study conducted by PRA the prize value randomly changed from trial to trial, and this added complexity might possibly have rendered it more difficult for the subjects to accurately perceive the two probabilities of winning.

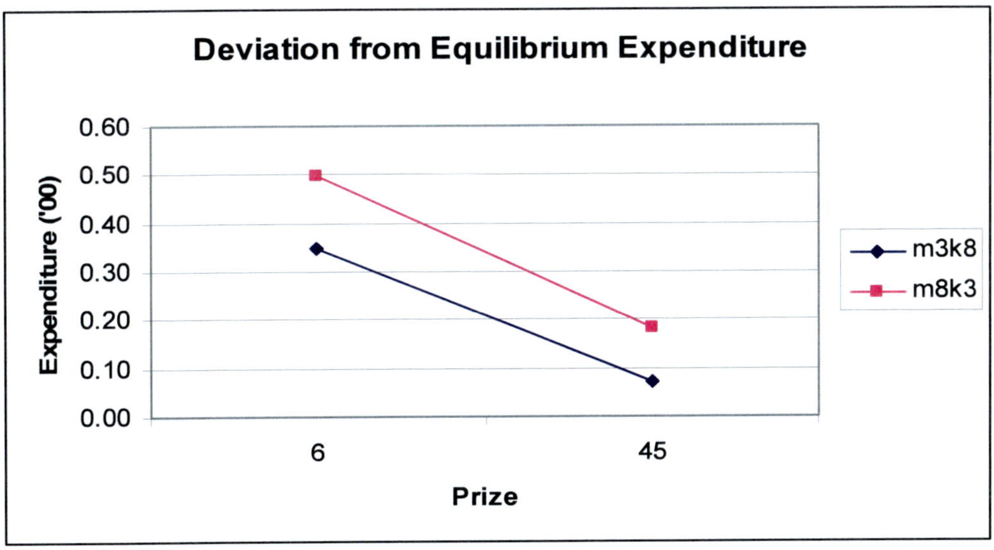

Figure 1. Deviation from Stage 1 Equilibrium Expenditure when $\omega = 0$.

Trends in Expenditure. The analyses reported above examined the mean expenditures. In an attempt to detect trends in the expenditure pattern of the subjects, we divided the 30 trials into 3 blocks of 10 trials each and then conducted an ANOVA to test for block effects.

Examination of the block means (Table 4) does not indicate an easily interpretable trend in the expenditure pattern. In fact, we cannot reject the null hypothesis of equality of the three block means ($F_{(2,184)}$=0.38, p>0.68). However, this result is qualified by a significant block by prize value interaction effect ($F_{(2,184)}$=12.59, p<0.001), as well as a significant interaction effect of prize value and group size ($F_{(2,184)}$=5.8, p<0.006), implying that the trends differ among the conditions.

Table 4. Trends in Stage 1 Expenditures

Condition	Reward	Group	Stage 1 Expenditure (a_h)		
			Block 1	Block 2	Block 3
m=3, k=8	r=6	Group 1	0.399	0.399	0.383
		Group 2	0.371	0.343	0.313
	r=45	Group 1	0.422	0.409	0.424
		Group 2	0.364	0.423	0.465
m=8, k=3	r=6	Group 1	0.716	0.705	0.693
		Group 2	0.528	0.519	0.497
	r=45	Group 1	0.667	0.764	0.752
		Group 2	0.620	0.574	0.609

Individual Differences. Further analyses show considerable individual differences. Figure 2 exhibits the frequency distributions of the mean stage 1 expenditures of the subjects for each of the four conditions. Each expenditure class in the figure accounts for 10% of the total budget. The observed empirical distributions of subjects by their mean expenditure are unimodal in all the four conditions. The mean expenditures of individual subjects in condition $m8r6$ ranges from 0.24 to 0.86 with a modal frequency in the expenditure class of 0.6-0.7.

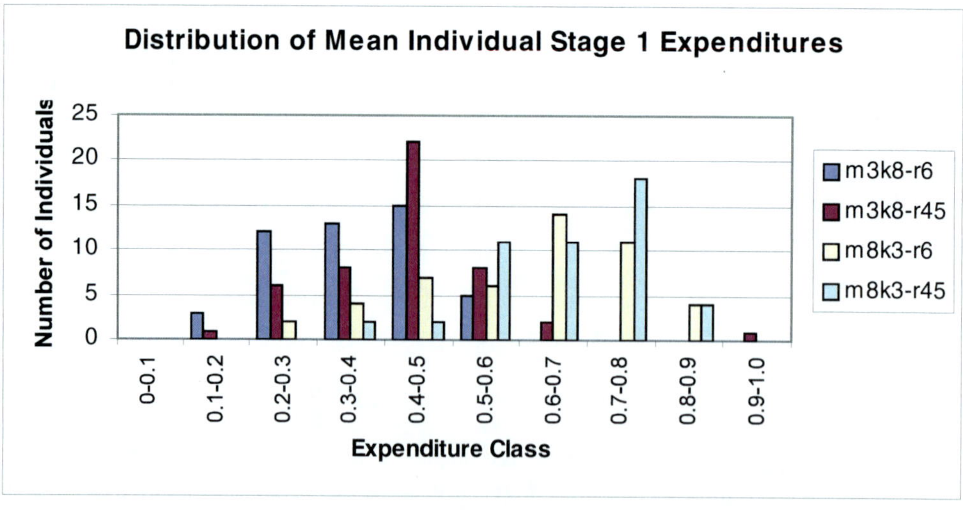

Figure 2. Distribution of Mean Expenditures of Individual Subjects.

Similarly, individual expenditures in condition *m8r45* range from 0.36 to 0.89 with the mode at 0.7-0.8. Likewise, the stage 1 expenditures of subjects in condition *m3r6* vary from 0.11 to 0.61, and the stage 1 expenditures of the subjects in condition *m3r45* range all the way from 0.1 to the highest expenditure class. We find very weak support for equilibrium play on the individual level.

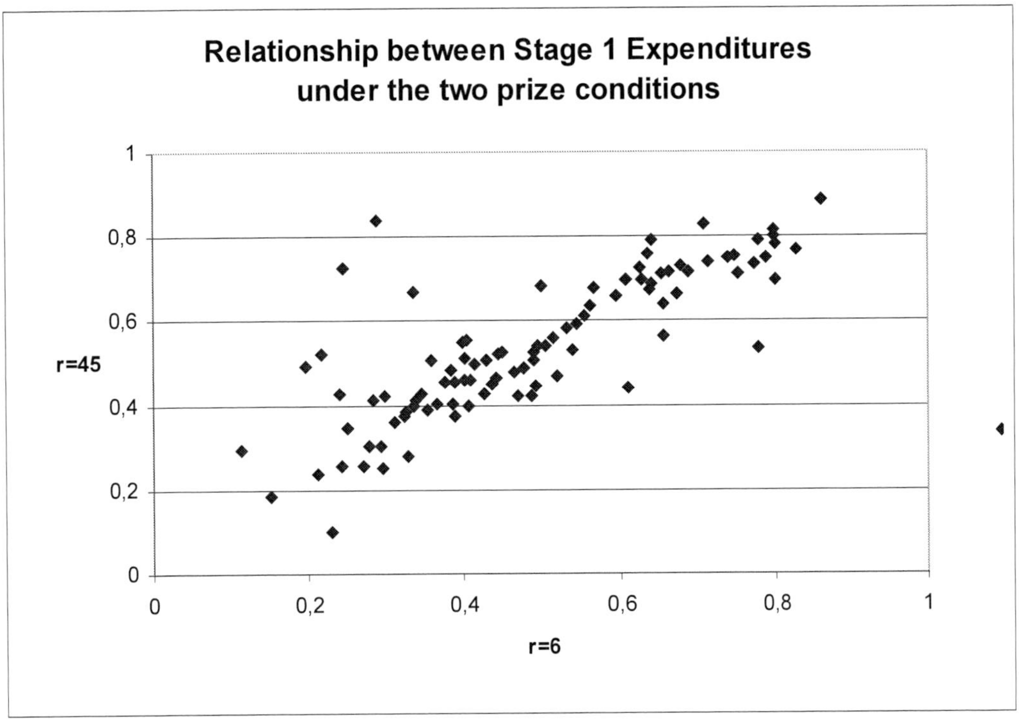

Figure 3. Relationship between Stage 1 Expenditures under the two Prize Conditions.

Table 5. Correlations between Mean Individual Expenditures in Stage 1 for the Two Prizes

Condition	Group	Correlation between Stage 1 Expenditures in the Two Prize Conditions*
$m=3, k=8$	Group 1	0.82
	Group 2	0.46
	Overall	0.61
$m=8, k=3$	Group 1	0.30
	Group 2	0.64
	Overall	0.65

* All correlations are positive and significant.

Since we used a within-, rather then between-subject design, we could explore whether there is any systematic pattern in the behavior of individual subjects across the two prize values, namely, *r=6* and *r=45*. Figure 3 exhibits the mean stage 1 expenditures of individual subjects simultaneously for the two prize values. It shows that subjects who spent more when *r=6* also spent more when *r=45*. Table 5 summarizes this relationship by reporting the correlation coefficients that range from 0.30 to 0.81 across the four conditions. Thus, the pattern of individual differences among subjects persists in both the prize values and both group sizes. A possible explanation of these systematic individual differences is that subjects differ from one another in their depth of reasoning; they are boundedly rational to varying levels and form beliefs that are not mutually consistent. We explore this explanation in the next section.

5. Cognitive Hierarchy Model

Our intent in this section is to examine whether the expenditure pattern of our subjects can be accounted for by differences in their depth of reasonong. We use the single-parameter Cognitive Hierarchy (CH) model of Camerer et al. (2004) to accomplish this goal. It is not our intention here to compare the CH model with other models of iterative thinking (e.g., Costa-Gomes & Crawford, 2004; Costa-Gomes, Crawford, & Broseta, 2001; Ho, Camerer, & Weigelt, 1998; Nagel, 1995).

The CH model assumes that players engage in iterative step-by-step reasoning. The iterative process starts with zero-step thinkers who make random choices. The one-step thinkers best respond to zero-step thinkers. In general, *k*-step thinkers assume that their opponents are distributed over zero to *k*-1 steps. So the *k*-step players fail to see the possibility that others could think as many steps as they do, if not more.

Let the strategy of player *h* spending a_h units in stage 1 of the game be s_h^a. Also let each expenditure class equal to 10% of the total budget. Zero-step thinkers are equally likely to choose any of the expenditure classes. Denote by $f(k)$ the frequency distribution of the *k*-step thinkers. Denote a *k*-step player's beliefs about the proportion of *z*-step thinkers by $g_k(z)$. As the *k*-step players only consider the possibility that opponents can consider $0 \le z \le k-1$ steps, the corresponding beliefs about competitors is given by $g_k(z) = f(z) / \sum_{l=0}^{k-1} f(l), \forall z < k$. It is useful to note that as *k* increases the deviation between actual frequencies $f(k)$ and beliefs $g_k(z)$ reduces.

Consistent with Camerer et al. (2004), we assume that subjects choose their stage-1 expenditure level, namely s_h^a, such that it maximizes their payoff given their beliefs $g_k(z)$ about their competitors. Further, the frequency distribution of *k*-step thinkers is given by a Poisson density function $f(k) = e^{-\tau}\tau^k / k!$, where the parameter τ is the mean of the distribution. Assuming that subjects derive additional utility from winning each stage of the game and that they do not correctly perceive probabilities as in PRA model, we estimate the value of τ for each of the four conditions by minimizing the mean root squared deviation between the observed mean expenditure and the CH model prediction.

We find that the mean predicted expenditures of the CH model are, in general, very closely aligned with the mean observed expenditures. The mean predicted stage 1 expenditures in conditions *m3r6*, *m3r45*, *m8r6*, *m8r45* are 0.368, 0.451, 0.610, and 0.644 respectively. In actuality, the corresponding expenditures are 0.368, 0.418, 0.610, and 0.664, respectively. We cannot reject the null hypothesis that the actual and predicted mean expenditures are the same in conditions *m3r6*, *m8r6*, and *m8r45* ($p>0.2$), but can in condition *m3r45* ($p<0.05$). The estimated values of τ in conditions *m3r6*, *m3r45*, *m8r6*, and *m8r45* were 0.75, 4.0, 1.32, and 3.39, respectively, implying that the majority of the subjects were thinking more than a single step. These estimates of τ are within the range of values reported in Camerer et al. for a wide variety of games (2004, p.878).

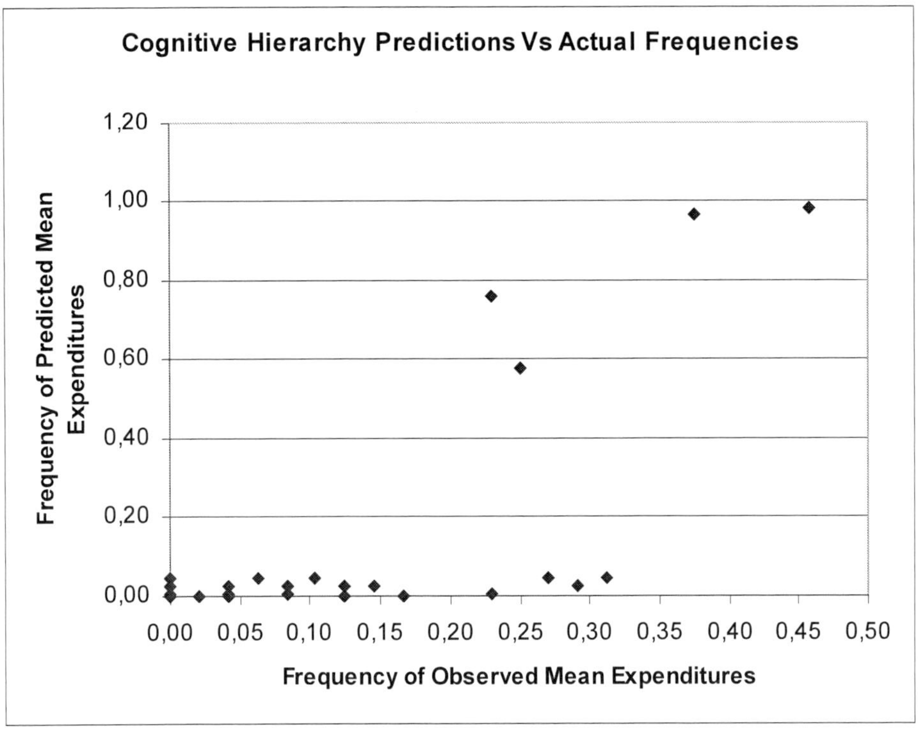

Figure 4. Relationship between Cognitive Hierarchy Model Predictions and the Actual Distribution of Mean Investments of Subjects in the Four Conditions.

Given that the CH model is surprisingly successful in accounting for the mean observed expenditures in each of the four treatments, we submit it to an even more stringent test. Now we compare the actual *distribution* of expenditures reported in Fig. 3 against the predicted expenditures in each of the expenditures classes. Figure 4 displays a scatter plot of the actual and predicted frequencies across all the four conditions. Now it is easy to see that CH model systematically over-predicts the high frequency events. For example, 46% of the subjects in condition *m3r45* spent 0.4-0.5 units, while the corresponding CH model prediction is 98%. Thus the CH model is successful in predicting the mean expenditures, not the entire distributions.

A related question is whether the CH model can track the observed mean expenditures in the absence of utility for winning? If the non-pecuniary utility for winning is zero ($\omega_1 = \omega_2 = 0$), then the estimated value of τ systematically reduces in conditions *m3r6* and *m8r45* so that the predicted mean expenditures move upward toward 0.5 (random choice prediction) and thereby better account for the high expenditure levels observed in these two conditions. For example, the estimated values of τ drop to 0.35 and 0.79 in conditions *m3r6* and *m8r45*, respectively. Recall that the mean actual expenditures in conditions *m8r6* and *m8r45* are 0.610 and 0.664, respectively. Even if τ =0 (random choice), the CH model cannot account for the excessive expenditures observed in these two conditions, as random choice leads to a mean expenditure of only 0.5. This analysis clarifies that the limited reasoning in itself cannot fully account for the excessive expenditures in conditions *m8r6* and *m8r45*.

Consistent with the analysis reported in Section 4, we note that misperception of probabilities is less important in our experiments. Allowing for accurate perception of probabilities ($\gamma = 1$) increases the estimated value of τ but does not hurt the predictive accuracy of CH model.

In summary, the excessive expenditures in stage 1 of the game cannot be accounted for by merely invoking poor reasoning on the part of our subjects. Further, the non-pecuniary utility of winning, not misperception of probabilities, plays a key role in the decisions of our subjects.

6. Conclusions

We report behavioral regularities in two-stage contests involving large number of groups and large group sizes and then explore whether the observed expenditure patterns can be accounted for by psychological factors such as non-pecuniary utility for winning and misperception of probabilities. Consistent with earlier experimental research on single-stage contests and two-stage contests with small groups, we observe that subjects tend to spend more than the equilibrium prediction in the first stage of the contest.

The excessive expenditures, though, cannot be explained by SR model, but can better be accounted by an equilibrium model proposed by PRA. The key reason for the superior performance of the PRA model is that it allows for non-pecuniary utility for winning. The PRA model accounts for the aggregate, but not the individual responses. We studied whether the individual differences can be accounted for by differences in the depth of reasoning. On fitting the Cognitive Hierarchy model to the data, we find that the non-pecuniary utility of winning also emerges here as a critical factor in explaining excessive expenditures in stage 1, and that the model accounts quite well for the mean expenditure across subjects but not for the distributions of mean individual expenditures.

Acknowledgement

We gratefully acknowledge financial support from the Hong Kong Research Grants Council to the Hong Kong University of Science and Technology (Grant HKUST6307/04H). We also thank William E. Stein for his invaluable support.

Appendix

This Appendix provides the instructions for the case where $m=3$ and $r=\{6,45\}$.

Subject Instructions

This experiment has been designed to study how people allocate their budget in a two-stage tournament. The instructions for this experiment are quite simple. If you follow them carefully and make good decisions, you may earn a considerable amount of money. Therefore, it is important that you try to do your best. A research foundation has contributed the funds to support this research.

General Description of the Tournament

Main features of the game. You are one of 24 players who volunteered to compete in a two-stage tournament with payoffs contingent on performance. Each tournament has the same structure. At the beginning of each tournament, players are randomly placed into EIGHT groups containing THREE players each. During the first stage, the three players within each group will compete against one another to determine a group winner. Each group winner (finalist) will then compete in the final match against the other seven group finalists. The winner of the final match will be awarded a prize. There will be two different prize values, one for trials 1-30 and the other for trials 31-60. The prize value will be publicly announced prior to each tournament.

The tournament consists of two stages:

- Stage 1 – "The Semi-finals". During stage 1 of the tournament, each player is given an initial budget of $1.00. Each player will be asked privately and independently to decide how much of her budget they wish to spend. Players can spend any amount between $0.00 and $1.00. The more a player spends, the greater her chances of winning the first stage. Once all three players within the group have decided how much to spend, the computer will determine the winner of that group. Simultaneously, the other twenty-one players assigned to the other seven groups will be competing against each other in a similar fashion. The rule used by the computer for choosing a winner in each of the eight semi-finals will be explained below.
- Stage 2 – "The Finals." After the computer determines the eight finalists (one winner from each group), they will compete against each other in a final match. The other sixteen players—two from each group—who did not win during the semi-finals--will no longer participate in this game. Just like in the semi-finals, each player in the finals will be asked to decide how much she wishes to spend to compete against the other seven players. No player can spend more than the amount remaining from the initial budget. Once all eight finalists have made their decisions in stage 2 of the contest, the computer will determine the Champion.

The Champion will receive a predetermined prize. The other twenty-three players will receive no prize at all. Regardless of whether or not a player receives the prize, each of the twenty-four players will keep whatever portion remains of her initial budget of $1.00 that was not spent during the tournament.

The diagram below depicts the basic structure of the game.

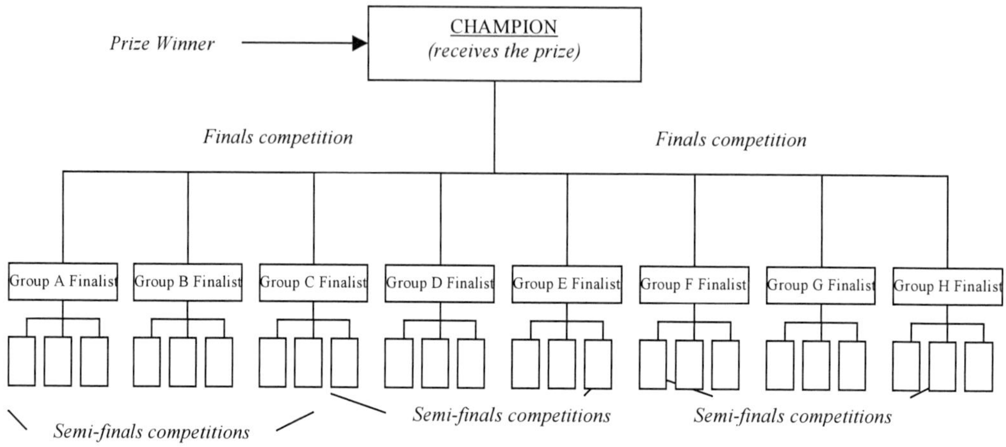

Calculating the "chances of winning" Each player's chances of winning the competition during either stage of the game are determined by how much she spends *in comparison to* how much all the players in her group spend. As long as a player spends something, then he will have some chance of winning. If a player spends nothing, then she "forfeits" the competition defaulting to the other players.

Once all three players of a group have made their decisions, the computer will calculate each player's chances of winning using the following formula:

$$\text{A player's chances of winning} = \frac{\text{Amount spent by Player}}{\text{(Amount spent by ALL the players in his/her group)}}$$

Example

Suppose that each of the three players in a group spends the following:

Player 1: $0.10
Player 2: $0.57 — Total amount spent: $1.42
Player 3: $0.75

Player 1 of Group A spends $0.10 (of his initial $1.00 budget) in the semi-finals. His/her chances of winning are given by:

$$\text{Player 1's chances of winning} = \frac{\$0.10}{\$1.42} = 0.071$$

So Player 1 has a 7.1% chance of winning the semi-final competition. The chances of winning of Player 2 are calculated in a similar way resulting in a 40.1% chance of winning.

On dividing each player's expenditure by the total group expenditures ($1.42), we have each player's chances of winning:

Player 1's chances of winning = $0.10 / $1.42 = 0.071 → 7.1% chance of winning
Player 2's chances of winning = $0.57 / $1.42 = 0.401 → 40.1% chance of winning
Player 3's chances of winning = $0.75 / $1.42 = 0.528 → 52.8% chance of winning

Determining the winner. Once the computer has calculated the chances of winning based on the amounts spent, it will assign "lottery ticket" numbers to each of the players in proportion to the total expenditures. The tickets are numbered from 1 to 1000. For example, if Player 1 has a 0.071 chance of winning, Player 2 has a 0.401 chance of winning, and Player 3 has a 0.528 chance of winning (the chances of all three players must always sum to 1.000), then Player 1 will be assigned ticket numbers 1 through 71 (71 tickets), Player 2 will be assigned ticket numbers 72 through 472 (401 tickets), and Player 3 will be assigned ticket numbers 473-1000 (528 tickets).

After each player has been assigned her tickets, the computer will randomly generate a number between 1 and 1000. The winner of the contest will be the player holding the lottery ticket number corresponding to the randomly drawn number. Thus, the greater the "chances of winning" are for a player, the more lottery tickets that will be assigned and thus, the more likely she is to be holding the winning ticket.

Note that under this rule it is possible for a player to win the contest as long as she holds at least 1 ticket. This rule for determining a winner is the same for both stages of the tournament.

Description of the Computer Displays

During each game you will have at most two decisions to make. At the beginning of each game (trial), you will be given $1.00. Your first decision is how much, if any, you wish to spend during the semi-finals. You will be presented with the following screen that will provide you with the following information:

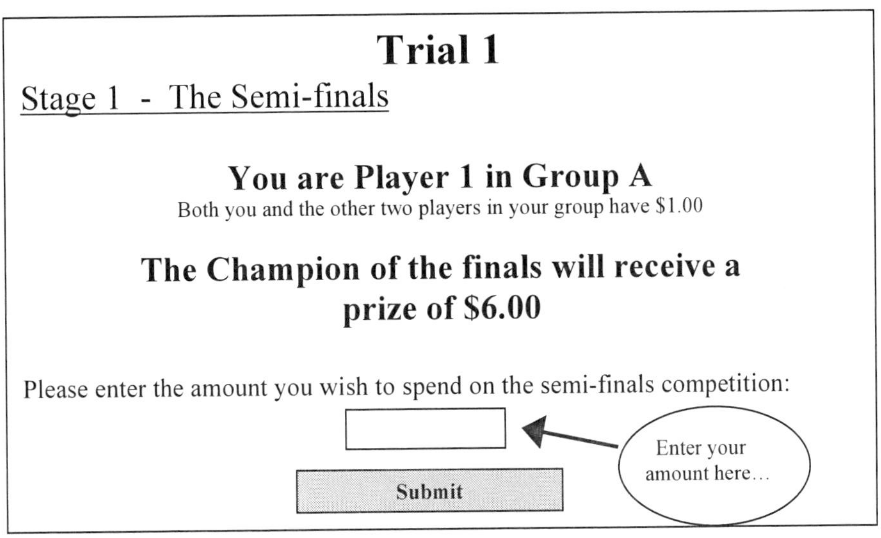

Assume that Player 1 of Group A decides to spend $0.95 (out of his $1.00 budget) during the semi-finals. After she types this amount into the white box, she has to click on the Submit button to confirm his decision. The remaining amount she has is $0.05 ($1.00 – $0.95). If she loses the semi-finals, then this will be her earnings for the trial. If she wins the semi-finals, this is the amount left for her to compete with in the finals.

Example. Suppose that each of the three players in Group A spent the following:

Player 1: $0.95
Player 2: $0.05
Player 3: $0.15

Then, the total amount spent by group members = $1.15

Player 1's chances of winning = $0.95 / $1.15 = 0.826 → 82.6% chance of winning
Player 2's chances of winning = $0.05 / $1.15 = 0.043 → 4.3% chance of winning
Player 3's chances of winning = $0.15 / $1.15 = 0.131 → 13.1% chance of winning

Total chances of winning: 1.000 → 100%

At the completion of stage 1, all players in Group A will be presented with the following screen:

	Starting Budget	Amount spent	Amount remaining	Chance of Winning	Ticket numbers
Player 1	$1.00	$0.95	$0.05	0.826	1-826
Player 2	$1.00	$0.05	$0.95	0.043	827-870
Player 3	$1.00	$0.15	$0.85	0.131	871-
	Total amount spent:	$1.15		1.000	

You are Player 1. The winning ticket is 283. You win!

In this example, Player 1 in Group A wins her semi-finals match and advances to the finals. Once she clicks on 'Continue' she will be shown the second-stage decision screen with the following information:

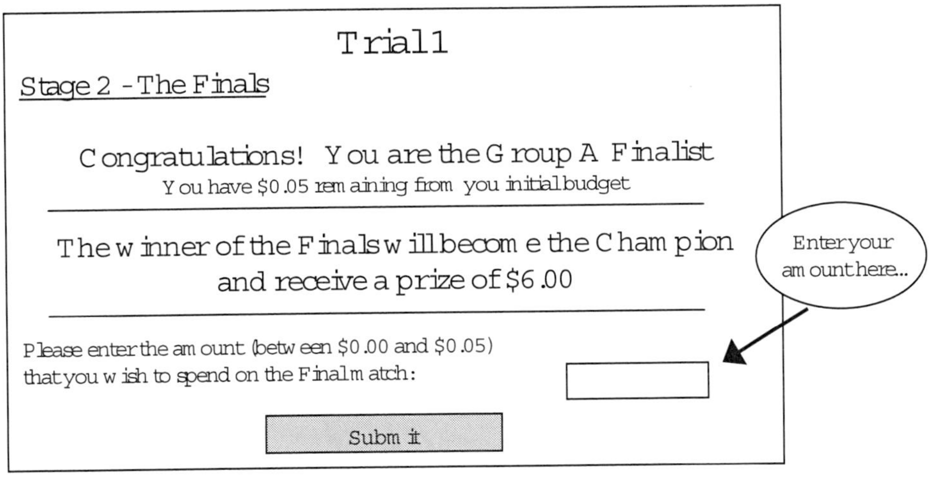

The Group A finalist can spend any amount she wishes for the Finals match up to the remaining amount of her budget. In this example, the Group A finalist has $0.05 remaining. She knows that each of the other seven group finalists also started with $1.00; however, she does not know how much any of her other opponents have at the start of the final match. In this example, suppose the Group A finalist spends all of her remaining $0.05.

Once she decides how much to spend, each player is asked to click on the Submit button. When all eight finalists have confirmed their decisions, the computer will compute their chances of winning the Final match exactly as in semi-finals (using the same formulas) and determine the tournament *Champion*. The finalists will then be shown the results of the final match.

	Starting budget	Amount spent	Amount remaining	Chance of Winning	Ticket numbers
Group A	$0.05	$0.05	$0.00	0.019	1 - 19
Group B	$0.87	$0.87	$0.00	0.332	20 - 351
Group C	$0.15	$0.15	$0.00	0.057	352 - 408
Group D	$0.10	$0.10	$0.00	0.038	409 - 446
Group E	$0.20	$0.20	$0.00	0.076	447 - 528
Group F	$0.75	$0.30	$0.40	0.115	529 - 638
Group G	$0.60	$0.40	$0.20	0.153	638 - 790
Group H	$0.55	$0.55	$0.00	0.210	791 -1000
Total amount spent:		$2.62		1.000	

The winning ticket is 942. You did not win. The Group H Finalist is the Champion and wins the $6.00 prize.

In this example, the Group F finalist spent less than half of her remaining amount ($0.30) while the Group A, B, C, D, E and H finalists spent all of their remaining endowments. As it turned out, the Group H finalist won the finals and became the *Champion* earning the prize of $6.00.

Once the tournament is over, all the players will view the same screen, which summarizes the final results of the game.

Once the prize value is announced, your primary consideration in this trial is to determine how much to spend in each stage of the tournament. The more you spend in the semi-finals, the greater your chances of winning this stage of competition. However, the more you spend during the semi-finals, the less money you have to spend in the finals, if you reach it, and the less money you have if you do not receive the prize. Recall that only the Champion receives a prize, and every player, including the Champion, keeps what remains of his or her initial budget.

View Earnings History. If you wish to review your results from the previous trial, you may do so by clicking on the button labeled 'Review previous results' at the bottom of the screen showing tournament diagram at the conclusion of the previous trial.

Summary

There are 24 individuals in this experiment who will participate in 60 trials. At the beginning of each trial, the computer will randomly divide the 24 individuals into 8 groups of 3 players each. Thus, at the beginning of each trial, you will be randomly matched with two other anonymous players. These group members will change from one trial to another. Each player will be given the same initial budget of $1.00. Earnings from previous trials cannot be used. Communication between the players during the experiment is strictly forbidden.

Once all players have made their decisions for stage 1, the computer will determine the winner of each of the eight groups. These group finalists will then make their stage 2 decisions. Once all eight finalists have entered their decisions, the computer will determine the Champion and award the prize to that player. The other players' earnings will only be that which they did not spend during stage 1 or stage 2.

Payment at the End of the Session

At the beginning of the experiment, 10 trials will be randomly selected out of 60 and you will be paid your earnings for these 10 trials. Therefore, it is important to do your best on each trial.

Your earnings will be paid to you in cash at the rate of two US dollars for every experimental dollars.

Please look up to indicate that you have completed reading the instructions. The supervisor will start the experiment in just a few minutes. Thank you for your participation.

References

Amegashie, J. A. (1999). The design of rent-seeking competitions: Committees, preliminary and final contests. *Public Choice*, **99**, 63-76.

Amegashie, J. A. (2000). Some results on rent-seeking contests with shortlisting. *Public Choice*, **105**, 245-253.

Anderson, L. R. and Stafford, S. L. (2003). An experimental analysis of rent seeking under varying competitive conditions. *Public Choice*, **115**, 199-216.

Camerer, C. F., Ho, T. –H., and Chong, J. –K. (2004). A cognitive hierarchy model of games. *Quarterly Journal of Economics*, **119**, 861-898.

Costa-Gomes, M., and Crawford, V. (2004). *Cognition and behavior in two-person guessing games: An experimental study*. University of California, San Diego, Working Paper.

Costa-Gomes, M., Crawford, V. and Broseta, B. (2004). Cognition and behavior in normal-form games: An experimental study, *Econometrica*, **59**, 1193-1235.

Che, Y.-K. and Gale. I. (1998). Standard auctions with financially constrained bidders. *Review of Economic Studies*, **65**, 1-21.

Che, Y.-K. and Gale. I. (2000). The optimal mechanism for selling to a bidder-constrained buyer. *Journal of Economic Theory*, **92**, 198-233.

Clark D. J. and Riis, C. (1996). A multi-winner nested rent-seeking contest. *Public Choice*, **87**, 177-184.

Davis, D. and Reilly, R. (1998). Do many cooks always spoil the stew? An experimental analysis of rent seeking and the role of a strategic buyer. *Public Choice*, **95**, 89-115.

Gradstein, M. (1995). Intensity of competition, entry and entry deterrence in rent seeking contest. *Economics and Politics*, **7**, 79-91.

Gradstein, M. (1998). Optimal contest design: volume and timing of rent seeking in contests. *European Journal of Political Economy*, **14**, 575-585.

Gradstein, M. and Konrad, K. A. (1999). Orchestrating rent seeking contests. *Economic Journal*, **109**, 536-545.

Hillman A. L. and Katz, E. (1984). Risk-averse rent-seekers and the social cost of monopoly power. *Economic Journal*, **94**, 104-110.

Hirshleifer, J. (1995). Anarchy and its breakdown. *Journal of Political Economy*, **103**, 26-52.

Ho, T. –H., Camerer, C. F., and Weigelt, K. (1998). Iterated dominance and iterated best-response in p-beauty Contests. *American Economic Review*, **88**, 947-969.

Millner E. L. and Pratt, M. D. (1989). An experimental investigation of efficient rent seeking society. *Public Choice*, **62**, 291-303.

Millner E. L. and Pratt, M. D. (1991). Risk aversion and rent seeking: An extension and some experimental evidence. *Public Choice*, **69**, 91-92.

Nagel, R. (1995). Unraveling in guessing games: An experimental study. *American Economic Review*, **85**, 1313-1326.

Nitzan, S. (1994). Modelling rent-seeking contests. *European Journal of Political Economy*, **10**, 41-60.

Önçüler, A. and Croson, R. (2004). Rent seeking for a risky rent: A model and experimental investigation. University of Pennsylvania, Wharton School, unpublished manuscript.

Parco, J. A., Rapoport, A., and Amaldoss, W. (2005). Two-stage contests with budget constraints: An experimental study. *Journal of Mathematical Psychology*, **49**, 320-338.

Potters, J. C., de Vries, C, and van Linden, F. (1998). An experimental examination of rational rent seeking. *European Journal of Political Economy*, **14**, 783-800.

Prelec, D. (1998) The probability weighting function. *Econometrica*, **66**, 497-527.

Rapoport, A. and Amaldoss, W. (2004). Mixed-strategy play in single-stage first-price all-pay auctions with symmetric players. *Journal of Economic Behavior and Organization*, **54**, 585-607.

Schmitt, P., Shupp, R., Swope, K., and Cardigan, J. (2004). Multi-period rent-seeking contests with carryover: Theory and experimental evidence. *Economics of Government*, **5**, 187-211.

Schmitt, D., Shupp, R., and Walker, J. M. (2003). *Resource allocation contests: Experimental evidence*. Federal Trade Commission, Bureau of Economics, unpublished manuscript.

Shogren, J. F. and Baik, K. H. (1991). Reexamining effective rent seeking in laboratory markets. *Public Choice*, **69**, 69-97.

Stein, W. E. and Rapoport, A. (2005). Symmetric multi-stage contests with budget constraints. *Public Choice,* **124**, 309-328.

Tversky, A., and Kahneman, D. (1992). Advances in prospect theory: Cumulative representation of uncertainty. *Journal of Risk and Uncertainty*, **5**, 297-323.

Tullock, G. (1967). The welfare costs of tariffs, monopolies, and theft. *Western Economic Review*, **5**, 224-232.

Tullock, G. (1980). Efficient rent seeking. In J. M. Buchanan, R. D. Tollison, and G. Tullock (Eds.), *Toward a Theory of the Rent-Seeking Society*. College Station: Texas A&M University Press, pp. 97-112.

Tversky, A., and Kahneman, D. (1974). Judgment under uncertainty: Heuristics and biases. *Science*, **185**, 1124-1130.

Vogt, C., Weimann, J., and Yang C. -L. (2002). Efficient rent-seeking in experiment. *Public Choice*, **110**, 67-78.

Weinmann, J., Yang, C., and Vogt, C. (2000). An experiment on sequential rent seeking. *Journal of Economic Behavior and Organization*, **41**, 405-426.

Wu, G. and Gonzalez (1996). Curvature of the probability weighting function. *Management Science*, **42**, 1676-1690.

In: Game Theory: Strategies, Equilibria and Theorems
Editors: Ingrid N. Haugen et al, pp. 267-287
ISBN: 978-1-60456-844-8

Chapter 10

THE EFFECTIVENESS OF SOPHISTICATED CAPITAL BUDGETING PRACTICES: EMPIRICAL EVIDENCE FROM THE NETHERLANDS

Frank H.M. Verbeeten *

RSM Erasmus University, Burg. Oudlaan 50, Room T 8-33, 3062 PA Rotterdam, The Netherlands

Abstract

This study examines the effectiveness of sophisticated capital budgeting practices. The theoretical applications of sophisticated capital budgeting practices (defined as the use of real option reasoning and/or game theory decision rules) have been well documented; however, empirical evidence on the effectiveness of these sophisticated capital budgeting practices is scarce. The empirical results from a survey of Dutch organizations suggest that the use of sophisticated capital budgeting practices is not necessarily associated with higher performance. However, the results also indicate that particular industries may benefit from sophisticated capital budgeting practices. Additional research may identify the specific characteristics that affect the effectiveness of sophisticated capital budgeting practices for individual companies.

Key words: capital budgeting practices; real option theory; game theory; performance.

1. Introduction

The investment decision is one of the most critical decisions made by top management (cf. McGrath *et al.*, 2004; Bowman & Hurry, 1993; Hofer & Schendel, 1978). Considerable attention has been devoted to theoretical explorations on the use sophisticated capital

* E-mail address: fverbeeten@rsm.nl. Tel. +31 10 4081817, fax +31 10 4089017. Correspondence address: RSM Erasmus University, Burg. Oudlaan 50, Room T 8-33, 3062 PA Rotterdam, Netherlands.
I would like to thank Rob Bannink and Bert Bettonvil for comments on previous versions of this chapter. Financial assistance for this project was provided by Deloitte.

budgeting practices, such as real option reasoning (Miller & Waller, 2003; Dixit & Pyndick, 1994; Trigeorgis, 1993; Brennan & Schwartz, 1992) and game theory decision rules (Zhu & Weyant, 2003; Smit, 2003). In addition, empirical studies have documented the methods and techniques that are used in evaluating and selecting investment projects (see, for example, Brounen et al, 2004; Graham & Harvey, 2001; and Segelod, 1998; Sangster, 1993; and Mukherjee & Henderson, 1987 for reviews). There is, however, only limited evidence on the firm-specific conditions that may influence the effectiveness of sophisticated capital budgeting practices (Chatterjee *et al.*, 2003; Farragher et al, 2001; Ho & Pike, 1998; Haka, 1987).

The purpose of this chapter is to investigate the effectiveness of sophisticated capital budgeting practices (SCBP, defined as the use of Real Option Reasoning and/or Game Theory). While theory recognizes that SCBP are essential in dealing with uncertainty, empirical evidence is scarce and mostly anecdotic (Nichols, 1994). Moreover, most empirical literature on capital budgeting practices has solely focused on the capital budgeting decision rule (exceptions are Farragher *et al.*, 2001 and Kim, 1982). Recent literature (Miller & Waller, 2003; Chatterjee *et al.*, 2003) indicates that the sophistication of capital budgeting practices must be considered in a wider context than solely the use of specific capital budgeting decision rules (such as decision criteria based net present value, payback period, real option pricing theory or game theory). In this study, I empirically investigate the relations amongst SCBP, firm-specific conditions, and performance in 189 Dutch organizations. To my knowledge, this is one of the first research projects that investigates the effectiveness of Real Option Reasoning, Game Theory and other associated capital budgeting tools in the investment process. The empirical results show that the use of sophisticated capital budgeting practices is not necessarily associated with higher performance. However, the results also indicate that particular industries may benefit from these SCBP; additional research may identify the specific industry characteristics that affect the effectiveness of SCBP.

The remainder of this chapter is organized as follows. The next section provides a review of literature on capital budgeting practices, firm-specific conditions that have been argued to affect the effectiveness of SCBP, and performance. Next, the research method and design are described, including the data collection methods and the sample selection. The fourth section provides the results of this study, as well as a discussion of the findings. Finally, some implications for theory and practice are discussed.

2. Literature Review

In this section, capital budgeting practices, firm-specific conditions and effectiveness are defined. In addition, the relevant literature on the relation between these two variables is discussed.

2.1. Capital Budgeting Practices

Capital budgeting practices are defined as the methods and techniques used to evaluate and select investment projects. Capital budgeting practices are designed to help managers select investment projects with the highest future cash flows at an acceptable 'risk of ruin'.

Literature has generally distinguished among advanced and simple (or naive) capital budgeting practices (eg. Haka, 1987; Haka *et al.*, 1985). Simple or naive capital budgeting practices (such as the payback and accounting rate of return) generally do not use cash flows, do not consider the time value of money and do not incorporate risk in a systematic manner. Advanced capital budgeting practices (discounted cash flow methods such as the internal rate of return, IRR, or the net present value, NPV) are those that consider cash flows, risk, and the time value of money.

Recent theoretical developments in financial management (Zhu & Weyant, 2003; Smit, 2003; Dixit & Pyndick, 1994; Trigeorgis, 1993; Brennan & Schwartz, 1992) have indicated that these discounted cash flow methods have serious shortcomings in analyzing investment projects when information concerning future investment decisions is not available. The application of Real Options Reasoning (ROR) and Game Theory (GT) principles can be used as analytical tools to evaluate such projects and to support the overall operating and investment strategy (Smit & Ankum, 1993). ROR has its roots in finance literature and frames investment projects in terms analogous to financial options (Miller & Waller, 2003). Option value may stem from the option to postpone, to grow or develop, to stage or sequence, to switch inputs or outputs, and/or to abandon a project (Busby & Pitts, 1997; Trigeorgis, 1993). GT, on the other hand, analyzes how strategic interactions among rational players produce outcomes with respect to the preferences (or utilities) of those players (Ross, 2003). While ROR stresses the value to delay an adoption decision until more information about the investment project is available, GT indicates that firms have an incentive to invest early in case of fear of pre-emption[1] (Zhu & Weyant, 2003; Smit, 2003). A firm runs the risk that another firm may pre-empt it by adopting the investment project first; the 'first-adopter' may eliminate the investment opportunities for all the other firms in that industry (Murto & Keppo, 2002). By integrating ROR with GT principles, firms can make a more complete assessment of strategic option value in an interactive competitive setting (Smit, 2003). For the purposes of this study, sophisticated capital budgeting practices (SCBP) are defined as the application of ROR and/or GT principles.

2.2. Firm-Specific Conditions

Historically, it has been argued that managers should use sophisticated capital budgeting practices in order to improve the capital budgeting decision-making process (cf Farragher et al, 2001). However, Haka (1987) indicates that the usefulness of sophisticated capital budgeting practices may depend on firm-specific conditions. That is, the application of SCBP is complicated and conceptually difficult to understand (cf. Bowman & Moskowitz, 2001); firms have to spend time and money on training of employees, acquisition of IT-tools, etc. (Miller & Waller, 2003; Busby & Pitts, 1997; Ho & Pike, 1996; Klammer et al, 1991). In determining the appropriate level of SCBP, firms will compare the net benefits of SCBP to their costs; the net benefit that firms may obtain from SCBP may be affected by firm-specific conditions. Based on previous literature (Farragher et al, 2001; Ho & Pike, 1992; Klammer et

[1] ROR may take the preemption effect in account by shortening the 'time to expiration' of the option (Bowman & Moskowitz, 2001). For example, in the case study by Bowman & Moskowitz (2001) on Merck, the firm using ROR set the maximum time to expiration at four years because after that, competing products would have entered the market, making entry by Merck unfeasible.

al, 1991; Haka, 1987; Haka et al, 1985; Kim, 1982; Schall & Sundem, 1980; Klammer, 1973), I distinguish amongst two firm-specific conditions that are likely to affect the effectiveness of SCBP: uncertainty and industry. Each of them will be discussed shortly in the next sections.

Uncertainty

Uncertainty is defined as the gap between the information currently available and the information required to make the decision (Galbraith, 1973). A condition of uncertainty usually exists in capital budgeting because investment decisions, by definition, involve uncertain outcomes that in the long run are important to firm survival and about which complete information is unavailable (Zhu & Weyant, 2003; Simerly & Li, 2000; Smit & Ankum, 1993). Generally, it is hypothesized that the application of ROR and GT principles helps to determine the trade-off between adopting investment projects early and waiting for more information (Zhu & Weyant, 2003; Murto & Keppo, 2002). Theory thus suggests that SCBP are most valuable in case of high uncertainty; in that situation, it is likely that the costs of SCBP are offset by additional gains from successful investment projects.

A core concept in the ROR literature is that there may be specific uncertainties (rather than uncertainty in general) that affect capital budgeting practices (Dixit and Pyndick, 1994). Also, game theory indicates that specific uncertainties can change the optimal investment criterion (Smit, 2003). Thus, theory suggests that specific uncertainties may need to be analyzed by using different capital budgeting tools (in other words, specific uncertainties have an impact on capital budgeting practices).

Previous empirical research also suggests that specific uncertainties may affect the usefulness of SCBP. For example, Harvey & Graham (2001) find a positive relation between foreign sales (i.e., exchange risk uncertainty) and the use of specific capital budgeting practices (IRR, NPV). Ho & Pike (1998) found a positive relation between socioeconomic uncertainty (governmental regulations, actions of trade unions and behavior of financial/capital markets) and the application of risk analysis techniques in capital budgeting practices, and no relation with actions of competitors and customer preferences. Thus, theory as well as previous empirical research suggests that specific uncertainties affect the usefulness of SCBP; however, few studies empirically investigate whether firms perform better if they adapt their capital budgeting practices to these specific uncertainties.

Industry

Some authors suggest that the application of real option or game theory analysis is more suitable for particular industries. For example, (mostly theoretical) studies have focused on the application of real option reasoning or game theory analysis in the high-tech industry (Billington *et al*., 2003; Zhu & Weyant, 2003), the pharmaceutical industry (McGrath & Nerkar, 2004; Bowman & Moskowitz, 2001; Nichols, 1994) and the extraction industry (see Trigeorgis, 1993; Brennan & Schwartz, 1992).

Previous empirical research also suggests that capital budgeting practices differ among industries (Verbeeten, 2006; Harvey & Graham, 2001; Ho & Pike, 1998; Haka *et al*., 1985; Aggarwal, 1980). One of the reasons that the application of SCBP differs is that there are significant differences in uncertainty across industries (Simerly & Li, 2000; Miller, 1992). As such, industry may be just another proxy for the (specific) uncertainties relevant to that industry. However, industry may also capture other effects such as the importance of identifying and exploiting investment projects (versus the importance of 'human capital' or 'relational capital'; Segelod, 1998), operating and financial leverage preferences, capital structure policies, regulation or other institutional characteristics (Brounen et al, 2004; Harvey & Graham, 2001). Again, while theory as well as previous empirical research suggests that industry affects the usefulness of SCBP there is little empirical evidence regarding the performance consequences of SCBP-practices in specific industries.

2.3. The Effectiveness of Sophisticated Capital Budgeting Practices

The academic community has tried to convince managers that SCBP can improve the capital budgeting decision-making process (cf Farragher et al, 2001). Over the years, many studies have documented the increasing use of SCBP (cf Brounen et al, 2004; Harvey & Graham, 2001; Pike, 1996). However, there is no clear evidence that more effective companies are more likely to employ SCBP relative to less effective companies (cf. Farragher et al, 2001).

Table 1. Overview studies on capital budgeting practices and performance

Author	Performance measure	Research method	Results
Farragher et al (2001)	Return on assets	Multiple regression: independent variables size, operating risk, capital intensity, focus and SCBP	No significant relation between ROA and SCBP (strategic analysis, search for investment, evaluation, implementation and post-audit)
Chen (1995)	Return on assets	Comparison of ROA of two groups (high-use or low-use) of capital budgeting decision rules	No significant differences between ROA for each of the capital budgeting decision rules (DCF, payback, ARR and nonfinancial)
Ho (1992)	Operating return measures	Matched pairs approach - matching variables size, risk and industry	No significant relationship between SCBP (DCF + formal risk analysis) and performance
Haka (1987)	Share price	Matched pairs approach - matching variables size, risk and industry	Positive relation between SCBP (DCF) and stock price performance when a firm's financial markets and competitors are more predictable
Haka et al (1985)	Share price	Matched pairs approach - matching variables size, risk and industry	No significant relationship between SCBP (DCF) practices and firm performance
Pike (1984)	Return on assets	Multiple regression: independent variables – risk, capital budgeting practices, size, capital intensity, industry variables	Negative relationship between SCBP (procedures, risk analysis and evaluation techniques) and return on assets
Kim (1982)	Average earnings per share	Multiple regression: independent variables- degree of sophistication of capital budgeting process, size, risk and capital intensity	Positive relationship between degree of SCBP process, DCF-methods, firm performance, size and risk
Klammer (1973)	Operating rate of return	Multiple regression: independent variables- capital budgeting techniques, size, risk, capital intensity	No significant relationship between profit performance and the use of SCBP (DCF)

Some studies have empirically investigated the effectiveness of various capital budgeting practices (see table 1 for an overview). It is important to notice several differences between the studies: for example, some projects have relied on cross-sectional comparisons of performance (Farragher et al, 2001; Kim, 1982; Klammer, 1973), while others have used a matched pairs approach (Ho, 1992; Haka et al, 1985). Also, different studies have employed different performance measures such as operating rate of return (Farragher et al, 2001; Klammer, 1973; Ho, 1992) and share price (Haka, 1987; Haka et al, 1985). Third, several definitions of capital budgeting practices have been used: some studies have defined the application of discounted cash flow method as advanced (Haka et al, 1985; Schall & Sundem, 1980; Klammer, 1973), while others argue that the application of other tools is necessary to qualify as 'sophisticated' (Farragher et al, 2001; Kim, 1982). Finally, to the author's knowledge there is no previous empirical evidence about the relation between ROR /GT, firm-specific conditions and performance.

Table 1 also indicates that previous studies have hardly found a direct relation between SCBP and performance. However, studies that have investigated a match between SCBP and firm-specific conditions appear to provide interesting avenues for research. For example, Haka (1987) concludes that the more predictable a firm's financial markets and competitors are, the more likely that the firm using NPV-methods will outperform a matching firm not using NPV-methods. The predictability of government regulations and the actions of labor unions, customers or suppliers did not have an impact on capital budgeting practices in her study. This suggests that SCBP should be matched to the firm-specific conditions in order to improve performance.

Summarizing, previous literature suggests that uncertainty and industry are two major determinants of capital budgeting practices. In addition, previous literature also suggests that a match between SCBP and firm-specific conditions should positively affect performance. Following literature and empirical findings discussed in the previous sections, I expect that a match between firm-specific conditions (specifically: uncertainty and industry) and SCBP is associated with higher performance (and that a mismatch is associated with lower performance). This expectation gives rise to the following hypotheses:

H1: Greater use of SCBP will have a positive impact on performance in the case of high-uncertainty firms relative to low-uncertainty firms

H2: Greater use of SCBP will have a positive impact on performance in specific industries.

2.4. Control Variables

Two control variables are included in the analysis: diversification and size.

Diversification

An organization's ability to strike options effectively is influenced by its structure (Bowman & Hurry, 1993). In addition, Schall & Sundem (1980) argue that it is not the level of uncertainty itself that affects capital budgeting methods, but rather the range of risk types across the different projects that a firm considers. When a firm is confronted with heterogeneous projects, the application of SCBP is helpful to distinguish among different types of projects (in terms of uncertainty). This situation is more likely to be present in diversified firms.

Size

Size is also included as a control variable. In general, large firms tend to adopt innovations for decision making (for example, SCBP) to a larger extent than small firms (Rogers, 1995). In addition, large firms have more at stake and are more likely to have the available resources to use SCBP (Ho & Pike, 1998). Previous empirical research has indicated that size has an impact on capital budgeting practices (Brounen et al, 2004; Graham & Harvey, 2001; Farragher *et al.*, 2001; Ho & Pike, 1996; Klammer *et al.*, 1991).

3. Method

3.1. Research Design

A sample of 704 large organizations operating in the Netherlands is used for this research project. To select the organizations, three criteria are used: (1) sales have to exceed €25m, (2) total assets have to exceed €20m and (3) costs for personnel have to exceed €15m. The survey is sent to the chief financial officers of the selected organizations since previous research indicates that the finance department is generally involved in the capital budgeting decision (cf. Brounen et al, 2004; Graham & Harvey, 2001; Northcott, 1992; Mukherjee & Henderson, 1987). These senior financial executives should be fully conversant both with the capital budgeting practices as well as with the factors hypothesized to affect the application of SCBP. The data collection process results in 189 (at least partially) useable responses (an actual response rate of 26.9%). Additional tests for non-response bias (i.e., comparing the characteristics and responses for early and late respondents; see Wallace & Mellor, 1988) reveal no particular problems.

3.2. Measurement of Variables

3.2.1. Sophisticated Capital Budgeting Practices

Similar to Farragher et al (2001), Pike (1984) and Kim (1982), I use a composite measure to proxy for the sophistication of capital budgeting practices. Respondents have been asked to indicate on a 5-point Likert scale (ranging from 1=never, to 5=always) to what extent they consider several capital budgeting techniques useful or important in the investment process. Factor analysis has been used again to verify whether the 3 capital budgeting practices recognized by theory (naive capital budgeting practices, NPV-capital budgeting practices and sophisticated capital budgeting practices, the latter including ROR and/or GT decision rules) are actually present. The factor analysis results in the 3 factors recognized by theory after eliminating one variable (profitability index; see table 2). Based on the previous factor analysis, three measures for capital budgeting practices are constructed by calculating the average score for the responses to the questions that load on the relevant factor[2]. The first factor is most relevant for this research project and is labeled Sophisticated Capital Budgeting Practices (SCBP); the other factors are not used in the remainder of this project. At 0.80, Cronbach's alpha for the SCBP measure is acceptable (Hair *et al.*, 1998).

Table 2. Results Factor Analysis for Sophistication of Capital Budgeting Practices

The next table presents the results for the factor analysis (eigenvalue > 1) on the capital budgeting factors. In addition, Cronbach's alpha and the number of observations are presented.

Component	Sophisticated Capital Budgeting Practices (SCBP)	NPV Capital Budgeting Practices (NPVCBP)	Naive Capital Budgeting Practices (NAIVCBP)
Monte Carlo simulations	***.732***	.117	-.035
Game theory decision rules (GT)	***.722***	-0.035	.185
Real option pricing (ROR)	***,706***	0.061	,193
Using certainty equivalents	***,664***	,152	,121
Decision trees	***.633***	.116	.190
CAPM analysis / ß analysis	***.598***	.385	-.263
Adjusting expected values	***.488***	.372	-.089
Sensitivity analysis/break-even analysis	.037	***.709***	.023
Scenario analysis	.152	***.703***	-.080
Adaptation of required return/discount rate	.010	***.589***	.203
Internal rate of return (IRR)	.047	***.582***	.350
Net present value (NPV)	.087	***.547***	.322
Uncertainty absorption in cash flows	.312	***.476***	.033
Payback period (PB)	-.010	.150	***.750***
Adaptation of required payback period	.194	.165	***.701***
Accounting rate of return (ARR)	.194	.018	***.530***
Cronbach's Alpha	*.80*	*.70*	*.52*
N	*183*	*186*	*187*

Extraction Method: Principal Component Analysis.
Rotation Method: Varimax with Kaiser Normalization.
a Rotation converged in 7 iterations.

The validity of the SCBP-measure is tested by correlating it with a measure for operational risk management practices; the results (non-tabulated) support the convergent validity of the SCBP-measure.

3.2.2. Uncertainty

Miller's (1992) uncertainty framework has been used to investigate which specific uncertainties have an impact on capital budgeting practices. Miller (1992) distinguishes among three types of uncertainty: general uncertainties (which affect all companies), industry-specific uncertainties

[2] For example, the score for Sophisticated Capital Budgeting Practices (labeled SCBP) is constructed by calculating the average score for the responses on the questions whether firms find Monte Carlo simulations, GT decision rules, ROR, certainty equivalents, decision trees, CAPM-/β-analysis and adjusting expected values more important and useful.

(which affect specific industries) and organization specific uncertainties (which affect one organization). Respondents have been asked to indicate on a 5-point Likert scale (ranging from 1=unimportant, to 5= very important) to what extent they consider a number of uncertainties relevant for the intended results of their organization, within the time frame of an investment decision. Factor analysis has been used to verify whether the 3 uncertainty categories mentioned by Miller (1992) are actually present. Factor analysis provides four separate uncertainty factors (eigenvalue>1) after eliminating four uncertainty variables (natural uncertainties, R&D uncertainties, credit uncertainties and behavioral uncertainties). I label these factors input uncertainty (INPUNCTY), financial uncertainty (FINUNCTY), social uncertainty (SOCUNCTY) and market uncertainty (MARKUNCTY). Considering that these factors are fairly similar to the variables that have been used in previous research (e.g. Ho & Pike, 1998; Haka, 1987), they have been used for this study. The cumulative responses on the questions that load on each factor are used as proxies for each measure[3]. With the exception of market uncertainty, Cronbach's alpha for each of these factors exceeds the lower limit of 0.60[4] (Hair *et al.*, 1998; see table 3 for details).

Table 3: Results Factor Analysis for Miller's Uncertainties Framework

The next table presents the results for the factor analysis (eigenvalue > 1) on the uncertainty factors. In addition, Cronbach's alpha and the number of observations are presented.

Component	INPUT UNC	FINAN UNC	SOCIO UNC	MARKT UNC
Raw material uncertainties	***.791***	.042	-.117	.045
Input market uncertainties	***.743***	.039	-.047	.009
Production uncertainties	***.729***	.154	.040	-.033
Labour uncertainties	***.635***	.153	.174	.239
Liability uncertainties	***.584***	-.069	.159	.046
Inflation uncertainties	.051	***.867***	.200	.057
Interest uncertainties	-.134	***.851***	.156	-.031
Exchange rate uncertainties	.190	***.732***	-.125	.106
Political uncertainties	.062	.114	***.809***	.042
Society uncertainties	.077	-.003	***.719***	.188
Policy uncertainties	-.009	.103	***.692***	-.278
Competitive uncertainties	-.094	.063	.057	***.838***
Output market uncertainties	.316	.051	-.056	***.730***
Cronbach's Alpha	0.74	0.76	0.62	0.51
N	188	188	184	189

Extraction Method: Principal Component Analysis.
Rotation Method: Varimax with Kaiser Normalization.
a Rotation converged in 5 iterations.

[3] For example, the measure INPUNCTY is calculated by summarizing the scores on the questions on raw material uncertainties, input uncertainties, production uncertainties, labor uncertainties and liability uncertainties.

[4] It is noticeable that the Cronbach's alpha for market uncertainty does not exceed the lower limit of .60 generally agreed upon (Hair *et al.*, 1998, p. 118). This is probably due to the fact that only two variables (competitive and output uncertainties) relate to this factor. Considering the fact that the inter-item correlation exceeds the .30 (another rule of thumb for judging the reliability of a scale), it has been decided to include this variable in the analysis.

The validity of the uncertainty measures is tested using CAPM-β data collected for a small subset of the sample. The survey included a question on 'additional remarks', where respondents could indicate whether they wanted to receive the final results of the study. Based on these data, it was possible to obtain the CAPM-β for 23 listed firms in the sample. CAPM-β is significantly correlated with three uncertainty factors (input, financial and market uncertainty; in all cases $\rho>0.45$, $p<0.04$), but not with social uncertainty ($\rho=0.21$, $p>0.34$). The results from the correlation analysis provide considerable confidence in the validity of the uncertainty measures used in the study.

3.2.3. Industry

Respondents have been asked to provide the SIC-codes for the industry in which their organization operates. The survey data on industry have been validated through archival data where possible (the survey is anonymous; respondents have provided results voluntarily). The industry codes are regrouped to obtain dummy variables for industry. I distinguish among extraction and manufacturing firms (EXTRMFTG), public utilities (water, electricity and gas companies, labeled PUBLUTIL), construction and building firms (CONSTRBU), transport and communication firms (TRANSPCO), wholesale and retail firms (WHOLRETA), financial services firms (FINSERV) and other service firms (NFSERV; non-financial services and government companies).

3.2.4. Effectiveness

Three measures are used to proxy for effectiveness: a subjective measure for 'overall effectiveness' (based on Gupta & Govindarajan, 1985) as well as two objective performance measures (return on assets and return on equity). The measure for perceived effectiveness, AGREFF, is based on the respondent's perception of the importance of several goals for

Table 4. Response structure: importance of goals, performance and effectiveness

	Importance of goals		**Performance**	
	Mean	**St.dev.**	**Mean**	**St.dev**
Profit, profit margin	4.48	.70	3.54	.84
Operational cash flows	3.83	.99	3.55	.78
Cost reduction	4.15	.71	3.24	.84
Sales growth	3.88	.91	3.37	.78
Market share	3.80	.92	3.34	.77
New product/new market development	3.70	.85	3.18	.76
Research & development	3.24	1.01	3.11	.76
Quality/customer service	4.27	.76	3.48	.75
Human capital/personnel development	3.69	.90	3.26	.77
Political public affairs	3.09	.95	3.23	.69
Ethical integrity	3.46	.96	3.62	.69
	Mean	St.dev.	Min	Max
Aggregate effectiveness	3.3681	.41458	2.28	4.86

several stakeholders as well as the perceived performance in 11 areas[5]. Each respondent was asked to indicate the rate of importance (1=not important at all, to 5=extremely important) attached by superiors to each of the 11 performance dimensions (profit, operational cash flows, cost reduction, sales growth, market share, development of new markets and products, research and development, quality, human capital, political/society effects and ethical integrity). Each respondent was also asked to rate the firm's performance on each of the performance dimensions (ranging from 1=not at all satisfactory to 5=outstanding). Using the data on dimensional importance obtained in the first question as weights, a weighted average effectiveness score is obtained for each firm. Summary data on the individual responses as well as the aggregate effectiveness measure are provided in table 4.

In addition to perceived performance, two measures for objective financial performance are obtained for a selected number of firms in the sample[6]. I use the average return on equity (ROE) as well as the average return on assets (ROA) over 5 years as financial performance measures. Due to their specific operations and the incomparability of ROE and ROA to other industries, the financial services industry is excluded from the analysis. The objective measures of performance are winsorized in order to reduce econometric problems.

3.2.5. Control Variables

Diversification

Respondents have also been asked to characterize their organization as (part of) a single business, (part of) a related diversified organization or (part of) an unrelated diversified organization. The distribution of this variable is mostly in accordance with other research in this area (see Christie *et al.*, 2003). The study includes a dummy variable for diversification which I label DIV (related diversified or unrelated diversified firms, coded 1; single business firms coded 0)[7].

Size

Three measures for size are included in this research project: total sales, total assets and number of employees of the organization. The Spearman correlation reveals that all indicators for size are highly correlated ($\rho>0.53$, $p<0.01$). To correct for skewness, the logarithm of the number of employees is determined (labeled SIZE) and used in the remainder of the analysis.

---

5 Originally, a question regarding shareholder value and dividends was included in the survey. However, this question has been excluded from the analysis due to the fact that several companies indicated that these goals did not apply to them.

6 The survey was anonymous; therefore, not all financial data could be obtained. In addition, some firms were part of larger organizations; under certain circumstances, these firms do not have to provide financial data. Therefore, our sample size is lower for the objective financial performance measures.

7 The validity of this measure is checked by comparing the responses to more objective information. Based on the data in the 'additional remarks' section of the survey, it has been possible obtain objective information on the number of industries (SIC-codes) in which 66 firms are operating. The number of industries in which the organization is operating may be considered as a proxy for diversification (see Pennings *et al*, 1994). The relation between the number of industries in which the firm is operating and the diversification dummy variable is mostly in the expected direction, yet not significant ($p>0.15$).

4. Results

4.1. Statistical Analysis

The descriptive statistics and the correlation matrix for all of the variables are included in table 5, respectively 6.

Table 5 indicates that there is considerable variation in the dispersion of the variables of interest, including SCBP. Additional analysis (non-tabulated) indicates that only about 5% of the respondents uses ROR and GT-principles 'often' or 'always' in investment decisions. In addition, 60% of the respondents uses ROR 'never' while 70% uses GT-principles 'never' in investment decisions. This suggests that the use of ROR and GT is not widespread and that these techniques may be used only for specific investment decisions.

Table 5. Descriptive statistics

The next table provides the mean, standard deviation, theoretical range, actual range and number of observations for the variables under consideration (excluding dummy variables).

	N	Min	Max	Mean	St.dev
AGGREFF	170	2.28	4.86	3.3681	.41458
ROE	66	-6.22	89.18	18.4253	23.02304
ROA	68	.37	29.11	8.1234	6.50745
SCBP	189	1.00	4.29	1.7828	.68875
INPUNCTY	189	1.20	5.00	3.0521	.80955
FINUNCTY	188	1.00	5.00	3.1543	.88318
SOCUNCTY	189	1.00	5.00	3.0467	.85574
MRKUNCTY	189	1.50	5.00	3.7989	.71593
DIV	188	.00	1.00	.5266	.50063
SIZE	188	.30	4.85	3.0053	.65953
EXTRMFTG	187	.00	1.00	.4332	.49684
PUBLUTIL	187	.00	1.00	.0749	.26388
CONSTRBU	187	.00	1.00	.0428	.20291
WHOLRETA	187	.00	1.00	.1016	.30294
TRANSPCO	187	.00	1.00	.0642	.24572
FINSERV	187	.00	1.00	.1337	.34123
OTHSERV	187	.00	1.00	.1497	.35777

AGGREFF = aggregate effectiveness; ROE = return on equity (5 year average); ROA = return on assets (5 year average); SCBP = sophisticated of capital budgeting practices (i.e., including game theory and real option pricing); INPUNCTY = input uncertainty; FINUNCTY = financial uncertainty; SOCUNCTY = social uncertainty; MRKUNCTY = market uncertainty; DIV = Diversification (dummy variable; related and unrelated diversified strategy=1, single business strategy =0); SIZE = size (log fte); EXTRMFTG = extraction and manufacturing industry (dummy variable); PUBLUTIL = public utilities (dummy variable); CONSTRBU = construction and building industry (dummy variable); TRANSPCO = transport and communication (dummy variable); FINSERV = financial services industry (dummy variable); OTHSERV = other (non-financial) services industry (dummy variable).

Table 6. Pearson Correlation Matrix

	1.	2.	3.	4.	5.	6.	7.	8.	9.	10.	11.	12.	13.	14.	15.	16.
1. AGREFF	1															
2. ROE	.46**	1														
3. ROA	.34**	.89**	1													
4. SCBP	.07	.09	-.02	1												
5. INPUNCTY	-.01	-.08	-.05	-.00	1											
6. FINUNCTY	.04	-.00	-.08	.27**	.18*	1										
7. SOCUNCTY	.00	.13	-.01	.08	.10	.15*	1									
8. MRKUNCTY	-.07	.15	.26*	.04	.22**	.13	.01	1								
9. SIZE	-.12	.08	.05	.19**	-.02	.08	-.04	.04	1							
10. DIV	-.08	-.03	.04	.11	.01	-.01	.03	.05	.12	1						
11. EXTRMFTG	-.03	-.22	-.06	-.16*	.22**	-.02	-.32**	.07	.04	.05	1					
12. PUBLUTIL	-.08	-.13	-.29*	.05	.05	-.11	.18*	-.22**	-.03	-.14	-.25**	1				
13. CONSTRBU	-.00	-.04	-.13	.07	-.05	.00	.03	-.03	.07	-.01	-.19*	-.06	1			
14. WHOLRETA	.03	.14	.23	-.14*	.07	-.05	.06	.20**	.00	.00	-.29**	-.10	-.07	1		
15. TRANSPCO	-.12	.05	-.09	.04	.03	.02	.10	-.00	.09	.03	-.23**	-.07	-.06	-.09	1	
16. FINSERV	.07	.a	.a	.19**	-.22**	.39**	.12	-.09	-.13	-.03	-.34**	-.11	-.08	-.13	-.10	1
17. OTHSERV	.08	.45**	.41**	.06	-.19*	-.24**	.07	.00	.00	.04	-.37**	-.12	-.09	-.14	-.11	-.17*

* and ** denote 5% and 1% significance levels (two-tailed), respectively.

a Cannot be computed because at least one of the variables is constant.

Explanation of variables:

AGGREFF = aggregate effectiveness; ROE = return on equity (5 year average); ROA = return on assets (5 year average); SCBP = sophisticated of capital budgeting practices (i.e., including game theory and real option pricing); INPUNCTY = input uncertainty; FINUNCTY = financial uncertainty; SOCUNCTY = social uncertainty; MRKUNCTY = market uncertainty; DIV = Diversification (dummy variable; related and unrelated diversified strategy=1, single business strategy =0); SIZE = size (log fte); EXTRMFTG = extraction and manufacturing industry (dummy variable); PUBLUTIL = public utilities (dummy variable); CONSTRBU = construction and building industry (dummy variable); TRANSPCO = transport and communication (dummy variable); FINSERV = financial services industry (dummy variable); OTHSERV = other (non-financial) services industry (dummy variable).

Table 7a. Multiple regression results
Dependent variable : AGREFF

Variable												
C	3.69***	3.70***	3.78***	3.78***	3.59***	3.96***	3.74***	3.68***	3.79***	3.71***	3.71***	3.63***
INPUNCTY	0.00	0.02	-0.01	0.02	0.02	0.02	0.02	0.02	0.01	0.02	0.02	0.02
FINUNCTY	0.01	0.02	0.02	-0.00	0.02	0.02	0.02	0.02	0.03	0.02	0.02	0.02
SOCUNCTY	-0.01	-0.00	-0.00	-0.00	0.03	-0.00	-0.00	-0.01	-0.01	-0.00	-0.00	-0.00
MRKUNCTY	-0.04	-0.05	-0.05	-0.05	-0.05	-0.12	-0.05	-0.05	-0.05	-0.05	-0.04	-0.05
SCBP	0.06	0.07	0.03	0.02	0.14	-0.08	0.04	0.09**	0.05	0.07	0.07#	0.10*
SIZE	-0.08**	-0.08**	-0.09**	-0.09**	-0.09**	-0.08**	-0.09**	-0.09**	-0.07	-0.08**	-0.09**	-0.09**
DIV	-0.06	-0.09#	-0.09#	-0.09	-0.09	-0.09#	-0.09	-0.10#	-0.11*	-0.09	-0.09#	-0.09#
EXTRMFTG		-0.10	-0.10	-0.10	-0.10	-0.10	-0.22	-0.10	-0.11	-0.10	-0.10	-0.09
PUBLUTIL		-0.30***	-0.30***	-0.30***	-0.31***	-0.30***	-0.30***	0.04	-0.30***	-0.30***	-0.30***	-0.30***
CONSTRBU		-0.10	-0.11	-0.10	-0.10	-0.11	-0.10	-0.11	-0.99***	-0.10	-0.10	-0.10
WHOLRETA		-0.03	0.03	0.03	0.03	-0.03	-0.04	-0.02	-0.04	-0.23	-0.03	-0.01
TRANSPCO		-0.27#	-0.27#	-0.27#	-0.27#	-0.27#	-0.27#	-0.27#	-0.27#	-0.27#	-0.90	-0.27#
FINSERV		-0.10	-0.10	-0.10	-0.10	-0.10	-0.09	-0.10	-0.10	-0.10	-0.10	0.18
SCBP*INPUNCTY			0.01									
SCBP* FINUNCTY				0.01								
SCBP* SOCUNCTY					-0.02							
SCBP* MRKUNCTY						0.04						
SCBP*EXTRMFTG							0.07					
SCBP* PUBLUTIL								-0.18				
SCBP* CONSTRBU									0.44***			
SCBP* WHOLRETA										0.14		
SCBP* TRANSPCO											0.35	
SCBP*FINSERV												-0.13*
R2	0.03	0.07	0.07	0.07	0.07	0.07	0.08	0.08	0.10	0.07	0.08	0.08
Adj R2	-0.01	-0.01	-0.01	-0.01	-0.01	-0.01	-0.01	-1.01	0.01	-0.01	-0.01	-0.00
F-stat	0.83	0.54	0.85	0.85	0.85	0.87	0.88	0.94	1.16	0.87	0.90	0.95

#, *, **, *** denote 15%, 10%, 5% and 1% significance levels respectively

Explanation of variables: AGGREFF = aggregate effectiveness; ROE = return on equity (5 year average); ROA = return on assets (5 year average); SCBP = sophisticated of capital budgeting practices (i.e., including game theory and real option pricing); INPUNCTY = input uncertainty; FINUNCTY = financial uncertainty; SOCUNCTY = social uncertainty; MRKUNCTY = market uncertainty; DIV = Diversification (dummy variable; related and unrelated diversified strategy=1, single business strategy =0); SIZE = size (log fte); EXTRMFTG = extraction and manufacturing industry (dummy variable); PUBLUTIL = public utilities (dummy variable); CONSTRBU = construction and building industry (dummy variable); TRANSPCO = transport and communication (dummy variable); FINSERV = financial services industry (dummy variable); OTHSERV = other (non-financial) services industry (dummy variable).

Table 7b. Multiple regression results
Dependent variable : ROE

Variable										
C	70.51**	79.11**	63.77	92.84**	102.54*	90.49**	77.63**	73.64**	68.66**	90.37**
INPUNCTY	-6.65**	-9.33	-6.61**	-6.80*	-6.29**	-6.04*	-6.94***	-7.00**	-7.01***	-7.65***
FINUNCTY	-5.10	-5.13	-3.39	-5.69	-5.57	-4.52	-5.23	-5.06	-5.00	-5.53
SOCUNCTY	6.94*	6.96*	7.03*	0.31	6.71*	6.40*	7.42*	6.75*	7.03*	4.79
MRKUNCTY	5.69	5.63	5.78	5.44	-2.35	6.18	4.88	5.74	6.06	2.56
SCBP	-5.00	-9.93	-1.28	-17.79	-20.79	-10.51#	-7.28	-5.60	-3.82	-3.63
SIZE	3.84	3.87	3.99	4.41	3.01	0.51	4.55	3.99	3.17	4.52
DIV	-7.53	-7.75	-7.44	-7.77	-7.62	-6.47	-7.07	-8.17	-7.38	-10.27*
EXTRMFTG	-58.56***	-57.94***	-58.86***	-58.89***	-59.12***	-88.92***	-61.12***	-59.36	-57.55***	-58.29***
PUBLUTIL	62.29***	-62.38***	-62.42***	-62.81***	-64.39***	-65.01***	-81.84***	-62.82	-61.37***	-65.25***
CONSTRBU	-62.71***	-62.85***	-62.77***	-65.29***	-62.73***	-65.30***	-66.08***	-76.56	-61.35***	-66.00***
WHOLRETA	-53.30***	-52.64***	-53.68***	-53.66***	-53.58***	-59.15***	-55.83***	-54.21	-39.40	-51.05***
TRANSPCO	-39.59	-38.74	-40.08	-39.46	-38.36	-39.08*	-41.93*	-40.03#	-38.75#	220.66*
SCBP*INPUNCTY		1.50								
SCBP* FINUNCTY			-1.09							
SCBP* SOCUNCTY				4.16						
SCBP* MRKUNCTY					4.58					
SCBP*EXTRMFTG						14.75#				
SCBP* PUBLUTIL							8.21			
SCBP* CONSTRBU								7.35		
SCBP* WHOLRETA									-8.16	
SCBP* TRANSPCO										-114.48**
R2	0.37	0.37	0.37	0.38	0.38	0.40	0.38	0.38	0.38	0.43
Adj R2	0.23	0.22	0.21	0.22	0.22	0.26	0.22	0.22	0.22	0.29
F-stat	2.58***	2.35**	2.35**	2.41**	2.42**	2.69***	2.42**	2.37**	2.39**	2.99***

#, *, **, *** denote 15%, 10%, 5% and 1% significance levels respectively

Explanation of variables:

AGGREFF = aggregate effectiveness; ROE = return on equity (5 year average); ROA = return on assets (5 year average); SCBP = sophisticated of capital budgeting practices (i.e., including game theory and real option pricing); INPUNCTY = input uncertainty; FINUNCTY = financial uncertainty; SOCUNCTY = social uncertainty; MRKUNCTY = market uncertainty; DIV = Diversification (dummy variable; related and unrelated diversified strategy=1, single business strategy =0); SIZE = size (log fte); EXTRMFTG = extraction and manufacturing industry (dummy variable); PUBLUTIL = public utilities (dummy variable); CONSTRBU = construction and building industry (dummy variable); TRANSPCO = transport and communication (dummy variable); FINSERV = financial services industry (dummy variable); OTHSERV = other (non-financial) services industry (dummy variable).

Table 7c. Multiple regression results
Dependent variable : ROA

Variable										
C	23.18***	21.78*	18.09	26.45**	28.23*	26.52***	27.20***	23.69***	22.63***	21.27**
INPUNCTY	-1.98**	-1.53	-1.94**	-1.99**	-1.93**	-1.86*	-2.21***	-2.03**	-2.08**	-1.89**
FINUNCTY	-1.26	-1.26	0.05	-1.35	-1.34	-1.18	-1.30	-1.26	-1.24	-1.20
SOCUNCTY	1.22	1.22	1.29	0.26	1.19	1.16	1.51#	1.19	1.26	1.37
MRKUNCTY	1.00	1.01	1.07	1.00	-0.26	1.03	0.63	1.00	1.09	1.22
SCBP	-1.64	-0.83	1.20	-3.51	-4.14	-2.57	-2.98#	-1.73	-1.29	-1.76
SIZE	2.42	2.41	2.50	2.49	2.27	1.92	2.78#	2.44	2.24	2.51
DIV	-1.83	-1.79	-1.76	-1.86	-1.83	-1.71	-1.53	-1.93	-1.82	-1.70
EXTRMFTG	-15.37***	-15.47***	-15.62***	-15.43***	-15.47***	-20.31***	-16.93***	-15.50***	-15.07***	-15.38***
PUBLUTIL	-20.43***	-20.42***	-20.56***	-20.50***	-20.70***	-21.01***	-31.52***	-20.53***	-20.18***	-20.21***
CONSTRBU	-20.93***	-20.91***	-20.97***	-21.30***	-20.93***	-21.44***	-22.89***	-23.21***	-20.57***	-20.78***
WHOLRETA	-13.66***	-13.76***	-13.95***	-13.72***	-13.72***	-14.65***	-15.21***	-13.80***	-9.43	-13.81***
TRANSPCO	-18.33***	-18.48***	-18.67***	-18.23***	-18.04***	-18.14***	-19.54***	-18.41***	-18.10***	-43.59
SCBP*INPUNCTY		-0.24								
SCBP* FINUNCTY			-0.83							
SCBP* SOCUNCTY				0.61						
SCBP* MRKUNCTY					0.73					
SCBP*EXTRMFTG						2.39				
SCBP* PUBLUTIL							4.81**			
SCBP* CONSTRBU								1.22		
SCBP* WHOLRETA									-2.49	
SCBP* TRANSPCO										10.63
R2	0.39	0.40	0.40	0.40	0.40	0.41	0.43	0.40	0.40	0.40
Adj R2	0.26	0.25	0.25	0.25	0.25	0.26	0.29	0.25	0.25	0.25
F-stat	2.94***	2.66***	2.68***	2.68***	2.68***	2.78***	3.03***	2.67***	2.72***	2.69***

#, *, **, *** denote 15%, 10%, 5% and 1% significance levels respectively

Explanation of variables:

AGGREFF = aggregate effectiveness; ROE = return on equity (5 year average); ROA = return on assets (5 year average); SCBP = sophisticated of capital budgeting practices (i.e., including game theory and real option pricing); INPUNCTY = input uncertainty; FINUNCTY = financial uncertainty; SOCUNCTY = social uncertainty; MRKUNCTY = market uncertainty; DIV = Diversification (dummy variable; related and unrelated diversified strategy=1, single business strategy =0); SIZE = size (log fte); EXTRMFTG = extraction and manufacturing industry (dummy variable); PUBLUTIL = public utilities (dummy variable); CONSTRBU = construction and building industry (dummy variable); TRANSPCO = transport and communication (dummy variable); FINSERV = financial services industry (dummy variable); OTHSERV = other (non-financial) services industry (dummy variable).

The correlation matrix in table 6 reveals that several variables appear to be related, yet that multicollinearity is not likely to be a problem. First of all, all performance variables (AGREFF, ROE and ROA) are correlated at the 1% level ($\rho>0.34$, $p<0.01$). Second, SCBP is not associated with (neither perceived nor objective) performance. Third, SCBP is positively correlated with financial uncertainties, size and the financial services industry ($p<0.01$) yet negatively with the extraction and manufacturing industry ($p<0.05$). Also, most uncertainties are correlated at the 10% level ($\rho>0.13$, $p<0.10$), with the exception of the correlations between social uncertainty and input uncertainty respectively market uncertainty. Some uncertainties are correlated with the industry dummy variables (for example, input uncertainty is positively correlated with extraction and manufacturing industry yet negatively correlated with financial services industry; financial uncertainty is positively correlated with financial services industries yet negatively correlated with non-financial services industry). Finally, size is positively related to a diversification strategy, yet this result is not significant ($p>0.10$).

Multiple linear regression is used to test the propositions stated previously. Table 7 presents the results for this analysis.

The hypothesis stated previously is only partially confirmed: the results of the analysis in table 7a indicate that SCBP in general is not associated with perceived performance. In other words, respondents do not feel that SCBP help to improve performance. However, SCBP is in some cases positively associated with performance if interaction effects are included (eg. SCBP has a positive effect on performance if the interaction for SCBP and primary/manufacturing, respectively financial services industry is included). In addition, the results suggest that the use of SCBP in the construction and building industry is positively associated with perceived performance while the use of SCBP in the financial services industry hampers performance (interaction effects are significant at the 10% level or below). Finally, a regression that includes all interaction effects (non-tabulated) provides similar results as those presented previously.

The results in table 7b and 7c indicate that objective performance (ROE and ROA) is mostly determined by industry. In addition, table 7b indicates that the use of SCBP is not (directly) associated with performance; however, the interaction effects suggest that SCBP is positively associated with ROE in the primary/manufacturing industry, yet negatively associated with ROE in the transport/communication industry. Finally, table 7b also indicates that SCBP is not directly associated with ROA; however, the use of SCBP in public utilities is positively associated with ROA. Jointly, these results suggest that the effectiveness of SCBP depends on industry characteristics (rather than on specific uncertainties).

5. Conclusion

Finance and strategic management theory suggest that organizations (should) react to specific uncertainties in the investment decision by adopting SCBP (i.e., ROR and/or GT principles; see Smit, 2003; Zhu & Weyant, 2003; Bowman & Hurry, 1993). In addition, previous literature suggests that SCBP may be more useful in specific industries (see McGrath & Nerkar, 2004; Zhu & Weyant, 2003; Nichols, 1994; Brennan & Schwartz, 1992). The findings in this study are only partially in line with these theoretic notions; one reason may be that the use of ROR and GT is not widespread. The empirical results indicate that there is hardly a direct relation between the use of SCBP and performance. In addition, the interaction effects between SCBP and uncertainties are not significant suggesting that a match between SCBP and uncertainties not necessarily increases performance. However, the results also indicate that industry characteristics may affect the effectiveness of SCBP: I find some effects of a positive relation between performance and the interaction effect of industry and SCBP. Specifically, SCBP appear to be effective in the primary/manufacturing industry, public utilities and building/construction industry yet seem to harm effectiveness in the transport/communications and financial services industry.

The impact of industry may relate to its characteristics (for example, production type or intensity of competition) that make administrative innovations like GT and/or ROR more suitable to implement (innovation diffusion theory; see for example Abrahamson, 1991). Another reason may be that these industries are used to dealing with uncertainty through option-like or game theory analyses (Billington *et al.*, 2003). Finally, the impact of investments in (fixed) assets on the financial performance and stability of the firm may determine the effectiveness of SCBP in these industries (cf. Segelod, 1998). Additional research may investigate the specific characteristics that affect the effectiveness of SCBP for (companies in) these industries.

Like all research projects, this study has several potential limitations. First of all, all problems associated with survey research (potentially skewed responses, inadequately informed respondents, etc) may affect the results; however, non-response bias tests should have mitigated these problems. Second, the control factors (industry and diversification) have been measured by rather crude measures while other measurement instruments have been adapted in order to explore the relations of interest. For example, the SCBP measure is 'extended' from the instrument used in previous studies (capital budgeting decision rules) to include uncertainty analysis and uncertainty adjustment tools. In case of uncertainty, the instrument has been adapted to capture aspects previously discussed in theory. Although these measures have been validated where possible, further research is necessary along these lines. Finally, the research project relies on perceptions rather than 'hard data'. As a result, we do not know whether the answers on the use of real options are based on using 'actual models' or 'intuitive adjustments' (Busby & Pitts, 1997).

There are a number of possibilities for future research that result from this project. First of all, case studies may reveal how companies use ROR and/or GT to deal with specific uncertainties in investment projects. Such case studies may reveal important insights on how specific companies profit from the application of such techniques, as well as on the specific firm-characteristics that may determine the effectiveness of these techniques. A second avenue for future research is to investigate for what specific investment decisions (eg. expansion, replacements, takeovers, etc) ROR and/or GT are used, and whether such practices are effective. Finally, empirical research may examine the interrelations between operational, financial and strategic risk management and capital budgeting practices. Such an analysis may reveal which management practices (insurance, capital budgeting, derivatives, etc.) are the most efficient for managing specific uncertainties.

References

Abrahamson, E., 1991. Managerial fads and fashions: the diffusion and rejection of innovations, *Academy of Management Review*, **16**, 586-612.

Aggarwal, R., 1980. Corporate Use of Sophisticated Capital Budgeting Techniques: A Strategic Perspective and a Critique of Survey Results, *Interfaces*, **10**, 2, 31-34.

Billington, C., B. Johnson & A. Triantis, 2003. A real options perspective on supply chain management in high technology, *Journal of Applied Corporate Finance*, **15**, 2, 32-43.

Bowman, E.H., & D. Hurry, 1993. Strategy through the option lens: an integrated view of resource investments and the incremental-choice process, *Academy of Management Review*, **18**, 4, 760-782.

Bowman, E.H., & G.T. Moskowitz, 2001. Real Options Analysis and Strategic Decision Making, *Organizations Science*, **12**, November-December, 772-777.

Brennan, M.J., & E.S. Schwartz, 1992. A New Approach to Evaluating Natural Resource Investments. From: *The Revolution in Corporate Finance*, 2nd edition. By J.M. Stern & D.H. Chew (editors). Oxford, UK, Blackwell Publishers, 107-117.

Brounen, D., A. De Jong, and K. Koedijk. 2004. Corporate finance in Europe: confronting theory with practice. *Financial Management* **33** (4):71-101.

Busby, J.S., & C.G.C. Pitts, 1997. Real options in practice: an exploratory survey of how finance officers deal with flexibility in capital appraisal, *Management Accounting Research*, **8**, 169-186.

Chatterjee, S., R.M. Wiseman, A. Fiegenbaum & C.E. Devers, 2003. Integrating Behavioural and Economic Concepts of Risk into Strategic Management: the Twain Shall Meet, *Long Range Planning*, **36**, 61-79.

Chen, S. 1995. An empirical examination of capital budgeting techniques: impact of investment types and firm characteristics. *Engineering economist* **40** (2):145-170.

Christie, A.A., M.P Joye & R.L. Watts, 2003. Decentralization of the Firm: Theory and Evidence, *Journal of Corporate Finance*, **9**, issue 1, 3-37.

Dixit, A.K., & R.S. Pindyck, 1994. *Investment under Uncertainty*. Princeton, New Jersey Princeton University Press.

Farragher, E.J., R.T. Kleiman & A.P. Sahu, 2001. The Association between the Use of Sophisticated Capital Budgeting Practices and Corporate Performance, *Engineering Economist*, **46**, 4, 300-311.

Graham, J. R., and C. R. Harvey. 2001. The theory and practice of corporate finance: evidence from the field. *Journal of Financial Economics* **60**:187-243.

Hair, J.F., jr, R.E. Anderson, R.L. Tatham & W.C. Black, 1998. *Multivariate Data Analysis*, 5th edition, Upper Saddle River, NJ, Prentice Hall.

Haka, S.F. 1987. Capital Budgeting Techniques and Firm Specific Contingencies: A Correlational Analysis, *Accounting, Organizations and Society*, **12**, 1, 31-48.

Haka, S.F., L.A. Gordon & G.E. Pinches, 1985. Sophisticated Capital Budgeting Techniques and Firm Performance, *Accounting Review*, **60**, 4, 651-669.

Ho, S. S. M. 1992. The impact of using risk analysis in capital budgeting on earnings performance: the UK experience. *International journal of accounting* **27**:1-14.

Ho, S., & R. Pike, 1996. Computer decision support for capital budgeting: some empirical findings of practice, *Journal of Information Technology*, 1996, 11, 119-128.

Ho, S.S.M., & R.H. Pike, 1998. Organizational Characteristics Influencing the Use of Risk Analysis in Strategic Capital Investments, *Engineering Economist*, **43**, 3, 247-268.

Hofer, C.W., & D. Schendel, 1978. *Strategy Formulation: Analytical Concepts*, St. Paul/New York, West Publishing Company.

Kim, S.H., 1982. An Empirical Study on the Relationship between Capital Budgeting Practices and Earnings Performance, *Engineering Economist*, **27**, 3, 185-196.

Klammer, T. 1973. The association of capital budgeting techniques with firm performance. *The accounting review* **48** (2):353-364.

Klammer, T., B. Koch & N. Wilner, 1991. Capital Budgeting Practices - A Survey of Corporate Use, *Journal of Management Accounting Research*, **3**, Fall, 113-130.

McGrath, R.G., & A. Nerkar, 2004. Real Options Reasoning and a new look at the R&D Investment strategies of pharmaceutical firms, *Strategic Management Journal*, **25**, 1-21.

McGrath, R.G., W.J. Ferrier & A.L. Mendelow, 2004. Response: Real Options as Engines of Choice and Heterogeneity, *Academy of Management Journal*, **29**, 1, 86-101.

Miller, K.D., & H.G. Waller, 2003. Scenarios, Real Options and Integrated Risk Management, *Long Range Planning*, **36**, 93-107.

Miller, K.D., 1992. A Framework for Integrated Risk Management in International Business, *Journal of International Business Studies*, **23**, 2, Second Quarter, 311-331.

Mukherjee, T.K., & G.V. Henderson, 1987. The Capital Budgeting Process: Theory and Practice, *Interfaces*, **17**, 2, March-April, 78-90.

Murto, P., & J. Keppo, 2002. A Game Model of Irreversible Investment under Uncertainty, *International Game Theory Review*, **4**, 2, 127-140.

Nichols, N.A., 1994. Scientific Management at Merck: An Interview with CFO Judy Lewent, *Harvard Business Review*, January-February, 89-99.

Northcott, D., 1992. *Capital Investment Decision-Making*, London, Academic Press, Harcourt Brace Jovanovich, Publishers.

Pennings, J.M., H. Barkema & S. Douma, 1994. Organizational Learning and Diversification, *Academy of Management Journal*, **37**, 3, 608-640.

Pike, R. H. 1984. Sophisticated capital budgeting systems and their association with corporate performance. *managerial and decision economics* **5** (2):91-97.

Rogers, E.M., 1995. *Diffusion of Innovations*, 4th Edition, New York, Free Press.

Ross, D., 2003. Game Theory. *The Stanford Encyclopedia of Philosophy (Summer 2003 Edition)*, Edward N. Zalta (ed.), http://plato.stanford.edu/archives/sum2003/entries/game-theory.

Sangster, A., 1993. Capital Investment Appraisal Techniques: A Survey of Current Usage, *Journal of Business Finance & Accounting*, **20**, 3, April, 307-332.

Schall, L.D., & G.L. Sundem, 1980. Capital Budgeting Methods and Risk: A Further Analysis, *Financial Management*, **9**, Spring, 7-11.

Segelod, E., 1998. Capital Budgeting in a Fast-Changing World, *Long Range Planning*, **31**, 4, 529-541.

Simerly, R.L., & M. Li, 2000. Environmental Dynamism, Capital Structure and Performance: A Theoretical Integration and an Empirical Test, *Strategic Management Journal*, **21**, 31-49.

Smit, H.T.J., & L.A. Ankum, 1993. A Real Options and Game-Theoretic Approach to Corporate Investment Strategy Under Competition, *Financial Management*, **22**, 3, Autumn, 241-250.

Smit, H.T.J., 2003. Infrastructure Investment as a Real Options Game: The Case of European Airport Expansion, *Financial Management*, **32**, 4, Winter 2003, 27-57.

Trigeorgis, L., 1993. Real Options and Interactions with Financial Flexibility, *Financial Management*, **22**, 3, Autumn 1993, p. 202-224.

Verbeeten, F. H. M. 2006. Do organizations adopt sophisticated capital budgeting practices to deal with uncertainty in the investment decision? A research note. *Management Accounting Research* **17**:106-120.

Wallace, R.S.O., & C.J. Mellor, 1988. Nonresponse Bias in Mail Accounting Surveys: A Pedagogical Note, *British Accounting Review*, **20**, 131-139.

Zhu, K., & J.P. Weyant, 2003. Strategic Decisions of New Technology Adoption under Asymmetric Information: A Game-Theoretic Model, *Decision Sciences*, **34**, 4, Fall, 643-675.

In: Game Theory: Strategies, Equilibria and Theorems
Editors: Ingrid N. Haugen et al, pp. 289-296
ISBN: 978-1-60456-844-8

Chapter 11

A New Type of Irresoluteness via Topological Ideals

***J. Bhuvaneswari*[1,a] *and N. Rajesh*[2,b]**
[1]Department of Computer Applications, Rajalakshmi Engineering College, Thandalam, Chennai-602 105, Tamilnadu, India
[2]Department of Mathematics, Kongu Engineering College, Perundurai, Erode-638 052, Tamilnadu, India

Abstract

The aim of this paper is to give a new class of functions called θ-β-I-irresolute functions in ideal topological space. Some characterizations and several basic properties of this class of functions are obtained.

Key words and phrases. Ideal topological spaces, β-I-open sets.

2000 Mathematics Subject Classification. 54C05.

1. Introduction

The subject of ideals in topological spaces has been studied by Kuratowski [4] and Vaidyanathasamy [7]. An ideal I on a topological space (X, τ) is a nonempty collection of subsets of X which satisfies (i) $A \in I$ and $B \subset A$ implies $B \in I$ and (ii) $A \in I$ and $B \in I$ implies $A \cup B \in I$. Given a topological space (X, τ) with an ideal I on X and if P(X) is the set of all subsets of X, a set operator $(.)^*$: $P(X) \rightarrow P(X)$, called the local function [7] of A with respect to τ and I, is defined as follows: for $A \subset X$, $A^*(I, \tau) = \{x \in X \mid U \cap A \notin I\}$ for every $U \in \tau(x)\}$ where $\tau(x) = \{U \in \tau \mid x \in U\}$. A Kuratowski closure operator $Cl^*(.)$ for a topology

[a] E-mail address: sai_jbhuvana@yahoo.co.in
[b] E-mail address: nrajesh_topology@yahoo.co.in

$\tau^*(I, \tau)$ called the *-topology, finer than τ is defined by $Cl^*(A) = A \cup A^*(I, \tau)$ when there is no chance of confusion, $A^*(I)$ is denoted by A^*. If I is an ideal on X, then (X, I, τ) is called an ideal space. By a space, we always mean a topological space (X, τ) with no separation properties assumed. If $A \subset X$, Cl(A) and Int(A) will denote the closure and interior of A in (X, τ), respectively. A subset S of an ideal space (X, τ, I) is said to be I-open if $S \subset Int(S^*)$. The family of all I-open sets of (X, τ, I) is denoted by IO(X). The aim of this paper is to give a new class of functions called θ-β-I-irresolute functions in ideal topological space. Some characterizations and several basic properties of this class of functions are obtained.

2. Preliminaries

A subset S of an ideal topological space (X, τ, I) is β-I-open [3] (resp. α-I-open [3], α-open [6]) if $S \subset Cl(Int(Cl^*(S)))$ (resp. $S \subset Int(Cl^*(Int(S)))$). The complement of a β-I-open set is called β-I-closed [3]. The intersection of all β-I-closed sets containing S is called the β-I-closure of S and is denoted by $_{\beta I}Cl(S)$. The β-I-Interior of S is defined by the union of all β-I-open sets contained in S and is denoted by $_{\beta I}Int(S)$. A subset S of an ideal space (X, τ, I) is said to be β-I-regular [9] if it is β-I-open and β-I-closed. The family of all β-I-regular (resp. β-I-open, β-I-closed, α-closed) sets of (X, τ, I) is denoted by β IR(X) (resp. βIO(X), βIC(X), αO(X)). The family of all β-I-regular (resp. β-I-open, β-I-closed) sets of (X, τ, I) containing a point $x \in X$ is denoted by βIR(X, x) (resp. βIO(X, x), βIC(X, x)). A point $x \in X$ is called the β-I-θ-cluster point of S if $_{\beta I}Cl(U) \cap S \neq \phi$ for every β-I-open set U of (X, τ, I) containing x. The set of all β-I-θ-cluster points of S is called the β-I-closure of S and is denoted by $_{\beta I}Cl_\theta(S)$. A subset S is said to be β-I-θ-closed set is said to be β-I-θ-open. A point $x \in X$ is called β-I-θ-interior point of S if there exists a β-I-regular set U of X containing x such that $x \in U \subset S$. The set of all β-I-θ-interior of S and is denoted by $_{\beta I}Int_\theta(S)$.

Definition 2.1 [8] An ideal space (X, τ, I) is said to be β-I-regular if for each closed set F and each $x \notin F$, there exist disjoint β-I-open sets U and V such that $x \in U$ and $F \subset V$.

Theorem 2.2 [8] For an ideal space (X, τ, I), the following properties are equivalent:

(i) X is β-I-regular;

(ii) For each open set U and each $x \in U$, there exists $V \in \beta IO(X)$ such that $x \in V \subset {_{\beta I}Cl(V)} \subset U$;

(iii) For each open set U and each $x \in U$, there exists $V \in \beta IR(X)$ such that $x \in V \subset U$.

Definition 2.3 A function f: (X, τ, I) → (Y, σ, J) is said to be β^*-I-irresolute [8] if $f^{-1}(V) \in \beta JO(X)$ for every $V \in \beta IO(Y)$.

3. θ-β-I-Irresolute Functions

We have introduced the following definition

Definition 3.1 A function f: (X, τ, I) → (Y, σ, J) is said to be θ-β-I-irresolute if for each x ∈ X and each V ∈ βIO(Y, f(x)), there exists U ∈ βJO(X, x) such that $f({}_{\beta I}Cl(U)) \subset {}_{\beta J}Cl(V)$.

Clearly, every β^*-I-irresolute function is θ-β-I-irresolute. But the converse is not true as shown by the following example.

Example 3.2 Let X = {a,b,c}, τ = {ϕ, {a}, X}, σ = {ϕ, {b}, Y} and I = {ϕ, {a}}. Clearly the identity function f: (X, τ, I) → (Y, σ, I) is θ-β-I-irresolute but not β^*-I-irresolute.

Theorem 3.3 For a function f: (X, τ, I) → (Y, σ, J) the following properties are equivalent:

(i) f is θ-β-I-irresolute;

(ii) ${}_{\beta I}Cl_\theta(f^{-1}(B)) \subset f^{-1}{}_{\beta J}l_\theta(B))$ for every subset B of Y;

(iii) $f({}_{\beta I}Cl_\theta(A)) \subset {}_{\beta J}Cl_\theta(f(A))$ for every subset A of X.

Proof. (i) ⇒ (ii): Let B be any subset of Y. Suppose that $x \notin f^{-1}({}_{\beta J}Cl_\theta(B))$. Then $f(x) \notin {}_{\beta J}Cl_\theta(B)$ and there exits V ∈ βJO(Y, f(x)) such that ${}_{\beta I}Cl(V) \cap B = \phi$. Since f is θ-β-I-irresolute, there exists U ∈ βIO(X, x) such that $f({}_{\beta I}Cl(U)) \subset {}_{\beta J}Cl(V)$. Therefore, we have $f({}_{\beta I}Cl(U)) \cap B = \phi$ and ${}_{\beta I}Cl(U) \cap f^{-1}(B) = \phi$. This shows that $x \notin {}_{\beta I}Cl_\theta(f^{-1}(B))$. Hence, we obtain ${}_{\beta I}Cl_\theta(f^{-1}(B)) \subset f^{-1}({}_{\beta J}Cl_\theta(B))$.

(ii) ⇒ (iii): Let A be any subset of X. Then we have ${}_{\beta I}Cl_\theta(A) \subset {}_{\beta I}Cl_\theta(f^{-1}(f(A))) \subset f^{-1}({}_{\beta J}Cl_\theta(f(A)))$ and hence $f({}_{\beta I}Cl_\theta(A)) \subset {}_{\beta J}Cl_\theta(f(A))$.

(iii) ⇒ (ii): Let B be a subset of Y. We have $f({}_{\beta I}Cl_\theta(f^{-1}(B))) \subset {}_{\beta J}Cl_\theta(f(f^{-1}(B))) \subset {}_{\beta I}Cl_\theta(B)$ and hence ${}_{\beta I}Cl_\theta(f^{-1}(B)) \subset f^{-1}({}_{\beta J}Cl_\theta(B))$.

(ii) ⇒ (i): Let x ∈ X and V ∈ βJO(Y, f(x)). Then we have ${}_{\beta J}Cl(V) \cap (Y - {}_{\beta J}Cl(V)) = \phi$ and $f(x) \notin {}_{\beta J}Cl_\theta(Y - {}_{\beta J}Cl(V))$. Hence, $x \notin f^{-1}({}_{\beta J}Cl_\theta(Y - {}_{\beta J}Cl(V)))$ and $x \notin {}_{\beta J}Cl_\theta(f^{-1}(Y - {}_{\beta J}Cl(V)))$. There exists U ∈ βIO(X, x) such that ${}_{\beta I}Cl(U) \cap f^{-1}(Y - {}_{\beta J}Cl(V)) = \phi$ and hence $f({}_{\beta I}Cl(U)) \subset {}_{\beta J}Cl(V)$. This shows that f is θ-β-I-irresolute.

Theorem 3.4 For a function f: (X, τ, I) → (Y, σ, J) the following properties are equivalent:

(i) f is θ-β-I-irresolute;

(ii) $f^{-1}(V) \subset {}_{\beta I}Int_\theta(f^{-1}({}_{\beta J}Cl(V)))$ for every V ∈ βJO(Y).

(iii) ${}_{\beta I}Cl_\theta(f^{-1}(V)) \subset f^{-1}({}_{\beta J}Cl(V))$ for every $V \in \beta JO(Y)$.

Proof. (i) ⇒ (ii): Suppose that $V \in \beta JO(Y)$ and $x \in f^{-1}(V)$. Then $f(x) \in V$ and there exists $U \in \beta IO(X, x)$ such that $f({}_{\beta I}Cl(U)) \subset {}_{\beta J}Cl(V)$. Therefore, $x \in U \subset {}_{\beta I}Cl(U) \subset f^{-1}({}_{\beta J}Cl(V))$. This shows that $x \in {}_{\beta I}Int_\theta(f^{-1}({}_{\beta I}Cl(V)))$. This shows that $f^{-1}(V) \subset {}_{\beta I}Int_\theta(f^{-1}({}_{\beta J}Cl(V)))$.

(ii) ⇒ (iii): Suppose that $V \in \beta JO(Y)$ and $x \notin f^{-1}({}_{\beta J}Cl(V))$. Then $f(x) \notin {}_{\beta J}Cl(V)$ and there exists $U \in \beta JO(Y, f(x))$ such that $U \cap V = \phi$ and hence ${}_{\beta J}Cl(U) \cap V = \phi$. Therefore, we have $f^{-1}({}_{\beta J}Cl(U)) \cap f^{-1}(V) = \phi$. Since $x \in f^{-1}(U)$, by (ii), $x \in {}_{\beta I}Int_\theta(f^{-1}({}_{\beta J}Cl(U)))$. There exists $W \in \beta IO(X, x)$ such that ${}_{\beta I}Cl(W) \subset f^{-1}({}_{\beta J}Cl(U))$. Thus, we have ${}_{\beta I}Cl(W) \cap f^{-1}(V) = \phi$ and hence $x \notin {}_{\beta I}Cl_\theta(f^{-1}(V))$. This shows that ${}_{\beta I}Cl_\theta(f^{-1}(V)) \subset f^{-1}({}_{\beta J}Cl(V))$.

(iii) ⇒ (i): Suppose that $x \in X$ and $V \in \beta JO(Y, f(x))$. Then, $V \cap (Y- {}_{\beta J}Cl(V)) = \phi$ and $f(x) \notin {}_{\beta J}Cl(Y- {}_{\beta J}Cl(V))$. Therefore, $x \notin f^{-1}({}_{\beta J}Cl(Y- {}_{\beta J}Cl(V)))$ and by (iii), $x \notin {}_{\beta I}Cl_\theta(f^{-1}(Y- {}_{\beta J}Cl(V)))$. There exists $U \in \beta IO(X, x)$ such that ${}_{\beta I}Cl(U) \cap f^{-1}(Y- {}_{\beta J}Cl(V)) = \phi$. Therefore, we obtain $f({}_{\beta I}Cl(U)) \subset {}_{\beta J}Cl(V)$. This shows that f is θ-β-I-irresolute.

Definition 3.5 A function $f: (X, \tau, I) \rightarrow (Y, \sigma, J)$ is said to be quasi β-I-irresolute [2] if for each point $x \in X$ and each $V \in \beta IO(Y, f(x))$, there exists $U \in \beta IO(X, x)$ such that $f({}_{\beta I}Cl(U)) \subset V$.

Theorem 3.6 Let (Y, σ, J) be a J-regular space. Then for a function $f: (X, \tau, I) \rightarrow (Y, \sigma, J)$ the following properties are equivalent:

(i) f is quasi β-I-irresolute;
(ii) f is β^*-I-irresolute;
(iii) f is θ-β-I-irresolute.

Proof. (i) ⇒ (ii): This is obvious.

(ii) ⇒ (iii): Suppose that $x \in X$ and $V \in \beta JO(Y, f(x))$. Since f is β^*-I-irresolute, $f^{-1}(V)$ is β-I-open and $f^{-1}({}_{\beta J}Cl(V))$ is β-I-closed in X. Now, put $U = f^{-1}(V)$. Then we have $U \in \beta IO(X, x)$ and ${}_{\beta I}Cl(U) \subset f^{-1}({}_{\beta J}Cl(V))$. Therefore, we obtain $f({}_{\beta I}Cl(U)) \subset {}_{\beta J}Cl(V)$. This shows that f is θ-β-I-irresolute.

(iii) ⇒ (i): Suppose that $x \in X$ and $V \in \beta JO(Y, f(x))$. Since Y is β-J-regular, there exists $W \in \beta JO(Y)$ such that $f(x) \in W \subset {}_{\beta I}Cl(W) \subset V$. Since f is θ-β-I-irresolute, there exists $U \in \beta IO(X, x)$ such that $f({}_{\beta I}Cl(U)) \subset {}_{\beta J}Cl(W) \subset V$. This shows that f is quasi β-I-irresolute.

Theorem 3.7 Let (X, τ, I) be a β-I-regular space. Then $f: (X, \tau, I) \rightarrow (Y, \sigma, J)$ is β-θ-I-irresolute if and only if it is weakly β-I-irresolute.

Proof. Suppose that f is weakly β-I-irresolute. Let $x \in X$ and $V \in \beta JO(Y, f(x))$. Then, there exists $U \in \beta IO(X, x)$ such that $f(U) \subset {}_{\beta J}Cl(V)$. Since X is β-I-regular, there exists $U_0 \in$

$\beta IO(X, x)$ such that $x \in U_0 \subset {}_{\beta I}Cl(U_0) \subset U$. Therefore, we obtain $f({}_{\beta I}Cl(U_0)) \subset {}_{\beta J}Cl(V)$. This shows that f is θ-β-I-irresolute.

Lemma 3.8 [8] Let A and X_0 be subsets of an ideal space (X, τ, I).

(i) If $A \in \beta IO(X)$ and $X_0 \in \alpha O(X)$, then $A \cap X_0 \in \beta IO(X_0)$;

(ii) If $A \in \beta IO(X_0)$ and $X_0 \in \alpha O(X)$, then $A \in \beta IO(X)$.

Lemma 3.9 Let A and X_0 be subsets of an ideal space (X, τ, I) such that $A \subset X_0 \subset X$. Let ${}_{\beta I}Cl_{X_0}(A)$ denote the β-I-closure of A with respect to the subspace X_0.

(i) If X_0 is α-open in X, then ${}_{\beta I}Cl_{X_0}(A) \subset {}_{\beta I}Cl(A)$;

(ii) If $A \in \beta IO(X_0)$ and $X_0 \in \alpha O(X)$, then ${}_{\beta I}Cl(A) \subset {}_{\beta I}Cl_{X_0}(A)$.

Proof. Follows from Lemma 3.8.

Theorem 3.10 If $f: (X, \tau, I) \to (Y, \sigma, J)$ is θ-β-I-irresolute and X_0 is an α-open subset of X, then the restriction $f_{|X_0}: X_0 \to Y$ is θ-β-$I_{|X_0}$-irresolute.

Proof. For any $x \in X_0$ and any $V \in \beta JO(Y, f(x))$, there exists $U \in \beta IO(X, x)$ such that $f({}_{\beta I}Cl(U)) \subset {}_{\beta J}Cl(V)$ since f is θ-β-I-irresolute. Let $U_0 = U \cap X_0$, then by Lemmas 3.8 and 3.9, $U_0 \in \beta IO(X_0, x)$ and ${}_{\beta I}Cl_{X_0}(U_0) \subset {}_{\beta I}Cl(U_0)$. Therefore, we obtain $(f_{|X_0})({}_{\beta I}Cl_{X_0}(U_0)) = f({}_{\beta I}Cl_{X_0}(U_0)) \subset f({}_{\beta I}Cl(U_0)) \subset f({}_{\beta I}Cl(U)) \subset {}_{\beta J}Cl(V)$. This shows that $f_{|X_0}$ is θ-β-$I_{|X_0}$-irresolute.

Theorem 3.11 A function $f: (X, \tau, I) \to (Y, \sigma, J)$ is θ-β-I-irresolute if for each $x \in X$ there exists an α-I-open set X_0 of X containing x such that the restriction $f_{|X_0}: X_0 \to Y$ is θ-β-$I_{|X_0}$-irresolute.

Proof. Let $x \in X$ and $V \in \beta JO(Y, f(x))$. There exists an α-I-open set X_0 of X containing x such that $f_{|X_0}: X_0 \to Y$ is θ-β-$I_{|X_0}$-irresolute. Thus, there exists $U \in \beta IO(X_0, x)$ such that $(f_{|X_0})({}_{\beta I}Cl_{X_0}(U)) \subset {}_{\beta J}Cl(V)$. By Lemmas 3.8 and 3.9, $U \in \beta IO(X, x)$ and ${}_{\beta I}Cl(U) \subset {}_{\beta I}Cl_{X_0}(U)$. Hence, we have $f({}_{\beta I}Cl(U)) = (f_{|X_0})({}_{\beta I}Cl(U)) \subset (f_{|X_0})\, {}_{\beta I}Cl_{X_0}(U)) \subset {}_{\beta I}Cl(V)$. This shows that f is θ-β-I-irresolute.

Corollary 3.12 Let $\{U_\alpha: \alpha \in \wedge\}$ be an α-I-open cover of an ideal space (X, τ, I). A function $f: (X, \tau, I) \to (Y, \sigma, J)$ is θ-β-I-irresolute if and only if the restriction $f_{|U_\alpha}: U_\alpha \to Y$ is θ-β-$I_{|X_0}$-irresolute for each $\alpha \in \wedge$.

Proof. The proof follows from Theorems 3.10 and 3.11.

Definition 3.13 A function f: (X, τ, I) → (Y, σ, J) is said to be weakly β-irresolute [8] if for each x ∈ X and each V ∈ βIO(Y, f(x)), there exists U ∈ βIO(X, x) such that f(U) ⊂ $_{\beta I}$Cl(V).

Theorem 3.14 Let f: (X, τ, I) → (Y, σ, I), g: (Y, σ, I) → (Z, η, I) be functions and g ° f: X → Z be the composition. Then the following properties hold:

(i) If f is θ-β-I-irresolute and g is θ-β-J irresolute;
(ii) If f is strongly β-I-irresolute and g is weakly β-I-irresolute, then g ° f is θ-β-I-irresolute;
(iii) If f is weakly β-I-irresolute and g is θ-β-I-irresolute, then g ° f is weakly β-I-irresolute;
(iv) If f is θ-β-I-irresolute and g is strongly β-I-irresolute, then g ° f is strongly β-I-irresolute.

Proof. The proof follows from the definitions.

Definition 3.15 An ideal space (X, τ, I) is said to be β-I-T_2 [8] if for each pair of distinct points x and y in X, there exists U ∈ βIO(X, x) and V ∈ βIO(X, y) such that $_{\beta I}$Cl(U) ∩ $_{\beta J}$Cl(V) = ϕ.

Recall that for a function f: (X, τ, I) → (Y, σ, J), the subset {(x, f(x)): x ∈ X} of X x Y is called the graph of f and is denoted by G(f).

Definition 3.16 The graph G(f) of a function f: (X, τ, I) → (Y, σ, J) is said to be strongly β-I-closed (resp. β-θ-I-closed) if for each (x, y) ∈ (X x Y)-G(f), there exists U ∈ βIO(X, x) and V ∈ βJO(Y, y) such that ($_{\beta I}$Cl(U) x V) ∩ G(f) = ϕ (resp. ($_{\beta I}$Cl(U) x $_{\beta J}$Cl(V)) ∩ G(f) = ϕ).

Lemma 3.17 The graph G(f) of a function f: (X, τ, I) → (Y, σ, J) is β-θ-I-closed in X x Y if and only if for each point (x, y) ∈ (X x Y) - G(f), there exist U ∈ βIO(X, x) and V ∈ βIO(Y, y) such that f($_{\beta I}$Cl(U)) ∩ $_{\beta J}$Cl(V) = ϕ.

Proof. The proof follows from the definitions.

Theorem 3.18 If f: (X, τ, I) → (Y, σ, J) is θ-β-I-irresolute and Y is β-J-T_2, then G(f) is β-θ-I-closed in X x Y.

Proof. Let (x, y) ∈ (X x Y)-G(f). It follows that f(x) ≠ y. Since Y is β-J-T_2, there exist β-J-open sets say V and W in Y containing f(x) and y, respectively, such that $_{\beta J}$Cl(V) ∩ $_{\beta J}$Cl(W) = ϕ. Since f is θ-β-I-irresolute, there exists U ∈ βIO(X, x) such that f($_{\beta I}$Cl(U)) ⊂ $_{\beta J}$Cl(V). Therefore, f($_{\beta I}$Cl(U)) ∩ $_{\beta J}$Cl(W) = ϕ and by Lemma 3.17, G(f) is β-θ-I-closed in X x Y.

Theorem 3.19 If f: (X, τ, I) → (Y, σ, J) is quasi β-I-irresolute and Y is β-J-T_2, then G(f) is β-θ-closed in X x Y.

Proof. Let $(x, y) \in (X \times Y) - G(f)$. It follows that $f(x) \neq y$. Since Y is β-J-T_2, there exist β-J-open sets V and W in Y containing f(x) and y, respectively, such that $V \cap W = \phi$ and hence $V \cap {}_{\beta J}Cl(W) = \phi$. Since f is strongly β-I-irresolute, there exists $U \in \beta IO(X, x)$ such that $f({}_{\beta I}Cl(U)) \subset V$. Therefore, $f({}_{\beta I}Cl(U)) \cap {}_{\beta J}Cl(W) = \phi$ and by Lemma 3.17, G(f) is β-θ-I-closed in X x Y.

Theorem 3.20 Let f, g: (X, τ, I) → (Y, σ, J) be functions. If G(f) is β-θ-I-closed and g is θ-β-I-irresolute, then the set $\{(x_1, x_2):f(x_1) = g(x_2)\}$ is β-θ-I^2-closed in the product space X x X.

Proof. Let $A = \{(x_1, x_2):f(x_1) = g(x_2)\}$. Suppose $(x_1, x_2) \notin A$. Then $f(x_1) \neq g(x_2)$ and hence $(x_1, g(x_2)) \notin G(f)$. Since G(f) is β-θ-I-closed, there exist $U \in \beta IO(X, x_1)$ and $W \in \beta IO(Y, g(x_2))$ such that $f({}_{\beta I}Cl(U)) \cap {}_{\beta J}Cl(W) = \phi$. Since g is θ-β-J-irresolute, there exists $U_0 \in \beta IO(X, x_2)$ such that $g({}_{\beta I}Cl(U_0)) \subset {}_{\beta J}Cl(W)$ and hence $f({}_{\beta I}Cl(U)) \cap g({}_{\beta I}Cl(U_0)) = \phi$. Therefore, we obtain $({}_{\beta I}Cl(U) \times {}_{\beta I}Cl(U_0)) \cap A = \phi$ and hence A is β-θ-I^2-closed.

Theorem 3.21 If f: (X, τ, I) → (Y, σ, J) is a θ-β-I-irresolute function and Y is β-I-T_2, then the subset $A = \{(x, y): f(x)=f(y)\}$ is β-θ-I-closed in X x X.

Proof. Since f is θ-β-I-irresolute and Y is β-I-T_2, by Theorem 3.18, G(f) is β-θ-I-closed. Therefore, by Theorem 3.20, A is β-θ-I-closed.

Definition 3.22 An ideal space (X, τ, I) is said to be:

(i) β-I-closed if every cover of X by β-I-open sets has a finite subcover whose β-I-closures cover X;
(ii) countably β-I-closed if every countable cover of X by β-I-open sets has a finite subcover whose β-I-closures cover X.

A subset K of an ideal space (X, τ, I) is said to be β-I-closed relative to X if for every cover $\{V_\alpha: \alpha \in \wedge\}$ of K by β-I-open sets of X, there exists a finite subset $\wedge_0$ of $\wedge$ such that $K \subset \bigcup\{{}_{\beta I}Cl(V_\alpha): \alpha \in \wedge_0\}$.

Theorem 3.23 If f: (X, τ, I) → (Y, σ, J) is θ-β-I-irresolute function and K is β-I-closed relative to X, then f(K) is β-f(I)-closed relative to Y.

Proof. Suppose that f: (X, τ, I) → (Y, σ, J) is θ-β-I-irresolute and K is β-I-closed relative to X. Let $\{V_\alpha: \alpha \in \wedge\}$ be a cover of f(K) by β-open sets of X. For each point $x \in K$, there exists $\alpha(x) \in \wedge$ such that $f(x) \in V_{\alpha(x)}$. Since f is θ-β-I-irresolute, there exists $U_x \in \beta IO(X, x)$ such that $f({}_{\beta I}Cl(U_x)) \subset {}_{\beta J}Cl(V_{\alpha(x)})$. The family $\{U_x: x \in K\}$ is a cover of K by β-I-open sets of X and hence there exists a finite subset K_1 of K such that $K \subset \bigcup_{x \in K_1} {}_{\beta I}Cl(U_x)$. Therefore, we obtain $f(K) \subset \bigcup_{x \in K_1} {}_{\beta J}Cl(V_{\alpha(x)})$. This shows that f(K) is β-f(I)-closed relative to Y.

Corollary 3.24 If f: $(X, \tau, I) \rightarrow (Y, \sigma, J)$ be a θ-β-I-irresolute surjection. Then the following properties hold:

(i) If X is β-I-closed, then Y is β-f(I)-closed;
(ii) If X is countably β-I-closed, then Y is countably β-f(I)-closed.

References

M. E. Abd El-Monsef, E. F. Lashien and A. A. Nasef, On I-open sets and I-continuous functions, *Kyungpook Math. J.,* **32**(1992), 21-30.

J. Bhuvaneswari and N. Rajesh, *A new type of irresolute functions via ideals* (preprint).

E. Hater and T. Noiri, On decomposition of continuity via idealization, *Acta Math. Hungar.,* **96**(2002), 341-349.

K. Kuratowski, *Topology*, Academic Press, New York, 1966.

N. Levine, A decomposition of continuity in topological spaces, *Amer. Math. Monthly*, **68** (1961), 44-46.

O. Njastad, On some classes of nearly open sets, *Pacific J. Math.*, **15**(1965), 961-970.

R. Vaidyanatahswamy, The localisation theory in set topology, *Proc. Indian Acad. Sci.*, **20** (1945), 51-61.

S. Yuksel, A. Acikgoz and E. Gursel, On weakly β-I-irresolute functions, *Far. East J. Math. Sci.*, **25**(1)(2007), 129-144.

S. Yuksel, A. Acikgoz and E. Gursel, *On β-I-Regular sets*, (submitted).

In: Game Theory: Strategies, Equilibria and Theorems
Editors: Ingrid N. Haugen et al, pp. 297-317
ISBN: 978-1-60456-844-8

Chapter 12

DIVERSITY BY TEMPORAL OSCILLATIONS IN PLANT COMMUNITIES WITH A DIFFERENTIAL TIMING OF REPRODUCTION

***Tessa E. Pronk*[1]*, Heinjo J. During*[2]*, Feike Schieving*[2] *and Marinus J.A. Werger*[2]**

[1] Integrative BioInformatics Unit / Computer Systems Architecture group, Faculty of Science, University of Amsterdam, Kruislaan 403, 1098 SJ Amsterdam, the Netherlands.

[2] Plant Ecology group, Utrecht University, Sorbonnelaan 16, 3584 CA Utrecht, the Netherlands.

Abstract

Background and Aims: Species can coexist at non-equilibrium circumstances, for instance by oscillations in population densities or chaos, caused by non-linear responses of species to their environment. We analyzed whether plant genotypes that vary in their timing of reproduction can coexist under equilibrium or non-equilibrium circumstances when competing for light.

Methods: We used a game theoretical approach, based on a biologically mechanistic model of plant growth.

Key Results: In our model, the genotype switching to reproduction slightly later than its competitor attained a higher fitness. This caused a succession from early switching genotypes to those switching later to reproductive investment. However, there were cyclic opportunities for extinct genotypes that switch early to reproduction to re-establish and grow into the community. The cause was that genotypes that switched very late produced relatively very little seed because of an overinvestment in vegetative growth; especially when competing against individuals of the same genotype. Because the very early switch genotypes could establish, circumstances were such that other extinct switch genotypes could re-enter the vegetation as well. In this way the diversity of genotypes was maintained over time by temporal oscillations of genotype abundances.

Conclusions: We show that within a model, an externally undisturbed plant community can produce its own temporal cyclic or chaotic disturbances to promote diversity, rather than

[1] E-mail address: tepronk@science.uva.nl. Tel.: 00 31 20 5257923. Corresponding Author.
[2] Fax: 00 31 30 2518366 Tel.: 00 31 30 2536843

converge to a stable equilibrium when competing for light. Cyclic fluctuations in species composition can occur in a model community of plants sharing the same growing season and that are limited just by light as a single resource.

Keywords: Game Theory, Light interception, Cyclic, Annuals, Coexistence, Temporal, Mechanistic Model, Oscillations, Reproduction

Introduction

In community ecology, many studies try to unravel the mechanisms by which species-rich communities overcome the competitive exclusion principle (Hardin, 1960). In most of these studies stable coexistence is considered, e.g. in communities or populations that are at equilibrium. As an alternative, however, the non-equilibrium view was proposed (Huisman and Weissing, 1999; Bauer et al., 2002; Nowak and Sigmund, 2004). It was predicted that it is possible that mechanisms that promote coexistence at non-equilibrium will prove to be most important for the maintenance of diversity (Hutchinson, 1961). Aside from regular disturbances that can create circumstances for species to coexist, species can coexist at non-equilibrium circumstances by non-linear responses that cause oscillations in population densities or chaos (Hutchinson, 1961; Huisman and Weissing, 1999; Bauer et al., 2002). So far, well-known examples can be found in the oscillating populations in trophic food-webs such as predator-prey systems (Peet et al., 2005), where predator, prey, and food source are dependent on the other for their survival. Aside from such dependence, non-hierarchical competitive relationships between species can cause species to coexist on a limited number of resources (Huisman and Weissing, 1999). Kerr et al. (2002), for instance, modeled and tested empirically the coexistence of a system of paper-rock-scissors (Maynard Smith, 1982). Species coexisted when interaction and dispersal was local and thus they emphasize, that spatially restricted patterns also may lead to coexistence of species. Similarly, species limited by one resource only may coexist indefinitely if they partition the growing season and each is active only in its "own" part of that season (Armstrong and McGehee, 1976). Huisman and Weissing (1999), on the other hand, found oscillating patterns in their model for well-mixed plankton species, on the basis of competition for three or more different resources. This was under the condition that species differed in their required resource ratio. Within oscillations, circumstances for growth differed and species with different requirements for growth found a window of opportunity. In their study, more species could coexist than there were limiting resources (Huisman and Weissing, 1999; Huisman et al., 2001).

We investigate whether or not in a system with one limiting resource coexistence of species is possible. We study this in virtual, competing, annual plants that differ in the timing of reproduction, just as species do in real communities. In this study we assume that in plants there is a trade-off between the production of vegetative parts and seed production. We also assume that the amounts of seed production and photosynthethic carbon gain are strongly correlated. Model studies have indicated that, to maximize seed mass yield, plants should show a sharp transition between the vegetative and reproductive phases of growth (e.g. Cohen, 1976; Iwasa, 2000). One would assume that the allocation scheme determining reproductive output is subject to strong selection because reproductive output is one of the main determinants of fitness (Stearns, 1992). Under strong selective pressure, all plants might be expected to evolve towards a single or at most a few optimal reproductive allocation schemes (Cohen, 1976; Schmid and Weiner, 1993). However, this is only the case in a system where reproductive performance is not affected by the presence of competitor plants. Competitive optimization sensu Anten (2005) leads to a delayed switch from growth to reproduction. But whether or not this leads to a single switch time from growth to reproduction is still an open question. In natural vegetation stands, however, often large numbers of plants with widely different timings of the transition to reproduction coexist (Reekie et al., 2002).

Because a large investment in seed mass implies a low investment in vegetative mass and hence competitive strength, reproduction and competitiveness are closely related. As plants mostly do grow in close proximity to each other, competition will often play a role in affecting the fitness of plants with different growth strategies (Weiner, 1988). Part of the variability in reproductive allocation between genotypes may therefore stem from the influence of competition.

When, as is the case in the direct competition for light (Anten, 2005), the success of a genotype depends on the strategies of the other genotypes present, game theory is an appropriate tool to assess the possibilities for different genotypes with their particular traits to persist (Falster and Westoby, 2003; Nowak and Sigmund, 2004). Also in the research of timing issues, game theory has often been successfully applied (Iwasa and Levin, 1995). Whereas most modeling studies on the selection for traits concentrate on the dynamics at plant or even population level, we explicitly include allocation patterns within a plant as the basis for competition. Plants thus grow on the basis of intrinsic allocation trade-offs. The size and leaf area distribution of a plant and that of neighboring plants during growth explicitly determines a plant's resource capture (Pronk et al., 2007a,b). This puts each plant into a direct game theoretical context. In addition, on a population level, plants on average will change their seed production as the total composition of the vegetation changes and thus the frequency of encounters with other plant types change. This puts the populations in a game theoretical context as well. Plants in our model are competing for light as a single limiting resource.

An important finding of this study is that competing annual plants that differ in the timing of reproduction do not necessarily converge towards a single evolutionary stable strategy without external disturbances. Populations of plants with different strategies are able to generate their own variability and show cyclic behavior. In this self-organizing model vegetation, multiple genotypes differing in their switch to reproduction find temporal opportunities to perform well so that the diversity of genotypes over a period of time is maintained.

Methods

The Plant Growth Model

To simulate the fitness and growth of competing plants that switch to reproduction at different moments in time, we used a mechanistic plant growth model. The model is described in detail in Appendix A (see also Pronk, 2004). Only a brief description of the most important features of the model is given here. Parameter settings are given in Table A1 in the Appendix.

Plants grow within a fixed area of ten by ten centimeters, referred to as a cell. Only plants that share such a cell compete (i.e. plants in separate cells do not compete). However, plants are able to disperse their seeds into other cells between growing seasons. Plants are annual and die at the end of the growing season. The growing season is 365 days, with a seasonal course of the light climate, as it would occur in a temperate region (Equation A14).

As we want to compare plants with respect to their timing of the switch to reproduction, other characteristics are held equal amongst plants. Each plant starts with an initial leaf area, root mass and height (Table A1 in the Appendix). All plants have a parabolic distribution of leaf area over their height (Caton et al., 1999; Pronk et al., 2007; Equation A1). Light interception is modeled following Beer's law (Monsi and Saeki, 1953; Equations A5-A7 Appendix) and the light response of photosynthesis is described by a non-rectangular hyperbola (Marchall and Biscoe, 1980; Equation A8). To obtain daily gross photosynthesis per plant, leaf photosynthesis is integrated over canopy depth and over the day (Equation A10). Part of the assimilate pool thus obtained is spent on the maintenance respiration of structural mass (Equation A9).

In the vegetative phase of growth, a plant invests a fixed proportion of its net photosynthetic production in height (i.e. stem mass). When growing in height, the leaves at the base of the plant are discarded while leaves at the top are newly produced, according to the leaf distribution (Equation A1). Whatever of the net photosynthetic production is left after height growth is invested in leaf and root mass, divided between these structures in some constant proportion. If at any time the carbon that is spent for maintenance and height growth exceeds the carbon income, leaf and root mass are shed and the retracted carbon from these leaves and roots is used to make the carbon balance equal to zero. If either leaf or root mass is zero grams or less, the plant is considered dead.

The Switch to Seed Production

The model plants have a clear switch from vegetative growth to seed mass production. After the switch, the plants no longer grow in height or leaf mass, but all available carbohydrates are invested in seeds. Seed production continues to the end of the growing season. All seeds stay on the plant until the end of the growing season. At the start of the new growing season, all old plants are removed, the seeds are redistributed over the cells and germinate. We assume, that switching time is genotype-specific. To evaluate the consequences of the timing of a switch to seed production, we distinguish ten genotypes which differ in their timing of this switch during the simulated growing season (top part of Table 1).

Simulations with Isolated Plants

First we simulate the growth and seed production of plants of the different switch genotypes (Table 1) without competition in a cell, for a single growing season.

Table 1. Top part of the table: Time step of the switch to the production of seed mass of plant genotypes. The timing of a switch to the production of seed mass is given in days after the start of the simulation. Lower part of the table: Seed output (gram per plant) of plant genotypes (rows) in pair-wise competition. The seed output of the genotype that is the fittest against a particular opponent genotype (columns) is marked in bold.

	Type 1	Type 2	Type 3	Type 4	Type 5	Type 6	Type 7	Type 8	Type 9	Type10
Switch (day)	5	25	45	65	85	105	125	145	165	185
					Opponent type					
Type	1	2	3	4	5	6	7	8	9	10
1	1.15	1.13	1.09	0.97	0.64	0.15	0.15	0.15	0.15	0.15
2	1.53	1.50	1.45	1.29	0.84	0.18	0.18	0.18	0.18	0.18
3	2.62	2.54	2.42	2.15	1.41	0.23	0.23	0.23	0.23	0.23
4	5.47	5.32	5.01	4.42	3.03	0.32	0.31	0.31	0.31	0.31
5	10.78	10.61	10.23	9.15	6.58	2.05	0.33	0.33	0.33	**0.33**
6	**12.37**	**12.41**	**12.49**	**12.52**	**10.66**	5.60	0.66	0.27	0.27	0.27
7	7.61	7.69	7.85	8.27	9.56	**7.96**	3.22	0.23	0.20	0.20
8	3.53	3.56	3.64	3.85	4.58	6.86	5.06	1.50	0.14	0.14
9	1.30	1.32	1.34	1.42	1.69	2.66	**5.33**	2.98	0.56	0.09
10	0.33	0.33	0.34	0.36	0.44	0.73	1.78	**4.16**	**1.50**	0.14

Simulations with Competing Plants

To simulate the competition between plants that grow up together, we let the different switch genotypes (described in Table 1) grow in pairs within a cell, in every possible combination of genotypes. The seed mass produced by the plants in each pair at the end of the 365-day simulation is put into a 'pay-off matrix' (Maynard Smith, 1974; Riechert and Hammerstein, 1983) (see Table 1).

Calculations on the Frequency Development of Plant Genotypes

Then, we extend the simulations to a situation where all different genotypes have an influence on performance of a genotype, using a replicator equation (Taylor and Jonker, 1978; Nowak and Sigmund, 2004). This can be visualized as an immense arena consisting of very many cells, each with two players. Often these players belong to two different genotypes, and the frequency of the various genotype combinations equals the product of the frequencies of both genotypes in the total seed pool. The success of a switch genotype now depends on its performances against every single switch genotype in the simulated community, and the frequency at which these encounters occur. For this purpose we sum the plant genotype's seed productions when competing with all the different opponents that are present:

$$P_i = f_i \cdot \sum_j p_{ij} \cdot f_j \quad (1)$$

Here P is the total pay-off (i.e. seed production) for a plant genotype and p_{ij} is the pay-off of a plant genotype i against an opponent plant genotype j, taken from the pay-off matrix (Table 1). The frequency at which the genotypes are present is f. The total pay-off is a measure of fitness for the plant genotype.

In addition, we calculate whether the genotype will increase or decrease in frequency f in the course of years, compared to other genotypes. Each year, the new frequency is the relative contribution of the target plant's previous-year pay-off to the total pay-off of all competing plants in that year:

$$f_{i(t+1)} = \frac{P_{i(t)}}{\sum P_{j(t)}} \tag{2}$$

With each repeated calculation, the relative frequencies of plant genotypes in the community will change and can be followed during simulation years. If the frequency of a genotype falls below a certain frequency (10^{-5} of the total community size) it is supposed to have gone extinct and its frequency is set to zero. This limit implicitly defines the order of magnitude of the system: ca 10^5 cells.

Reintroduction of Plant Genotypes

The selective pressure changes with the composition of the community. It is possible that some genotypes have gone extinct before conditions occur that would benefit them. The switch genotype, or genotypes, that finally become dominant are therefore not necessarily evolutionary stable strategies (Maynard Smith, 1982). To investigate this, the estimated threshold frequency of 10^{-5} is removed, allowing each genotype to remain in the vegetation at (infinitely) low frequencies. This can also be seen as a probability that such an 'extinct' genotype re-immigrates into the system. We explore the question whether switch genotypes can successfully re-establish themselves from a low frequency at some point during the development of the simulated community.

Results and Interpretation

Isolated Plants

Plants that switch to seed mass production early have a long period for seed mass production (Iwasa and Levin, 1995). However, these plants have not yet produced much leaf area and gross photosynthetic production is small, resulting in a low production of seed mass. Plants that switch to seed mass production at a very late stage have produced much leaf area and have a large gross photosynthetic production. However, the end of the simulation period is near and the period for seed mass production is short. Secondly, such plants have more leaf area and have a higher degree of self-shading, resulting in more non-productive leaves. Thirdly, being larger, the plant will have high costs of maintenance and not much carbon will be left to invest in seed mass. As a result, the optimal switch time for isolated plants is somewhere in the middle of the season (see also Iwasa, 2000; Widen, 1991).

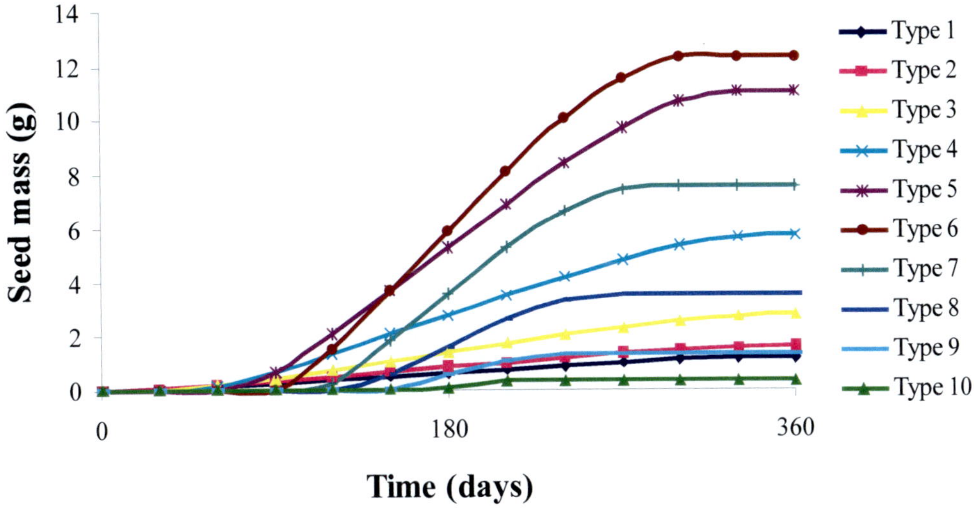

Pronk et al. 2006

Figure 1. The seed mass production during one simulation year of isolated plants switching to seed mass production at different times. See the top part of Table 1 for a description of the different switch genotypes.

Figure 1 shows the produced seed mass per switch genotype (see Table 2) for isolated plants. Seed mass is highest for a medium switch genotype. The highest seed mass yield is achieved when switching to the production of seed mass at 105 days after the onset of growth (this is switch genotype 6). The further (both earlier or later) the switch is from this 105 days, the more the production of seed mass declines.

Simulations on Competing Plants

Table 1 shows the pay-offs of the different switch genotypes (rows) in pair-wise competition with opponent switch genotypes (columns). Within a column, the switch genotype that has the highest production of seed mass in competition against a particular opponent switch genotype is indicated in bold lettering.

When switch genotypes compete with weakly competitive opponent genotypes such as genotype 1 to 5 (left columns in Table 1), genotype 6 achieves the highest seed mass. Genotype 6 also showed the highest seed mass as a single plant. When switch genotypes have to compete with genotypes that have a later switch (columns more to the right in Table 1), the latter genotypes have a higher seed mass production. This implies that genotypes, above a certain optimal switch, should delay their investment in seed mass slightly when competing with genotypes that are late in the production of seed mass as well.

If a plant starts earlier than its competitor with the production of seed mass, it will become shaded. This obviously has negative effects on its seed production. The early switcher can even become so severely shaded that it cannot intercept enough light to pay for maintenance. This forces the plant to shed leaves and roots, leading to a further reduction of its ability to intercept light. Consequently, a switch to seed production should be as close as possible to the switch time that gave the highest seed mass in isolated plants, but also later than the switch of the neighboring plant.

Calculations on the Frequency Development of Plant Genotypes

As the performance of a switch genotype depends on the genotype of the competitor (Table 1) the total pay-off per switch genotype will differ with each change in the overall composition of the community.

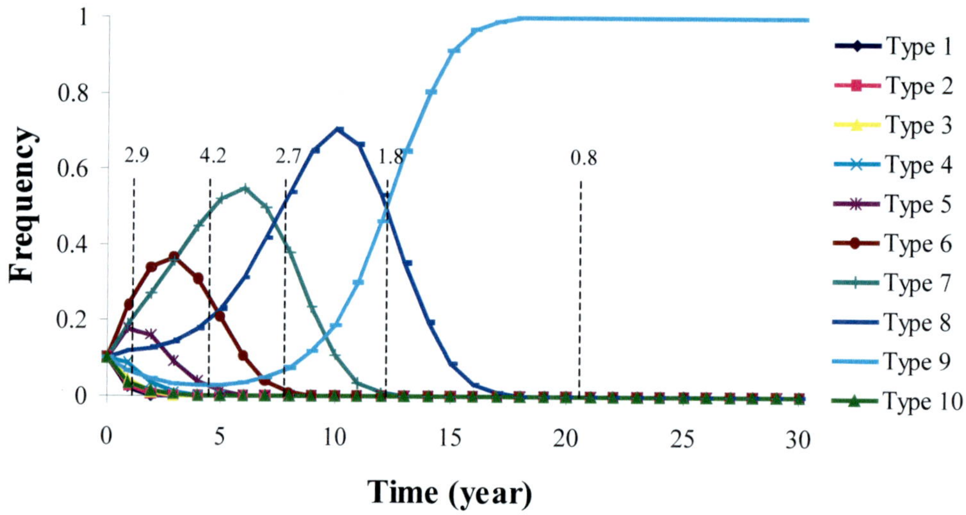

Figure 2. The frequency development of competing switch genotypes. The frequencies of switch genotypes are relative to each other, e.g. the frequencies of all genotypes combined add up to 1 at each point in time. For a description of switch genotypes, see the top part of Table 1. The numbers above the vertical dashed lines in the figure depict the seed mass produced on average per plant in the particular composition at the vertical dashed line.

Figure 2 shows the frequency development over several years (Equations 1 and 1) of switch genotypes 1 to 10 under pair-wise competition (Table 1). It shows that there is an initial shift in dominance during the simulation. Initially, when all plants start with equal frequency, switch genotype 6 is dominant. During the subsequent simulation years, switch genotypes 1 to 5 that start investing in the production of seed earlier (and thus lose from switch genotype 6 in competition) go extinct. Genotype 6 now encounters relatively more late-switching plants, which have a larger competitive vigor. These more competitive genotypes increasingly suppress genotype 6 and its seed production declines. The next genotype, which is genotype 7, invests almost as much in seed mass but has a slightly better competitive vigor than genotype 6. It thus replaces genotype 6. As genotype 6 starts to decline in frequency, genotype 7 encounters more competitive genotypes, and in its turn declines also. As can be seen in Figure 2, genotype 6, 7, 8 and 9 are successively dominant.

Genotype 10 has gone extinct already at the beginning of the simulations because of its relatively small seed production. The genotype that eventually becomes all dominant is genotype 9. This genotype 9 switches to seed mass production at day 165.

In our results we see a clear succession from the switch genotype that has the highest fitness in isolated plants, towards genotypes that switch later. In Figure 2 the average seed production per plant during the frequency development is indicated. Initially, the average seed production increases slightly. This is because the very early switch genotypes go extinct. These genotypes produce a low amount of seed because they start allocation to reproduction before they have produced sufficient leaf area, and additionally are suppressed by all other genotypes. After the initial increase, the average seed production gradually decreases again to a low level. This is because the later switch genotypes that become dominant produce a very low amount of seed, as a result of their late switching time. The seed production of the final population thus is lower than in earlier stages of the succession and may even reach values which are insufficient for self-replacement of the remaining genotype (Fig. 2).

Reintroduction of Plant Genotypes

Figure 3 shows that, when formerly extinct switch genotypes are allowed to re-establish from a low frequency, the abundance of switch genotypes shows a regular pattern of switch genotypes decreasing and increasing in frequency in the course of years. Apparently, there are cyclic opportunities for the reintroduction of switch genotypes (Figure 3).

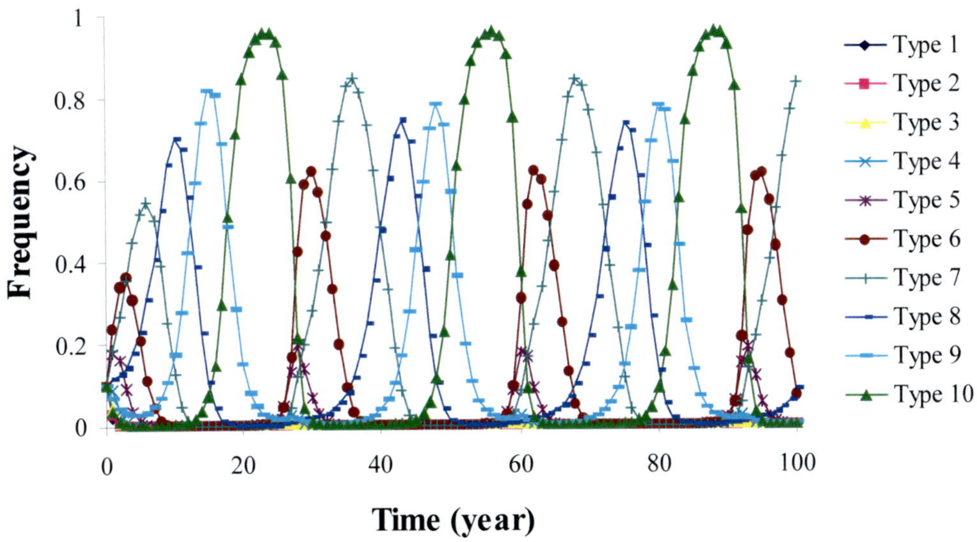

Pronk et al. 2006

Figure 3. The frequency development of competing switch genotypes, when reintroduction of switch genotypes is allowed. The frequencies of switch genotypes are relative to each other, e.g. the frequencies of all genotypes combined add up to 1 at each point in time. See the top part of Table 1 for a description of the switch genotypes.

In the previous simulation, genotype 10 plants were already extinct, but with reintroduction they can make a re-entry at the point when genotype 8 and 9 are most abundant (Figure 3). The plants of genotype 10 slowly exclude the plants of genotype 9. However, plants of genotype 10 grow vegetative for such a long period of time, that by the time they become reproductive they spend most of their photosynthetic production on maintenance of vegetative biomass. Genotype 10 already produces very little seed mass, and even less when competing against another plant of genotype 10 (see Table 1). Therefore, as the frequency of genotype 10 increases, the seed production lessens. Against plants of genotype 8 or 9 it performed better, for the counter-intuitive reason that these genotypes hamper its growth, which resulted in the plant being further from steady state as the switch to reproduction occurs. Genotype 5 has a comparable seed mass production to genotype 10 when competing against it (see Table 1) and a better seed mass production when it encounters con-specifics. As it increases in frequency, its overall seed production increases (Figure 3). Because genotype 5 is present, genotype 6 can perform well. Genotype 6 is followed by genotype 7, 8, 9, and 10, and the cycle, as explained in the case without reintroduction, repeats itself.

So far, we have taken only 10 genotypes that differed 20 days from each other in their switch to the production of seed mass. If the genotypes are taken closer together, even more genotypes can coexist in a cyclic manner (results not shown). The ability of genotypes to invade and exclude other genotypes also can be calculated formally. Appendix B gives a formal calculation for the possibilities for any switch genotype to invade the population of another switch genotype and/or to exclude that switch genotype.

Conclusion

In this paper, the aim was to investigate the adaptive values of different seed investment strategies for annual plants under competition and to find whether they could coexist at equilibrium or non-equilibrium. Initially, we found a shift towards delayed reproduction over the course of subsequent growing seasons. With the shift to delayed reproduction, we also found a classic example of a 'tragedy of the commons' as introduced by Hardin (1968). In a simulated community where all genotypes were present initially, there was a clear succession from highly fecund genotypes (that switch early) towards a community consisting of genotypes that switched later and, as a consequence, produced an excessive vegetative mass (Donald, 1968; Kawecki, 1993; Zhang et al., 1999) at the disadvantage of seed production. It is known that strong selection for growth is in general one of the key processes causing delayed reproduction (Rees and Rose, 2002). In other studies, selection for growth was also found for separate plant traits rather than whole plant size. For instance, Gersani et al. (2001) showed a tragedy of the commons for plants competing for root-space (and thus nutrients and water). Schieving and Poorter (1999) showed that species should increase specific leaf area in a competitive game, at the expense of carbon gain.

When we allowed the different switch genotypes to be reintroduced after extinction, we found stable cyclic species replacement patterns. Cycles were caused by the succession from early to late switch genotypes that in turn could be reinvaded by the early switch genotypes. These cyclic patterns were generated not by external factors, but by the process of competition itself. The strongest competitor invested in vegetative growth for such a long period of time that it could produce only a small amount of seeds. A switch genotype that switched to seed production earlier in the season, made relatively more seed and was able to increase its frequency compared to the strongest competitor. Thus, as the composition of plants developed, the circumstances for growth changed. Switch genotypes that were excluded earlier in community development were able to re-enter the community.

Empirical evidence for the occurrence of large-scale fluctuations in species composition such as we found in our model study, will not be easy to trace in real plant communities. However, it is clear that populations often show cyclic or chaotic behavior rather than converging to a stable equilibrium (Kendall et al., 1999; Nowak and Sigmund, 2004) and that this also applies to plant populations (Bauer et al., 2002). Silvertown et al. (2002) for instance, argue that cyclic fluctuations might be common phenomena in herbaceous plant communities, even those otherwise thought to be stable. They put forward the 'Park Grass experiment' (van den Bergh, 1979; Tilman, 1996) which is a long-term experimental set of equilibrium plant communities. Although monitoring showed the gross composition of the vegetation was at equilibrium over a 60-year period, annual records showed that individual species exhibited a range of dynamics. Because long-term records are not available for most other experimental sites, these cycles or the chaotic behavior would go undiscovered.

For plants, seasonality or drought are among the recorded mechanisms that can cause oscillations. Also, oscillations can be caused by density dependent effects that affect life-history traits, such as overcompensation (Crone and Taylor, 1996; Bauer et al., 2002). Overcompensation can, theoretically, cause a stand of two coexisting plant species to oscillate (Damgaard, 2004). Furthermore, oscillations can be caused by delayed feedback mechanisms such as can occur in plant – nutrient interactions (Daufresne and Hedin, 2005).

While the occurrence of periodically cycling populations has been documented (Kendall et al., 1999), there has not been much evidence whether these cycles can permit coexistence of populations of species that would otherwise exclude each other. Huisman and Weissing (1999) found that species may coexist by oscillating patterns on the basis of competition for three different resources, in their model for well-mixed plankton species. An additional species also found opportunity to grow within these cycles.

Our simulation model results show that the coexistence of plants can be a consequence of the oscillatory behavior of different genotypes. This exemplifies that cycling populations are not necessarily restricted to multi-resource systems (Huisman and Weissing, 1999) but can also occur in systems where individuals compete for light as a single resource. We propose that a simple life-history trade-off between growth and reproduction, mediated by competition, can in theory generate oscillatory dynamics within a plant community that maintain diversity, even without an environmental trigger. Of course, although model results can give clues to what the mechanisms behind plant growth and competition are, only the confirmation or rejection of these clues in an experimental setting will be conclusive. Because the assumptions for the individual plant growth in this model are explicitly formalized, they can be tested and falsified. Phenotypic or genetic manipulations, or the comparison of closely related plant genotypes that differ slightly in the expression of their traits, will give the best system to test model predictions.

Obviously, the model results may be partly determined by its simplicity and rigidity. Thus, the population is conceived as a collection of cells with competition restricted to just two plants per cell and a mean-field approach to the distribution of seeds at the beginning of each new growing season. We suggest, however, that the presence of genotypes with only very few seeds can be interpreted as probabilities of re-entry of extinct genotypes rather than presence at extremely low densities.

Another restriction in our analysis is the uniform distribution of seeds over the cells. A Poisson distribution over the cells would be more realistic, but would make it far more difficult to determine the frequencies of competing genotype pairs (cf. Geritz (1995), who introduces simplicity in a different way, i.e., by allowing only the largest seed in a cell to survive).

Furthermore, the conditions of the model are highly determinate and so, the strategies are discrete, without stochasticity or mutational drift. Studies explicitly comparing games with discrete and continuous strategies suggest, that the results may be qualitatively different (e.g., Maynard Smith, 1974; Bishop and Cannings, 1978), and that continuous strategies may lead to one stable equilibrium in games that allow several coexisting strategies if these are discrete (e.g., Broom et al., 2005). It would be fascinating to test, whether introduction of stochasticity or evolutionary change in switching dates of our strategies would similarly turn our oscillating system into a system with one stable equilibrium strategy. We expect, however, that oscillations will be maintained because of a defining characteristic of our system: selection will favour later-switching genotypes until the winning genotype has so little time left for seed production, that it can no longer produce sufficient numbers of seeds to fill all cells, after which an earlier-switching genotype can re-immigrate.

Acknowledgements

We thank Niels Anten for his valuable comments and suggestions for improvement of this study.

Appendix A. The Model

1. Shape of the Model plant

A plant is confined to grow in a predetermined surface area, referred to as a cell. More than one plant can inhabit this cell. Each plant has a specified smooth distribution of infinitesimally small leaf elements over its height. The leaf area distribution in leaf area per unit height $\lambda(h)$ per cell basal area is described as (Caton et al., 1999):

$$\lambda(h) = p_0 \cdot \frac{L_t}{h_{pt}} \cdot \left(\frac{h}{h_{pt}}\right)^{p_1} \cdot \left(1 - \frac{h}{h_{pt}}\right)^{p_2} \qquad \text{(A1)}$$

in which h_{pt} is the height of the plant at time t, L_t is the total leaf area of the plant at time t, h is the height at which the leaf density is calculated and p_0, p_1, p_2 are shape parameters. Here, the shape parameter p_0 is given in terms of p_1 and p_2 by the integral:

$$p_0 = \frac{1}{\int_0^1 dx \cdot x^{p_1} \cdot (1-x)^{p_2}} \quad \text{with } x = \frac{h}{h_{pt}} \tag{A2}$$

By changing the shape parameters p_1 and p_2 different leaf area distribution curves can be achieved.

The relation between height h_{pt} and stem volume S_t at time t is given by a standard allometric equation (Stearns, 1992). In this equation, α and β are constants.

$$S_t = \alpha \cdot h_{pt}{}^{\beta} \tag{A3}$$

Although a root system is present in the model plant, it merely acts as a carbon sink for the carbohydrates from photosynthesis. The root mass is a linear function of the leaf area:

$$R_t = \sigma \cdot L_t \tag{A4}$$

Here σ is a constant factor depicting the ratio between leaf area and root mass.

2. The Light Climate within the Vegetation

We assume a light climate in the vegetation cell in which the photons flow vertically downward. At each depth h the interception of light by the plants is modeled by Beer's law (Monsi and Saeki, 1953). The horizontal light intensity $i_h(h)$ is given for each height h in the cell:

$$i_h(h) = i_h(h_v) \cdot \exp\left[- \int_h^{hv} \sum_{j=1}^{n} dh \cdot \lambda_j(h) \cdot \cos\alpha \cdot a \right] \tag{A5}$$

where $i_h(h_v)$ is the horizontal light intensity above the vegetation and h_v is the height of the vegetation. The light intensity $i_h(h_v)$ over a season is described further on. Cos α is the leaf inclination and a is the absorption coefficient of leaves. For simplicity, we take these parameters equal for all plant genotypes. The rate of photon absorption for the leaf elements at height h is calculated from the light climate at height h, the leaf inclination and absorption coefficient of a plant. It is given by

$$i_{al}(h) = i_h(h) \cdot \cos\alpha \cdot a \tag{A6}$$

3. Leaf and Plant Photosynthetic Rates

The leaf photosynthetic rate is calculated on the basis of a non-rectangular hyperbolic relation between maximum photosynthetic capacity, light interception and quantum yield (Marchall and Biscoe, 1980). Leaf photosynthetic rate is given as:

$$P_{gl}(h) = P_{ml} \cdot \frac{(1+x) - \sqrt{(1+x)^2 - 4 \cdot \theta \cdot x}}{2 \cdot \theta} \quad with \; x = \frac{\Phi \cdot i_{al}(h)}{P_{ml}} \tag{A7}$$

Here P_{ml} is the photosynthetic capacity of the leaves, Φ the quantum yield per unit absorbed light, θ a curvature factor and $i_{al}(h)$ is the rate of photon absorption of the leaves at height h (see 2.4). For reasons of simplicity, we assume that capacity P_{ml}, quantum yield Φ and curvature θ do not vary with height or time.

For each plant the total photosynthetic rate is given by the integral over the plant's height, of the leaf photosynthetic rate times the leaf area. This gross photosynthesis is in μmol carbon per unit time.

$$P_g = \int_0^{hp} dh \cdot P_{gl}(h) \cdot \lambda(h) \tag{A8}$$

4. Carbon Balance and the Allocation Program

The growth of plant organs brings about carbon costs. These costs are balanced at any time by the carbon production rate of the plant. Firstly, the plant allocates carbon to account for maintenance costs.

$$P_n = P_g - R_m \text{ with } R_m = r_{ml} \cdot c_l \cdot L + r_{ms} \cdot c_s \cdot S + r_{mr} \cdot c_r \cdot R \tag{A9}$$

R_m is the maintenance rate and is based on the weights of the plant parts. The parameters r_{ml}, r_{ms}, r_{mr} are the maintenance respiration rates in gram carbon per unit carbon mass per unit time. The conversion factor c_l is in g carbon per m^2 leaf, c_s is in g carbon per volume stem, c_r is in g carbon per gram roots. After maintenance, plants invest the net photosynthetic production in the growth of their vegetative parts or, if they are in reproductive mode, invest it in seed mass.

The carbon balance we use in the model is according to the following set of rules: 1. The consumption of carbon should be, at any time, equal to the carbon gain from net photosynthesis P_n. 2. Height growth rate dh_p / dt is always non-negative. 3. Leaf area rate dL / dt can be either positive or negative. If the carbon balance at any moment is negative, because the plant consumes more carbon than comes from photosynthetic production, leaf area is shed and carbon from this leaf area is used to account for the shortage. If the carbon balance is positive, leaf area is produced. 4. If the leaf area is equal to zero, the plant is considered dead. The carbon balance is applicable for any specific allocation program the plant may follow.

$$P_n = [\ldots]\frac{dL}{dt} + \left[(c_s + r_s)\frac{dS}{dh_p} + (c_l + r_l)\int_0^{hp} dh \cdot POS\left(\frac{d\lambda}{dh_p}\right) - c_{rl}\int_0^{hp} dh \cdot NEG\left(\frac{d\lambda}{dh_p}\right)\right]\frac{dh_p}{dt} \tag{A10}$$

In case of a positive leaf production rate ($dL / dt > 0$) or a negative leaf production rate ($dL / dt < 0$) we write for the leaf production rate, respectively:

$$\left[(c_l + r_l) + \sigma(c_r + r_r)\right]\frac{dL}{dt} \text{ or } \left[-(c_{rl} + \sigma \cdot c_{rr})\right]\frac{dL}{dt} \quad \text{(A11)}$$

In these equations A10 and A11, c_l, c_r, c_s, c_f are the structural carbon imbedded in leaves, roots, stem and seed respectively. Parameters r_l, r_r, r_s and r_f are the growth respiration with leaf growth, root growth and stem growth. If a plant grows in height h_p with rate dh_p / dt this requires costs for the breakdown and production of leaves to maintain the crown shape as dictated by the parameters p_1 and p_2 in Equation A1. *POS* ($d\lambda / dh_p$) and *NEG* ($d\lambda / dh_p$) respectively depict the leaf area that is produced with growth in height and the leaf area that is cast away with growth in height. Parameters c_{rl} and c_{rr} depict the carbon that can be retrieved if leaves and roots, respectively, are cast away.

The investment in stem is always a fixed fraction of P_n. The actual height growth dh_p / dt is calculated from the relation between stem volume and height, net photosynthesis P_n and the allocation of carbon to height growth a_s.

$$\frac{dS}{dt} = f(P_n) = \frac{a_s \cdot P_n}{c_s} \text{ and as } \frac{dh_p}{dt} = \frac{f(P_n)}{dS / dh_p}$$

the height growth can be written as

$$\frac{dh_p}{dt} = \frac{(a_s \cdot P_n) / c_s}{\alpha \cdot \beta \cdot (h_p)^{\beta - 1}} \quad \text{(A12)}$$

The denominator of the equation is the allometric height-volume relation, derived from Equation A3.

In conclusion, two state variables determine the state of the plant when it is not in reproductive mode. They are total leaf area L (with associated roots) and total height h_p. The allocation to the latter is an imposed control; it determines the strategy of the plant. It is a fraction of the net photosynthetic production. In contrast, the allocation to the leaf area and roots can be determined from the carbon balance as written in Equation A10 and A11.

After the plant switches to seed production, all of the net photosynthetic production goes to the production of seed. If net photosynthesis is negative or zero, the allocation rate to seed is zero.

$$\frac{dF_m}{dt} = P_n \quad \text{(A13)}$$

Here F_m is the seed mass that is produced. It is assumed that maintenance of current biomass is a priority. This is paid from net photosynthetic production. This means the plant is able to maintain its shape when it has switched to seed mass production. Leaf area and consequent light interception changes only if the plant is forced to shed leaf area as a result of a negative carbon balance, caused by a diminishing light climate.

5. The Seasonal Light Climate

We assume a seasonal fluctuation in the yearly light climate, as we would expect it to occur in a temperate region (Kirschbaum, 1999):

$$\textit{Daily Incidence } I_{veg} = I_{mean} + I_{amp} \cdot \sin\left[\Pi \cdot \left(\frac{t - 0.5 \cdot year + 0.25 \cdot year}{0.5 \cdot year}\right)\right] \quad \text{(A14)}$$

Here I_{veg} is the daily incident light above the canopy at day t in the simulation year, I_{mean} is the mean incident light above the canopy during a year and I_{amp} is the amplitude of the fluctuation in incident light.

Table A1. Parameters settings. Some parameters have been taken as an approximate average of several literature sources. Sources of the parameters: 1. Poorter (1991) 2. Goudriaan and Van Laar (1994) 3. Anten and Hirose (2003) 4. Calibrated parameters (this study) 5. Caton et al. (1999) 6. Kastner-Maresch and Mooney (1994) 7. Kirschbaum (1999) 8. Lieth and Reynolds (1988)

Symbol	Unit	Value	Source
L (initial)	m^2	0.00165	1
H (initial)	m	0.01	1
c_s	$g\ C/\ m^3$	0.000045	4,1
c_l	$g\ C/\ m^2$	15	1,2
c_r	g C/ g mass	0.45	1
σ	g C root/ m^2 leaf	7.5	1,6
α, β	-	0.018, 1.4	4
p_0, p_1, p_2	-	30, 2, 2	4
P_{ml}	$\mu mol\ CO_2/\ m^2/\ s$	16	3
$\cos \alpha$	-	0.5	2

Table A1. Continued

Symbol	Unit	Value	Source
Φ	μmol CO_2/ μmol ppfd	0.05	3
θ	-	0.7	3,8
r_s, r_l, r_r	g C/ g C	0.213	2,8
r_{ms}, r_{ml}, r_{mr}	g C/ g C plant/ day	0.0235	2,5
c_{rl}, c_{rr}	g C/ g C	0.6	4
I_{amp}	μmol/ m2/ s	1360	7
I_{mean}	μmol/ m2/ s	1700	7
Day length	s/ day	25200	4

Appendix B. Invasion and Coexistence

The possibility for a pair of plant genotypes to coexist, invade each other or exclude each other can be calculated as followed.

We consider a pair of competing genotypes '1' and '2'. The seed production 'A' of genotype 1 will be different depending on whether it is competing against conspecific plants (A_{11}) or heterospecific plants (A_{12}), this also applies to the other genotype (A_{22} and A_{21}). Examples of such seed productions can be found in the lower part of Table 2.

The frequencies 'x' of the two genotypes add up to unity. The frequency of genotype 2 can thus be expressed in terms of the frequency of genotype 1:

$$x_2 = 1 - x_1 \tag{B1}$$

Because the total produced seed mass 'S' of the pair-wise competing genotypes is dependent on the frequency of the heterospecific and the conspecific plants we write:

$$S_1 = x_1(A_{11} \cdot x_1 + A_{12} \cdot x_2) = x_1(A_{11}x_1 + A_{12}(1 - x_1)) \tag{B2}$$

$$S_2 = x_2(A_{22} \cdot x_2 + A_{21} \cdot x_1) = (1 - x_1)(A_{22}(1 - x_1) + A_{21}x_1) \tag{B3}$$

The frequency at the next time step for genotype 1 can be written as the produced seed mass (S_1) divided by the total produced seed mass for both genotypes ($S_1 + S_2$):

$$x_1^+ = \frac{x_1(A_{11}\cdot x_1 + A_{12}\cdot(1-x_1))}{x_1(A_{11}\cdot x_1 + A_{12}(1-x_1)) + (1-x_1)\cdot(A_{22}\cdot(1-x_1)+A_{21}\cdot x_1)} \quad \text{(B4)}$$

The sign of $x_1^+ - x_1$ determines the increase or decrease of genotype 1. If $x_1^+ - x_1$ is positive, genotype 1 will increase in frequency. If it is negative, it will decrease in frequency.

$$x_1^+ - x_1 = \frac{x_1(A_{11}x_1 + A_{12}(1-x_1)) - x_1(x_1(A_{11}x_1 + A_{12}(1-x_1)) + (1-x_1)(A_{22}(1-x_1)+A_{21}x_1))}{x_1(A_{11}x_1 + A_{12}(1-x_1)) + (1-x_1)\cdot(A_{22}(1-x_1)+A_{21}x_1)}$$

(B5)

The numerator of the formula will determine the sign of the change in frequency. To clean up the formula (B5), we multiply all terms by A_{22} / A_{22} and by doing so scale the pay-off to A_{22} so that

$$\alpha = A_{11} / A_{22} \qquad \beta = A_{12} / A_{22} \qquad \gamma = A_{21} / A_{22} \quad \text{(B6)}$$

This gives:

$$x_1^+ - x_1 = \frac{x_1(x_1 - 1)\cdot(\alpha x_1 + \beta(1-x_1) - (1-x_1) - \gamma x_1)}{[....]} \quad \text{(B7)}$$

If the sign of the formula is positive near $x_1 = 0$, it means a genotype 1 is able to invade a population consisting of the competitor genotype 2. If the sign is negative near $x_1 = 1$, it means a population of the genotype 1 can be invaded by a competitor of genotype 2.

We can derive what the conditions in pair-wise competition are for either genotype 1 to win, genotype 2 to win, or to get a stable or unstable coexistence of the genotypes.

There is an unstable coexistence if $\beta - 1 < 0$ near $x_1 = 0$ and $\alpha - \gamma > 0$ near $x_1 = 1$

There is a stable coexistence if $\beta - 1 > 0$ near $x_1 = 0$ and $\alpha - \gamma < 0$ near $x_1 = 1$.

Genotype 1 cannot invade if $\beta - 1 < 0$ near $x_1 = 0$ and $\alpha - \gamma < 0$ near $x_1 = 1$

Genotype 1 cannot be invaded if $\beta - 1 > 0$ near $x_1 = 0$ and $\alpha - \gamma > 0$ near $x_1 = 1$

References

Anten NPR (2005). Optimal photosynthesis characteristics of individual plants in vegetation stands and implications for species coexistence. *Annals of Botany* **95**: 495-506.

Anten NPR, Hirose T (2003). Shoot structure, leaf physiology, and daily carbon gain of plant species in a tall grass meadow. *Ecology* **84**: 955-968.

Armstrong RA, McGehee R (1976). Coexistence of species competing for shared resources. *Theoretical Population Biology* **9**: 317-328.

Bauer S, Berger U, Hildenbrandt H, Grimm V (2002). Cyclic dynamics in simulated plant populations. *Proceedings of the Royal Society, London, B Biology* **269**: 2443-2450.

Bergh van den JP (1979). Changes in the composition of mixed populations of grassland species. In: Werger MJA, ed. *The study of vegetation.* The Hague: Dr W. Junk Publishers, pp 59-80.

Broom M, Speed MP, Ruxton GD (2005). Evolutionarily stable investment into secondary defenses. *Functional Ecology* **19**: 836-843

Caton BP, Foin TC, Hill JE (1999). A plant growth model for integrated weed management in direct-seeded rice. III. Interspecific competition for light. *Field Crops Research* **63**: 47-61.

Cohen D (1976). The optimal timing of reproduction. *American Naturalist* **110**: 801-807.

Crone EE, Taylor DR (1996). Complex dynamics in experimental populations of an annual plant, *Cardamine pensylvanica*. *Ecology* **77**: 289-299.

Damgaard C (2004). Dynamics in a discrete two-species competition model: coexistence and over-compensation. *Journal of Theoretical Biology* **227**: 197-203.

Daufresne T, Hedin LO (2005). Plant coexistence depends on ecosystem nutrient cycles: Extension of the resource-ratio theory. *Proceedings National Academy of Science* **102**: 9212-9217.

Donald CM (1968). The breeding of crop ideotype. *Euphytica* **17**: 385-403.

Falster DS, Westoby M (2003). Plant height and evolutionary games. *Trends in Ecology and Evolution* **18**: 337-343.

Geritz SAH (1995). Evolutionarily stable seed polymorphism and small-scale spatial variation in seedling density. *American Naturalist* **146**: 685-707.

Gersani M, Brown JS, O'Brien EE, Maina GM, Abramsky Z (2001). Tragedy of the commons as a result of root competition. *Journal of Ecology* **89**: 660-669.

Goudriaan J, van Laar HH (1994). *Modelling Potential Crop Growth Processes.* The Netherlands: Kluwer Academic Publishers.

Hardin G (1968). The tragedy of the commons. *Science* **162**: 1243-1248.

Huisman J, Weissing FJ (1999). Biodiversity of plankton by species oscillations and chaos. *Nature* **402**: 407-410.

Huisman J, Johansson AM, Folmer EO, Weissing FJ (2001). Towards a solution to the paradox of the plankton: the importance of physiology and life-history. *Ecology Letters* **4**: 408-411.

Iwasa Y (2000). Dynamic optimization of plant growth. *Evolutionary and Ecological Research* **2**: 437-455.

Iwasa Y, Levin SA (1995). The timing of life-history events. *Journal of Theoretical Biology* **172**: 33-42.

Karstner-Maresch AE, Mooney HA (1994). Modeling optimal plant biomass partitioning. *Ecological Modeling* **75**: 309-20.

Kawecki TJ (1993). Age and size at maturity in a patchy environment - fitness maximization versus evolutionary stability. *Oikos* **66**: 309-317.

Kendall BE, Briggs CJ, Murdoch WW, Turchin P, Ellner SP, McCauley E, Nisbet RM, Wood SN (1999). Why do populations cycle? A synthesis of statistical and mechanistic modeling approaches. *Ecology* **80**: 1789-1805.

Kerr B, Riley MA, Feldman MW, Bohannan BJM (2002). Local dispersal promotes biodiversity in a real-life game of rock-paper-scissors. *Nature* **418**: 171-174.

Kirschbaum MUF (1999). CenW, a forest growth model with linked carbon, energy, nutrient and water cycles. *Ecological Modeling* **118**: 17-59.

Lieth JH, Reynolds JF (1988). A plant-growth model for controlled-environment conditions incorporating canopy structure and development - application to Snap bean. *Photosynthetica* **22**: 190-204.

Marchall B, Biscoe PV (1980). A model for C3 leaves describing the dependence of net photosynthesis on irradiance. *Journal of Experimental Botany* **31**: 29-39.

Maynard-Smith, J (1974). The theory of games and the evolution of animal conflicts. *Journal of Theoretical Biology* **47**: 209-221.

Maynard-Smith J (1982). *Evolution and the Theory of Games.* Cambridge: Cambridge University Press.

Monsi M, Saeki T (1953). Ueber den Lichtfaktor in den Pflanzengeselschaften und seine Bedeutung fur die Stoffproduktion. *Japanese Journal of Botany* **14**: 22-52.

Nowak MA, Sigmund K (2004). Evolutionary dynamics of biological games. *Science* **303**: 793-799.

Peet AB, Deutsch PA, Peacock-López E (2005). Complex dynamics in a three-level trophic system with intraspecies interaction. *Journal of Theoretical Biology* **232**: 491-503.

Poorter H (1991). *Interspecific Variation in the Relative Growth Rate of Plants: the Underlying Mechanisms.* PhD Thesis, University of Utrecht, The Netherlands.

Pronk TE (2004). *The role of plant traits in the regulation of diversity: a modelling study.* http://www.library.uu.nl/digiarchief/dip/diss/2004-0224-135436/inhoud.htm PhD thesis, University of Utrecht, The Netherlands.

Pronk TE, During HJ, Schieving F (2007a). Coexistence by temporal partitioning of the available light in plants with different height and leaf investments. *Ecological Modeling* **204**: 349-358.

Pronk TE, Schieving F, Anten NPR, Werger MJA. (2007b). Plants that differ in height investment can coexist if they are distributing non-uniformly within an area. *Ecological Complexity* **4**: 182-191.

Reekie EG, Budge S, Baltzer JL (2002). The shape of the trade-off function between reproduction and future performance in *Plantago major* and *Plantago rugelii*. *Canadian Journal of Botany* **80**: 140-150.

Rees M, Rose KE (2002). Evolution of flowering strategies in *Oenothera glazioviana*: an integral projection model approach. *Proceedings of the Royal Society, London, B Biology* **269**: 1509-1515.

Riechert SE, Hammerstein P (1983). Game theory in the ecological context. *Annual Review of Ecology and Systematics* **14**: 377-409.

Schieving F, Poorter H (1999). Carbon gain in a multispecies canopy: the role of specific leaf area and photosynthetic nitrogen-use efficiency in the tragedy of the commons. *New Phytologisy* **143**: 201-211.

Schmid B, Weiner J (1993). Plastic relationships between reproductive and vegetative mass in *Solidago altissima*. *Evolution* **47**: 61-74.

Silvertown J, McConway KJ, Hughes Z, Biss P, Macnair M, Lutman P (2002). Ecological and genetic correlates of long-term population trends in the park grass experiment. *American Naturalist* **160**: 409-420.

Stearns SC (1992). *The evolution of life histories*. Oxford: Oxford University Press.

Taylor P, Jonker L (1978). Evolutionarily stable strategies and game dynamics. *Mathematical Bioscience* **40**: 145-156.

Tilman D (1996). Biodiversity: Population versus ecosystem stability. *Ecology* **77**: 350-363.

Weiner J (1988). The influence of competition on plant reproduction. In: Lovett Doust J, Lovett Doust L, eds. *Plant reproductive ecology: patterns and strategies.* New York: Oxford University Press, pp 228-245.

Widen B (1991). Phenotypic selection on flowering phenology in *Senecio integrifolius*, a perennial herb. *Oikos* **61**: 205-215.

Zhang DY, Sun GJ, Jiang XH (1999). Donald's ideotype and growth redundancy: a game theoretical analysis. *Field Crops Research* **61**: 179-187.

In: Game Theory: Strategies, Equilibria and Theorems ISBN 978-1-60456-844-8
Editors: Ingrid N. Haugen et al, pp. 319-335

Chapter 13

A GAME-THEORETIC ANALYSIS OF THE 'TIPPING POINT' PHENOMENON: INFORMATIONAL PHASE TRANSITIONS IN SOCIAL NETWORKS

Juan M.C. Larrosa [*] ***and Fernando Tohmé*** [†]
CONICET - Department of Economics, Universidad Nacional del Sur,
San Juan y 12 de Octubre
Bahia Blanca, Buenos Aires 8000, Argentina

Abstract

A well-known phenomenon (popularized by Malcolm Gladwell's book) in real world social networks is the existence of 'tipping points'. That is, thresholds that once surpassed lead to a whole new configuration of the social structure. Political upheavals, sudden fads and the fast adaptation of innovations are just examples of this. We intend to provide some clues on how this might happen, but instead of assuming that it is an unintended consequence of random actions we consider here a game theoretic framework in which rational agents make decisions aimed to maximize their payoffs.

Starting with a framework very much like Bala & Goyal's (2000) we consider a finite society in which agents are endowed with some amount of a private but reproducible good (*information*) that upon contact can be copied or transmitted from one agent to another. While there is a cost of establishing a connection, there are also gains in accessing new information. The difference between these two yields the payoff of a connection. Rational agents will behave strategically and the Nash equilibria will provide the *network architecture*.

As it is well known from Erdös and Renyi (1959) seminal treatment of random graphs, new connections may lead to phase transitions in the density of the graph. That is, jumps in the number of clusters from many to a single major one. While for social networks the framework of random graph is not quite cogent, similar results may arise varying the nature of the probability distribution on potential connections (Newman et al., 2002).

In this paper we will show how the same is true in our non-probabilistic, game-theoretic framework. By slight changes in the information carried by individual agents

[*]E-mail address: jlarrosa@criba.edu.ar
[†]E-mail address: ftohme@criba.edu.ar

(representing the influence of non-social sources) the equilibrium networks may vary suddenly. We will see that in the end, if each agent has an information endowment larger than the cost of establishing connections to her, a minimally connected network becomes the unique outcome. That means that a highly organized structure arises when everyone is "valuable". On the other hand, if the value is too low for every agent (i.e. there is no gain in connecting to others) the only efficient outcome is the empty network. In the middle, we will show, there exist some critical agents to which most of the others will want to establish contact and yield components in a disconnected network.

1. Introduction

In the last decade, the study of the structure of interactions among agents in different contexts has grown to the point to constitute the core of a new area in the social sciences. The tools of this field, the study of *social networks*, are currently being applied in different disciplines, ranging from sociology to anthropology. and of course in economics.

The mathematics behind the analysis of social networks is provided by *graph theory*. Here the nodes are interpreted as individual agents and the edges as the links over which the agents carry out exchanges (Wasserman and Faust (1994)). Game theory has been also applied to the analysis of networks, being the fundamental tool in the study of their normative properties, particularly their stability or efficiency (Jackson and Wolinsky (1996), Bala and Goyal (2000), Dutta and Jackson (2000)). The strategic approach focuses on the individual strategies available to the agents as well as on the corresponding individual payoff functions. The decisions the agents have to make concern their connection to other agents. The rational choices of the agents lead to Nash equilibria, which may support various types of networks.

A modeling primitive is the representation of the links between agents as directed or non-directed. While the latter are important to represent situations in which the direction of flow of utility goods is less important or just irrelevant (Dutta and Mutuswami (1997)), the former are important for distinguishing the agents that initiate connections as well as the direction of flow in the ensuing network (Bala and Goyal (2000), Dutta and Jackson (2001)).

In this paper we consider networks only as directed graphs with one-way flows. We denote by "information" the good that flows in the networks. Each agent is endowed with some amount of information which, up to communication, can be reproduced by other agents. Contrary to Bala and Goyal, the amount of goods to which an agent has access is also relevant and not only the amount of agents that provide them. By establishing links to other agents she can obtain the information held by them, but she has to pay a small fee to establish those links. Since the "distance" among agents varies, the strategic goal of each rational agent is to maximize the access to the information hold by the others, at the lowest cost possible. Nash equilibria yield the structures from which no individual agent may deviate to increase her benefit.

The main goal of our analysis is to provide a game-theoretic foundation of the *tipping point* phenomenon, popularized by Gladwell (2000). While it lacks a precise definition, it is commonly understood as the existence of a critical value of a parameter that, once surpassed, leads to a substantive change in a social system. Tipping point processes are widely found in the sociological analyses embodied in *critical mass theory* (See Marwell et al. (1988)). The analogy to well-known phenomena studied in statistical physics is obvious (Durlauf (1997)). Therefore, the tools developed in the study of criticality in collective phenomena have been also applied to modeling tipping point processes.

One of the most influential approaches has been to see the social structure as a network that may arise from random interactions among individuals. Changes in the probability of encounters may lead to drastic changes in the final structure that obtains. In mathematics this is a well-known result, originally presented in Erdös and Renyi groundbreaking study on *random graphs* (1959). They showed that if the average number of links established by any agent is slightly increased in a small neighborhood of 1, a complete disconnected graph becomes a completely connected one. Newman et al. (2001) have generalized this result for generic probability distributions, showing that intermediate phase transitions exist, for which new components arise in a graph.

Our idea is to reproduce a similar result, but instead of attributing it to random encounters between agents, they will arise from the intentional behavior of rational agents. In consonance with Gladwell's point of view, one key parameter will be the relative presence of "mavens", i.e. agents sought for connection. We represented this in terms of their amount of private information.

We will see that if the amount of information held by the agents increases (due to external sources) several structures arise as equilibria, from the empty network to a completely connected one. The key observation is that the final topology depends on the agents that become mavens.

The organization of the paper is as follows. Section 2 introduces our game-theoretic model. In section 3, the equilibria are determined, particularly two extreme cases, one in which information is so scarce that the only equilibrium is the empty net while in the other, each agent is so richly endowed with information that in equilibrium every pair of agents becomes connected. In section 4 we analyze how increasing amounts of information that monotonically increase the number of links among agents lead to phase transitions, all of them due to the presence of agents that connect previously separated components. Finally, section 5 discusses these results.

2. Networks and Games

Let $N = (1, \ldots, n)$ be a set of agents. To avoid trivial results we will always assume that $n \geq 3$. If i and j are two typical members of N, a link among them, without intermediaries, originated by i and ending in j will be represented as ij. The interpretation of ij is that i establishes a contact with j that allows i to get acquainted with both the information possessed by j as well as connect to j's network of contacts. Each agent $i \in N$ has some information of her own, $I_i \in \Re^+$, (i.e. represented as a nonnegative real number). As said i can have access to more information by forming links with other agents. The formation of links is costly, in time, resources and effort, but we will assume that a link ij has cost of $c\, l_{ij}$ where c is the cost per unit of length (measured in units of utility of information), while

l_{ij} is the *social distance* between i and j.[1] For simplicity, we assume that $l : N \times N \rightarrow Z^+$, i.e. that the length of each link is a non-negative integer and that $l_{ij} = l_{ji}$ (each i is trivially connected to itself through a link of length 0).

The agents will try to maximize the utility of the information available to them as well as to minimize the cost of connecting to other agents. In order to do this, they will be endowed with a set of *strategies*. Each strategy for $i \in N$ is a $(n-1)$-dimensional vector $g_i = \langle g_{i,1}, \ldots, g_{i,i-1}, g_{i,i+1} \cdots g_{i,n} \rangle$ where each $g_{i,j}$, for $j \neq i$, is either 0 or 1. This is interpreted as meaning that i establishes a direct link with j if $g_{i,j} = 1$ while if $g_{i,j} = 0$ there is no such direct link. The set of all i's strategies is denoted as G_i. Since we restrict our analysis to only pure strategies, $|G_i| = 2^{n-1}$. Finally, $G = G_1 \times \ldots \times G_n$ denotes the set of strategy profiles in the interaction among the agents in N.

The existence of a direct link ij indicates an asymmetric communication between i and j. That is, $g_{i,j} = 1$ indicates that i establishes a communication with j that permits i to access to j's information but not vice versa (the symmetry between i and j is restored if also $g_{j,i} = 1$). Structures with this feature are called **one-way flow** networks.

In one-way flow networks a strategy profile can be represented as a directed graph $g = (g_1 \ldots g_n)$ over N. That is, in the directed graph the elements of N are the *nodes* while any established link like $g_{i,j} = 1$ is represented by an arrow beginning in j with its head pointing to i.[2] That is, arrowheads always point toward the agent who establishes the link. It follows immediately that:

Proposition 1 *There exists a one-to-one map between directed graphs among n nodes and strategy profiles in G.*

Proof: *A directed graph with n nodes is such that for each node i there exists at most one incoming arrow from each $j \neq i$ (and none from itself). Then, for each j define $g_{i,j}$ equal to 1 if there exists an incoming arrow from j, and 0 otherwise. This defines $g_i = \langle g_{i,1}, \ldots, g_{i,i-1}, g_{i,i+1}, \ldots, g_{i,n} \rangle$ for each $i \in N$. That is, it defines a $g = \langle g_1, \ldots, g_i, \ldots, g_n \rangle \in G$. Conversely, given g, a directed graph can be obtained by just adding an arrow from j to i if $g_{i,j} = 1$. Since $g_{i,i}$ is not defined, the graph is loop-less, and since $g_{i,j}$ has only two possible values, there exist either one or zero links between them.*

Example 1: *consider a group of four agents, $N = \{a, b, c, d\}$. Each link among them is assumed to have length 1. A joint strategy $g = \langle g_a, g_b, g_c, g_d \rangle$ can be represented as a table:*

Strategy	a	b	c	d
g_a	X	1	0	0
g_b	0	X	1	0
g_c	0	0	X	1
g_d	0	0	0	X

[1]The meaning of *social distance* here should not be confused with the definition in Akerlof (1997). While there it is an emergent property of the social structure, the social distance in this paper is a constant, given previously to the emergence of the social network.

[2]In order to represent the idea that when i establishes a link with j, the information flows *from j to i*.

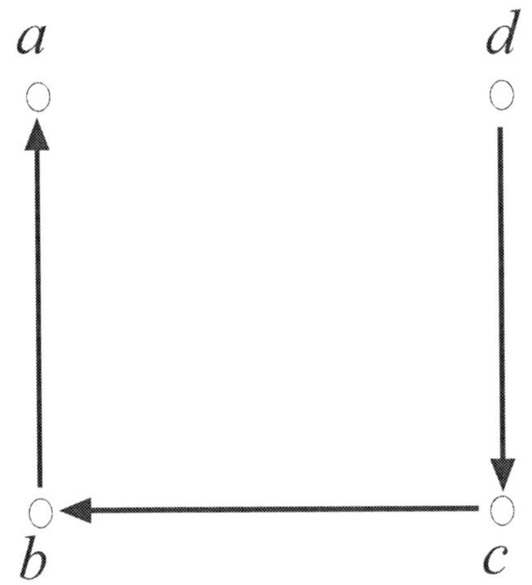

Figure 1.

Each row is the strategy chosen by one of the agents. Columns correspond to the agents. An entry 1 *in row* i *and column* j *means that the strategy of agent* i *prescribes to establish a link with agent* j. *Entries in the diagonal are crossed out since agents are trivially connected to themselves by links of length* 0. *In Figure 1 we can see the directed graph that corresponds to* g.

We define $N^{g_i} = \{k \in N | g_{i,k} = 1\}$ as the set of agents to whom i establishes a direct link according to her strategy g_i. We say that there exists a *path* of information flow from j to i according to $g \in G$ if there exists a sequence of different[3] agents $j_0 \dots j_m$ (with $i = j_0$ and $j = j_m$) such that $g_{j_0,j_1} = \dots = g_{j_{m-1},j_m} = 1$. In other words, given the joint strategy g, we have that $j_1 \in N^{g_{j_0}}$, $j_2 \in N^{g_{j_1}}$, ..., $j_m \in N^{g_{j_{m-1}}}$. The length of the path from $j = j_m$ to $i = j_0$, denoted as $j \to_g i$, is the sum of the lengths of links in the path, $\sum_{j=0}^{m-1} l_{j(j+1)}$. Notice that a direct link is a path of length 1.

Example 1 revisited: *Given the strategy* $g = \langle g_a, g_b, g_c, g_d \rangle$, *we have that* $N^{g_a} = \{b\}$, $N^{g_b} = \{c\}$ *and* $N^{g_c} = \{d\}$ *while* $N^{g_d} = \emptyset$. *This sequence establishes a path from* d *to* a *of length* 3.

We denote the set of agents accessed (directly and otherwise) by i as $N^{i;g} = \{k \in N | k \to_g i\} \cup \{i\}$. We include i in $N^{i;g}$ to indicate that i knows her own valuation through a link, of length 0, from i to herself. Let $\mu_i : G \to \{0, ..., n \times (n-1)\}$ be the number of links in all paths that end in i, originated by agents in $N^{i;g}$ under any given joint strategy: $\mu_i(g) = \sum_{(j,k) \in \mathcal{L}^{i;g}} l_{jk}$, where $\mathcal{L}^{i;g} = \{(j,k) \in N \times N : g_{j,k} = 1$, and $\exists l \in N^{i;g}$ and $l \to_g i$ with $j, k \in l \to_g i\}$.[4]

Example 2: *Assume that we have* $N = \{1, 2, 3, 4, 5\}$ *and the links among agents are given*

[3]To avoid cycles.

[4]Notice that there may be more than one path from j to i.

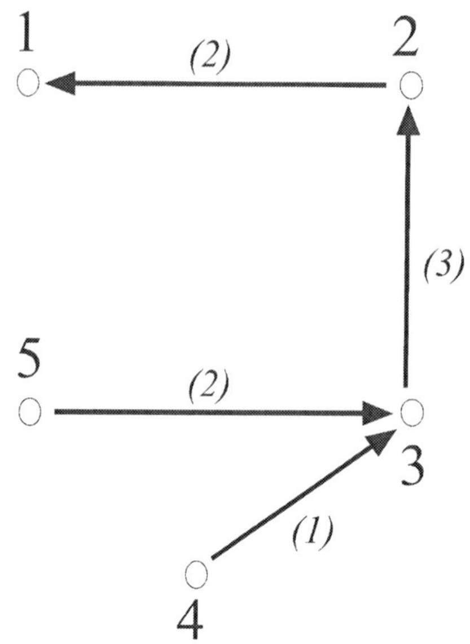

Figure 2.

in the following matrix, where the entry in the ith row and jth column indicates the length l_{ij} of the link from i to j (∞ means that the length is excessively large, compared to the finite lengths):

	1	**2**	**3**	**4**	**5**
1	0	2	∞	∞	1
2	2	0	3	∞	∞
3	∞	3	0	1	2
4	∞	∞	1	0	∞
5	1	∞	2	∞	0

Consider strategy $g = \langle g_1, g_2, g_3, g_4, g_5 \rangle$ given by the following table:

Strategy	**1**	**2**	**3**	**4**	**5**
g_1	X	1	0	0	1
g_2	0	X	1	0	0
g_3	0	0	X	1	1
g_4	0	0	0	X	0
g_5	0	0	0	0	X

Figure 2 shows the corresponding network. We have that $N^{1;g} = \{1,2,3,4,5\}$, $N^{2;g} = \{2,3,4,5\}$, $N^{3;g} = \{3,4,5\}$ while $N^{4;g} = \{4\}$ and $N^{5;g} = \{5\}$. That is, under g we have that 1 can access to the information of all the agents while 4 and 5 have access only to their own information. The links in the path towards 1 are described by $\mathcal{L}^{1,g}$=$\{(1,2),(2,3),(3,4),(3,5)$,$(1,5)\}$. For the other agents we have that

$\mathcal{L}^{2,g}=\{(2,3),(3,4),(3,5)\}$, $\mathcal{L}^{3,g}=\{(3,4),(3,5)\}$, *while* $\mathcal{L}^{4,g}=\mathcal{L}^{5,g}=0$.[5] *Then, the total lengths of the links required to obtain the information are* $\mu_1(g) = 9$, $\mu_2(g) = 6$ *and* $\mu_3(g) = 3$, *while* $\mu_4(g) = \mu_5(g) = 0$.

To endow this framework with a game-theoretic structure, we have to define the payoffs to agents. Let $\Pi_i : G \to R$ be the payoff function:

$$\Pi_i(g) \equiv \sum_{j \in N^{i;g}} I_j - c\mu_i(g)$$

where c is the cost of establishing each link. That is, i's payoff is just the sum of all the information that can be accessed by her, less the total cost of the paths reaching her that are established according to g. The intuition here is that i gets a payoff from accessing to more information but at the same time she has to pay a "fee" for each of the links on the paths to the sources of information.

Example 2 revisited: *Suppose the information owned by the agents is:* $I_1 = 2$, $I_2 = 2$, $I_3 = 4$, $I_4 = 3$ *and* $I_5 = 3$. *Then, under strategy* g *and* $c = 0.5$ *we have:*

$$\Pi_1(g) = I_1 + \cdots + I_5 - \mu_1(g) = 2+2+4+3+3 \; - \; 0.5 \times 9 = 9.5$$

$$\Pi_2(g) = I_2 + \cdots + I_5 - \mu_2(g) = 2+4+3+3 \; - \; 0.5 \times 6 = 9$$

$$\Pi_3(g) = I_3 + \cdots + I_5 - \mu_3(g) = 4+3+3 \; - \; 0.5 \times 3 = 5.5$$

$$\Pi_4(g) = I_4 - \mu_4(g) = 3 \; - \; 0 = 3$$

$$\Pi_5(g) = I_5 - \mu_5(g) = 3 \; - \; 0 = 3.$$

We can notice here that, for example, if $g_{1,5} = 0$, 1 *could improve her payoff (i.e. obtaining* 10 *instead of* 9.5*) because she would still have access to* I_5 *but using one link less.*

For each $g \in G$, agent i obtains a structure $N^{i;g}$ and her payoff depends critically on the type of directed graph that corresponds to $N^{i;g}$ as summarized in the following proposition:

Proposition 2 *Given two joint strategies* g *and* g', $\Pi_i(g) \geq \Pi_i(g')$ *iff the corresponding graphs* $N^{i;g}$ *and* $N^{i,g'}$ *are such that:*

$$\sum_{j \in N^{i;g}} I_j - \sum_{j \in N^{i;g'}} I_j \geq c(\mu_i(g) - \mu_i(g')).$$

Proof: *Trivial.*

This result conveys the intuition that the goal of a rational agent is to get as much information as possible traversing the shortest possible paths.

[5]Links from agents to themselves do not count, since they do not, by definition, belong to paths.

3. Nash Equilibria

Given a network $g \in G$,[6] let g_{-i} be the directed graph obtained by removing all of agent i's direct links. Then, g can be written as $g = g_i \oplus g_{-i}$ where $\oplus$ indicates that g is formed by the union of the links of g_i and those in g_{-i}. A strategy g_i is said the *best response* of agent i to g_{-i} if

$$\Pi_i(g_i \oplus g_{-i}) \geq \Pi_i(g_i^{'} \oplus g_{-i})$$

for all $g_i^{'} \in G_i$.

Example 3: *Consider again the case of* $N = \{1, 2, 3, 4, 5\}$, *where* $I_1 = 2$, $I_2 = 2$, $I_3 = 4$, $I_4 = 3$ *and* $I_5 = 3$. *The social distances among agents are now:*

	1	**2**	**3**	**4**	**5**
1	0	2	5	5	1
2	2	0	3	∞	1
3	5	3	0	2	2
4	5	∞	2	0	∞
5	1	1	2	∞	0

Let g_{-1} *be described by the following table:*

Strategy	1	2	3	4	5
g_2	0	X	1	0	1
g_3	0	0	X	1	0
g_4	0	0	0	X	0
g_5	0	0	0	0	X

See Figure 3 for the situation faced by 1. She has to decide to whom establish a connection, assuming that $c = 0.5$. *A possibility is to remain isolated, but that would give her a payoff of only 2. Alternatively, she could connect to as many of the other agents as she likes. But some connections may be redundant in terms of the gain in information. Such redundancy, in turn, would mean a higher cost for the same information. So, if 1's strategy were, for instance, to connect both to 3 and 4, it would ensure her access to the information of 3 and 4, i.e.* $\mathcal{L}^{1,g}$=$\{(1,3), (1,4), (3,4)\}$. *The payoff is then* $2 + 4 + 3 - 0.5 \times (5 + 5 + 2)$=3. *She could, instead, connect only to 3, since she would still get hold of the information of 3 and 4 but it would require only 2 links, i.e., her payoff would be* $2 + 4 + 3 - 0.5 \times (5 + 2)$=5.5. *A bit of reflection shows that the best answer for 1 would be to connect only to the agent with the higher payoff under* g_{-1}. *That is, to agent 2, who has a payoff of* $2 + 4 + 3 + 3 - 0.5 \times (1 + 2 + 3)$=9. *Then, 1 will reach the information of 2, 3, 4 and 5, through the links* $(1, 2)$, $(2, 3)$, $(2, 5)$ *and* $(3, 4)$. *That is, her payoff would be of 10. Figure 3 shows the resulting network.*

[6]According to Proposition 1 we identify a joint strategy g with its corresponding directed graph.

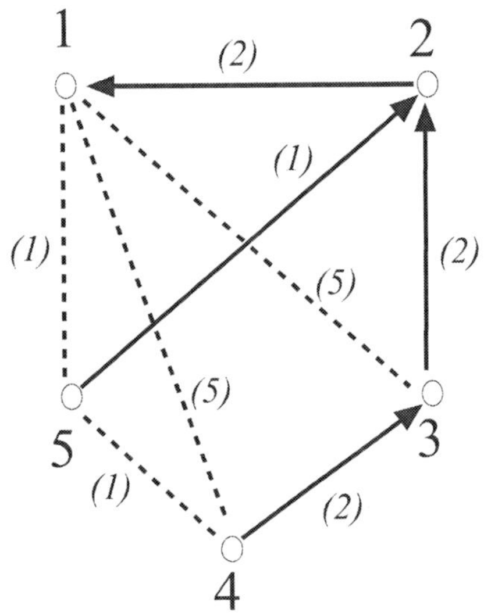

Figure 3.

The set of best responses to g_{-i} is $BR_i(g_{-i})$. A network $g = \langle g_1, \ldots, g_n \rangle$ is said to be a *Nash network* if for each i, $g_i \in BR_i(g_{-i})$ i.e. if g (as a joint strategy) is a Nash equilibrium. In order to determine the structure of Nash networks let us give a few more definitions that will allow us to describe some additional properties of networks.

Given a network g, a set $C \subset N$ is called a (directed) *component* of g if for every pair of agents i and j in C ($i \neq j$) we have that $j \in N^{i;g}$ and $i \in N^{j;g}$, while if just there exists a path from i to j or vice versa these nodes are said to belong to the same (undirected) component. Furthermore, if there does not exist C', $C \subset C'$ for which this is true, C it said to be *minimal*: C is not a component anymore once a link $g_{i,j} = 1$ between two agents i and j in C is cut off, i.e. if $g_{i,j} = 0$.

Example 4:*If $N = \{1, 2, 3, 4\}$, consider the following network g, represented in Figure 4:*

Strategy	1	2	3	4
g_1	X	1	0	0
g_2	0	X	1	0
g_3	0	0	X	1
g_4	0	1	0	X

Clearly $C = \{2, 3, 4\}$ is a component, since $N^{2;g} = N^{3;g} = N^{4;g} = \{2, 3, 4\}$ and if we consider $C' = C \cup \{1\}$, it is easy to see that C' is not a component, since 1 does not belong to $N^{2;g}$, $N^{3;g}$ or $N^{4;g}$. On the other hand, C is minimal, since if we cut off any of the links $(2, 3)$, $(3, 4)$ or $(4, 2)$ some of the agents will no longer be reachable for at least one agent in C. So, for instance, if $(2, 3)$ is cut off, in the new network g' we have that $N^{2,g'} = \{2\}$.

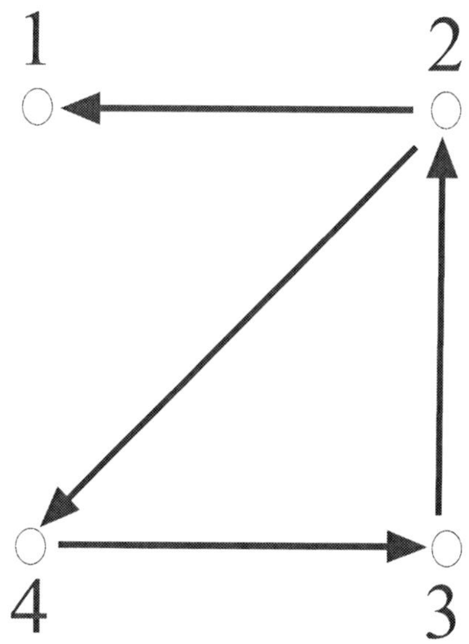

Figure 4.

A network g is said to be *connected* if it supports a single component. If that component is minimal, g is *minimally connected.* A network that is not connected is *disconnected.*

Then, with all these elements at hand we can state the following results:

Lemma 1 *A strict Nash network is empty if for each i, $I_i < c$. If, instead, for each i, I_i is larger than the cost of accessing i from every $j \neq i$, the strict Nash network is minimally connected.*

Proof: *Let us consider $\Pi_i : G \to \Re^+$, for each $i \in N$. If we show that there exists a unique (up to isomorphism) $g^* \in G$ that maximizes Π_i for each i, given that the others choose g^*_{-i} we would establish that there exists only one Nash equilibrium in the game. In fact, recalling Proposition 2 it is easy to show that for every i the corresponding payoff should be:*

$$\Pi_i(g^*) = \sum_{j \in N_{>c}^{i;g^*_{-i}}} I_j - c\bar{\mu}_i(g^*)$$

where $N_{>c}^{i;g^*_{-i}} = \{j : I_j > c\bar{\mu}_i(i \to_{g_i;g^*_{-i}} j)\} \cup \{i\}$. Here $\bar{\mu}_i(j \to_{g_i;g^*_{-i}})i$ is the shortest length of a path from j to i, for any g_i while keeping fixed g^*_{-i}. In turn, $\bar{\mu}_i(g^*)$ is $\sum_{(j,k)\in \mathcal{G}L^{i;g^*}} l_{jk}$, for $\mathcal{G}L^{i;g^*}$, the class of shortest paths *between i and any other node in $N^{i;g^*}$.*

In words: the maximum of information that can be reached is the sum of all the information that outperforms the costs of access, while the minimum sum of distances that make that amount of information available is $\bar{\mu}_i(g^*)$. Notice that $\mu_i(g^*) \geq |N^{i;g^*}| - 1$ since, otherwise, there would be at least one node $j \in N^{i;g^*}$ that will not be accessed, yielding a contradiction.

It is easy to check that if all agents other than i choose g_j^*, i's choice will be g_i^*. Suppose by way of contradiction that i has chosen $g_i^{'} \neq g_i^*$ such that $\Pi_i(g_i^{'}, g_i^*) > \Pi_i(g^*)$. That is, $\Pi_i(g_i^{'}, g_i^*) > \sum_{j \in N_{>c}^{i;g_{-i}^*}} I_j - c\bar{\mu}_i(g^*)$. This can happen either by accessing a node $j \notin N_{>c}^{i;g_{-i}^*}$ or, if $N_{>c}^{i;g_{-i}^{'}} = N^{i;(g_i^{'}, g_i^*)}$, by covering less distance to access the same information. This last possibility must be discarded since $\bar{\mu}_i(g^*)$ obtains summing the distances in the shortest paths that connect i with the other nodes. Therefore, there cannot exist shorter paths towards the same nodes. With respect to the possibility of accessing a $j \notin N_{>c}^{i;g_{-i}^*}$, this implies that $I_j \leq c\bar{\mu}_i(i \rightarrow_{g_i^{'};g_{-i}^*} j)$. Therefore, the gain of I_j does not compensate the cost of accessing it. Contradiction.

Now, if $I_j < c$ for $j = 1, \ldots, n$, it follows that the optimal choice of every agent is described by $N_{>c}^{i;g_{-i}^*} = \{i\}$, since, otherwise, for any $g^{'} \neq g^*$, $c\bar{\mu}_i(g^{'}) > \sum_{j \in N_{>c}^{i;g_{-i}^{'}}} I_j$. That is, the network is empty.

To the contrary, if $I_j > c\bar{\mu}_i(i \rightarrow_g j)$ for every j and every $g \in G$ it follows that every i will try to access, directly or indirectly each other j. This means that in equilibrium there will exist a path from every i to every j, i.e. $N^{i;g^*} = N$ for every i. Therefore, the network will be connected. On the other hand, in equilibrium the network will be minimal, since the cost of accessing any j from any i cannot be lowered without severing a link that disconnects the graph.

Example 5: *Consider three agents, a, b and c. The distances are $l_{ab} = 3$, $l_{ac} = 1$ and $l_{bc} = 2$ while $c = 0.5$. Consider two cases:*

- *$I_i = 0.1$ for $i = a, b, c$. Let us determine the behavior that can be expected of a. If she accesses both b and c, she obtains 0.3 of total information at a cost of $0.5 \times (3 + 2)$. Then, this strategy yields a negative payoff. The same is true if a would connect just to b or c. In the last case the total information available to a would be 0.2 while the cost would be 0.5×1. It is easy to see that this argument applies to each of the tree agents. Therefore, the Nash strategy prescribes to stay isolated obtaining a positive payoff of 0.1.*

- *$I_i = 6$ for $i = a, b, c$. Then, the Nash network will be constituted by the links (a, b), (b, c) and (c, a) (an alternative Nash equilibrium, yielding the same payoffs, obtains reversing the direction of the edges).[7] To see why this is so, consider the decision of a when she faces (b, c) and (c, a). She could try to access just c, obtaining a payoff $(6 + 6) - 0.5 \times 1$, which yields a lower result than trying to access also b. Now, a could try to establish two independent links, one to b and the other to c, obtaining $(6+6+6) - 0.5 \times (3+2+1)$ (since b also accesses c, a would have to pay for the two paths that lead to c). But instead she could get the same information at a lower cost by just establishing a link towards b, which yields a payoff of $(6+6+6) - 0.5 \times (3+2)$. A similar analysis shows that indeed the circular network is a Nash equilibrium.*

[7] This pattern is sometimes called a *circular* or *wheel* network.

4. Phase Transitions

It is immediate that, from Lemma 1, changes in the information carried by the agents lead to *phase transitions*. More precisely, consider a function $I : N \rightarrow \Re^+$, the *information distribution* among the agents. Lemma 1 has shown that there exists two different distributions, say I^1 (such that for each i, $I_i < c$) and I^2 (with, for each i, I_i larger than the cost of accessing i from every j) such that I^1 shields a totally disconnected graph as equilibrium while I^2 a minimally connected one.

I^1 can be continuously transformed into I^2. To see this consider two particular instances of I^1 and I^2, say $\hat{I}^1$ and $\hat{I}^2$ such that $\hat{I}^k_i = \hat{I}^k_j$, for every pair $i, j \in N$ and for $k = 1, 2$. Let us define a transformation $t : \Re^+ \times \mathcal{I} \rightarrow \mathcal{I}$, where $\mathcal{I}=\{I | I : N \rightarrow \Re^+\}$ defined as, for any i, $t(r, \hat{I}^1)_i = \hat{I}^1_i + r$. This transformation is continuous in r. It is obvious that there exists $r \in \Re^+$ such that $t(r, \hat{I}^1) = \hat{I}^2$: it suffices to take any $i \in N$ and $r = \hat{I}^2_i - \hat{I}^1_i$.

Then, there exist $\bar{r}, \tilde{r} \in \Re^+$, $\bar{r} < \tilde{r}$ such that for $r \leq \bar{r}$, $I = t(r, \hat{I}^1)$ yields an empty network, while for $r \geq \tilde{r}$, $I = t(r, \hat{I}^1)$ is a minimally connected network. We say that at $\bar{r}$ and $\tilde{r}$ the network undergoes phase transitions.

Before we generalize this argument, let us introduce some notions that will be used in what follows.

A directed graph, corresponding to a Nash equilibrium g^*, can be fully described by means of its *Laplacian matrix* $L(g^*) = D(g^*) - A(g^*)$. It obtains as the difference between the degree matrix $D(g^*)$ and the adjacency matrix $A(g^*)$. $D(g^*)$ is a diagonal $n \times n$ matrix, in which for each i, the ii entry is $\sum_{\{j: g^*_{i,j} \neq 0\}} l_{ij}$. $A(g^*)$ is a $n \times n$ matrix in which each entry ij is l_{ij} if $g^*_{ij} = 1$ and 0 otherwise.

Example 5 revisited: *Consider the circular network obtained when* $I_i = 6$ *for* $i = a, b, c$. *Then, the degree matrix will be:*

	a	b	c
a	3	0	0
b	0	2	0
c	0	0	1

while the adjacency matrix is:

	a	b	c
a	0	3	0
b	0	0	2
c	1	0	0

Instead, when $I_i = 0.1$ *for* $i = a, b, c$, *the Laplacian is the null matrix, since there are no adjacencies in the graph and thus no node has a non-zero degree.*

The main properties of $L(g^*)$ are (Mohar 1991),(Wu 2005):

- The eigenvalues of $L(g^*)$, $\lambda_1, \lambda_2, \ldots, \lambda_n$ have all non-negative real parts.

- Furthermore, $Re(\lambda_1) \leq Re(\lambda_2) \leq \ldots \leq Re(\lambda_n)$ are such $|\{k : Re(\lambda_k) = 0\}|$ is the number of undirected[8] components of g^*.

- In the case that g^* is an empty network, $Re(\lambda_k) = 0$ for every $k = 1, \ldots, n$, while if g^* is connected, $Re(\lambda_k) > 0$ for $k = 2, \ldots, n$.

Example 5, continued: *When $I_i = 6$ for $i = a, b, c$ the Laplacian matrix*

	a	b	c
a	3	−3	0
b	0	2	−2
c	−1	0	1

has three eigenvalues, 0, $3 + 1.41i$ and $3 - 1.41i$. Since only one of them has zero real part, it verifies that the network has a single component. On the other hand, for $I_i = 0.1$ for $i = a, b, c$ the Laplacian is a null matrix and therefore the three eigenvalues are all zero, indicating that the corresponding network has tree components (the three isolated nodes).

Now consider the space $\mathcal{H}^n$ of the Laplacian matrixes of directed graphs with n nodes and distances given by $\{l_{ij}\}_{i,j=1,\ldots,n}$. It can be endowed with a partial order $\preceq$, such that $L^i, L^{ii} \in \mathcal{H}^n$ are such that $L^i \preceq L^{ii}$ iff all the eigenvalues of $L^{ii} - L^i$ have non-negative real parts. That is, $L^{ii} - L^i$ corresponds to a direct graph (not necessarily to an equilibrium). This, in turn means that the graph corresponding to L^{ii} includes the undirected edges of L^i as a subgraph.

We can define a transformation $t : N \cup \{0\} \times \mathcal{I} \to \mathcal{I}$, where N is the class of natural numbers, defined as follows:

- For any information distribution I, $I = t(0, I)$.

- For $k \geq 1$, $t(k+1, I) = I'$ such that I' is an information distribution with a Nash graph, $g^*(I')$ that includes all the links in the Nash graph of $t(k, I)$. Furthermore, $t(k+1, I)$ yields a Nash graph that *strictly* includes the graph of $t(k, I)$.[9] Otherwise, if no such graph exists, $t(k+1, I) \in t(k, I)$.

We finally have that:

Theorem 1 *Consider distributions I^1 and I^2 in the conditions of Lemma 1 and the transformation t. Then, there exist two parameters $\bar{k}, \tilde{k} \in N$ such that $\bar{k} < \tilde{k}$ verifying that for $k \leq \bar{k}$, $t(k, I^1)$ yields an empty network, while for $k \geq \tilde{k}$, $t(k, I^1)$ is a minimally connected network.*

Proof: *First of all, notice that t determines an increasing sequence in $\mathcal{H}^n$. This is immediate from the fact that $t(k, I)$ yields a class of information distributions such that the corresponding Nash graphs all have the undirected Nash graph of I as a subgraph. This in turn means that the Laplacian matrixes will verify that $L^I \preceq L^{t(k,I)}$.*

[8]That is, the components that obtain by disregarding the direction of edges.

[9]It is immediate that t is a correspondence, since there might exist several minimal graphs in this condition.

Since the number of eigenvalues with zero real part of a Laplacian is the number of components of its graph, and $L^I \preceq L^{t(k,I)}$ *it follows that, since a graph corresponding to* $t(k, I)$ *keeps the links of the graph for* I*, the number of components in* $t(k, I)$ *has to be greater or equal than the amount of components for* I*. Then, the number of eigenvalues with zero real part must either remain the same or decrease from* L^I *to* $L^{t(k,I)}$*.*

Therefore, $t(0, I^1)$ *yields the empty network, with the maximal number of components. So, trivially,* $\bar{k} = 0$ *since for any* $k > 0$*,* $t(k, I^1)$ *has already at least one link among two nodes. On the other hand, considering the ordering among Laplacian matrixes, there must exist* L^{I^2}*, corresponding to* I^2*. It is clear that* $I^2 \in t(k, I^1)$ *for some* k*, since the minimally connected graph for* I^2 *includes the empty graph as subgraph. Take* $\tilde{k}$ *as the minimal* k *that verifies that the information distribution supports a minimally connected graph. It follows that for every* $k > \tilde{k}$*,* $t(k, I^1)$ *will also be an information distribution that corresponds to a minimally connected graph.*

The proof of this claim has the key for a further understanding of this process. An increasing transformation on the partial ordering of $\mathcal{H}^n$ will determine chains of increased connection in the corresponding graph. This means that the number of eigenvalues with null real part will decrease. Furthermore, the process described in Theorem 1 goes through all the intermediate Nash networks between the totally disconnected and a completely disconnected one, as seen as follows.

Example 5, continued: *Consider the Nash network that obtains if* I_a *jumps from* 0.1 *to* 6*, while* I_b *and* I_c *are kept fixed at* 0.1*:*

Strategy	a	b	c
g_a	X	0	0
g_b	1	X	0
g_c	1	0	X

The corresponding Laplacian, call it L^1 *is:*

	a	b	c
a	0	0	0
b	-3	3	0
c	-1	0	1

while the Laplacian for $I_i = 6$ *for all three agents, call it* L^2*, is such that* $L^2 - L^1$ *is:*

	a	b	c
a	3	-3	0
b	3	-1	-2
c	0	0	0

which has three eigenvalues with non-negative real parts (0*,* $1 + 2.23i$ *and* $1 - 2.23i$*), corresponding to the fact that the undirected edges of the former network are included in the latter. Also, if we call* L^0 *the null matrix corresponding to the totally disconnected* *network that arises when $I_i = 0.1$ for the three agents, we have that $L^1 - L^0 = L^1$ has three eigenvalues with non-negative real parts (0, 1 and 3).*[10]

This means that $L^0 \preceq L^1 \preceq L^2$. Furthermore, the increase of I_a from 0.1 to 6 determines a phase transition. Another phase transition in this chain obtains when I_b and I_c also increase to 6.

An important feature of each of the chains determined by t is that each one will be determined by the nodes that receive heavier loads of information. This is expressed in the concept of *eigenvector centrality*, which should not be confused with the mere number of links pointed towards the central nodes. More broadly, it indicates their relative importance. To determine the centrality of each node, we have to consider the lowest non-zero eigenvalue of the Laplacian matrix and compute the corresponding eigenvector. Formally, given an equilibrium g^*, and the family of eigenvalues of $L(g^*)$, $Re(\lambda_1) \leq Re(\lambda_2) \leq \ldots \leq Re(\lambda_n)$, let k be the least such that $Re(\lambda_k) > 0$. Then let $\{\bar{x}^j\}_{j \geq k}$ be the eigenvectors corresponding to positive eigenvalues (Newman 2008). Then, the centrality of the i node is $\frac{1}{\sum_{j \geq k} \bar{x}^j_i}$.

Example 5, continued: *Consider the case in which $I_a = 6$, while $I_b = I_c = 0.1$. The eigenvalues are 0, 1 and 3. The eigenvectors corresponding to the two positive eigenvalues are $\langle 0, 0, 1.41 \rangle$ and $\langle 0, 1.22, 0 \rangle$, respectively. Summing their components we get $\langle 0, 1.22, 1.41 \rangle$. Therefore, the centrality degree of a is $\frac{1}{0}$, while for b and c are $\frac{1}{1.22}$ and $\frac{1}{1.41}$. The fact that the centrality of a is infinite is consistent with its salient role in the network.*

Formally, we have the following result:

Proposition 3 *For each chain of information distributions between I^1 and I^2 induced by t, the critical parameters $\{k_m\}_{m=1}^n$ that determine the transition from Nash graphs with $m+1$ components to Nash graphs with m components, are associated with different centrality values for different values of m. These values will be such that for each m a few nodes will become central.*

Proof: *Consider two critical values k_m and k_{m-1}. They will be associated to eigenvalues $\{\lambda^m_j\}_{j=1}^n$ and $\{\lambda^{m-1}_j\}_{j=1}^n$, respectively. Notice that since the real parts of the eigenvalues are ordered, $Re(\lambda^m_{m-1}) = 0$ and $Re(\lambda^m_m) > 0$. In turn, $Re(\lambda^{m-1}_{m-2}) = 0$ while $Re(\lambda^{m-1}_{m-1}) > 0$. Therefore, the corresponding least non-zero eigenvalues are $Re(\lambda^m_m) > 0$ and $Re(\lambda^{m-1}_{m-1}) > 0$ and the corresponding eigenvectors are different. The centrality of nodes will differ accordingly. But even so, notice that, since each Nash graph contains the undirected Nash graph of the previous stage in the chain, if the number of components becomes reduced from m to $m-1$ is because two components become linked. That is because some nodes acquire enough information to makes them central for the two previous components.*

For each transition, in particular for $\bar{k}$ and $\tilde{k}$, the critical parameters for the transition between non-connectedness and full connection, the central nodes play the role

[10]Notice that only one of the eigenvalues is zero. Therefore, the corresponding network has only one undirected component, which is not a component in terms of the directed edges.

of "mavens", i.e. highly connected individuals that facilitate the connection between previously separated agents.

5. Conclusion

We presented in this paper a model of network formation as a non-cooperative game where agents decide to whom to link by comparing the net benefits from their actions. We found that equilibria depend on the distribution of information among the agents. Two extreme cases arise, one in which no agent is linked to any other, while in the other case every pair of agents is connected. Between them there exist a large number of intermediate cases, which can be ordered in terms of their corresponding Laplacian matrixes. We defined a transition correspondence which is increasing in that ordering. It follows that transitions are determined by the increase in the number of eigenvalues of the Laplacian matrixes with non-zero parts. They indicate that some components that arose with less information have been joined and became a larger component. In that sense, some individuals whose information amount increases achieve a central role in the reorganization of the network. This is exactly the type of phenomenon described as a "tipping point" effect. That is, some agents contribute, by their influence (described here by the amount of their information loads), in substantial rearrangements of the network.

References

Akerlof, G. (1997), "Social Distance and Social Decisions", *Econometrica* **65**:1005–1027.

Alam, R. and S. Bora (2003), "Effect of Linear Perturbations on Spectra of Matrices", *Linear Algebra and Applications* **368**: 329-342.

Bala, V. and S. Goyal (2000), "A Noncooperative Model of Network Formation", *Econometrica* **68**:1181-1229.

Durlauf, S.N. (1997), "Statistical Mechanics Approaches to Socioeconomic Behavior", in W.B. Arthur, S.N. Durlauf, and D.A. Lane, (eds.) *The Economy as a Complex Evolving System II*, Addison-Wesley, Redwood City, CA.

Dutta, B. and M. Jackson (2000), "The Stability and Efficiency of Directed Communication Networks", *Review of Economic Design* **5**:251-272.

Dutta, B. and M. Jackson (2001), "On the Formation of Networks and Groups", in B. Dutta and M. Jackson (eds), *Models of the Strategic Formation of Networks and Groups*, Springer-Verlag, N.Y.

Dutta, B., A. van den Nouweland, and S. Tijs (1998), "Link Formation in Cooperative Situations", *International Journal of Game Theory* **27**:245-255.

Dutta, B. and S. Mutuswami (1997), "Stable Networks", *Journal of Economic Theory* **76**:322-344.

Erdös, P. and Renyi, A. (1959), "On Random Graphs", *Publicationes Mathematicae* **6**: 290-291.

Gladwell, M. (2000), *The Tipping Point: How Little Things can Make a Big Difference*, Little, Brown and Co., N.Y.

Jackson, M. and A. Wolinsky (1996), "A Strategic Model of Social and Economic Networks", *Journal of Economic Theory* **71**:44-74.

Marwell, G., P.E. Oliver, and R. Prahl. (1988) "A Theory of the Critical Mass. III. Social Networks and Collective Action." *American Journal of Sociology.* **94**:502-534.

Mohar, B. (1991), "The Laplacian Spectrum of Graphs", in Y. Alavi, G. Chartrand, O. Öllemann and A. Schwenk (eds.), *Graph Theory, Combinatorics and Applications* **2**, Wiley, N.Y.

Newman, M. (2008), "Mathematics of Networks", in L. E. Blume and S. N. Durlauf (eds.), *The New Palgrave Encyclopedia of Economics* 2nd edition, Palgrave Macmillan, Basingstoke, forthcoming.

Newman, M., Strogatz, S. and Watts, D. (2001), "Random Graphs with Arbitrary Degree Distributions and their Applications", *Physical Review E* **64** 026118.

Newman, M., Watts, D. and Strogatz, S. (2002), "Random Graph Models of Social Networks", *Proceedings of the National Academy of Sciences* **99**: 2566-2572.

Slikker, M. and A. van den Nouweland (2001), "A One-Stage Model of Link Formation and Payoff Division", *Games and Economic Behavior* **34**:153-175.

Wasserman, S. and K. Faust, *Social Networks Analysis*, Cambridge University Press, N.Y., 1994.

Watts, D. (2002), "Networks, Dynamics, and the Small-World Phenomenon", *American Journal of Sociology* **105**: 493–527.

In: Game Theory: Strategies, Equilibria and Theorems
Editors: Ingrid N. Haugen et al, pp. 337-367
ISBN: 978-1-60456-844-8

Chapter 14

GAME THEORY AS A GENERAL DECISION MAKING TOOL IN ELECTRIC POWER INDUSTRY - EXAMPLES OF APPLICATION TO ELECTRICITY MARKET AND CYBER SECURITY OF SCADA

Dong-Joo Kang[1,a], Balho H. Kim[2,b] and Tai-hoon Kim[3,c]

[1]Electric Power Research Lab, Korea Electro-technology Research Institute, Korea
[2]Department of Electrical Engineering, Hong-ik University, Korea
[3]Department of Multimedia Engineering, Hannam University, Korea

Abstract

Game theory has been applied to many different kinds of fields including economics, politics, even engineering since it was developed and proposed by John Von Neumann and John Forbes Nash. It was originally applied to economic problems to analyze the interactions between players in competitive situations, but it has evolved as a general tool which is useful for modeling various kinds of systems including not just social but also physical ones if they have distributed schemes composed of interactive components. On the aspect of application, this paper introduces two different examples analyzed and modeled by game theoretic approach. Two examples have totally different qualities although both are the problems in electric power industry. One is a typical application of game theory to economic problems in electricity market representing social system, and the other is the application to SCADA communication and its security problem in the area of physical system which does not have independent decision making entities. By comparing two different cases based on different systems we intended to show the flexible application of game theory to solving the problems.

[a] E-mail address: djkang@keri.re.kr.
[b] E-mail address: bhkim@hongik.ac.kr.
[c] E-mail address: taihoon@hannam.ac.kr.

1. Introduction

Game theory is a useful theory in various fields by providing analytical tools for examining strategic interactions among two or more participants' decision making process under changing conditions which could be competitors or dynamic environments. The decision making is processed considering those variables, in which how to make a choice in the process is called as strategy. Most of models in game theory are formulized based on mathematics, so the strategy is expressed in mathematical description. In the beginning stage of game theory, mini-max theory was proposed for solving zero-sum game by John Von Neumann and Oskar Morgenstern in [1]. However mini-max theory couldn't solve the prisoner's dilemma which is non-zero sum game and was framed by Merrill Flood and Melvin Dresher working of RAND Corporation and mathematically formulized by Albert W. tucker of Princeton University in 1950. Prisoner's dilemma was explained by Nash equilibrium proposed by John Forbes Nash, and following studies by many scholars have made the game theory richer to be able to explain various kinds of situations and fields. Today game theory has been applied to politics, biology, mechanics, communication engineering, etc as well as economics. In this chapter game theory is shown as a general tool for decision making on two different fields in electric power industry based on similar theoretical principles.

2. Game Theory as a Tool for Describing Relationships and Interactions

There is a famous English proverb, "The actions of men, being guided by their antecedents, are in reality never inconsistent, but, however capricious they may appear, only form part of one vast scheme of universal order." This proverb talks about the autonomy of each member in a community, but the autonomy is affected by the ideas of former members and also adjusted by the relationship with other members in the community. Game theory is the research field to analyze the whole system as a group of many components interacting with each other. Considering the terms in game theory, the idea could be mapped into human's rationality in social science or the forces of nature affecting all of natural phenomenon in fundamental level, and the relationship could be described with many sub-theories of game theory like non-cooperative game or cooperative game.

Most of events and situations could be projected by the eye of game theory on the aspect that there always exist conflicts and cooperation on the interactions in them. Game theory was originated form economics, one of social science, but it applies to not just social system but also the realm of nature. For example we could see the molecular world of materials with game theoretic viewpoint. Molecules in material are connected or interacted with each other by several different kinds of physical forces like metallic bond, ionic bond, van der Waals forces, etc. These different forces are interpreted as relationship policies between molecules for the objective function like organizing the material in stable way. In social system, which is the origin field giving birth to game theory, the relationship is built between many different entities or participants pursuing their own profits within the common good of the entire social system. Based on these examples we could also see game theory as the theory on relationship which affects each member's status. Fig 1 shows us the relationship building process between players independent decision makers under constraints of and interactions with entire system. Each player has its own objective function which might correspond with or opposed to other player's. Considering physical system, the term, 'player' could be replaced by 'component' or 'sub-system' composing the overall system.

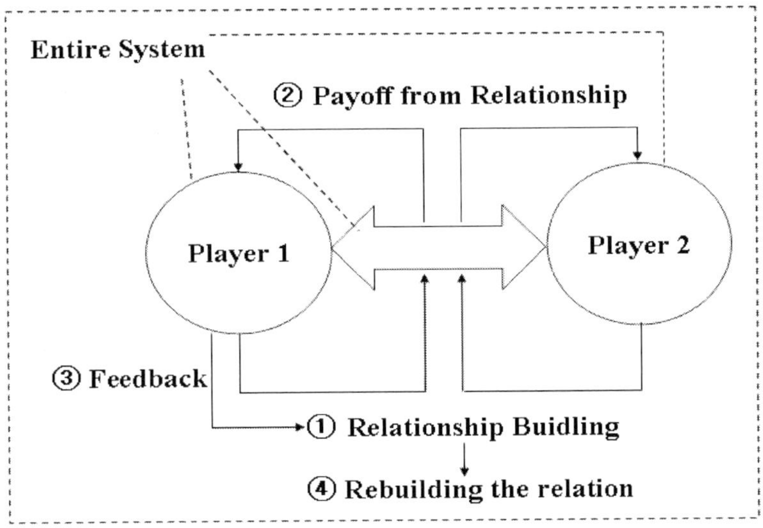

Figure 1. Game Theory for Relationship Building.

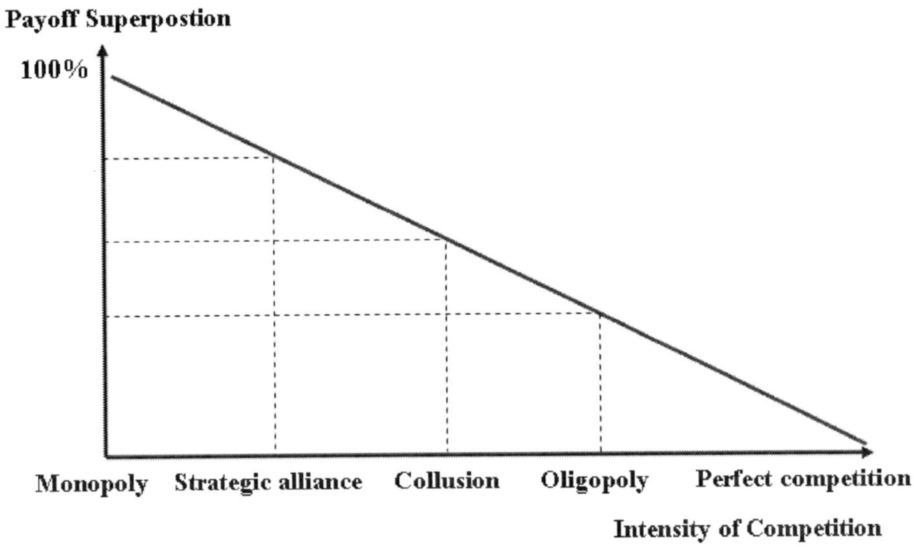

Figure 2. Functional relationship between payoff superposition and intensity of competition.

The quality of relationship is determined based on the superstition of the objective functions of two players. When the objective functions 100% correspond with each other, the game would be full cooperative game. On the other hand, when the interests of two players are completely opposed to each other, the game would be zero-sum game and non-cooperative game. Dependent on the quality of game, the strategy of player should be changed for the maximization of its profit. We just mentioned two extreme cases of cooperative and non-cooperative games, but there exist countless number of relationships as shown in Fig 2. The terms in Fig 2 are quoted as ones in economics. Payoff superposition is inversely proportional to intensity of competition, which determines the quality of the relationship and thereby the game.

Synthetically game theory is considered as a study on the relationship between the parts of a whole system, thereby huge and complex problem could be transformed into multiple sub-problems easier to be solved. On this aspect, game theory is more proper to be applied to the distributed control systems to interact with each other within whole system boundary. Totally different two fields are introduced in this article while there are lots of other fields to which game theory is applied.

One is the case in competitive electricity market, and the other is cyber security problem in communication engineering. One in electricity market is a typical case of game theory application to economics. Electric power industry has been also transformed from vertically integrated monopoly system into deregulated market like many other commodity markets. Under vertically integrated system there is only one decision maker to optimize the resource allocation on system operation. However there have come multiple players performing independent decision making like electricity generating companies, transmission companies, distribution companies, and regulatory entities. This is a typical situation appropriate to be applied by game theoretic solution by analyzing and formulating each player's objective function and the relationship between those players.

The other example is to solve the problems faced in cyber security field which has also many detail categories like decision making for investment on security facilities, hacking pattern analysis and defense strategy, etc. The pattern cyber attacks has generally been one-dimensional type representing as DoS (denial of service), viruses or worms, and unauthorized access, but it has recently diversified into multi-stage and multi-dimensional attacks with a variety of tools and technologies [2].

From next section two different fields are introduced with game theoretic approach respectively, and in last section it is concluded with finding common or general principles on game theory application to the different fields.

3. Application to the Competition in Electricity Market

3.1. Multiple Decision Makers in Deregulated Electricity Market

As the electric power industry has evolved into deregulated electricity markets, there have emerged multiple decision makers pursuing their own objective. Generation company (GENCO), transmission company (TRANSCO), distribution company (DISTCO), load serving entity (LSE), market operator (MO) and other stakeholders are the independent decision makers in electricity market. Electricity market is quite complicated to be analyzed and forecasted because there are too many complexities. One of the complexities is incurred by different layers not existing in a similar dimension like the relation between physical system and abstract market architecture as shown in Fig 3 [3]. The balance between supply and demand, system stability, and load flow on transmission system are the primary requirements of physical dimension, while transactions in pool market and bilateral contract market are the secondary requirements in the facet of business area. Business contracts should be made only unless the contracts violate the constraints of the system to be stable and secure. Therefore many players stated above interacted with each other multi-dimensional environment composed of physical system and market architecture.

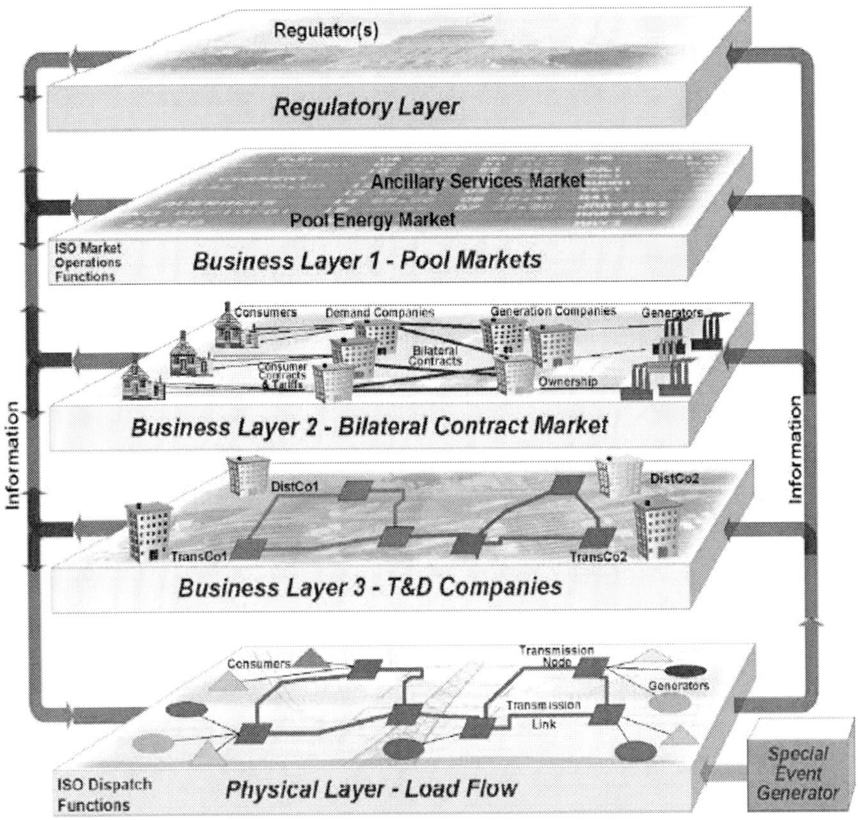

Figure 3. Layers of Electricity Market [3].

Among many players in competitive electricity market, GENCO is a representative entity owning generating facilities and participating in the market with the sole objective of maximizing its benefit, without concern for the system unless there is an incentive for it. This strategic behavior of multiple GENCO is one of the representative characteristics differentiated from the vertically integrated monopoly in which one entity makes the decision to optimize the resource allocation on the aspect of whole system operation. Though the GENCO's planning activities differ with the market structure in which it is operating, electric power sellers in pool markets are required to submit bids (called sometimes offers) to the market operator for the amounts of power that they are willing to trade in the market and the associated price. They would compete for the right to supply energy to the grid, and not for specific customers. If a market participant bids too high, it may not be able to sell [4]. Most evidently, individual bidding strategies are of the essence in interactions where the actions of participants alter one another's possible outcomes. Therefore understanding the strategic behaviors of market participants is very important for all members of market. When we model the GENCOs' strategic behaviors, the important thing is how to define the behavior as a model for mathematical formulation. Recent years witnessed many academic research tries to model these kinds of bidding strategic activities mainly based on Cournot and supply function model. Including these two models game theory related methods have been preferred in many previous researches. On general aspect game theoretic approaches are interested in the interaction between market equilibrium and strategic behaviors of market participants in competitive market environment. Nash equilibrium is one of representative concept in game theory and has been applied to finding out the equilibrium which is different from one expected by traditional economics.

3.2. Game Theoretic Approach to Competitive Market Model

There have been many applications of game theories to electric power transactions in deregulated market. One of the popular models is Cournot model which has been widely used for analyzing oligopoly market in many different kinds of oligopoly markets. Cournot model takes the supply (offer) quantity as independent variable which determines the market price by interacting with demand function which has price elasticity while offer (bid) price is strategic variable in Bertrand model. Lagrangian relaxation method is one of methods used to solve the bidding and self-scheduling problems within a simplified game theoretical framework. As stated above, the idea of Nash equilibrium is geared towards providing reasonable explanation for price clearing in deregulated energy marketplaces where each pool participant has incomplete information about its competitors' strategy. The concept of linear bid sensitivities is employed to determine the optimal bids taking into account not merely individual suppliers' profit maximization but system security as well. Many kinds of network optimization technique has been applied to building Nash equilibrium bidding strategies for generators on transmission network, and a step-by-step algorithm to Nash equilibrium in auction-based multi-period electricity markets.

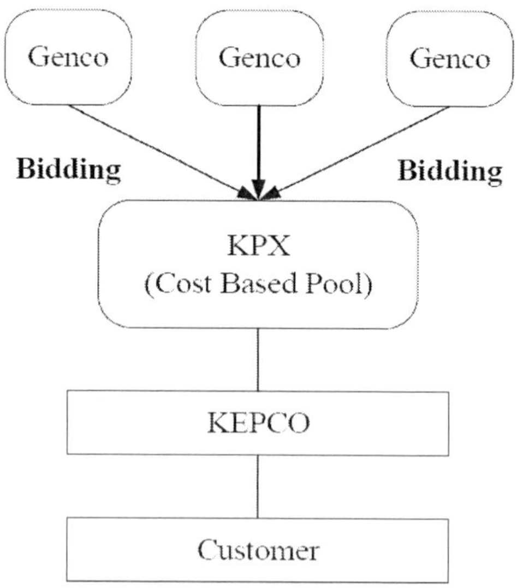

Figure 4. Structure of CBP market.

Competition in electricity market is done based on bidding process. There are many electricity markets around the world, and each market has its own characteristics and rules. However most of them are based on similar bidding process, so we just introduce Korea's electricity market, one of competitive electricity market. At present, the Korea's electric power industry is in the midst of restructuring transition from a vertically integrated structure to deregulated competitive market scheme. The three stages of transition were designed for Korea Electric Power Corporation (KEPCO) which is the vertically integrated monopoly company. First stage is Cost Based Pool (CBP) market in which all GENCOs offer their electricity with generation cost reviewed by public cost evaluation committee. All generators are required to provide details of their production costs, which are then independently checked and approved. Second stage is Price Based Pool (PBP) is similar with CBP on the aspect of one-way bidding from generation side but different from that GENCOs determine the offer price by themselves considering their bidding strategies and profit maximization. In the TWBP market, market prices are determined by bids from generation companies, disco/retailers and wholesale customers. Generation companies and purchasers are able to trade with each other according to market rules. Fig 4 and 5 shows the structure of CBP (PBP) and TWBP market respectively [5].

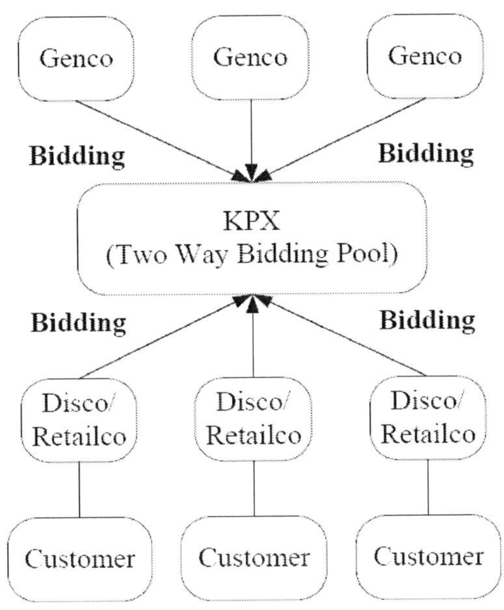

Figure 5. Structure of TWBP market.

After offering from GENCOs and bidding form DISTCOs the price is determined by optimization formulation. The general objective function of electricity market is to maximize the social welfare subject to satisfying the operating and resource constraints. Fig 6 shows the process of market clearing price (MCP) in bidding process [5].

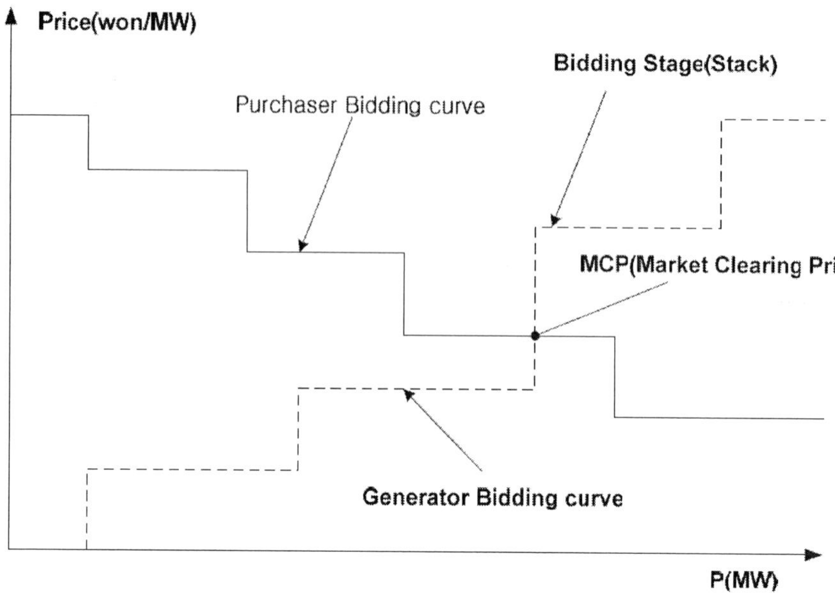

Figure 6. Determination of MCP.

3.3. Strategic Bidding Model of GENCO Based on Offer Stack

Considering the bidding process mentioned in 3.2 GENCOs' strategic behaviors could be simply modeled with two independent variables of offer quantity and offer price. Offer quantity means the generation quantity which a GENCO is willing to supply to the market, and offer price means the least price the GENCO is willing to supply the offered quantity.

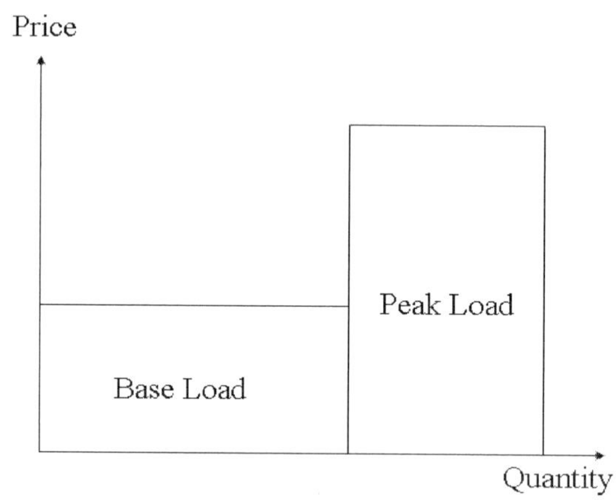

Figure 7. An example of GENCO's stack curve.

Offer quantity means the generation quantity which a GENCO is willing to supply to the market, and offer price means the least price the GENCO is willing to supply the offered quantity. Fig 7 shows a stack curve submitted to the pool. We assume a simplified 2-tuple offer curve composed of base and peak load while the stack curve could be generally illustrated as N-tuple.

When the offer stack of Fig 1 is considered of a GENCO, the available behavior pattern could be defined with limited degree of freedom of strategic variables, which is illustrated as in Fig 2 under the assumptions as follows;

i) GENCO is willing to offer its available capacity fully to the market.
ii) Offer stack is standardized with 2-tuple staircase form.
iii) GENCO has two strategic variables, offer quantities of base load and peak load
iv) GENCO has other two strategic variables, offer prices of base load and peak load.
v) Offer price is determined at least above the cost curve.

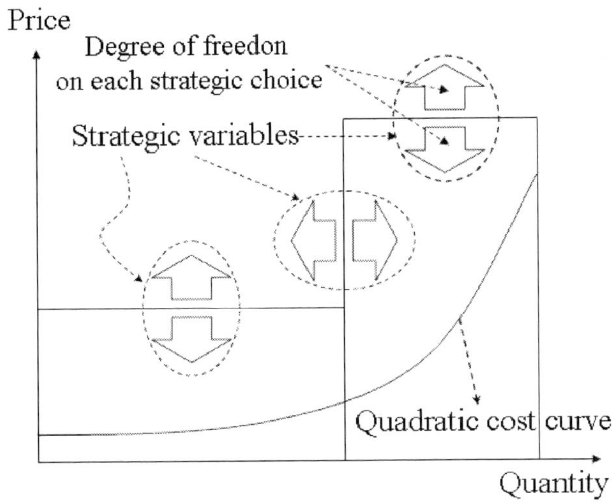

Figure 8. Strategic options on offer stack.

We indicate the notations of GENCO's strategy and profit as S_{GEN} and PF_{GEN} respectively. S_{GEN} is composed of three strategic variables of the ratio and two offer prices of two offer quantity bands. Offer prices of base load and peak load are represented as P_{Offer1} and P_{Offer2}, Offer quantities of base load and peak load as Q_{Offer1} to Q_{Offer2}. Fig 9 can be referred to for understanding these symbols better. The profit of GENCO is the function of offer strategic variables, which can be represented as $PF_{GEN} = f\ (S_{GEN})$. Here, $S_{GEN} = \{(Q_{Offer1}, P_{Offer1}), (Q_{Offer2}, P_{Offer2})\}$. This an example of offer stack model when we model the offer stack having two bands.

If a GENCO is more interested in market share than net profit it would offer the lower price than expected marginal price because its priority lies in the increase of generation quantity supplied to the pool. Fig 10 shows the strategy of lowering the offer price to raise the market share. For determining the price level we could use the Bertrand model mentioned in 3.2.

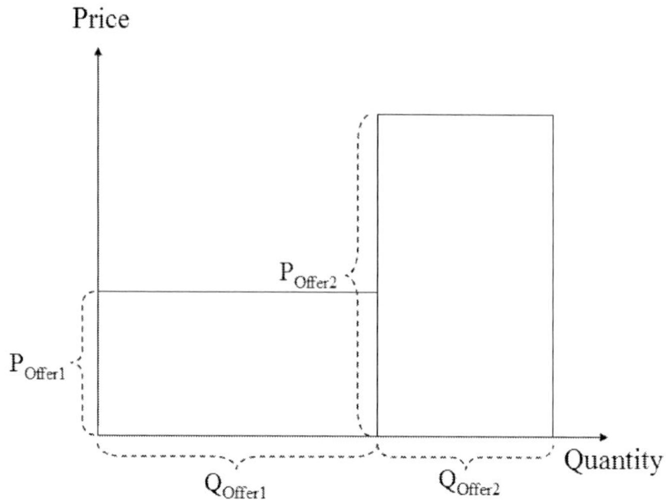

Figure 9. Independent variables in offer strategy.

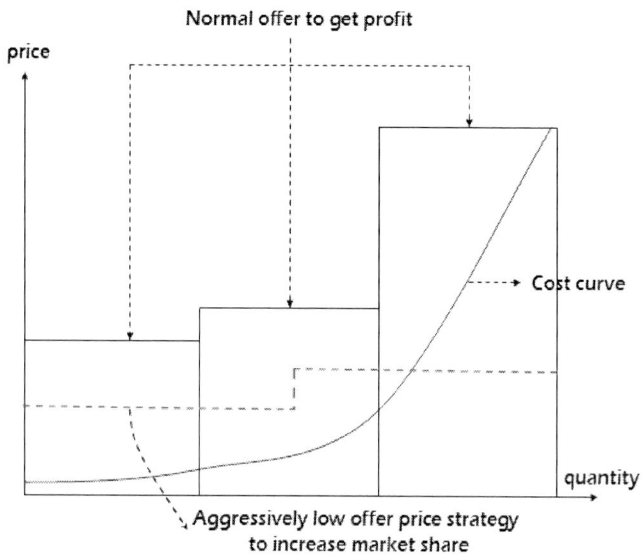

Figure 10. Offer strategy change by objective function.

Genco can choose the offer strategy as seen in Fig 11 if it prefers the risky strategy to pay back higher returns. If the Genco has the marginal generator, it will offer its generation with higher price than expected market price to raise the market clearing price. This is a typical phase of economic withholding in exercising market power.

Nevertheless the GENCO should take a risk of being cut in its offer quantity partly or entirely, leading to the decreased net profit. Accordingly, the offer strategies of GENCOs are influenced by their objective functions and strategic preference on offer or bid. For example, we could assume two GENCOs on competitive bidding process with offer prices and variable system demand in Fig 12 and 13 respectively [6].

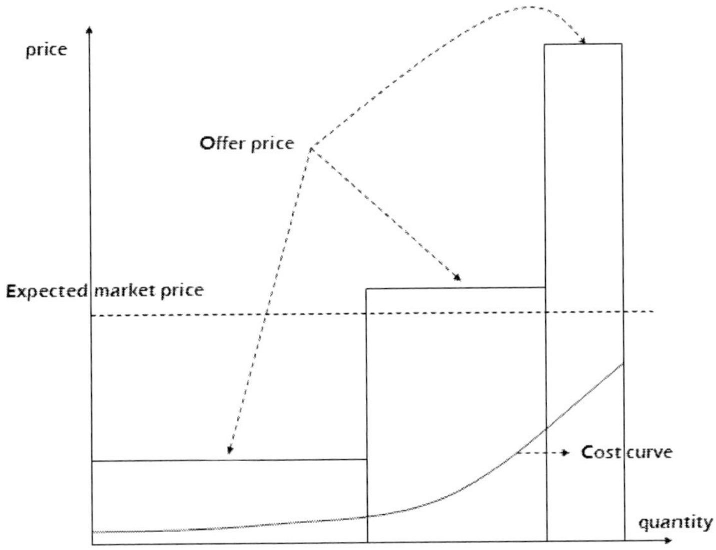

Figure 11. High offer price strategy expecting higher return.

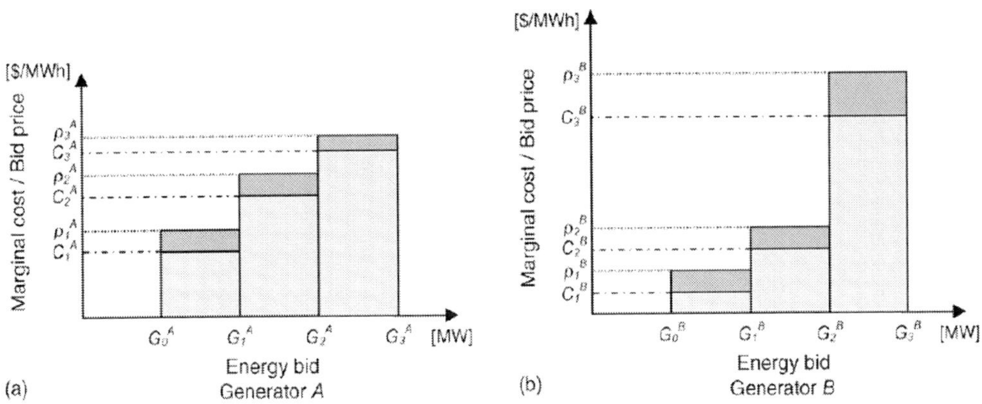

Figure 12. Examples of generation cost and strategic offer price stack of two GENCOs.

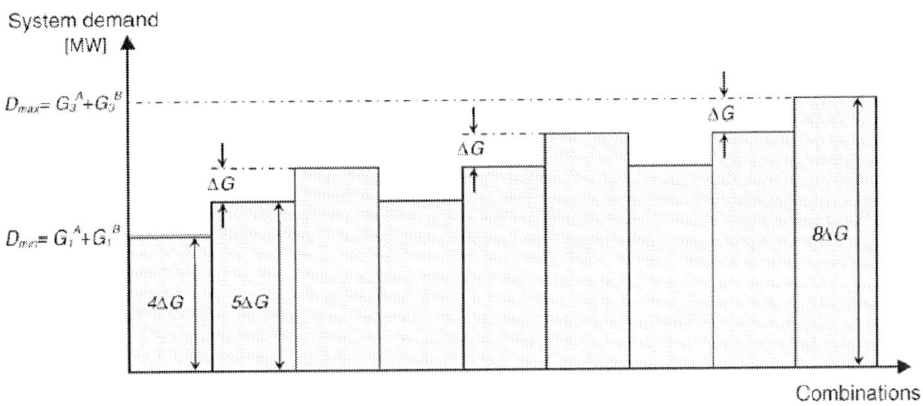

Figure 13. Variable system demand.

If we assume the market status of Fig 12 and 13 we could get the possible scenarios of market equilibriums as shown in Fig 14. And the decision making on how to choose a strategy on the table is differentiated by each GENCO's strategic preference or Nash equilibrium concept as shown in Fig 15, which could be one example of game theoretic approach to bidding process [6].

D_{R5}=8ΔG	D_{f1}^B	D_{f2}^B	D_{f3}^B	D_{f4}^B	D_{f5}^B

D_{R4}=7ΔG	D_{f1}^B	D_{f2}^B	D_{f3}^B	D_{f4}^B	D_{f5}^B

D_{R3}=6ΔG	D_{f1}^B	D_{f2}^B	D_{f3}^B	D_{f4}^B	D_{f5}^B

D_{R2}=5ΔG	D_{f1}^B	D_{f2}^B	D_{f3}^B	D_{f4}^B	D_{f5}^B

D_{R1}=4ΔG	D_{f1}^B	D_{f2}^B	D_{f3}^B	D_{f4}^B	D_{f5}^B				
						8ΔG	8ΔG	8ΔG	8ΔG
D_{f1}^A	4ΔG, 4ΔG	4ΔG, 5ΔG	4ΔG, 6ΔG	4ΔG, 7ΔG	4ΔG, 8ΔG	8ΔG	8ΔG	8ΔG	8ΔG
D_{f2}^A	5ΔG, 4ΔG	5ΔG, 5ΔG	5ΔG, 6ΔG	5ΔG, 7ΔG	5ΔG, 8ΔG	8ΔG	8ΔG	8ΔG	
D_{f3}^A	6ΔG, 4ΔG	6ΔG, 5ΔG	6ΔG, 6ΔG	6ΔG, 7ΔG	6ΔG, 8ΔG	8ΔG	8ΔG		
D_{f4}^A	7ΔG, 4ΔG	7ΔG, 5ΔG	7ΔG, 6ΔG	7ΔG, 7ΔG	7ΔG, 8ΔG	8ΔG			
D_{f5}^A	8ΔG, 4ΔG	8ΔG, 5ΔG	8ΔG, 6ΔG	8ΔG, 7ΔG	8ΔG, 8ΔG				

Figure 14. All conceivable scenarios with the system demand in the real market.

		Generator *B*		
		s_1^B	s_2^B	s_3^B
	s_1^A	(PF_{11}^A, PF_{11}^B)	(PF_{12}^A, PF_{12}^B)	(PF_{13}^A, PF_{13}^B)
Generator *A*	s_2^A	(PF_{21}^A, PF_{21}^B)	(PF_{22}^A, PF_{22}^B)	(PF_{23}^A, PF_{23}^B)
	s_3^A	(PF_{31}^A, PF_{31}^B)	(PF_{32}^A, PF_{32}^B)	(PF_{33}^A, PF_{33}^B)

Figure 15. Equilibrium determination on game matrix of two generators' bidding competition.

3.3. Computerized Tool for Strategic Bidding Based on Game Theory

The strategic bidding is usually very complicated because there are too many variables considered in real electricity market. Therefore it has required a computerized tool to perform the market simulation. For fulfilling the demand for market simulation, many global venders like GE, Henwood, Drayton Analytics, CRA, etc. have developed and provided electricity market simulators. Most of these simulators are based on the optimization formulation which has been used mainly for the least cost resource planning in the centralized power system planning and operation. From this standpoint, it seems somehow inevitable to face many challenges on modeling competitive market based on the method of traditional market simulators. In this paper, we propose a kind of new method, which is MAS based market simulation. The agent based model has already been introduced in EMCAS, one of commercial market simulators, but there may be various ways of modeling agent. This paper, in particular, seeks to introduce a model for MAS based market simulator.

There are two ways of modeling market simulators according to the approaches of handling the problem. One is the analytic method using the strict mathematical formulation with optimization techniques. The other is the empirical method based on trial and error concept using heuristic methods, for example genetic algorithm, tabu search, simulated annealing, and so on. Analytic approach has been widely used in current market simulators, whose original formulation was made for the resource planning model based on production cost minimization or social welfare maximization in a vertically integrated power industry. Empirical approach has been being introduced for making up the weak points of analytic model on formulating electricity market environment consisting of many decision makers, which is entirely different from traditional industry environment in which there is only one decision maker. The characteristics of two methods are summarized in Tab 1.

Table 1. Characteristics of Two Simulation Models

	Analytic model	Empirical model
Simulation method	Mathematical Optimization	Trial and error based on heuristic approach
Outcomes	Deterministic value on each scenario	Stochastic value having pdf
Commercial Simulators	PLEXOS, CeMOS	EMCAS

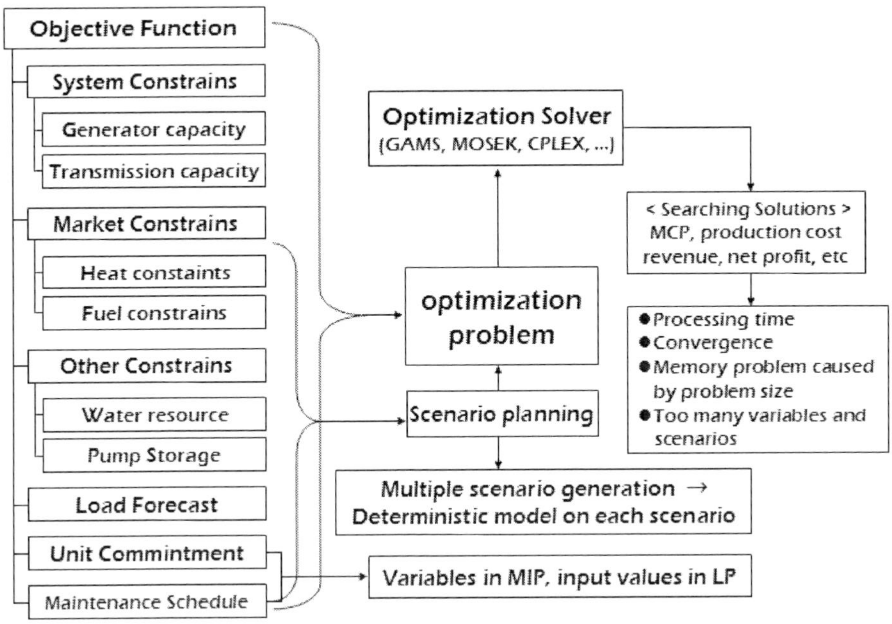

Figure 16. Structure of traditional market simulator.

Traditional simulators have applied the optimization formulation to market simulation. Linear programming has been preferred among many kinds of different optimization methods like linear programming, non-linear programming, combinatorial optimization, etc. because of its assurance on convergence and simplicity of modeling. Generally, generation cost minimization and social welfare maximization may be the objective functions of optimization problem. Physical capacities of generators and transmission lines are considered as the constraints of the optimization problem. Optimization was a good method for the centralized resource planning and operation scheduling problem of vertically integrated utility in the electric power industry since there is only one decision maker which is the best case in considering the simulation as an optimization problem. But the formulation as one optimization problem of overall system or market tends to make the processing time longer and sometimes even shut down the computer by a critical memory problem due to the big problem size. Both PLEXOS of Drayton Analytics and CeMOS of CRA were created on the basis of the structure shown in Fig 11. Also, they adopt commercial optimization solvers like MOSEK and GAMS for solving the formulated optimization problem.

3.4. MAS (Multi-agent System) Based Market Simulator

The concept of "agent" has been introduced in artificial intelligence when we describe the program which performs some tasks specialized in some field or very complicated jobs to be done instead of a human. Multi-agent is a collection of more than one agent and has common characteristics like autonomy, social ability, intelligence, etc. Autonomy means that agent can judge by itself and performs some tasks based on the judgment without the orders from humans or other programs. For example, if we apply agent technology to information search engine, then agents are able to collect, analyze, and arrange the information by itself even when there is no commands for searching from humans, while traditional information search engines do their job simply when they get orders. Social ability means the agent's capability to cooperate with other agents to perform some tasks or accomplish some objects. Agent is not a stand-alone program but a member of society composed of a number of agents who cooperate with each other for the objective or a common good of the society. Each agent has its own unique role, and some agent has the role of coordinating the agent. Intelligence means that agents perform tasks or accomplish objective according to their own reasoning and judging process not just based on codes or program already made by human, implying that agent can have the creativity or evolution on problem solving process.

The MAS based market simulator could have multiple agents each of which is in charge of their unique roles for the operation of power system and electricity market. For instance, we can establish a society composed of multiple agents and define their relationships using the game theory. Fig 9 is the MAS based version of traditional optimization model like in Fig 11.

System agent is normally responsible for power system operation related to load flow and voltage stability subject to the generator installed capacity and transmission line capacity. Market agent tries to perform the market operation for the cost minimization and social welfare maximization under the heat constraints and fuel constraints. Energy constraint agent is obligated to manage limited energy resources like water flow, pump-storage, renewable energy sources. Agent may be also regarded as individual market participant like GENCO, TRANSCO, DISTCO, etc., which is the different aspect from the traditional market simulator having only one decision-making entity. Each agent represents each market participant having its own utility function and pursues the maximization of the utility function. Profit maximization can be another objective function of GENCO and DISTCO. Social welfare maximization may be the objective function of MO (market operator). Relationship between agents is duly defined using game theory.

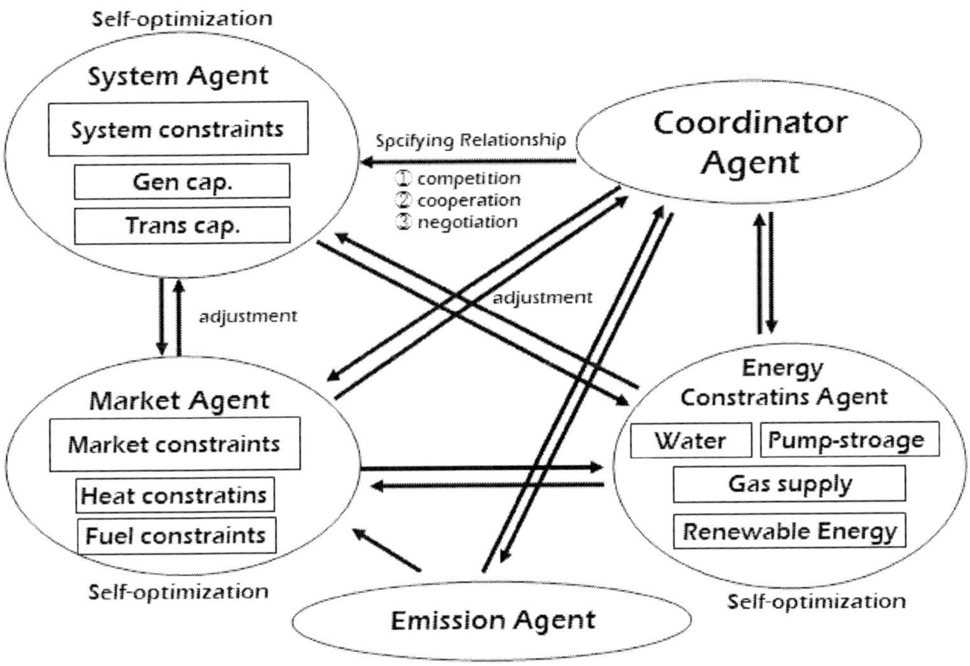

Figure 17. MAS based on Functional Composition.

A's action

A's payoff

	R	P	S
R	0	−1	+1
P	+1	0	−1
S	−1	+1	0

B's payoff

	R	P	S
R	0	+1	−1
P	−1	0	+1
S	+1	−1	0

Figure 18. Payoff Matrix.

Game theory provides us with three different kinds of gaming situation classified as cooperative game, non-cooperative game, and negotiation (or bargaining) game. In the same fashion, each game situation has a few sub-models like Cournot model, Bertrand model, Stackelberg game, Nash bargaining game, etc. A game is specified as n players, their strategies and payoff matrix as shown in Fig 6. If two payoff matrices of two players are the same or proportional, the game of two players is a fully cooperative game. If the payoff matrices are completely different from each other, the game would be a non-cooperative game.

The strategic choice for the relationship between two agents is wholly determined by the judgment of agents. The choices of agents may be different from each other even in same situation dependent on the objective function or the strategic preference of each agent. Each GENCO has its own offer stack differentiated by its objective function and strategic preference. Fig 13 shows us an example of GENCO agent.

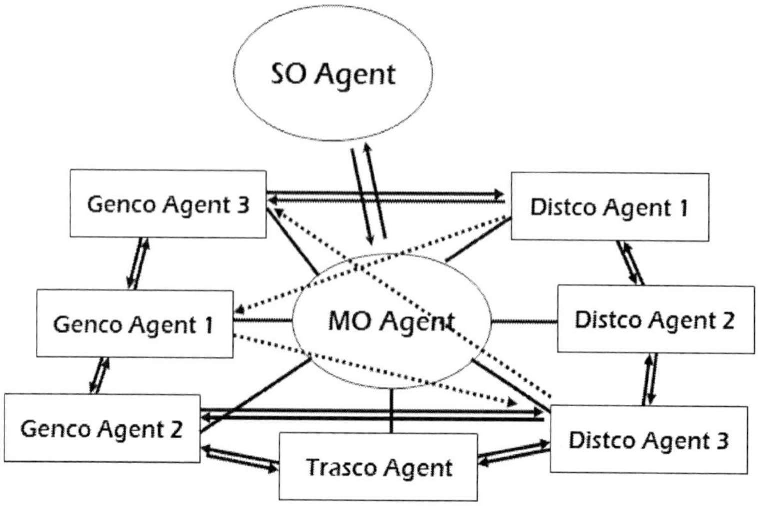

Figure 19. Agent Representing Market Participant.

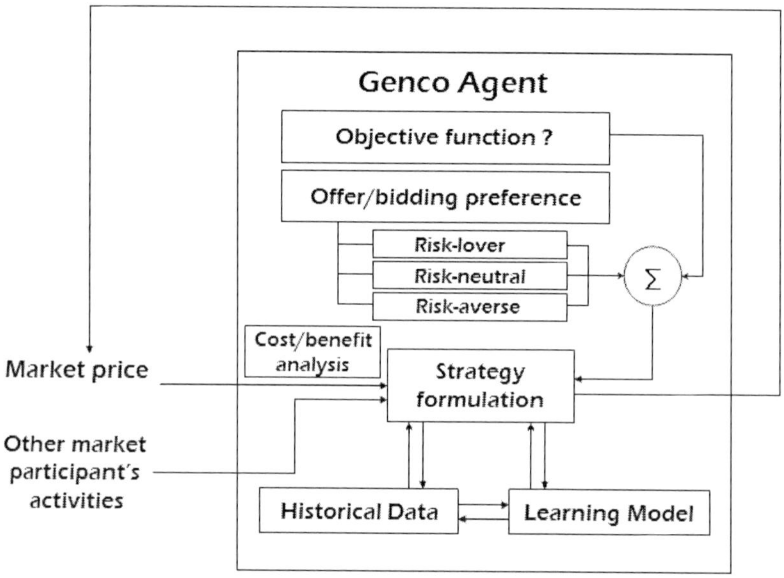

Figure 20. An Example of Agent based GENCO model.

The MAS based market simulator reflects these kinds of different tendencies and strategic evolutions based on intelligence or learning ability. And it makes the model more realistic in the fact that each GENCO has its own decision making independency. Traditional market simulators focused on the mathematical formulation of an optimization problem and the algorithm for searching the solution. For mathematical formulation, the problem model was required to have a deterministic objective function and related constraints, which leads the model too deterministic and static. It has been of the highest importance to find out an optimal solution for investment on generators and transmission lines under vertically integrated environment since there is only one decision maker for the investment. Under the competitive market the equilibrium replaces the concept of an optimal point in the steady-state. And the dynamism of market also gets importance in that the market players always try to unbalance the equilibrium for increasing their profit while the market operator try to maintain it. Even if we find out the equilibrium of the market at a specific moment, it would be negligible by the reactions of market players within a short time-horizon and require a new equilibrium to be found. Capturing these characteristics of market environment, the emphasis of the MAS based simulator is on making the problem model itself more realistic as close as possible to the real situation rather than solving the problem and finding out the optimal solution. Although there is no obvious evidence which method is better for competitive market simulation, the MAS based market simulator is expected to overcome several weak points of traditional market simulators.

4. Application to Cyber Security of SCADA

4.1. SCADA System Introduction

As the power industry relies increasingly on information to operate the power system, two infrastructures must now be managed: not only the Power System Infrastructure, but also the Information Infrastructure. The management of the power system infrastructure has become reliant on the information infrastructure as automation continues to replace manual operations, as market forces demand more accurate and timely information, and as the power system equipment ages. Therefore, the reliability of the power system is increasingly affected by any problems that the information infrastructure might suffer [7].

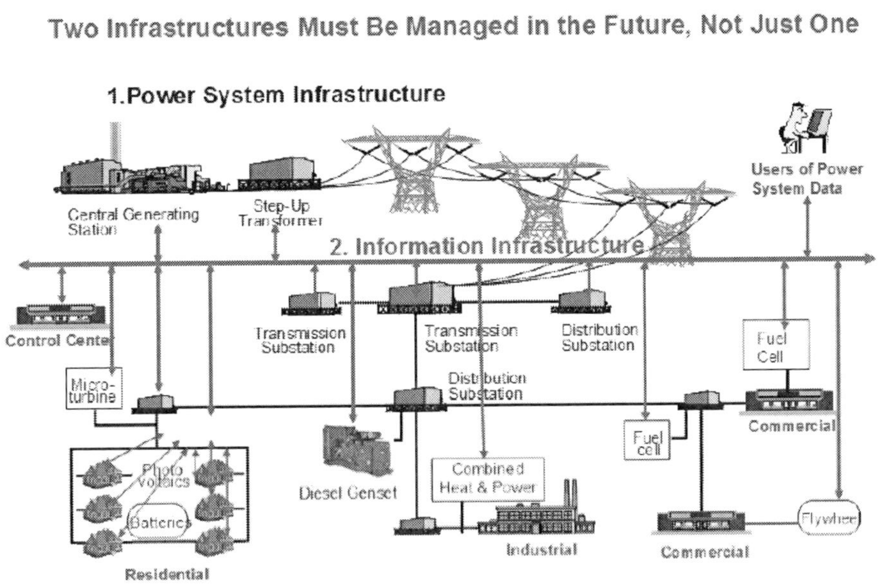

Figure 21. Two infrastructures in power system [7].

SCADA (Supervisory Control and Data Acquisition) system is a system operation with coded signals over communication channels so as to provide control of RTU (Remote Terminal Unit) equipment [7]. Recently Intelligent Electronic Device (IED) which is control unit having communication function with master station is replacing the role of RTU. The communication system links the control center with IEDs. Common methods of communication include radio, leased line, landline, and digital and analog microwave. SCADA system has been used for remote measurement and control on the critical infrastructures such as electric power, gas and oil as well as modern industrial facilities such as chemical factories, manufacturing facilities. More recently analog and digital cellular communication has been introduced. For remote service, satellite communication is sometimes employed. SCADA security in communication typically refers to the ability to perform error correction, rather than authentication or encryption [8].

4.2. Cyber Security Problem in SCADA System

SCADA network has been exposed to general cyber attacks with IT advancement and network growth. Especially, SCADA systems of energy industry such as electric power, gas and oil are vulnerable to targeted cyber attack and terrorism because the attacks to these infrastructures could cause huge loss in an entire social system. According to a DOE report released at 2005 year, the portion of energy industry takes almost 70% in main infrastructures being attacked in the States from 2002 to 2004 year. Several trends have led to an environment in which the security of SCADA system has become more vulnerable. The representative weaknesses are caused by the use of common operating systems such as Microsoft Window and Unix in SCADA and control system platforms, the increased use of TCP/IP communications and the demand from corporate users for operational data on a near-real-time basis.

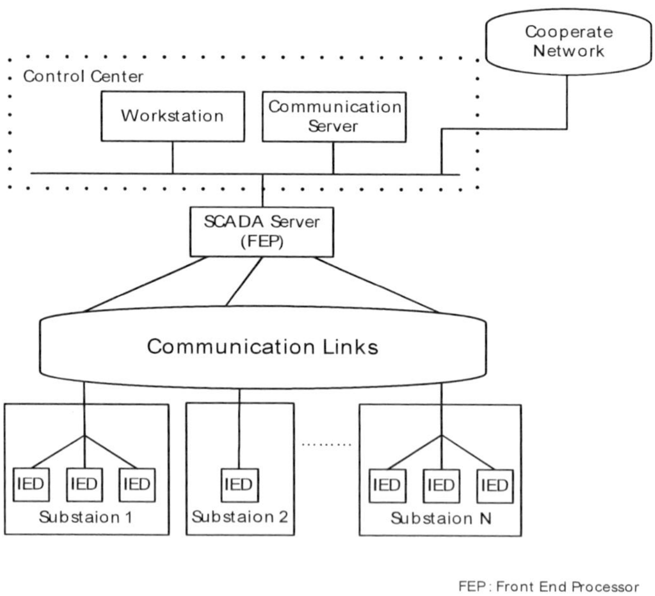

Figure 22. SCADA system configuration.

Recently, research efforts to solve the problems have been progressed in SCADA network security. There are many research challenges such as access control, firewall, intrusion detection system, protocol vulnerability assessment, cryptography, key management, devices/OS security and security management for SCADA networks [9]. Cyber attacks could also cause the system fault similar with physical attack. A coordinated attack on major power plants or substations could trigger a cascading blackout with major social and economic impacts [10]. For example, computer hacker could destroy a substation transformer by sending the transformer overload signals, causing it to rapidly overheat and explode like doing it by a bomb or setting a fire [11].

Figure 23. Attack to Substation [11].

Measurement data and control signal is exchanged on the communication line between RTU and Master Station like in Fig 24. When we assume there is no internal approaches from inside RTU or SCADA master station, one of the probable methods to crack the system is to tap a communication line directly because the network is closed to other networks and thereby allows no detour route to access the SCADA system. To protect the system against this kind of intrusion it is shown to encrypt the information by cryptography in next section.

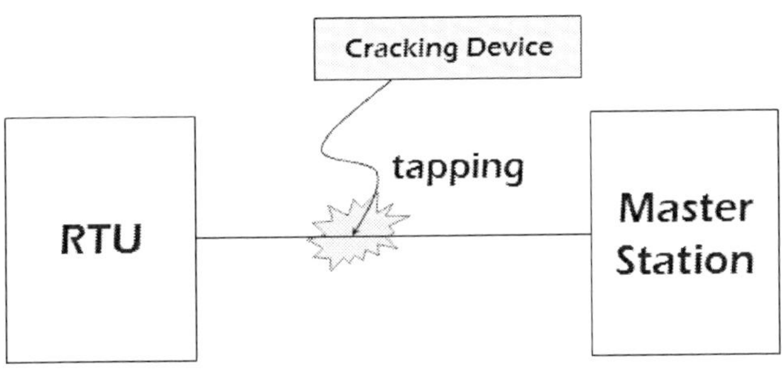

Figure 24. Intrusion into a communication link.

4.3. Encryption Methods - Symmetric and Asymmetric Encryption Methods

Encryption is the most common and powerful approach to securing the system. There are two fundamental alternatives for the location of encryption gear or device: link encryption and end-to-end encryption [12]. However end-to-end encryption is enough for radial network type of Korea's SCADA network. There are various cryptographic methods already developed for the encryption and decryption of information. Those cryptographic algorithms could be categorized into two different groups, which are symmetric and asymmetric encryption. Symmetric encryption algorithm could be characterized by the fact that the decryption key is identical to the encryption key. Symmetric encryption, also referred to as conventional encryption or single-key encryption was the only type of encryption in use prior to the development of public key encryption in the 1970s [13]. The key must be exchanged in advance between sender and receiver in a secure manner and must be kept secret [13]. Fig 25 illustrates the basic process of symmetric encryption.

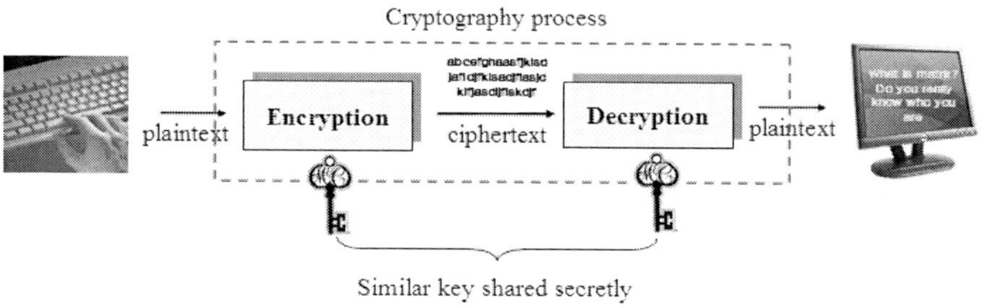

Figure 25. Symmetric encryption.

The concept of public key cryptography evolved from an attempt two of the most difficult problems associated with symmetric encryption. Public key (other name of asymmetric key) algorithms have different encryption and decryption keys, and the latter cannot be derived from the former by any efficient algorithm [12]. Fig 26 shows us the illustrated concept of asymmetric key based encryption method.

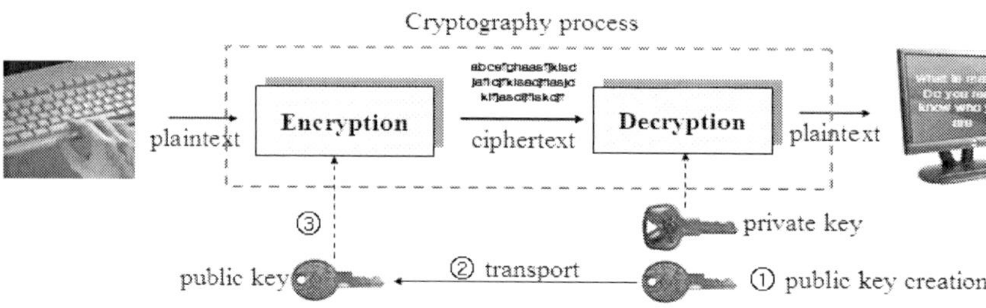

Figure 26. Asymmetric encryption.

4.4. Encryption Method for SCADA Communication

We focus on the encryption method although there are many kinds of methods for strengthening the cyber security of SCADA system. Encryption strengthens the security by protecting the information on the network from attack and thereby decreasing the vulnerability of network. However the encryption itself is always exposed to the danger of being cracked. So we change the secret key of encryption periodically and the level of danger would increase as the time duration of key distribution period lasts longer. But if we make the period too short, it could cause the inefficiency and load increase of network and key distribution server. Symmetric encryption is proposed to be applied to SCADA system because of next two reasons.

The first reason is the number of communication combinations in SCADA network, which is very different from the one in usual networks. Assuming N hosts, there are [N(N-1)]/2 cases for the pairs of communication in usual networks like Ethernet or Internet in which each host can communicate with any other hosts in the network. As increasing the number of hosts in general network or internet, the number of communication combination increases (N-1)N/2 when we assume the number of hosts as N, which is a quadratic functional increase of N. However on SCADA network one additional RTU adds only one communication combination to previous combinations. So it increases linearly as the number of RTUs increase because all RTUs only communicate with only on master station. So, it seems reasonable to apply symmetric encryption to SCADA communication information when considering the number of keys to be shared.

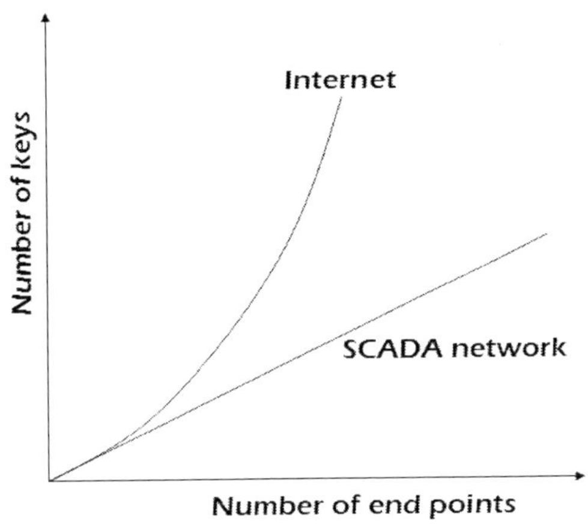

Figure 27. Number of keys as a function of number end points.

The second reason is the process speed required in SCADA communication. SCADA deals with mass size data in a very short period. Even when there is a system failure, there is a possibility of traffic jam on network which might be worse by the encryption process. So it is recommended to reduce the time taken in encryption process as short as possible. Considering the time consumed for encryption simply, the symmetric encryption is the better method for SCADA communication. However the symmetric encryption is more vulnerable to attack compared to asymmetric encryption. Complementary measures are required in this aspect when we use the symmetric encryption.

For symmetric encryption to work, two parties involved in communication must share the same key, and that key protected by access by others [12]. We could think of two kinds although there are several ways of key distribution methods. First one is that the communication initiator makes the key and sends it to the responder as shown in Fig 28. We call this method decentralized key distribution. There is no key distribution center on this method. Initiator A requests B to send the session key at ①process, and B responds to A with the key encrypted with master key already shared with A at ②. And finally A confirms the key distribution process at ③.

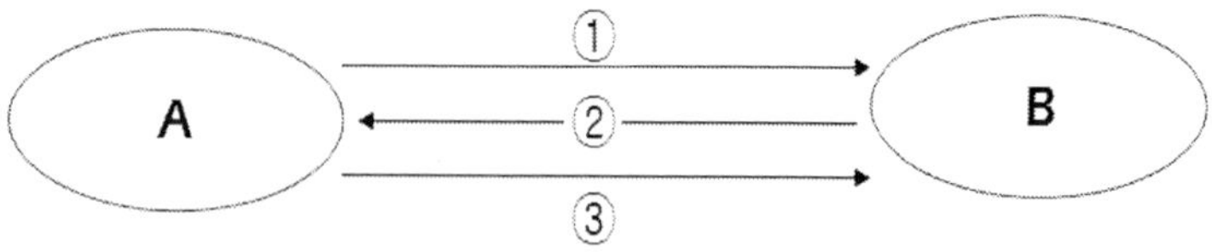

Figure 28. Decentralized Key Distribution.

Second one is that the third party makes the key and distributed to the initiator or both of them, which is called as centralized key distribution in Fig 28. The flow notated with solid lines and Arabic numbers indicate the key distribution to only communication initiator A, while the one with dotted line and alphabets indicate the key distribution to both parties involved in communication. ①, ② and ③ are same processes also corresponding to ⓐ and ⓑ as in the decentralized key distribution of Fig 29. ④ and ⑤ are authentication processes for the session key shared by two parties.

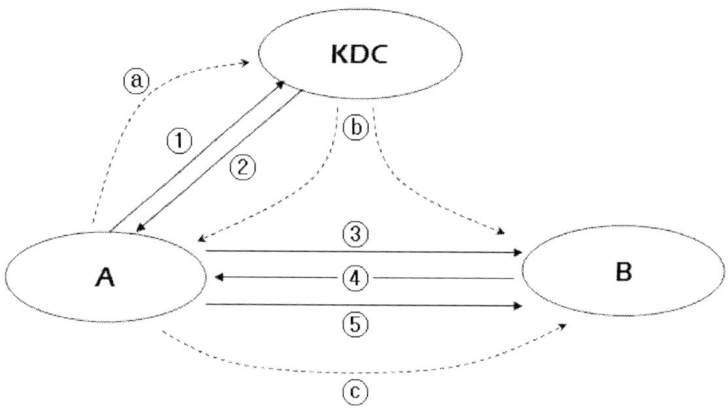

Figure 29. Centralized Key Distribution.

In this case we need KDC (Key Distribution Center) for key management. Considering Internet network it is better to have KDC to manage key distribution process, because there are lots of hosts in the network and also lots of communication combinations. In this aspect we need some kind of key distribution policy to keep or increase the security level of encryption while maintaining the efficiency of the system, and we propose a flexible key distribution scheme based on multi-agent concept and security assessment method.

4.5. MAS[1] Applied Key Distribution Process

Encryption strengthens the security by protecting the network from attack and thereby decreasing the vulnerability of network. However the encryption itself is always exposed to the danger of being cracked as stated in previous section. So we change the secret key of encryption periodically and the level of danger would increase as the time duration of key distribution period lasts longer. But if we make the period too short, it could cause the

[1] Multi-agent system

inefficiency and load increase of network and key distribution server. In this aspect we need some kind of key distribution policy to keep or increase the security level of encryption while maintaining the efficiency of the system, and we propose a flexible key distribution scheme based on multi-agent concept and security assessment method.

MAS based system makes it possible to allocate many functions centrally controlled to local agents, which is expected to increase the efficiency and flexibility of the system.

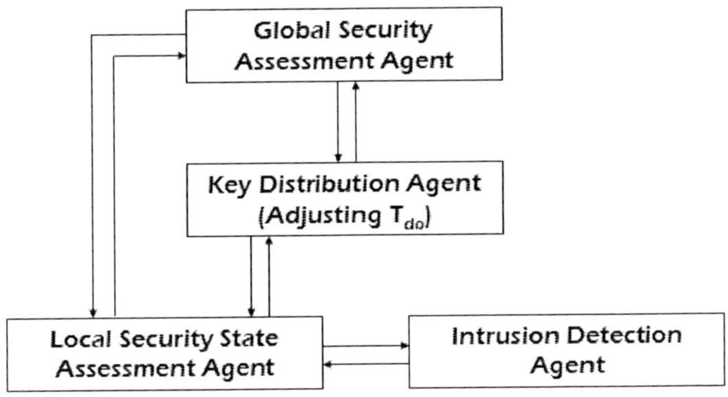

Figure 30. Multi-Agent based flexible key distribution.

Fig 30 shows us MAS based key distribution process. T_{dp} stands for key distribution period. IDA (Intrusion detection agent) monitors whether intrusion is or not periodically on a specific node or an area which it is in charge of. IDA has previously input knowledge on many types of intrusion patterns and has capability of learning and analyzing new patterns of attacking. IDA could reside in each communication node or have control of a local area composed of more than one node. Fig 31 shows node based IDA. Each IDA stays in its responsible node continuously and watches if there is intrusion at the node.

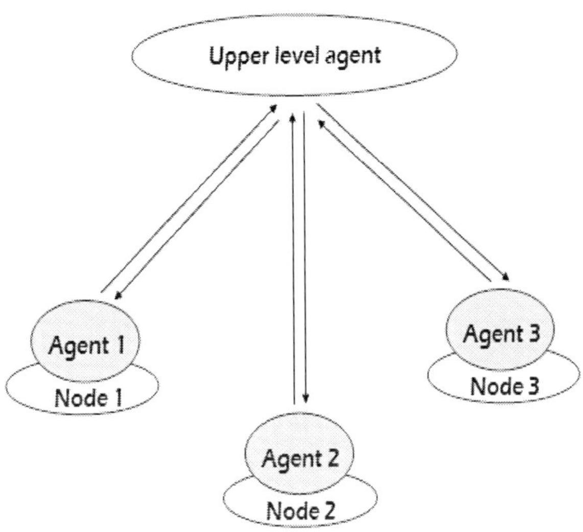

Figure 31. Node based ID agent.

Fig 32 shows area based IDA. When there are too many nodes in the system, it is impossible or inefficient to dispatch all agents to all nodes. In this case we could divide the whole system into several areas by grouping those nodes. Agent is in charge of its nodes in its responsible area.

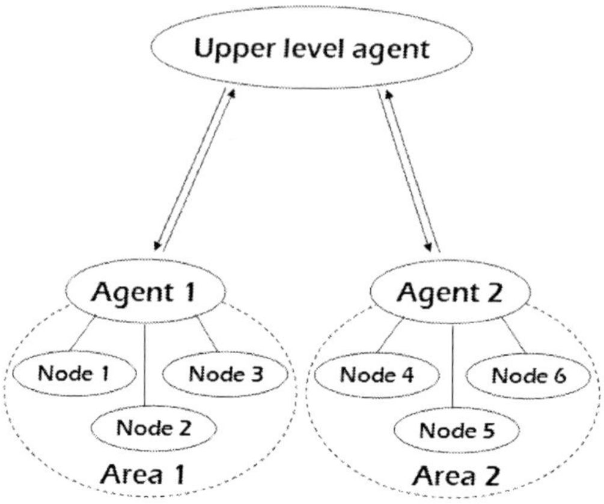

Figure 32. Area based agent.

We introduce two kinds of agent to manage the security of nodes, areas and entire system. Local Security State Assessment (LSSA) Agent performs the analysis on each node or each area, and thereby they check local based security level or vulnerability based on mathematical model using random variable distribution model. Global Security Assessment (GSA) agent performs the assessment global security state or vulnerability of whole system. When the security key is renewed based on whole system, key distribution (KD) agent communicates with VIA agent to adjust the period of key distribution according to the result of security assessment of GSA agent. When this key distribution is done at each area, key distribution agent communicates with LSSA agent directly to determine the new period of key distribution. It depends on security or key distribution policy, and Fig 33 shows this communication and calculation flow between agents dependent on the policy.

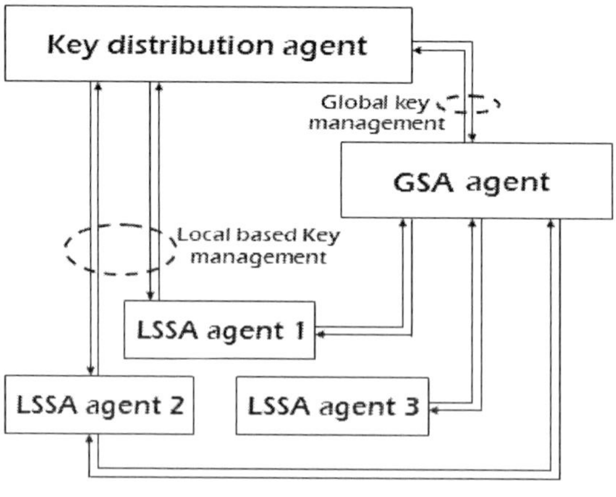

Figure 33. Communication between agents.

4.6. Relationship Modeling between Agents Using Game Theory

We introduced 4 agents in Fig 33, which are KDA, IDA, GSA agent, and LSSA agent. Basically all agents in the system are required to cooperate with each other. They have common object to accomplish the good performance and security of the whole system. Therefore the relationship between agents could be modeled using cooperative game theory. However there might be also a conflict between agents like GSA and LSSA agents. GSA covers whole system and is responsible for the reliability of entire system, while LSSA is a local agent in charge of security and performance monitoring in its node or region. Here we define the reliability as the overall index induced by the security and performance level of the system. When we consider the local and global reliabilities at the same time, it is required to be balanced between local and global aspects. Considering this situation, we could assume the trade-off between local and global reliability as bargaining process between LSSA and GSA agents, and thereby the NBS (Nash bargaining solution) could be an appropriate solution to model the relationship between two agents.

4.6.1. Nash Bargaining Solution

Nash bargaining solution is used for modeling cooperative games of two players. When there are two bargainers, A and B seeking to split a total value v which they can achieve if and only if they agree on a specific division, they would get their final outcomes x for A and y for B respectively. Because two bargainers are supposed to share the common outcome v, the equation $x + y = v$ should be fulfilled. If no agreement is reached, A will get a and B will get b, each by acting alone or in some other way acting outside of this relationship. Thinking differently, a and b could be considered as the initial cost of investment for obtaining the outputs, x and y, or the opportunity cost of x and y. Here, a and b are named as their backstop payoffs or, in the jargon of the Harvard Negotiation Project, BATNAs (best alternative to a negotiated agreement) [14]. Of course $a + b < v$ should be satisfied. When we assume two bargainers divide the common surplus v with the fraction of α for A and β for B, two bargainer's outcomes x and y could be also expressed respectively as the sum of each BATNA and backstop as follows:

$$x = a + \alpha(v - a - b) \rightarrow x - a = \alpha(v - a - b) \tag{1}$$

$$y = b + \beta(v - a - b) \rightarrow y - b = \beta(v - a - b) \tag{2}$$

If we divide (1) by (2), we could get another form of equation as follows:

$$\frac{x - a}{y - b} = \frac{\alpha}{\beta} \tag{3}$$

There are three principles prerequisite to model cooperative game as Nash bargaining solution [14].

i) The outcome should be invariant when the scale of payoffs changes linearly.

ii) The outcome should be efficient, which means $x + y = v$, therefore there is no unexploited gain.

iii) The outcome should be independent from other variables except x and y we are considering.

When three assumptions are fulfilled, the bargaining game can be modeled as following Nash bargaining formula.

$$\underset{x,y}{Max}\ (x-a)^{\alpha}(y-b)^{\beta} \tag{4}$$

subject to $y=f(x)$

Besides three assumptions mentioned above, Nash originally imposed fourth assumption $\alpha=\beta$ which means both parties share the outcome equally, thereby equation (4) could be re-written on the unique case.

$$\underset{x,y}{Max}\ (x-a)^{1/2}(y-b)^{1/2} \tag{5}$$

4.6.2. Application of NBS to Relationship Model between Agents

When we consider Fig 33, the agents are in cooperative relation and need to interact with each other to increase the reliability of entire system. The reinforcement of local security generally increases the global security of the entire system. However there is always the problem of limited resources and is required to allocate the limited resources in optimal way.

What is the limited resource in network communication and encryption process? One of the resources would be time. Encryption process causes network traffic increase and thereby communication time delay which result in the reliability degradation. Therefore we need to find a balance between security strength and network performance, which is the role of agents like LSSA or GSA agents in Fig 33. The second resource would be the communication network shared by many entities or agents. The network is also limited to be fully used by each of all entities on communication.

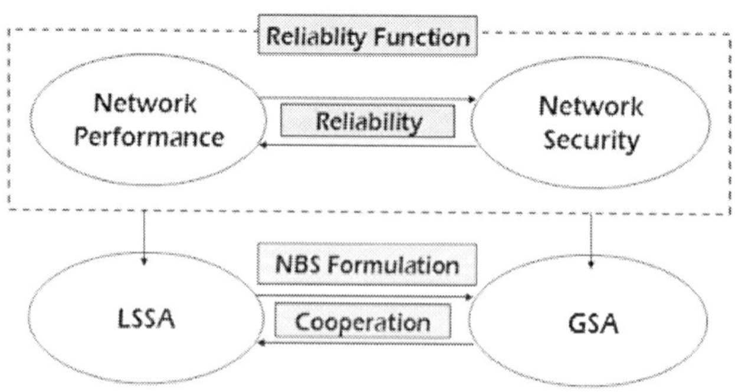

Figure 34. NBS application to Agents' relationship.

Considering two aspects mentioned above there are two bargaining dimensions on the problem. One is the bargaining between security and performance. The other is the bargaining between LSSA and GSA agents for optimizing the resource allocation. This concept is illustrated in Fig 34. Security and performance are co-optimized in the reliability function and thereafter agents adjust for the optimal balance on the entire system aspect.

Flexible key management scheme based on MAS is proposed for enhancing security level of SCADA network. And it could be modeled based on NBS which is one of game theories for cooperative game. There are many different kinds of alternatives for game theories, which model applied would be dependent on network or communication characteristics and objective function. We use the NBS for cooperative game model between agents because we intend to find out a balance between performance and security. So there could be more flexibilities and varieties for the key management process using game theory applied MAS.

4.7. Block Matrix Model for Decision Making on Choosing Security Methods

As network systems become more complex, the implementation of the security policy should be considered on many aspects like the strength level of security, economic problem, etc. It has been always an important issue to balance between technology and economics not just in security field but also in most of engineering fields. Security engineering is focused on the security requirements for securing the system on both aspects of software or hardware. The scope of security engineering could be elaborated as follows [15]:

- the security engineering activities for a secure software or a trusted system addressing the complete lifecycle of: concept definition, analysis of customer's requirements, high level design and low level design, development, integration, installation and generation, operation, maintenance end de-commissioning;
- requirements for product developers, secure systems developers and integrators, organizations that develop software and provide computer security services and computer security engineering;
- application to all types and sizes of security engineering organizations from commercial entities to governmental organizations.

Implementation of any security policy is subject to limited technology level and economic resource. If security investment is not sufficient to prevent the threats expected, the existence of the countermeasures can not be regarded as real countermeasures expected and just considered as like waste. On the other hand if security countermeasures are built with overinvestment to the risks owners really have, it would be the waste of their limited economic resource which could have been invested to other fields. Considering the characteristics of security field, minimum investment at every node is required to maintain some level of security of entire system as shown in Fig 35, because just one vulnerable node exposed to attacks could cause the danger of entire system. This kind of trade-off problem could be modeled using NBS as shown in Fig 34.

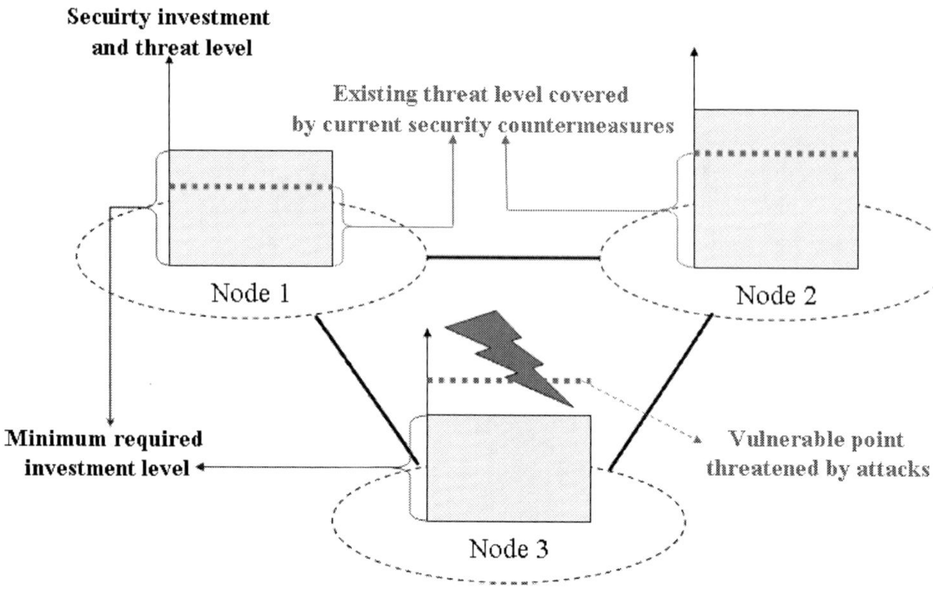

Figure 35. Security investment against existing threats on network system.

It is assumed IT system is divided into 4 parts which could be differentiated by system environments, and each part may have three common components of product, technique and operation as shown in Fig 36. It means the system consists of 4 different parts and security countermeasures are composed of 3 aspects such as product, technique, and operational and human problem. Therefore there are totally 12 (4*3) blocks for security investment.

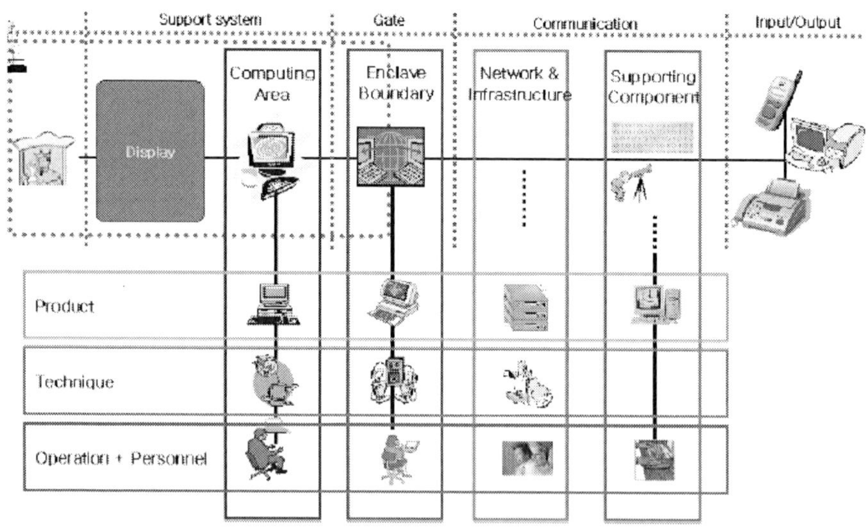

Figure 36. Block matrix [15].

Security methods or investments could be various in many different kinds of ways considering which factors to be considered as shown in Fig 37. For example security method A indicates the investment done at computing area and enclave boundary on the aspect of personal and operation factors. Fig 38 is the 3-dimensional version of Fig 36 informing of the security assurance level by security methods.

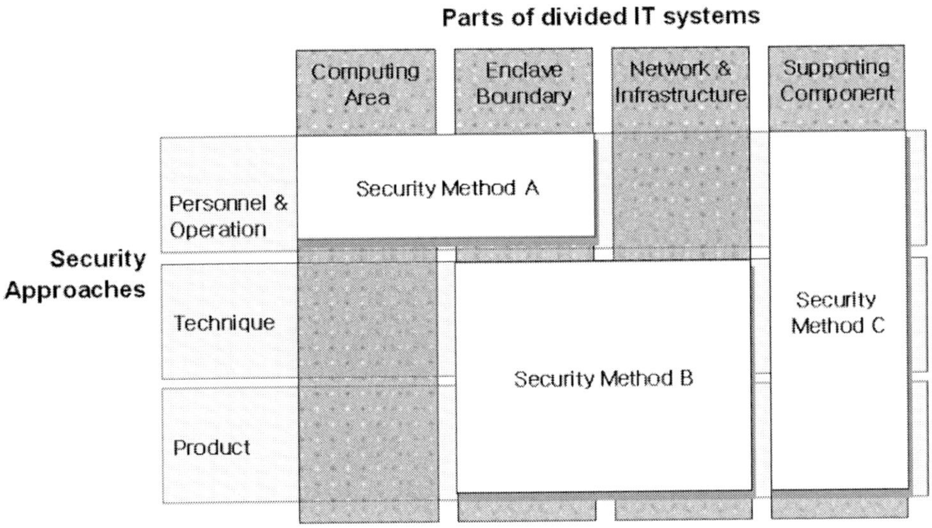

Figure 37. Block matrix model for security methods [15].

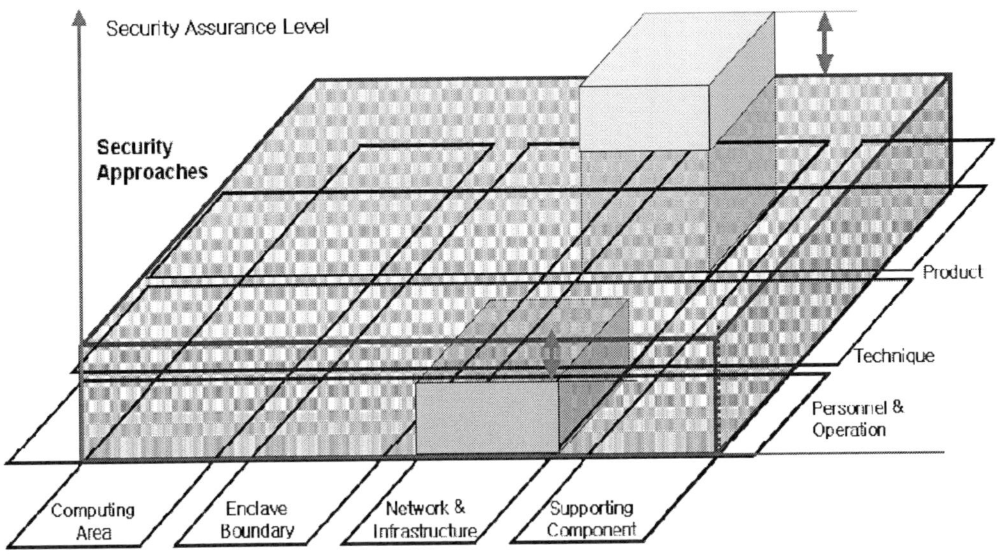

Figure 38. Security assurance improved by each security method application [15].

Conclusively security countermeasures should be chosen with the consideration of applicable threats and security solutions subject to limited investment resource. The proposed matrix model is useful for the decision making process required to reflect multiple factors, which could be also regarded as an extended version of the game matrix model and realized by MAS based NBS formulation of game theory.

5. Conclusions

We looked over several applications of game theories to decision making process on two different problems in electric power industry. Actually game theory itself has also many sub-models in it, therefore is not defined clearly as a certain formulation. As we have learned till now, game theory is applied to the analysis of many different situations in many different fields. For example, when game theory is applied to market behavior analysis, it could be applied to oligopoly market status as well as perfect competition. Cournot model is used for modeling oligopoly market while NBS model for strategic alliance as explained in 4.6.1. Physical system could be also designed based on game theoretic approach as shown in an example of security system. However there is a representatively common philosophy shared by many different models, which is the viewpoint on problems. As we already mentioned game theory is the study about the relationship. The relationships between players in a market, components in a system, sub-problems in a big problem, etc are all good materials for game theory to be applied. By applying game theory to those problems we could get a new insight on them we have not had before. Fig 38 illustrates the concept of the common insight to many different problems based on game theory, which indicates that everything is a part of a whole thing or the whole thing is composed of many small parts.

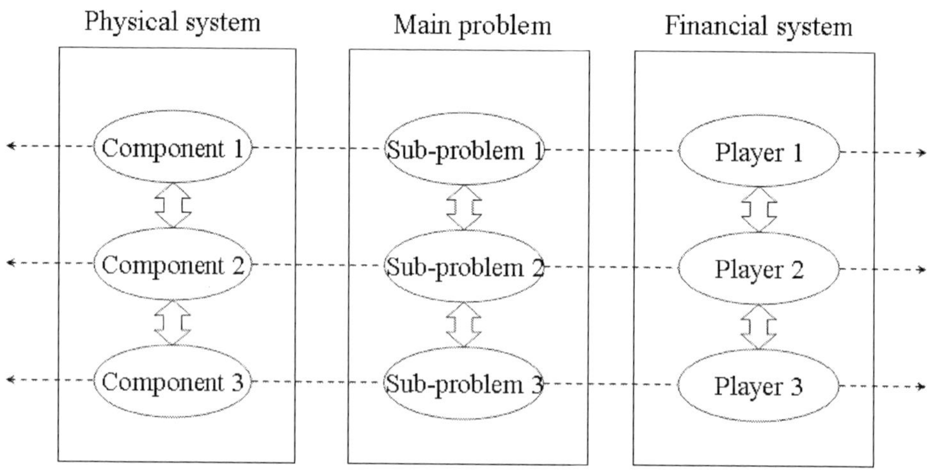

Figure 39. Common approach into many different problems by game theory.

Game theory has not been so long, about 60 years ago, since it was introduced for the first time in [1]. It is still young and on the making, but it has spread out to so many different study fields and given us new insights on many problems. Furthermore it is expected to be more advancement on many fields through game theory application. There is a Korean proverb "*You shall know while you walk through the path, and you shall get the truth while you keep practicing.*" We hope that game theory will be applied to many different fields and improved on each application with many studies and upgrades, which will lead us to having deeper insights on the world.

References

[1] John Von Neumann and Oskar Morgenstern, *Theory of Games and Economic Behavior*, Sixtieth-anniversary edition, Princeton University Press, 2004.

[2] Dan Shen, Genshe Chen, Jose B. Cruz Jr, Leonard S. Haynes, Martin Kruger, and Erik Blasch, *A Markov game approach to cyber security*, Newsroom 10.1117/ 2.1200707.0783, SPIE, 2007

[3] Daniel M. Reeves, "*Generating Trading Agent Strategies : Analytic and Empirical Methods for Infinite and Large Games*", A dissertation submitted in partial fulfillment of the requirements for the degree of Doctor of Philosophy at Michigan State University, 2005

[4] M. Shahidehpour, H. Yamin, Z. Li, *Market operations in electric power systems: forecasting, scheduling, and risk management*, 1st ed., Wiley, New York, 2002.

[5] Jin Hur, Dong-Joo Kang, Young-Hwan Moon, Development of an Integrated Power Market Simulator for the Korean Electricity Market, *KIEE International Transactions on Power Engineering*, Vol. 5-A, No. 4, pp. 416~424, 2005

[6] Dong-Joo Kang, Balho H. Kimb, Don Hur, Supplier bidding strategy based on non-cooperative game theory concepts in single auction power pools, *Electric Power System Research*, **77** (2007) 630–636, 2007

[7] A.B. Smith, IEEE Std C37.1-1994, IEEE Standard Definition, Specification, and Analysis of Systems Used for Supervisory Control, Data Acquisition, and Automatic Control, *IEEE Power Engineering Society, Sponsored by the Substations Committee, Institute of Electrical and Electronics Engineers*, Inc., New York, 1994.

[8] Rolf Carlson, *Sandia SCADA Program: High-Security SCADA LDRD Final Report*, Sandia Report, SAND2002-0729, April 2002.

[9] V.M. Igure, S.A. Laughter, R.D. Williams, Security issues in SCADA networks, *Computer & Society*, Vol. 25, 2006, pp. 498-506.

[10] Eric Byres, Justin Lowe, *The Myths and Facts behind Cyber Security Risks for Industrial Control Systems*, British Columbia Institute of Technology, PA Consulting Group

[11] John Douglas, *Grid Security in the 21-th Century*, EPRI Journal, 2005

[12] William Stallings, *Cryptography and Network Security – Principles and Practices*, Pearson International Edition, 2006

[13] Dacfey Dzung, Mario Crevatin, Security for Industrial Communication Systems, *2005 Proceedings of the IEEE*

[14] Avinash Dixit, Susan Skeath, *Games of Strategy*, W. W. Norton

[15] Tai-hoon Kim, Seok-soo Kim and Gil-cheol Park, Block Model Application for Decision Support Systems, *2007 ECSIS Symposium on Bio-inspired, Learning, and Intelligent Systems for Security (BLISS 2007)*, pp 25~28, IEEE Computer Society, 2007

INDEX

A

B

C

F

G

H

I

O

P

Q

R

T

U

V

W

Y

Z